Harnessing High-Performance Computing and AI for Environmental Sustainability

Arshi Naim
King Khalid University, Saudi Arabia

I0894745

IGI Global
PUBLISHER of TIMELY KNOWLEDGE

A volume in the Advances in Systems Analysis, Software Engineering, and High Performance Computing (ASASEHPC) Book Series

Published in the United States of America by
IGI Global
Engineering Science Reference (an imprint of IGI Global)
701 E. Chocolate Avenue
Hershey PA, USA 17033
Tel: 717-533-8845
Fax: 717-533-8661
E-mail: cust@igi-global.com
Web site: http://www.igi-global.com

Library of Congress Cataloging-in-Publication Data

ISBN: 979-8-3693-1794-5
eISBN: 979-8-3693-1795-2

This book is published in the IGI Global book series Advances in Systems Analysis, Software Engineering, and High Performance Computing (ASASEHPC) (ISSN: 2327-3453; eISSN: 2327-3461)

British Cataloguing in Publication Data
A Cataloguing in Publication record for this book is available from the British Library.

For electronic access to this publication, please contact: eresources@igi-global.com.

Advances in Systems Analysis, Software Engineering, and High Performance Computing (ASASEHPC) Book Series

Vijayan Sugumaran
Oakland University, Rochester, USA

ISSN:2327-3453
EISSN:2327-3461

Mission

The theory and practice of computing applications and distributed systems has emerged as one of the key areas of research driving innovations in business, engineering, and science. The fields of software engineering, systems analysis, and high performance computing offer a wide range of applications and solutions in solving computational problems for any modern organization.

The **Advances in Systems Analysis, Software Engineering, and High Performance Computing (ASASEHPC) Book Series** brings together research in the areas of distributed computing, systems and software engineering, high performance computing, and service science. This collection of publications is useful for academics, researchers, and practitioners seeking the latest practices and knowledge in this field.

Coverage

- Metadata and Semantic Web
- Computer Networking
- Network Management
- Computer Graphics
- Engineering Environments
- Performance Modelling
- Software Engineering
- Distributed Cloud Computing
- Storage Systems
- Parallel Architectures

The Advances in Systems Analysis, Software Engineering, and High Performance Computing (ASASEHPC) Book Series (ISSN 2327-3453) is published by IGI Global, 701 E. Chocolate Avenue, Hershey, PA 17033-1240, USA, www.igi-global.com. This series is composed of titles available for purchase individually; each title is edited to be contextually exclusive from any other title within the series. For pricing and ordering information please visit http://www.igi-global.com/book-series/advances-systems-analysis-software-engineering/73689. Postmaster: Send all address changes to above address.

Titles in this Series

For a list of additional titles in this series, please visit: http://www.igi-global.com/book-series/advances-systems-analysis-software-engineering/73689

Big Data Quantification for Complex Decision-Making
Chao Zhang (Shanxi University, China) and Wentao Li (Southwest University, China)
Engineering Science Reference • © 2024 • 312pp • H/C (ISBN: 9798369315828) • US $315.00

Digital Technologies in Modeling and Management Insights in Education and Industry
G. S. Prakasha (Christ University, India) Maria Lapina (North-Caucasus Federal University, Russia) Deepanraj Balakrishnan (Prince Mohammad Bin Fahd University, Saudi Arabia) and Mohammad Sajid (Aligarh Muslim University, India)
Information Science Reference • © 2024 • 409pp • H/C (ISBN: 9781668495766) • US $250.00

Serverless Computing Concepts, Technology and Architecture
Rajanikanth Aluvalu (Chaitanya Bharathi Institute of Technology, India) and Uma Maheswari V. (Chaitanya Bharathi Institute of Technology, India)
Engineering Science Reference • © 2024 • 310pp • H/C (ISBN: 9798369316825) • US $300.00

Developments Towards Next Generation Intelligent Systems for Sustainable Development
Shanu Sharma (ABES Engineering College, Ghaziabad, India) Ayushi Prakash (Ajay Kumar Garg Engineering College, Ghaziabad, India) and Vijayan Sugumaran (Oakland University, Rochester, USA)
Engineering Science Reference • © 2024 • 327pp • H/C (ISBN: 9798369356432) • US $385.00

Technological Advancements in Data Processing for Next Generation Intelligent Systems
Shanu Sharma (ABES Engineering College, Ghaziabad, India) Ayushi Prakash (Ajay Kumar Garg Engineering College, Ghaziabad, India) and Vijayan Sugumaran (Oakland University, Rochester, USA)
Engineering Science Reference • © 2024 • 357pp • H/C (ISBN: 9798369309681) • US $300.00

Advanced Applications in Osmotic Computing
G. Revathy (SASTRA University, India)
Engineering Science Reference • © 2024 • 370pp • H/C (ISBN: 9798369316948) • US $300.00

Omnichannel Approach to Co-Creating Customer Experiences Through Metaverse Platforms
Babita Singla (Chitkara Business School, Chitkara University, Punjab, India) Kumar Shalender (Chitkara Business School, Chitkara University, India) and Nripendra Singh (Pennsylvania Western University, USA)
Engineering Science Reference • © 2024 • 223pp • H/C (ISBN: 9798369318669) • US $270.00

IGI Global
PUBLISHER of TIMELY KNOWLEDGE

701 East Chocolate Avenue, Hershey, PA 17033, USA
Tel: 717-533-8845 x100 • Fax: 717-533-8661
E-Mail: cust@igi-global.com • www.igi-global.com

Table of Contents

Preface xvi

Acknowledgment xxi

Chapter 1
A Study on AI-ML-Driven Optimizing Energy Distribution and Sustainable Agriculture for Environmental Conservation........ 1
J Barnabas Paul Glady, Department of Electrical and Electronic Engineering, Sathyabama Institute of Science and Technology, Chennai, India
Sonia Maria D'Souza, Department of Artificial Intelligence and Machine Learning, New Horizon College of Engineering, Bengaluru, India
A. Parvathi Priya, Department of Chemistry, RMK Engineering College, Chennai, India
K. Amuthachenthiru, Department of Artificial Intelligence and Data Science, Ramco Institute of Technology, Rajapalayam, India
G. Vikram, Karunya School of Management, Karunya Institute of Technology and Sciences, Coimbatore, India
S. Boopathi, Mechanical Engineering, Muthayammal Engineering College, Namakkal, India

Chapter 2
AI-Integrated Biosensors and Bioelectronics for Agriculture 28
Pankaj Bhambri, Guru Nanak Dev Engineering College, Ludhiana, India
Alex Khang, Global Research Institute of Technology and Engineering, USA

Chapter 3
Machine Learning and Deep Learning Solutions for Green Computing 46
Nagendra Singh Yadav, Engineering College, Bikaner, India
Vishal Kumar Goar, Engineering College, Bikaner, India

Chapter 4
Machine Learning-Integrated Sustainable Engineering and Energy Systems: Innovations at the Nexus 74
M. Sarat Chandra Prasad, Department of Computer Science and Engineering, Aditya College of Engineering, Surampalem, India
M. Dhanalakshmi, Department of Computer Science and Engineering, New Horizon College of Engineering, Bangalore, India
M. Mohan, Department of Computer Science and Engineering, KPR Institute of Engineering and Technology, Arasur, India
B. Somasundaram, School of Mechanical Engineering, REVA University, Bengaluru, India
R. Valarmathi, Department of S&H (Mathematics), RMK Engineering College, Kavaraipettai, India
Sampath Boopathi, Mechanical Engineering, Muthayammal Engineering College, Namakkal, India

Chapter 5
The Impact of AI on Sustainability 99
Arshiya Begum, King Khalid University, Saudi Arabia
Arshi Naim, King Khalid University, Saudi Arabia
Asfia Sabahath, King Khalid University, Saudi Arabia

Chapter 6
Applications of Blockchain in Building Marketing Framework: Approach to Environmental Sustainability 114
Arshi Naim, King Khalid University, Saudi Arabia

Chapter 7
Internet of Things Framework for Sustainable Agriculture in Smart Farming 139
Pratima Bais, Dr. C.V. Raman University, India
Preeti Singh Bahadur, Amity University, India

Chapter 8
Circular Economy Digital Practices for Ethical Dimensions and Policies for Digital Waste Management 166
V. Vijaya Lakshmi, Humanities and Mathematics Department, G. Narayanamma Institute of Technology and Science, Hyderabad, India
Manoj Mishra, Department of Basic Science and Humanities, Institute of Technology and Management, Gwalior, India
Jitendra Singh Kushwah, Department of Information Technology, Institute of Technology and Management, Gwalior, India
Umaya Salma Shajahan, Department of Management Studies, Sona College of Technology, Salem, India
M. Mohanasundari, Department of Management Studies, Kongu Engineering College, Erode, India
Sampath Boopathi, Mechanical Engineering, Muthayammal Engineering College, Namakkal, India

Chapter 9
Applications of Machine Learning Techniques for Achieving Financial Sustainability 194
Arshi Naim, King Khalid University, Saudi Arabia

Chapter 10
Sustainable Practices for Green Computing and Digital E-Waste Management 214
K. R. Kalphana, Department of Agricultural Engineering, Mahendra Engineering College (Autonomous), Namakkal, India
K. Madu Mathi, Department of Electrical and Electronics Engineering, Mahendra Institute of Technology (Autonomous), Namakkal, India
A. Ranjithkumar, Department of Mechanical Engineering, AVS College of Technology, Salem, India
Sanchita Banerji, Department of HRM, DSIMS, Mumbai, India
S. Anandakumar, Department of Civil Engineering, KPR Institute of Engineering and Technology, Coimbatore, India

Chapter 11
Next Generation Materials for Sustainable Development.. 236
Vijay Bhutani, Amity University, India
Preeti Singh Bahadur, Amity University, India
Sunil Kumar Sansaniwal, The Energy and Resources Institute, New Delhi, India

Chapter 12
Leveraging High-Performance Computing and Artificial Intelligence in Climate Modeling and Prediction 266
Asfia Sabahath, King Khalid University, Saudi Arabia
Arshiya Begum, King Khalid University, Saudi Arabia
Arshi Naim, King Khalid University, Saudi Arabia

Chapter 13
Energy Optimization and Smart Grids: IoT-Based Smart Grid Solution and Smart Grids Applications 278
Sana Dahmani, Independent Researcher, USA

Chapter 14
Cloud Computing Adoption for Small and Medium Enterprises in Engineering and Environmental Aspects 305
S. Abirami, Department of Artificial Intelligence and Data Science, Jai Shriram Engineering College, Tirupur, India
Santosh K. C., Department of Computer Science and Engineering, Bapuji Institute of Engineering and Technology, Davangere, India
R. Somasundaram, Department of Management Studies, Kongu Engineering College, Erode, India
Kavitha K. S., Department of Computer Science and Engineering, Dayananda Sagar College of Engineering, Bangalore, India
Harshita Gangadhar Patil, Department of Computer Science and Engineering, S.B. Jain

Institute of Technology, Management, and Research, Nagpur, India
Sureshkumar Myilsamy, Bannari Amman Institute of Technology, India

Chapter 15
Assessment of Water Quality Using Multi-Layered Mamdani Fuzzy Inference Expert System........ 334
Gulzar Ahmad, Minhaj University, Lahore, Pakistan
Sehrish Aqeel, University of Malaysia, Sarawak, Malaysia
Zobia Zafar, University of South Asia, Pakistan
Kanza Fatima, University of South Asia, Pakistan

Compilation of References 353

About the Contributors 396

Index 399

Detailed Table of Contents

Preface.......... xvi

Acknowledgment xxi

Chapter 1
A Study on AI-ML-Driven Optimizing Energy Distribution and Sustainable Agriculture for Environmental Conservation.......... 1

J Barnabas Paul Glady, Department of Electrical and Electronic Engineering, Sathyabama Institute of Science and Technology, Chennai, India
Sonia Maria D'Souza, Department of Artificial Intelligence and Machine Learning, New Horizon College of Engineering, Bengaluru, India
A. Parvathi Priya, Department of Chemistry, RMK Engineering College, Chennai, India
K. Amuthachenthiru, Department of Artificial Intelligence and Data Science, Ramco Institute of Technology, Rajapalayam, India
G. Vikram, Karunya School of Management, Karunya Institute of Technology and Sciences, Coimbatore, India
S. Boopathi, Mechanical Engineering, Muthayammal Engineering College, Namakkal, India

The chapter examines how machine learning (ML) and artificial intelligence (AI) could be used to solve environmental problems throughout the world. It emphasizes how crucial AI and ML are to optimizing energy distribution, including energy demand forecasting, improving smart grid performance, and incorporating renewable energy sources. The chapter also covers the use of AI and ML methods to sustainable agriculture, emphasizing predictive analytics for pest management, soil health monitoring, and precision farming. It highlights the effectiveness of resource use and encourages actions that are ecologically friendly. The chapter also covers ethical issues, societal ramifications, legal systems, and the synergies between energy and agricultural solutions. It imagines a day when advances led by AI and ML will be essential to a sustainable and environmentally balanced planet.

Chapter 2
AI-Integrated Biosensors and Bioelectronics for Agriculture 28

Pankaj Bhambri, Guru Nanak Dev Engineering College, Ludhiana, India
Alex Khang, Global Research Institute of Technology and Engineering, USA

The integration of artificial intelligence (AI) alongside biosensors and bioelectronics has introduced a new age in precision agriculture, enabling the enhancement of agricultural processes. This book chapter examines the utilization of AI, biosensors, & bioelectronics in the domains of agriculture as well as aquaculture. The chapter commences with an examination of the agricultural sector's preoccupations, such as constraints on resources, environmental sustainability, and productivity. Subsequently, the text examines how biosensors and bioelectronics might effectively tackle these challenges by continuously monitoring and collecting agricultural data in real-time. A significant portion of the chapter focuses on the integration of artificial intelligence with biosensors. Farmers can make well-informed judgments about crop management, disease detection, and yield optimization by utilizing effective data analysis, predictive modeling, and decision-making. AI-powered biosensors have enhanced agricultural efficiency in several case studies and real-world applications.

Chapter 3
Machine Learning and Deep Learning Solutions for Green Computing 46
Nagendra Singh Yadav, Engineering College, Bikaner, India
Vishal Kumar Goar, Engineering College, Bikaner, India

In recent years, the field of green computing has gained significant attention due to the growing concerns about the environmental impact of information technology. As the demand for computing resources continues to rise, it becomes imperative to develop sustainable solutions that reduce energy consumption, minimize electronic waste, and lower carbon emissions. This proposed chapter aims to explore the application of machine learning and deep learning algorithms as innovative solutions for addressing these challenges within the domain of green computing.

Chapter 4
Machine Learning-Integrated Sustainable Engineering and Energy Systems: Innovations at the Nexus 74
M. Sarat Chandra Prasad, Department of Computer Science and Engineering, Aditya College of Engineering, Surampalem, India
M. Dhanalakshmi, Department of Computer Science and Engineering, New Horizon College of Engineering, Bangalore, India
M. Mohan, Department of Computer Science and Engineering, KPR Institute of Engineering and Technology, Arasur, India
B. Somasundaram, School of Mechanical Engineering, REVA University, Bengaluru, India
R. Valarmathi, Department of S&H (Mathematics), RMK Engineering College, Kavaraipettai, India
Sampath Boopathi, Mechanical Engineering, Muthayammal Engineering College, Namakkal, India

Machine learning (ML) has revolutionized various fields, including engineering, energy systems, and sustainability. This abstract explores the synergies between ML and these domains, focusing on its optimization in predictive maintenance, energy consumption efficiency, and smart grids. ML's role in renewable energy forecasting, building energy management, and materials science is also explored. It also highlights its impact on supply chain optimization, environmental monitoring, and sustainability assessments. The holistic approach extends to smart city initiatives and infrastructure development, paving the way for intelligent urban planning. ML enhances decision-making processes, enabling more resilient, efficient, and sustainable practices in engineering and energy systems. This exploration serves as a beacon for researchers, practitioners, and policymakers seeking innovative solutions at the intersection of ML, engineering, and sustainability.

Chapter 5
The Impact of AI on Sustainability 99
Arshiya Begum, King Khalid University, Saudi Arabia
Arshi Naim, King Khalid University, Saudi Arabia
Asfia Sabahath, King Khalid University, Saudi Arabia

Artificial intelligence (AI) has been increasingly popular in various businesses in recent years. AI has proven to be an effective tool for enhancing efficiency and decision-making across industries, from healthcare to finance. However, sustainability is the area in which AI is having a huge impact is sustainability. AI has the potential to transform the way we think about sustainability. AI can find trends and make predictions based on massive amounts of data, allowing us to understand better and handle environmental challenges. For example, artificial intelligence (AI) can analyze weather patterns and predict natural disasters, allowing for improved disaster preparedness and response. It can also be utilized in industries like manufacturing and transportation to optimize energy usage and eliminate waste.

Chapter 6
Applications of Blockchain in Building Marketing Framework: Approach to Environmental Sustainability 114
Arshi Naim, King Khalid University, Saudi Arabia

Firms apply various technological applications to achieve their businesses objectives. Developing marketing framework (MF) has been a challenge for firms in all types of products and services. This is because firms are not sure of which MF can best suit their business practices and attain environmental sustainability. Another important reason is that there are some marketing models (MM) which are confused with the MF that leads to misinterpretation of applications. This research chapter is qualitative in nature and has two parts. The first part presents the definitions of MM and MF and presents a comparative analysis between them. The second part presents the application of blockchain for developing successful marketing framework for environmental sustainability. Results are presented by using observation methods. Analysis and are based on the firm's experience in the use of blockchain in building MF for environmental sustainability.

Chapter 7
Internet of Things Framework for Sustainable Agriculture in Smart Farming 139
Pratima Bais, Dr. C.V. Raman University, India
Preeti Singh Bahadur, Amity University, India

The global human population is constantly increasing whereas the agricultural land resources are declining for reasons of urban settlements and industrial growth. This has certainly strained the agricultural sector for increased annual yields, especially in the changing climate. To accomplish the rising demand of consumers for food, feed, and fiber, up to 70% upsurge in crop production is farsighted by 2050. The agricultural industry is getting more data-centric and requires precise, more advanced data and technologies than before, despite being familiar with agricultural processes. The agriculture industry is being advanced by various information and advanced communication technologies, such as the internet of things (IoT) and cloud computing are anticipated to inspire growth and initiate the use of robots and artificial intelligence in farming. The rapid emergence of these advanced technologies has restructured almost all other industries, as well as advanced agriculture, which has shifted the industry from a statistical approach to a quantitative one.

Chapter 8
Circular Economy Digital Practices for Ethical Dimensions and Policies for Digital Waste Management 166
V. Vijaya Lakshmi, Humanities and Mathematics Department, G. Narayanamma Institute of Technology and Science, Hyderabad, India
Manoj Mishra, Department of Basic Science and Humanities, Institute of Technology and Management, Gwalior, India
Jitendra Singh Kushwah, Department of Information Technology, Institute of Technology and Management, Gwalior, India
Umaya Salma Shajahan, Department of Management Studies, Sona College of Technology, Salem, India
M. Mohanasundari, Department of Management Studies, Kongu Engineering College, Erode, India
Sampath Boopathi, Mechanical Engineering, Muthayammal Engineering College, Namakkal, India

Digital technology has revolutionized our lives, but it also presents a growing challenge: digital waste. This issue involves electronic, data, and obsolete devices, causing significant environmental and social consequences. To tackle this, a multi-faceted approach is needed, including ethical considerations, policy interventions, technological innovations, and consumer behavior change. Ethical frameworks, strengthened regulatory frameworks, eco-design, producer responsibility, and circular economy principles are essential. Industry best practices and technological innovations can reduce digital waste, while consumer education and behavior change initiatives encourage informed digital habits. Sustainable digital practices require a concerted effort from all stakeholders.

Chapter 9
Applications of Machine Learning Techniques for Achieving Financial Sustainability 194
Arshi Naim, King Khalid University, Saudi Arabia

In the contemporary business landscape, enterprises grapple with the pursuit of marginal profits, recognizing that the bedrock of their success lies in financial sustainability. Consequently, the mitigation or prevention of business financial risks (BFR) emerges as a pivotal factor in enhancing firms' profitability prospects. Various technological strategies are employed to address the challenges posed by financial risks, with machine learning (ML) standing out as a particularly pertinent and effective solution across diverse business sectors. In the current milieu, the significance and efficacy of ML are keenly acknowledged, with businesses leveraging this cutting-edge technology to tackle challenges associated with BFR.

Chapter 10
Sustainable Practices for Green Computing and Digital E-Waste Management 214
K. R. Kalphana, Department of Agricultural Engineering, Mahendra Engineering College (Autonomous), Namakkal, India
K. Madu Mathi, Department of Electrical and Electronics Engineering, Mahendra Institute of Technology (Autonomous), Namakkal, India
A. Ranjithkumar, Department of Mechanical Engineering, AVS College of Technology, Salem, India
Sanchita Banerji, Department of HRM, DSIMS, Mumbai, India
S. Anandakumar, Department of Civil Engineering, KPR Institute of Engineering and Technology, Coimbatore, India

The chapter explores the integration of circular economy principles into digital operations, focusing on green computing and email waste management. Green computing uses eco-friendly technologies to reduce digital infrastructure's environmental footprint. It involves energy-efficient hardware, virtualization techniques, and cloud computing to reduce energy consumption, carbon emissions, and electronic waste. Email waste management is crucial to address digital waste, which contributes to carbon emissions, energy consumption, and resource depletion. Implementing email optimization, efficient storage practices, and behavioural interventions can help organizations mitigate environmental impacts while improving productivity and information management efficiency.

Chapter 11
Next Generation Materials for Sustainable Development.. 236
Vijay Bhutani, Amity University, India
Preeti Singh Bahadur, Amity University, India
Sunil Kumar Sansaniwal, The Energy and Resources Institute, New Delhi, India

In a rapidly evolving world, sustainability has emerged as a central focus, and the exploration of next-generation materials for sustainable development and engineering has become a pivotal endeavour. The exploration of next-generation materials for sustainable development and engineering is a vital pursuit in our quest for a more sustainable future. This book chapter delves into the critical era of "Next Generation Materials for Sustainable Development or Sustainable Engineering." providing an in-depth examination of the evolution of materials, their characteristics, benefits, challenges, and innovative applications across various industries.

Chapter 12
Leveraging High-Performance Computing and Artificial Intelligence in Climate Modeling and Prediction 266
Asfia Sabahath, King Khalid University, Saudi Arabia
Arshiya Begum, King Khalid University, Saudi Arabia
Arshi Naim, King Khalid University, Saudi Arabia

Climate change is a pressing global challenge that demands improved understanding, accurate predictions, and effective mitigation strategies. High-performance computing (HPC) combined with artificial intelligence (AI) presents a transformative opportunity for climate modeling and prediction. HPC and AI for environmental sustainability have the potential to revolutionize our efforts in addressing pressing environmental challenges. By combining the computational power of HPC with the analytical capabilities of AI, we can accelerate research and develop innovative solutions to promote sustainability.

Chapter 13
Energy Optimization and Smart Grids: IoT-Based Smart Grid Solution and Smart Grids Applications 278
Sana Dahmani, Independent Researcher, USA

Grid computing, utilizing distributed resources, achieves goals in handling workloads in a distributed system. Unlike traditional high-performance computing, grid computing assigns tasks to nodes, showcasing heterogeneity without physical coupling. Grids, with variable sizes based on applications, use general-purpose middleware. In modern energy systems, integrating smart grids and grid computation is vital. Emphasizing resource efficiency, grid computation enhances smart grid performance; by enabling real-time monitoring, rapid adjustments for optimization, and outage prevention. Grid computation supports demand response strategies for electricity management and integrates distributed generation resources, such as solar panels and wind turbines, including autonomous microgrid supervision.

Chapter 14
Cloud Computing Adoption for Small and Medium Enterprises in Engineering and Environmental Aspects 305
S. Abirami, Department of Artificial Intelligence and Data Science, Jai Shriram Engineering College, Tirupur, India
Santosh K. C., Department of Computer Science and Engineering, Bapuji Institute of Engineering and Technology, Davangere, India
R. Somasundaram, Department of Management Studies, Kongu Engineering College, Erode, India
Kavitha K. S., Department of Computer Science and Engineering, Dayananda Sagar College of Engineering, Bangalore, India
Harshita Gangadhar Patil, Department of Computer Science and Engineering, S.B. Jain Institute of Technology, Management, and Research, Nagpur, India
Sureshkumar Myilsamy, Bannari Amman Institute of Technology, India

Cloud computing is revolutionizing small and medium enterprises (SMEs) by offering cost savings, scalability, collaboration, and security. However, it also presents challenges like security concerns, regulatory compliance, and migration complexities. The chapter explores the adoption of cloud technologies in engineering domains like CAD/CAM, simulation, project management, and remote monitoring, highlighting their role in enhancing efficiency and innovation. It also discusses the environmental impact of cloud computing, highlighting its potential to reduce carbon emissions through energy-efficient data centers and sustainable practices. The chapter provides case studies and best practices for SMEs to embrace cloud computing while navigating environmental responsibilities and identifies future trends in integrating cloud computing with emerging technologies and sustainable infrastructure.

Chapter 15
Assessment of Water Quality Using Multi-Layered Mamdani Fuzzy Inference Expert System........ 334
Gulzar Ahmad, Minhaj University, Lahore, Pakistan
Sehrish Aqeel, University of Malaysia, Sarawak, Malaysia
Zobia Zafar, University of South Asia, Pakistan
Kanza Fatima, University of South Asia, Pakistan

In this chapter, a new multi-layered mamdani fuzzy inference system (ML-MFIS) is proposed for the assessment of water quality (AWQ). The proposed AWQ-ML-MFIS expert system can categorize the level of water quality into excellent, normal, or dangerous for health (polluted). AWQ-ML-MFIS expert system for drinking water is developed by the guidelines from the World Health Organization (WHO) and Pakistan's Punjab environmental quality standard. AWQ-ML-MFIS expert system uses input water parameters such as bacterial, physical, chemical, and radioactive for different layers. The chemical parameters are further divided into essential inorganics, toxic inorganics, and organics. This chapter also analyses the intensities of the parameters and the results achieved by using the proposed AWQ-ML-MFIS expert system. All these parameters and results are discussed with the experts of the Pakistan Council of Scientific and Industrial Research (PCSIR), Lahore. The accuracy of the proposed AWQ-ML-MFIS Expert System is more accurate as compared to others approaches used for it.

Compilation of References 353

About the Contributors 396

Index 399

Preface

The world is addressing the insistent challenge of climate change, and the need for innovative solutions has become paramount. In this period of technical developments, artificial intelligence (AI) has emerged as a powerful instrument with enormous prospects to combat climate change and other environmental subjects. AI's ability to process vast amounts of data, identify patterns, and make intelligent predictions offers unprecedented opportunities to tackle this global crisis. High-Performance Computing (HPC) or super-computing environments address these large and complex challenges with individual nodes (computers) working together in a cluster (connected group) to perform massive amounts of computing in a short period. Creating and removing these clusters is often automated in the cloud to reduce costs. Computer networks, communication systems, and other IT infrastructures have a growing environmental footprint due to significant energy consumption and greenhouse gas emissions. To address this seemingly self-defeating conundrum, and create a truly sustainable environment, new energy models, algorithms, methodologies, platforms, tools, and systems are required to support next-generation computing and communication infrastructures.

Harnessing High-Performance Computing and AI for Environmental Sustainability navigates through AI-driven solutions from sustainable agriculture and land management to energy optimization and smart grids. It unveils how AI algorithms can analyze colossal datasets, offering unprecedented insights into climate modeling, weather prediction, and long-term climate trends. Integrating AI-powered optimization algorithms revolutionizes energy systems, propelling the transition towards a low-carbon future by reducing greenhouse gas emissions and enhancing efficiency.

This book is ideal for educators, environmentalists, industry professionals, and researchers alike, and it explores the ethical dimensions and policies surrounding AI's contribution to environmental development. It tackles the unintended consequences of AI, providing a comprehensive guide to harnessing AI for sustainable outcomes. Readers will discover how AI can catalyze sustainable farming practices, optimize energy distribution, and combat deforestation. Moreover, the text probes the profound impacts of AI on the Earth, both guided and unguided, offering a balanced perspective on its potential and challenges.

Its chapters cover diverse facets, from the role of HPC and open climate data in water and natural resource management to the design principles for environmentally sustainable computing. It unravels the implications of global warming on data center power consumption, introduces intelligence models for computing resource management, and explores the applications of blockchain in energy management and trading agricultural products.

Chapter 1, *A Study on AI-ML Driven Optimizing Energy Distribution and Sustainable Agriculture for Environmental Conservation,* The chapter examines how machine learning (ML) and artificial intelligence (AI) could be used to solve environmental problems throughout the world. It emphasizes how crucial AI and ML are to optimizing energy distribution, including energy demand forecasting, improving smart grid performance, and incorporating renewable energy sources. The chapter also covers the use of AI and ML methods to sustainable agriculture, emphasizing predictive analytics for pest management, soil health monitoring, and precision farming. It highlights the effectiveness of resource use and encourages actions that are ecologically friendly. The chapter also covers ethical issues, societal ramifications, legal systems, and the synergies between energy and agricultural solutions. It imagines a day when advances led by AI and ML will be essential to a sustainable and environmentally balanced planet.

Chapter 2, *AI-Integrated Biosensors and Bioelectronics for Agriculture*, The integration of artificial intelligence (AI) alongside biosensors and bioelectronics has introduced a new age in precision agriculture, enabling the enhancement of agricultural processes. This book chapter examines the utilization of AI, biosensors, & bioelectronics in the domains of agriculture as well as aquaculture. The chapter commences with an examination of the agricultural sector's preoccupations, such as constraints on resources, environmental sustainability, and productivity. Subsequently, the text examines how biosensors and bioelectronics might effectively tackle these challenges by continuously monitoring and collecting agricultural data in real-time. A significant portion of the chapter focuses on the integration of artificial intelligence with biosensors. Farmers can make well-informed judgments about crop management, disease detection, and yield optimization by utilizing effective data analysis, predictive modeling, and decision-making. AI-powered biosensors have enhanced agricultural efficiency in several case studies and real-world applications.

Chapter 3, *Machine Learning and Deep Learning solutions for Green Computing*, In recent years, the field of green computing has gained significant attention due to the growing concerns about the environmental impact of information technology. As the demand for computing resources continues to rise, it becomes imperative to develop sustainable solutions that reduce energy consumption, minimize electronic waste, and lower carbon emissions. This proposed chapter aims to explore the application of machine learning and deep learning algorithms as innovative solutions for addressing these challenges within the domain of green computing.

Chapter 4, *Machine Learning integrated Sustainable Engineering and Energy Systems: Innovations at the Nexus*, Machine Learning (ML) has revolutionized various fields, including engineering, energy systems, and sustainability. This abstract explores the synergies between ML and these domains, focusing on its optimization in predictive maintenance, energy consumption efficiency, and smart grids. ML's role in renewable energy forecasting, building energy management, and materials science is also explored. It also highlights its impact on supply chain optimization, environmental monitoring, and sustainability assessments. The holistic approach extends to smart city initiatives and infrastructure development, paving the way for intelligent urban planning. ML enhances decision-making processes, enabling more resilient, efficient, and sustainable practices in engineering and energy systems. This exploration serves as a beacon for researchers, practitioners, and policymakers seeking innovative solutions at the intersection of ML, engineering, and sustainability.

Chapter 5, *The Impact of AI on Sustainability*, Artificial intelligence (AI) has been increasingly popular in various businesses in recent years. AI has proven to be an effective tool for enhancing efficiency and decision-making across industries, from healthcare to finance. However, sustainability is the area in which AI is having a huge impact is sustainability. AI has the potential to transform the way we think about sustainability. AI can find trends and make predictions based on massive amounts of data, allowing us to understand better and handle environmental challenges. For example, artificial intelligence (AI) can analyze weather patterns and predict natural disasters, allowing for improved disaster preparedness and response. It can also be utilized in industries like manufacturing and transportation to optimize energy usage and eliminate waste.

Chapter 6, *Applications of Blockchain in building Marketing Framework -Approach to Environmental Sustainability*, Firms apply various technological applications to achieve their businesses objectives. Developing Marketing Framework (MF) has been a challenge for firms in all types of products and services. This is because firms are not sure of which MF can best suit their business practices and attain environmental sustainability. Another important reason is that there are some Marketing models (MM) which are confused with the MF that leads to misinterpretation of applications. This research paper is qualitative in nature and has two parts. First part presents the definitions of MM and MF and presents a comparative analysis between them. Second part presents the application of blockchain for developing successful marketing framework for environmental sustainability. Results are presented by using observation methods. Analysis and are based on the firm's experience in the use of blockchain in building MF for environmental sustainability.

Chapter 7, *Internet of Things Framework for Sustainable Agriculture in Smart Farming: Internet of Things Framework for Sustainable Agriculture in Smart Farming,* The global human population is constantly increasing whereas the agricultural land resources are declining for reasons of urban settlements and industrial growth. This has certainly strained the agricultural sector for increased annual yields, especially under the changing climate. To accomplish the rising demand of consumers for food, feed and fiber, up to 70% upsurge in the crop production is farsighted by 2050. The agricultural industry is getting more data-centric and requires precise, more advanced data and technologies than before, despite being familiar with agricultural processes. The agriculture industry is being advanced by various information and advanced communication technologies, such as the Internet of Things (IoT) and cloud computing are anticipated to inspire growth and initiate the use of robots and artificial intelligence in farming. The rapid emergence of these advanced technologies has restructured almost all other industries, as well as advanced agriculture, which has shifted the industry from a statistical approach to a quantitative one.

Chapter 8, *Circular Economy Digital Practices for Ethical Dimensions and Policies for Digital Waste Management*, Digital technology has revolutionized our lives, but it also presents a growing challenge: digital waste. This issue involves electronic, data, and obsolete devices, causing significant environmental and social consequences. To tackle this, a multi-faceted approach is needed, including ethical considerations, policy interventions, technological innovations, and consumer behavior change. Ethical frameworks, strengthened regulatory frameworks, eco-design, producer responsibility, and circular economy principles are essential. Industry best practices and technological innovations can reduce digital waste, while consumer education and behavior change initiatives encourage informed digital habits. Sustainable digital practices require a concerted effort from all stakeholders.

Chapter 9, *Applications of Machine Learning Techniques for achieving Financial Sustainability*, In the contemporary business landscape, enterprises grapple with the pursuit of marginal profits, recognizing that the bedrock of their success lies in financial sustainability. Consequently, the mitigation or prevention of business financial risks (BFR) emerges as a pivotal factor in enhancing firms' profitability prospects. Various technological strategies are employed to address the challenges posed by financial risks, with Machine Learning (ML) standing out as a particularly pertinent and effective solution across diverse business sectors. In the current milieu, the significance and efficacy of ML are keenly acknowledged, with businesses leveraging this cutting-edge technology to tackle challenges associated with BFR.

Chapter 10, *Sustainable practices for Green Computing and digital Waste Management*, The chapter explores the integration of circular economy principles into digital operations, focusing on green computing and email waste management. Green computing uses eco-friendly technologies to reduce digital infrastructure's environmental footprint. It involves energy-efficient hardware, virtualization techniques, and cloud computing to reduce energy consumption, carbon emissions, and electronic waste. Email waste management is crucial to address digital waste, which contributes to carbon emissions, energy consumption, and resource depletion. Implementing email optimization, efficient storage practices, and behavioural interventions can help organizations mitigate environmental impacts while improving productivity and information management efficiency.

Chapter 11, *Next Generation Materials for Sustainable Development: Next Generation Materials for Sustainable Development*, In a rapidly evolving world, sustainability has emerged as a central focus, and the exploration of next-generation materials for sustainable development and engineering has become a pivotal endeavour. The exploration of next-generation materials for sustainable development and engineering is a vital pursuit in our quest for a more sustainable future. This book chapter delves into the critical era of "Next Generation Materials for Sustainable Development or Sustainable Engineering." providing an in-depth examination of the evolution of materials, their characteristics, benefits, challenges, and innovative applications across various industries.

Chapter 12, *Leveraging High-Performance Computing and Artificial Intelligence in Climate Modeling and Prediction*, Climate change is a pressing global challenge that demands improved understanding, accurate predictions, and effective mitigation strategies. High-performance computing (HPC) combined with artificial intelligence (AI) presents a transformative opportunity for climate modeling and prediction. HPC and AI for environmental sustainability have the potential to revolutionize our efforts in addressing pressing environmental challenges. By combining the computational power of HPC with the analytical capabilities of AI, we can accelerate research and develop innovative solutions to promote sustainability.

Chapter 13, *Energy Optimization and Smart Grids: IoT-based Smart Grid Solution and Smart Grids applications*, Grid computing, utilizing distributed resources, achieves shared goals in handling workloads in a distributed system. Unlike traditional high-performance computing, grid computing assigns tasks to nodes, showcasing heterogeneity without physical coupling. Grids, with variable sizes based on applications, use general-purpose middleware. In modern energy systems, integrating smart grids and grid computation is vital. Emphasizing resource efficiency, grid computation enhances smart grid performance, enabling real-time monitoring, rapid adjustments for optimization, and outage prevention. Grid computation supports demand response strategies for electricity management and integrates distributed generation resources, such as solar panels and wind turbines, including autonomous microgrid supervision.

Chapter 14, *Cloud Computing Adoption for Small and Medium Enterprises in Engineering and Environmental Aspects*, Cloud computing is revolutionizing Small and Medium Enterprises (SMEs) by offering cost savings, scalability, collaboration, and security. However, it also presents challenges like security concerns, regulatory compliance, and migration complexities. The chapter explores the adoption of cloud technologies in engineering domains like CAD/CAM, simulation, project management, and remote monitoring, highlighting their role in enhancing efficiency and innovation. It also discusses the environmental impact of cloud computing, highlighting its potential to reduce carbon emissions through energy-efficient data centers and sustainable practices. The chapter provides case studies and best practices for SMEs to embrace cloud computing while navigating environmental responsibilities and identifies future trends in integrating cloud computing with emerging technologies and sustainable infrastructure.

Chapter 15, *Assessment of Water Quality Using Multi-Layered Mamdani Fuzzy Inference Expert System,* In this article, a new Multi-Layered Mamdani Fuzzy Inference System (ML-MFIS) is proposed for the Assessment of Water Quality (AWQ). The proposed AWQ-ML-MFIS Expert System can categorize the level of water quality into Excellent, Normal, or Dangerous for Health (Polluted). AWQ-ML-MFIS Expert System for drinking water is developed by the guidelines from the World Health Organization (WHO) and Pakistan's Punjab Environmental Quality Standard. AWQ-ML-MFIS Expert System uses input water parameters such as bacterial, physical, chemical, and radioactive for different layers. The chemical parameters are further divided into Essential Inorganics, Toxic Inorganics, and Organics. This article also analyses the intensities of the parameters and the results achieved by using the Proposed AWQ-ML-MFIS Expert System. All these parameters and results are discussed with the experts of the Pakistan Council of Scientific and Industrial Research (PCSIR), Lahore. The accuracy of the proposed AWQ-ML-MFIS Expert System is more accurate as compared to others approaches used for it.

Arshi Naim
King Khalid University, Saudi Arabia

Acknowledgment

The completion of this book, *Harnessing High-Performance Computing and AI for Environmental Sustainability*, would not have been possible without the collaborative efforts and support of numerous individuals and organizations.

I extend my deepest gratitude to the researchers, scientists, and experts in the fields of high-performance computing and artificial intelligence who have dedicated their time and expertise to advancing environmental sustainability. Your contributions have laid the foundation for innovative solutions that address pressing environmental challenges.

I would like to thank the reviewers for their insightful feedback and constructive criticism, which have significantly improved the quality and relevance of the content.

Special appreciation goes to my colleagues and peers who have provided encouragement, guidance, and inspiration throughout the writing process. Your encouragement has been invaluable in keeping us motivated and focused on our goal.

Last but not least, I express my heartfelt appreciation to my family for their unwavering love, patience, and understanding during the course of this project. Your support has been the cornerstone of our perseverance and success.

Thank you to everyone who has contributed to this endeavor. Your collective efforts have made this book a reality and will hopefully inspire further advancements in leveraging technology for the betterment of our planet.

Chapter 1
A Study on AI-ML-Driven Optimizing Energy Distribution and Sustainable Agriculture for Environmental Conservation

J Barnabas Paul Glady
Department of Electrical and Electronic Engineering, Sathyabama Institute of Science and Technology, Chennai, India

K. Amuthachenthiru
Department of Artificial Intelligence and Data Science, Ramco Institute of Technology, Rajapalayam, India

Sonia Maria D'Souza
Department of Artificial Intelligence and Machine Learning, New Horizon College of Engineering, Bengaluru, India

G. Vikram
Karunya School of Management, Karunya Institute of Technology and Sciences, Coimbatore, India

A. Parvathi Priya
Department of Chemistry, RMK Engineering College, Chennai, India

S. Boopathi
https://orcid.org/0000-0002-2065-6539
Mechanical Engineering, Muthayammal Engineering College, Namakkal, India

ABSTRACT

The chapter examines how machine learning (ML) and artificial intelligence (AI) could be used to solve environmental problems throughout the world. It emphasizes how crucial AI and ML are to optimizing energy distribution, including energy demand forecasting, improving smart grid performance, and incorporating renewable energy sources. The chapter also covers the use of AI and ML methods to sustainable agriculture, emphasizing predictive analytics for pest management, soil health monitoring, and precision farming. It highlights the effectiveness of resource use and encourages actions that are ecologically friendly. The chapter also covers ethical issues, societal ramifications, legal systems, and the synergies between energy and agricultural solutions. It imagines a day when advances led by AI and ML will be essential to a sustainable and environmentally balanced planet.

DOI: 10.4018/979-8-3693-1794-5.ch001

INTRODUCTION

The integration of Artificial Intelligence (AI) and Machine Learning (ML) in environmental conservation is a significant shift in our approach to tackling environmental issues. AI focuses on creating intelligent machines that mimic human cognition, while ML enables systems to learn and improve from experience. AI's key role in environmental conservation is data analysis, pattern recognition, and predictive modeling, enabling valuable insights for understanding and mitigating climate change, biodiversity loss, and resource management challenges (Javaid et al., 2022).

AI and ML are being utilized to enhance energy distribution systems, thereby reducing inefficiencies and unpredictability in traditional grids. This predictive analytics enables efficient resource allocation and reduced wastage.AI enhances the adaptability and responsiveness of smart grids by incorporating machine learning algorithms for real-time monitoring, consumption analysis, and seamless integration of renewable energy sources. This maximizes clean energy utilization, optimizes distribution networks, minimizes transmission losses, and ensures a more resilient and sustainable energy infrastructure(Nishant et al., 2020).

AI and ML are promising for transforming agricultural practices towards sustainability, addressing challenges like resource depletion, climate change, and food security. AI-powered precision farming, utilizing drones, sensors, and data analytics, offers hope in optimizing crop cultivation. Machine Learning algorithms process data to provide precise recommendations for farmers, enabling informed decisions on planting, irrigation, and pest control. This minimizes resource wastage and chemical usage while maximizing yields. Precision agriculture minimizes environmental footprint. The integration of AI and ML in environmental conservation offers a holistic approach to mitigating degradation. By harnessing these technologies, we can create a sustainable future with innovative solutions to address the planet's challenges(Galaz et al., 2021).

The integration of Artificial Intelligence and Machine Learning in environmental sustainability is a significant step towards tackling the interconnected challenges of our planet. These technologies not only enhance our understanding of environmental dynamics but also drive the development of innovative solutions for mitigating degradation and promoting sustainable practices. AI's importance in environmental sustainability lies in its ability to process and extract actionable insights from diverse datasets. ML algorithms, powered by these datasets, identify patterns, correlations, and trends that conventional methods cannot. This enables us to understand complex environmental phenomena like climate change, biodiversity loss, and ecosystem disruptions, enabling informed conservation efforts(Wu et al., 2022).

ML algorithms are crucial for predictive modeling, predicting environmental changes and their impacts. They help policymakers, conservationists, and stakeholders devise proactive strategies to mitigate risks and protect ecosystems. AI and ML are also essential in optimizing resource management practices, ensuring efficient utilization and conservation. In energy sustainability, AI-driven solutions optimize energy production, distribution, and consumption. ML algorithms enable the integration of renewable energy sources, enhance power grid efficiency, and manage demand, reducing reliance on fossil fuels and minimizing carbon footprint(G. D. Sharma et al., 2020). AI and ML are revolutionizing agriculture by optimizing resource usage and reducing environmental impact. Precision agriculture uses AI algorithms to process data from soil moisture, weather patterns, and crop health, providing real-time insights to farmers. This empowers them to make informed decisions about irrigation, fertilizer application, and pest control, leading to increased yields and reduced environmental impact. The integration of AI and ML in environmental sustainability initiatives fosters innovation and efficiency(Ben Ayed & Hanana, 2021).

Energy distribution faces numerous challenges, including reliance on conventional energy sources and inefficiencies in existing systems. The use of fossil fuels contributes to greenhouse gas emissions and energy security due to resource depletion. Aging infrastructure and centralized grids suffer from transmission losses and instability, leading to wastage and reliability issues. These issues highlight the need for cleaner, more efficient energy systems(Boopathi, 2024b; Rahamathunnisa et al., 2024; Revathi et al., 2024a). Challenges in the energy sector present opportunities for innovation and transformation. Renewable energy sources like solar, wind, and hydroelectric power offer a sustainable alternative to fossil fuels. Smart grids, powered by AI and ML, enable decentralized energy distribution, empowering consumers to generate their own energy. This shift promotes resilience, reduces transmission losses, and promotes a more sustainable energy landscape(Pallathadka et al., 2023).

The agricultural sector faces numerous challenges, including climate change-induced disruptions, soil degradation, water depletion, and excessive use of agrochemicals. However, AI and ML-driven precision agriculture technologies can revolutionize practices by optimizing resource usage and enhancing crop yields. Real-time monitoring of soil health, crop growth patterns, and pest infestations using sensors, drones, and data analytics allows for precise interventions like targeted irrigation, optimized fertilization, and integrated pest management. This reduces environmental impact and ensures food security(Benos et al., 2021).

The integration of AI and ML in energy distribution and agriculture presents a potential for a sustainable future. By harnessing renewable energy surplus, precision agriculture technologies can be developed, promoting self-sufficiency and sustainability. This also allows for energy demand predictions to be optimized. AI-driven solutions can also create 'agro-energy' systems, converting agricultural residues into biofuels or energy generation. This circular economy approach minimizes waste and maximizes resource efficiency, paving the way for a more interconnected and sustainable future(Boopathi, 2024a; Pachiappan et al., 2024a; Rebecca et al., 2024a). Technological advancements, especially in AI and ML, present opportunities for transformative solutions in energy distribution and agriculture, potentially revolutionizing these sectors and significantly contributing to environmental sustainability(Tzachor et al., 2022).

ENERGY DISTRIBUTION OPTIMIZATION USING AI AND ML

Current Energy Distribution Challenges

The contemporary energy distribution landscape grapples with a multitude of challenges, stemming from traditional centralized energy grids reliant on fossil fuels. One prominent issue involves the inefficiencies inherent in these systems, resulting in transmission losses, grid instability, and compromised reliability. Additionally, the demand for energy continues to escalate, posing challenges in meeting fluctuating demands while ensuring a stable supply. Moreover, the integration of renewable energy sources introduces complexities due to their intermittency, necessitating innovative solutions for seamless incorporation into existing grids(Ali & Choi, 2020).

Role of AI in Energy Demand Forecasting

AI plays a pivotal role in mitigating the challenges associated with energy distribution, particularly through its capability in energy demand forecasting. Machine Learning algorithms, when fed with historical consumption data, weather patterns, economic indicators, and other pertinent variables, enable accurate predictions of future energy demands. These predictive models offer insights into consumption patterns, allowing energy providers to anticipate peak loads and fluctuations, thereby optimizing resource allocation, reducing wastage, and enhancing grid stability. AI-driven demand forecasting empowers energy planners and grid operators to make informed decisions in resource management, ensuring a reliable and efficient supply of energy(Ahmad et al., 2021).

Machine Learning for Smart Grids and Energy Efficiency

Smart grids, empowered by Machine Learning, stand as a cornerstone in the quest for energy efficiency and sustainability. ML algorithms embedded within smart grids facilitate real-time monitoring and analysis of energy consumption patterns. These systems employ advanced analytics to detect anomalies, predict system failures, and optimize distribution in response to changing demands. By leveraging AI, smart grids enable demand-side management, allowing consumers to actively participate in energy conservation efforts through load shifting or demand response programs. Furthermore, AI-driven optimization algorithms enhance the integration of renewable energy sources into the grid, ensuring their efficient utilization while maintaining grid stability(Rangel-Martinez et al., 2021).

The utilization of AI and ML in energy distribution optimization not only addresses the current challenges but also fosters a more resilient, efficient, and sustainable energy ecosystem. These technologies pave the way for adaptive and responsive energy grids capable of meeting evolving demands while embracing renewable energy sources, ultimately contributing to a more sustainable and environmentally conscious energy distribution system(Boopathi, 2024a, 2024b, 2024c).

AI APPLICATIONS IN RENEWABLE ENERGY INTEGRATION

Artificial Intelligence (AI) is revolutionizing the renewable energy sector by offering innovative solutions that enhance efficiency, reliability, and scalability(Saeed et al., 2020). The figure 1 depicts the use of AI in integrating renewable energy sources.

Figure 1. AI applications in renewable energy integration

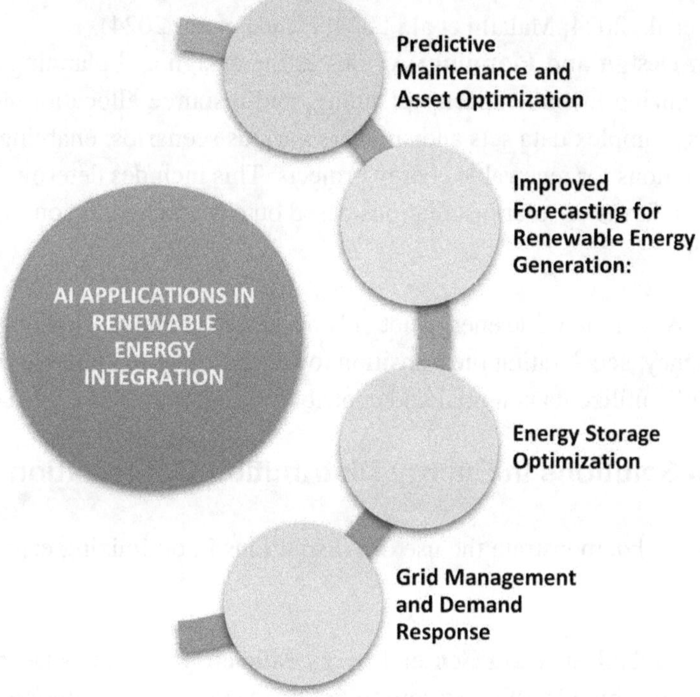

- **Predictive Maintenance and Asset Optimization:** AI-driven predictive maintenance utilizes machine learning algorithms to analyze data from sensors embedded in renewable energy infrastructure. By monitoring equipment performance and detecting anomalies, AI predicts potential failures before they occur, enabling proactive maintenance. This approach optimizes asset utilization, minimizes downtime, and extends the lifespan of renewable energy assets, ensuring consistent energy production(Paul et al., 2024).
- **Improved Forecasting for Renewable Energy Generation:** AI algorithms, particularly machine learning models, enhance the accuracy of renewable energy generation forecasts. These models analyze historical weather data, satellite imagery, and real-time inputs to predict solar irradiance, wind speeds, and other environmental variables. Such precise forecasts aid energy grid operators and utilities in effectively managing the intermittency of renewable sources, optimizing energy distribution, and integrating them seamlessly into the grid(Boopathi, 2024b; Das et al., 2024; Rahamathunnisa et al., 2024).
- **Energy Storage Optimization:** AI plays a crucial role in optimizing energy storage systems, addressing the intermittency challenge of renewable sources. Machine learning algorithms analyze patterns in energy demand and generation, determining the most efficient storage solutions. AI-driven optimization ensures the effective charging and discharging of batteries or other storage systems, enhancing grid stability and facilitating a more reliable supply of renewable energy.
- **Grid Management and Demand Response:** AI-powered grid management systems utilize predictive analytics to balance energy supply and demand. Machine learning algorithms analyze data on consumption patterns, weather forecasts, and market dynamics to optimize energy distribu-

tion. Additionally, AI enables demand response programs, allowing consumers to adjust their energy usage based on real-time pricing or grid conditions, thus promoting more efficient energy utilization(Das et al., 2024; Malathi et al., 2024; Nanda et al., 2024).

- **Energy System Design and Planning:** AI aids in the design and planning of renewable energy systems by optimizing layout, capacity planning, and resource allocation. Machine learning algorithms process complex data sets and simulate various scenarios, enabling the identification of optimal configurations for renewable energy projects. This includes determining the most suitable locations for wind farms or solar installations based on environmental conditions and energy generation potential.

The integration of AI in renewable energy not only addresses technical challenges but also promotes innovation and efficiency, accelerating the transition towards a more sustainable energy landscape, enabling the sector to fully utilize its potential and contribute significantly to global clean energy.

Case Studies: AI Solutions in Energy Distribution Optimization

The case studies provided demonstrate the use of AI solutions in optimizing energy distribution(Yang et al., 2020).

- **Google's DeepMind AI for Data Center Energy Efficiency:** Google's DeepMind, an AI subsidiary, developed a machine learning system to optimize energy consumption in their data centers. The AI system, known as "DeepMind for Google," utilized historical data on cooling systems, power usage, and environmental factors to predict future energy consumption patterns. By adjusting parameters in real-time, such as temperature and airflow, the AI system achieved a 30% reduction in energy usage for cooling, contributing to substantial energy savings and improved efficiency in data center operations.(Rebecca et al., 2024a)
- **Pacific Gas and Electric (PG&E) - AI for Grid Optimization:** PG&E, a leading energy company in the United States, implemented AI solutions to optimize grid operations. Using machine learning algorithms, PG&E analyzed massive amounts of data from smart meters, weather forecasts, and grid sensors. The AI system predicted electricity demand patterns more accurately, enabling proactive grid management. This optimization reduced inefficiencies in energy distribution, minimized transmission losses, and enhanced grid reliability, ensuring a stable energy supply for consumers(Das et al., 2024; Nanda et al., 2024).
- **Siemens - AI for Predictive Maintenance in Wind Turbines:** Siemens, a global leader in energy technology, integrated AI-driven predictive maintenance in their wind turbines. By collecting and analyzing sensor data from turbines, machine learning models predicted potential failures and maintenance needs. This proactive approach allowed Siemens to schedule maintenance precisely when needed, reducing downtime and optimizing turbine performance. As a result, Siemens achieved higher operational efficiency, increased turbine reliability, and reduced maintenance costs.
- **National Grid - AI-enabled Energy Demand Forecasting:** National Grid, a major energy transmission company in the UK, employed AI algorithms for accurate energy demand forecasting. Using machine learning models, National Grid analyzed historical consumption data, weather patterns, and market trends. The AI-driven forecasts improved accuracy in predicting energy demand

peaks and troughs, allowing National Grid to optimize energy distribution, avoid overloads, and better manage the grid, ensuring a more reliable energy supply for consumers(Boopathi, 2024c; Pachiappan et al., 2024a).

- **ENEL - AI in Smart Grid Management:** ENEL, a multinational energy company, implemented AI in smart grid management. By leveraging machine learning algorithms, ENEL analyzed data from smart meters and sensors across the grid. The AI system enabled real-time monitoring, anomaly detection, and predictive analysis. This proactive approach optimized energy flow, reduced losses, and allowed for better integration of renewable energy sources into the grid, contributing to a more resilient and sustainable energy infrastructure(Naval & Yusta, 2021).

SUSTAINABLE AGRICULTURE THROUGH AI AND ML

Sustainable Agriculture and its Importance

Sustainable agriculture aims to meet current agricultural needs without compromising the ability of future generations to meet their own needs. It emphasizes environmentally friendly practices, conservation of resources, and the promotion of economic viability for farmers. By prioritizing soil health, water conservation, biodiversity preservation, and minimizing chemical inputs, sustainable agriculture seeks to ensure long-term food security while mitigating environmental impacts such as soil degradation, water pollution, and loss of biodiversity(Agrawal et al., 2023; Pachiappan et al., 2023; Sankar et al., 2023).

AI for Precision Farming and Crop Monitoring

AI technologies, particularly in precision farming, revolutionize agricultural practices by offering precise and data-driven solutions. Through the use of AI-powered drones, sensors, and satellite imagery, farmers can collect detailed information about their fields, including soil moisture levels, crop health, pest infestations, and growth patterns. Machine learning algorithms analyze this vast array of data, providing actionable insights and enabling farmers to make informed decisions in real-time. This facilitates targeted interventions such as precise irrigation, optimal fertilizer application, and timely pest control, thereby minimizing resource wastage and maximizing crop yields(Boopathi, 2024c; Pachiappan et al., 2024a).

ML Algorithms for Soil Health and Nutrient Management

Machine learning algorithms are instrumental in soil health assessment and nutrient management. By processing data collected from soil sensors, historical soil data, and environmental factors, ML models predict soil quality, nutrient deficiencies, and soil erosion risks. These insights assist farmers in implementing customized soil management practices, such as tailored fertilization schedules and soil conservation measures, optimizing soil health while minimizing the environmental impact of excessive chemical usage(Boopathi et al., 2023a, 2023b; Jeevanantham et al., 2022).

Sustainable Soil-Less Agriculture

The integration of sustainable soil-less agriculture, also known as hydroponics or aeroponics, with AI & ML is a promising area of agricultural innovation(Koshariya, Kalaiyarasi, et al., 2023; Koshariya, Khatoon, et al., 2023; Vanitha et al., 2023). The figure 2 depicts the use of AI in integrating renewable energy sources.

Figure 2. Integration of sustainable soil-less agriculture

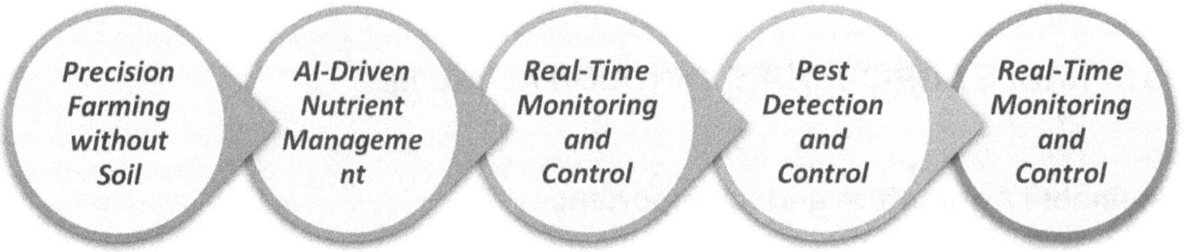

Precision Farming Without Soil

- Hydroponics and Aeroponics: Soil-less cultivation methods that rely on nutrient-rich water or mist for plant growth, offering controlled environments for optimal crop production(Boopathi et al., 2023b; Gnanaprakasam et al., 2023; P. R. Kumar et al., 2023).
- AI-ML Integration: AI algorithms optimize nutrient delivery, monitor plant health, and predict growth patterns in these controlled environments, maximizing yields without soil dependence.

AI-Driven Nutrient Management

- Nutrient Optimization: AI algorithms analyze plant growth data, environmental conditions, and nutrient solutions, precisely regulating nutrient delivery to match crop requirements.
- ML for Growth Patterns: Machine learning models forecast plant growth based on historical data, adjusting nutrient solutions dynamically for optimal growth without soil.

Real-Time Monitoring and Control

- Sensors and AI Integration: AI-powered systems leverage sensor data to monitor pH levels, humidity, temperature, and nutrient concentrations in hydroponic or aeroponic setups.
- Automated Control: AI systems regulate environmental factors in real time, ensuring ideal conditions for plant growth without the need for soil-based adjustments(Revathi et al., 2024b).

Pest Detection and Control

- Early Detection: AI-ML models analyze patterns in plant health data to detect early signs of pests or diseases in soil-less environments.
- Precision Pest Management: AI enables targeted interventions, such as precise application of organic pesticides or alterations in environmental conditions to deter pests without soil disruption.

Resource Efficiency and Sustainability

- Water and Resource Optimization: AI-ML algorithms optimize water usage in hydroponic systems, reducing waste and ensuring efficient resource utilization.
- Reduced Environmental Footprint: Soil-less agriculture, coupled with AI, minimizes the need for large land areas and conserves water, making it an environmentally sustainable option.

Adaptive Crop Cultivation

- Adaptive Growing Conditions: AI systems adapt cultivation parameters based on changing environmental factors or crop requirements, ensuring consistent and optimized growth.
- Crop Diversification: AI-ML models facilitate experimentation with a wide range of crops and growth conditions, expanding cultivation possibilities in soil-less setups.

The integration of AI and ML in soil-less agriculture is revolutionizing traditional farming methods, enhancing resource efficiency, minimizing environmental impact, and promoting year-round crop production, thus enhancing food security and agricultural resilience.

Predictive Analytics in Pest and Disease Control

AI and ML play a pivotal role in predictive analytics for pest and disease control in agriculture. By leveraging historical data on pest occurrences, weather patterns, crop types, and environmental conditions, machine learning models predict the likelihood and severity of pest infestations or disease outbreaks(Sathish et al., 2023; M. Sharma et al., 2024; Sundaramoorthy et al., 2024). These predictive models enable farmers to implement preventive measures proactively, such as targeted application of pesticides or deploying natural predators, precisely when and where they are needed. This proactive approach minimizes the use of chemical inputs, reduces crop losses, and safeguards crop health, contributing to sustainable and environmentally friendly pest management practices(Boopathi, 2023a).

Integrating AI-Driven Solutions for Efficient Water Usage

AI-driven solutions are being utilized to optimize water usage, a crucial aspect of sustainable agriculture, particularly in regions facing water scarcity or erratic rainfall patterns(Kumara et al., 2023).

- **Precision Irrigation**: AI-enabled systems analyze soil moisture data, weather forecasts, and plant water requirements to deliver precise amounts of water directly to the root zone. This targeted irrigation minimizes water wastage, reduces runoff, and ensures optimal plant hydration, enhancing crop yields while conserving water resources.
- **Crop-specific Water Management**: Machine learning algorithms process data related to different crop varieties, growth stages, and their specific water needs. This allows for tailored water management strategies, optimizing irrigation schedules and water application methods based on the individual requirements of different crops within a field.
- **Smart Water Monitoring**: AI-powered sensors and IoT devices monitor water usage in real-time, providing insights into water consumption patterns. Machine learning models analyze this data to identify inefficiencies and suggest improvements, enabling farmers to make informed decisions for water conservation and efficient usage.

AI-driven pest and disease control solutions and water optimization in agriculture can reduce environmental impact, conserve resources, and improve sustainability. These technologies enable farmers to adopt more precise, resource-efficient, and environmentally conscious practices, contributing to a more sustainable agricultural ecosystem(Kumara et al., 2023). The integration of AI-driven solutions for efficient water usage involves utilizing technology to enhance water management across various sectors as shown in Figure 3.

Precision Irrigation in Agriculture: AI-powered smart irrigation systems use algorithms to analyze soil moisture, weather forecasts, crop type, and evapotranspiration rates for precise watering. AI adjusts irrigation schedules and volumes, reducing waste from runoff and evaporation.

Leak Detection and Pipeline Management: AI-enabled sensors and data analytics are used for leak detection in water supply systems, while predictive maintenance optimizes pipeline maintenance schedules and prevents water loss due to infrastructure issues.

Smart Water Metering and Consumption Analysis: AI-powered smart meters monitor and analyze real-time water consumption patterns, enabling consumers to make informed decisions and conserve water through behavioral analysis.

Water Quality Monitoring and Treatment: AI is being used for water quality assessment, predicting contamination and impurities, and optimizing treatment processes to ensure efficient removal of contaminants and reduce wastage.

Urban Planning and Management: AI models aid urban planning by optimizing water distribution networks and reservoir management, and predict demand patterns, enabling authorities to plan infrastructure upgrades or demand-responsive pricing strategies.

Climate Adaptation and Water Resource Forecasting: AI enhances climate modeling and reservoir optimization, predicting rainfall patterns and droughts, promoting proactive water resource management and sustainable water availability.

Figure 3. Integration of AI-driven solutions for efficient water usage

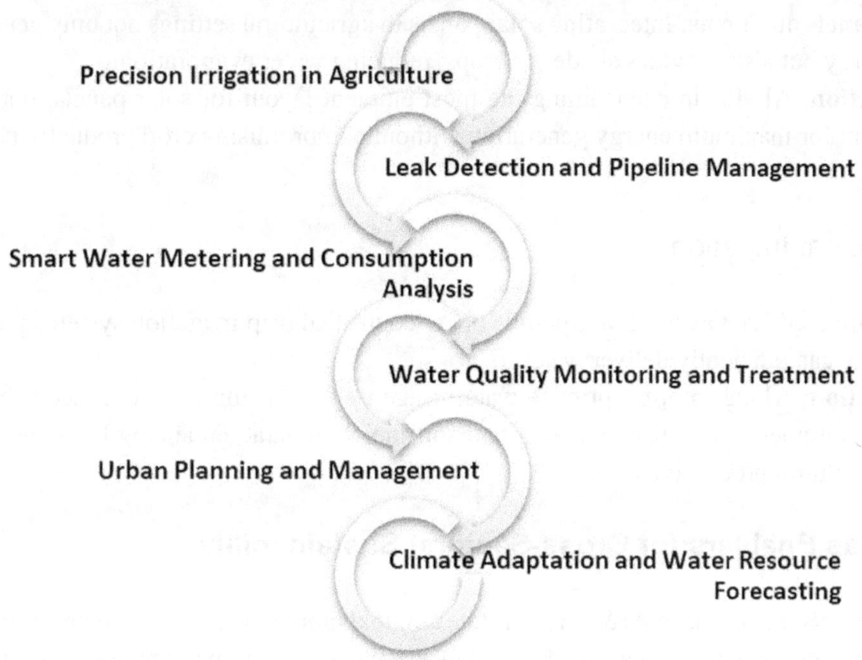

Integrating AI-driven solutions for efficient water usage spans across agricultural, urban, and industrial domains. By harnessing the power of AI to optimize water management, conserve resources, and enhance sustainability, these solutions pave the way for more responsible and efficient water usage across various sectors.

SYNERGIES AND INTERSECTIONS

Overlapping Solutions: Energy and Agriculture

Energy from Agricultural Waste

- **Anaerobic Digestion**: Agricultural waste, such as crop residues or animal manure, can be utilized in anaerobic digesters to produce biogas, a renewable energy source.
- **Intersection**: AI & ML algorithms optimize the process by predicting optimal conditions for biogas production, managing feedstock composition, and enhancing biogas yield through predictive analytics(Boopathi, 2023a, 2023b).

Solar-Powered Agriculture

- **Solar Panels on Farms**: Integrating solar panels in agricultural settings not only generates renewable energy but also provides shade for crops, reducing water evaporation.
- **Intersection**: AI aids in determining the most efficient layout for solar panels, optimizing their placement for maximum energy generation without compromising crop productivity.

Energy-Efficient Irrigation

- **Solar-Powered Irrigation**: Solar pumps or AI-controlled drip irrigation systems powered by solar energy can efficiently deliver water to crops.
- **Intersection**: AI algorithms optimize water usage by forecasting irrigation needs based on crop water requirements, weather forecasts, and soil moisture data, enhancing both energy efficiency and agricultural productivity.

AI and ML as Enablers for Cross-Sectoral Sustainability

Data-Driven Decision-Making: AI & ML facilitate the amalgamation of data from energy and agriculture sectors, enabling cross-sectoral analysis. For instance, weather data impacting both energy demand and crop growth can be jointly analyzed for more accurate predictions(A. Sharma et al., 2020).

Circular Economy Approach: AI-enabled systems in both sectors promote a circular economy by optimizing resource utilization. For instance, AI can predict surplus energy from renewable sources, which could then be channeled to power AI-driven precision agriculture technologies.

Resilience through Integration: AI & ML foster interconnectivity between energy grids and smart agriculture systems. Excess energy could power agricultural operations during low-demand periods, while agricultural data could inform energy demand predictions.

Innovation in Sustainability: Lessons learned from AI applications in one sector can be adapted to the other. For instance, predictive maintenance models developed for energy infrastructure can be applied to optimize maintenance in agricultural machinery(Ingle et al., 2023; Nishanth et al., 2023; Ugandar et al., 2023).

The integration of AI & ML-driven solutions fosters synergies between the energy and agriculture sectors, leading to shared innovations that enhance sustainability, resource efficiency, and resilience across both domains. These intersections exemplify the potential for cross-sectoral collaboration, highlighting how advancements in one sector can be harnessed to drive sustainability in another.

Challenges in Integrating Energy and Agriculture Solutions

Technological Compatibility

- **Diverse Systems**: Energy and agriculture solutions often utilize different technologies, creating compatibility challenges when integrating systems for combined applications.

- **Data Standardization**: Varied data formats and standards between sectors hinder seamless data exchange and integration for joint decision-making using AI & ML(Ragazou et al., 2022).

Resource Allocation and Prioritization

- **Competing Resource Demands**: Both sectors require resources like land, water, and funding, often leading to conflicts in resource allocation for energy and agricultural purposes.
- **Policy Misalignment**: Inconsistent or conflicting policies across energy and agriculture sectors impede integrated planning and implementation efforts.

Knowledge and Adoption Barriers

- **Skill Set Requirements**: Implementing AI & ML solutions demands specialized skills and expertise, posing a challenge in fostering a workforce capable of managing integrated systems.
- **Adoption Rate**: Resistance to change and a lack of awareness about the benefits of integrated solutions hinder widespread adoption.

Case Studies: Successes in Combined Approaches

AgriVoltaics - Fraunhofer Institute, Germany

- **Integration**: Fraunhofer Institute deployed solar panels above agricultural fields, harnessing land for both energy generation and crop cultivation.
- **Outcome**: This combined approach increased land use efficiency, protected crops from extreme weather, reduced water evaporation, and enhanced solar panel efficiency through natural cooling(Ragazou et al., 2022; A. Sharma et al., 2020; Spanaki et al., 2022).

Wind-Solar Energy Farming - Tamil Nadu, India

- **Integration**: Farmers in Tamil Nadu combined wind turbines and solar panels on their lands, generating renewable energy while continuing agricultural activities.
- **Outcome**: The integrated approach provided dual revenue streams, supplemented farmers' incomes, and contributed to local renewable energy production.

Smart Water-Energy Nexus - Arizona, USA

- **Integration**: AI-controlled irrigation systems were powered by solar panels in agricultural fields in Arizona.

- **Outcome**: This combination optimized water usage, improved crop yields, and reduced energy costs, showcasing the synergy between efficient water usage and renewable energy.

Biomass Energy and Agricultural Waste - Sweden

- **Integration**: Sweden utilized agricultural waste for bioenergy production, employing AI-based models to predict feedstock availability and optimize bioenergy generation.
- **Outcome**: This approach reduced waste, provided renewable energy, and enhanced the economic viability of both the agricultural and energy sectors.

Case studies showcase successful integration of energy and agriculture solutions, highlighting innovative approaches that maximize synergies. Despite challenges, these examples demonstrate potential for sustainability, resource optimization, and economic benefits.

ETHICAL AND SOCIAL IMPLICATIONS

Ethical Considerations in AI and ML Applications for Environment

Ethical considerations in AI & ML applications are crucial for responsible and sustainable deployment, focusing on key ethical aspects(Boopathi & Khang, 2023). Ethical considerations are crucial in AI & ML applications for the environment to maximize their positive environmental impact, minimize risks, and ensure accountability and fairness in their deployment.

Transparency and Accountability

- **Algorithms**: Ensuring transparency in AI & ML models used for environmental applications is crucial. Understanding how these algorithms make decisions is essential for stakeholders to trust the results.
- **Decision-Making**: Clearly defining responsibility for AI-driven actions in environmental conservation is essential to address potential errors or unintended consequences.

Bias and Fairness

- **Data Bias in Models**: AI & ML models can inherit biases from historical data, potentially perpetuating environmental disparities or inaccuracies.
- **Ensuring Fairness**: Rigorous testing and monitoring are necessary to identify and rectify biases, ensuring fair and equitable outcomes in environmental applications.

Privacy and Data Security

- **Protection of Sensitive Data**: AI systems often require access to sensitive environmental data. Striking a balance between access for analysis and protecting individual privacy and sensitive information is crucial.
- **Data Governance**: Implementing robust data governance measures to safeguard against unauthorized access, misuse, or breaches is essential for maintaining public trust.

Environmental Impact of AI Systems

- **Energy Consumption**: AI models, especially deep learning algorithms, consume significant computational resources. Minimizing the environmental impact of these systems through energy-efficient hardware or optimization techniques is imperative.
- **Lifecycle Impact**: Assessing the environmental footprint of AI infrastructure and hardware throughout their lifecycle, from production to disposal, is crucial.

Public Engagement and Ethical Guidelines

- **Inclusive Decision-Making**: Involving diverse stakeholders, including communities affected by environmental changes, in the development and deployment of AI & ML solutions fosters inclusive decision-making.
- **Ethical Guidelines and Standards**: Establishing clear ethical guidelines and standards specific to AI applications in environmental conservation ensures responsible and ethical use of these technologies.

Global Cooperation and Governance

- **International Collaboration**: Environmental issues transcend borders. International cooperation and governance frameworks are crucial to ensure standardized ethical practices globally.
- **Policy and Regulation**: Governments and international bodies need to develop and enforce policies and regulations that guide the ethical use of AI & ML in environmental applications.

Social Impact: Access, Equity, and Adoption

AI and ML applications in environmental conservation must consider social impact, prioritize equitable access, and effectively engage communities for a more inclusive and sustainable future(Myilsamy & Sampath, 2017; Rahamathunnisa et al., 2023; Reddy et al., 2023).

Access to Technology

- **Digital Divide**: Access to AI & ML technologies is unevenly distributed, leading to disparities in access between regions or communities.
- **Addressing Barriers**: Efforts to bridge the digital divide through initiatives such as public-private partnerships, community programs, and infrastructure development are crucial for equitable access to AI-driven solutions for environmental conservation.

Equity in Implementation

- **Ensuring Inclusivity**: Implementing AI & ML solutions for environmental conservation should consider the needs of marginalized communities or regions, ensuring that benefits reach everyone equitably.
- **Addressing Socio-economic Disparities**: Consideration of socio-economic disparities in the deployment of these technologies is essential to prevent further marginalization.

Community Engagement and Empowerment

- **Community Involvement**: Engaging local communities in the development and deployment of AI & ML solutions fosters a sense of ownership and ensures that solutions align with their needs.
- **Capacity Building**: Providing training and capacity-building programs empowers communities to utilize and benefit from AI & ML technologies for sustainable environmental practices.

Ethical Use for Social Good

- **Social Impact Assessment**: Conducting thorough social impact assessments before deploying AI & ML solutions helps understand potential consequences and tailor implementations to maximize positive impact.
- **Accountability and Transparency**: Ensuring that AI & ML applications are accountable, transparent, and aligned with societal values is crucial for social acceptance and trust.

Education and Awareness

- **Raising Awareness**: Educating the public about the benefits and implications of AI & ML in environmental conservation fosters acceptance and promotes informed decision-making.
- **Promoting Ethical Literacy**: Encouraging ethical literacy regarding AI & ML technologies and their implications for society, particularly in environmental contexts, empowers individuals to engage meaningfully in discussions and decision-making processes.

Policy and Regulation

- **Policy Frameworks**: Governments and organizations should develop policies and regulations that prioritize equitable access and ensure that AI & ML applications benefit society at large(Saravanan et al., 2022).
- **Ethical Guidelines**: Establishing ethical guidelines and standards that emphasize equitable access and social inclusion in AI & ML deployments for environmental conservation is imperative.

Regulatory Frameworks and Policies for Responsible AI in Conservation

Data Governance and Privacy Protection

- **Data Collection and Use**: Policies should outline guidelines for ethical data collection, ensuring that environmental data is obtained responsibly and used transparently.
- **Privacy Protection**: Regulations must safeguard individuals' privacy rights when collecting and utilizing environmental data for AI-driven conservation efforts.

Ethical Guidelines and Standards

- **Ethical Principles**: Establishing ethical guidelines specific to AI applications in conservation, emphasizing environmental responsibility, fairness, and accountability.
- **Standards Compliance**: Enforcing compliance with ethical standards through certification processes and audits to ensure adherence to responsible AI practices.

Transparency and Explain-ability

- **Algorithm Transparency**: Policies should mandate transparency in AI algorithms used for conservation to ensure stakeholders understand how decisions are made.
- **Requirement**: Regulations may require AI systems to provide explanations for decisions or recommendations made in environmental conservation contexts.

Accountability and Liability

- **Defining Responsibility**: Clear delineation of responsibility for AI-driven actions in conservation to ensure accountability for outcomes or unintended consequences.
- **Liability Frameworks**: Policies should address liability issues in cases of AI-related environmental damage or negative impacts caused by faulty algorithms.

Bias Mitigation and Fairness

- **Bias Assessment**: Mandating thorough assessments to identify and mitigate biases in AI models used for conservation to ensure fair and equitable outcomes.
- **Fairness Guidelines**: Establishing guidelines to ensure that AI applications do not perpetuate or exacerbate environmental disparities.

International Collaboration and Standards

- **Global Cooperation**: International collaboration on regulatory frameworks to ensure consistency and coherence in ethical standards and responsible AI practices.
- **Standardization Efforts**: Participation in global efforts to develop standardized guidelines and best practices specific to AI in environmental conservation.

Continuous Monitoring and Adaptation

- **Ongoing Evaluation**: Regular assessment and adaptation of regulatory frameworks to keep pace with technological advancements and evolving ethical concerns.
- **Monitoring Implementation**: Monitoring the implementation of regulations to ensure compliance and address emerging ethical challenges in AI-driven conservation.

By establishing robust regulatory frameworks and policies, governments and organizations can ensure that AI applications in conservation align with ethical principles, prioritize environmental responsibility, and contribute positively to sustainable conservation efforts while safeguarding societal interests.

FUTURE DIRECTIONS AND CONCLUSION

Emerging Trends in AI and ML for Environmental Conservation

Emerging trends highlight the growing importance of AI and ML in addressing environmental challenges, promoting data-driven, sustainable conservation practices. Their integration into conservation efforts can enhance ecosystem understanding and promote proactive environmental management, as these technologies continue to evolve(M. Kumar et al., 2023; Pachiappan et al., 2024b; Rebecca et al., 2024b).

- **Autonomous Environmental Monitoring:** Integration of sensor networks with AI algorithms for real-time environmental monitoring, allowing autonomous data collection, analysis, and early detection of ecological changes or threats.
- **Generative Models for Environmental Simulation:** Utilization of GANs and other generative models to simulate environmental scenarios, aiding in understanding potential impacts of changes like deforestation or climate shifts.

- **AI-Enhanced Climate Modeling:** AI & ML algorithms augmenting climate models to enhance accuracy in predicting complex climate patterns, aiding in long-term climate forecasting and resilience planning(Pachiappan et al., 2024b; Rebecca et al., 2024b; Revathi et al., 2024b).
- **AI-Driven Precision Conservation:** AI-enabled precision mapping and analysis to identify optimal areas for conservation efforts, maximizing impact by focusing on high-priority ecological zones.
- **Reinforcement Learning for Ecosystem Management:** RL applications in optimizing ecosystem management strategies, facilitating dynamic decision-making to conserve biodiversity or manage natural resources effectively(BOOPATHI, 2022; Sampath et al., 2021).
- **AI-Based Wildlife Monitoring and Protection:** AI-powered computer vision systems for wildlife monitoring and anti-poaching efforts, enabling efficient surveillance and protection of endangered species.
- **AI-ML for Sustainable Agriculture:** AI-driven solutions optimizing sustainable agricultural practices, such as precision farming, smart irrigation, and crop disease detection for minimized environmental impact.
- **AI-ML in Circular Economy Solutions:** AI & ML applications to optimize resource reuse, recycling, and waste management, fostering a circular economy approach for minimal environmental footprint(Dhanya et al., 2023; Mohanty et al., 2023).
- **Ethical AI Frameworks for Conservation:** Growing focus on developing ethical frameworks specific to AI applications in conservation, ensuring responsible and accountable deployment of AI & ML technologies.
- **Federated Learning for Global Collaboration:** Collaborative AI approaches like federated learning enabling global data sharing while preserving data privacy, fostering international cooperation in environmental research.

Figure 4. Potential innovations and unexplored avenues

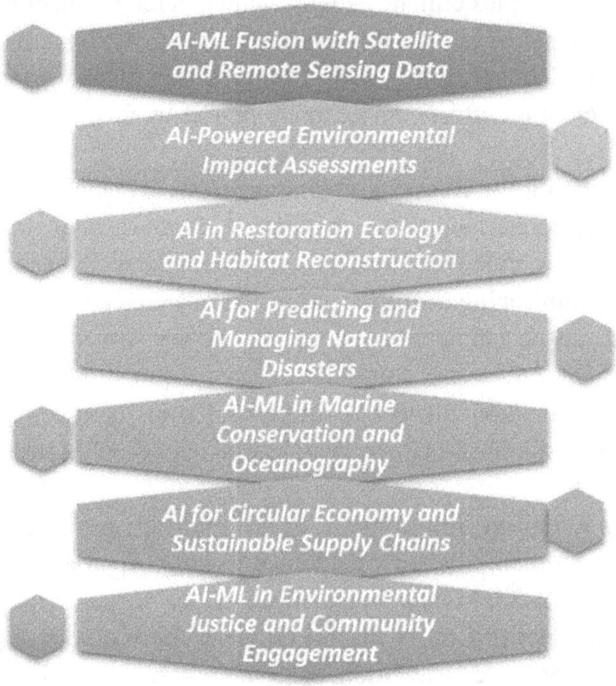

Potential Innovations and Unexplored Avenues

Leveraging AI & ML technologies in environmental conservation holds immense potential for addressing challenges, fostering sustainability, and creating a more resilient future for our planet(M. Kumar et al., 2023; Sundaramoorthy et al., 2024). Figure 4 presents potential innovations and unexplored avenues for growth and explained below.

AI-ML Fusion with Satellite and Remote Sensing Data

- **High-Resolution Imaging**: Integrating AI techniques with satellite imagery and remote sensing for higher resolution environmental monitoring and ecosystem analysis.
- **Real-Time Global Observation**: Exploring AI-powered systems for real-time global observation to track changes in ecosystems, biodiversity, and climate at a more granular level(Mishra & Mishra, 2023; Shaikh et al., 2022).

AI-Powered Environmental Impact Assessments

- **Automated Impact Assessment**: Utilizing AI for rapid and automated environmental impact assessments in development projects, aiding in quick decision-making with minimal environmental footprint.

- **Predictive Impact Modeling**: Developing predictive models using AI to forecast potential environmental impacts of policy decisions or human activities.

AI in Restoration Ecology and Habitat Reconstruction

- **Ecosystem Restoration Strategies**: AI-driven restoration plans and simulations for rebuilding degraded ecosystems or restoring habitats to support biodiversity.
- **Predictive Habitat Mapping**: AI algorithms predicting optimal locations for habitat reconstruction to aid in biodiversity conservation efforts.

AI for Predicting and Managing Natural Disasters

- **Disaster Prediction and Response**: Advanced AI models predicting natural disasters like wildfires, floods, or landslides, aiding in better disaster preparedness and response strategies.
- **AI-Driven Emergency Management**: AI-powered systems for real-time disaster response, optimizing resource allocation and rescue efforts during crises.

AI-ML in Marine Conservation and Oceanography

- **Marine Life Monitoring**: Developing AI solutions for monitoring marine ecosystems, tracking endangered species, and identifying illegal fishing activities.
- **Ocean Health Assessment**: AI-enabled analysis of oceanographic data to assess and predict ocean health, including acidification levels, temperature changes, and plastic pollution.

AI for Circular Economy and Sustainable Supply Chains

- **Supply Chain Optimization**: AI-driven solutions for designing sustainable supply chains, minimizing waste, and promoting circular economy practices in industries(Mohanty et al., 2023; Verma et al., 2024).
- **Life Cycle Analysis**: Utilizing AI to conduct comprehensive life cycle analyses for products and services to reduce environmental impacts across their entire life cycle.

AI-ML in Environmental Justice and Community Engagement

- **Community-Centered Conservation**: AI applications fostering community participation in conservation efforts, considering local knowledge and promoting environmental justice.
- **Ethical AI for Indigenous Knowledge**: Integrating indigenous knowledge with AI algorithms to preserve and leverage traditional conservation practices.

CONCLUSION

The chapter discusses the integration of AI and ML in environmental conservation, highlighting their potential to address complex ecosystem challenges. It discusses their applications in optimizing energy distribution, promoting sustainable agriculture, and fostering responsible environmental stewardship. However, it also highlights challenges like data privacy, bias mitigation, and ethical considerations, emphasizing the need for responsible deployment and regulatory frameworks in AI applications for conservation.

AI and ML technologies are transforming the energy and agriculture sectors, offering synergy and optimizing resource usage. Emerging trends like autonomous environmental monitoring and predictive impact modeling highlight the potential for further innovation. However, collaboration, ethical guidelines, and equitable access to these technologies are crucial for harnessing their full potential for environmental conservation. By embracing these innovations responsibly and focusing on inclusivity, we can create a more sustainable, resilient, and harmonious planet.

REFERENCES

Agrawal, A. V., Magulur, L. P., Priya, S. G., Kaur, A., Singh, G., & Boopathi, S. (2023). Smart Precision Agriculture Using IoT and WSN: In T. Murugan & N. E. (Eds.), Advances in Information Security, Privacy, and Ethics (pp. 524–541). IGI Global. https://doi.org/ doi:10.4018/978-1-6684-8145-5.ch026

Ahmad, T., Zhang, D., Huang, C., Zhang, H., Dai, N., Song, Y., & Chen, H. (2021). Artificial intelligence in sustainable energy industry: Status Quo, challenges and opportunities. *Journal of Cleaner Production*, *289*, 125834.

Ali, S. S., & Choi, B. J. (2020). State-of-the-art artificial intelligence techniques for distributed smart grids: A review. *Electronics (Basel)*, *9*(6), 1030.

Ben Ayed, R., & Hanana, M. (2021). Artificial intelligence to improve the food and agriculture sector. *Journal of Food Quality*, *2021*, 1–7.

Benos, L., Tagarakis, A. C., Dolias, G., Berruto, R., Kateris, D., & Bochtis, D. (2021). Machine learning in agriculture: A comprehensive updated review. *Sensors (Basel)*, *21*(11), 3758. PMID:34071553

Boopathi, S. (2022). Effects of Cryogenically-treated Stainless Steel on Eco-friendly Wire Electrical Discharge Machining Process. Preprint : Springer.

Boopathi, S. (2023a). Internet of Things-Integrated Remote Patient Monitoring System: Healthcare Application. In A. Suresh Kumar, U. Kose, S. Sharma, & S. Jerald Nirmal Kumar (Eds.), (pp. 137–161). Advances in Healthcare Information Systems and Administration. IGI Global. doi:10.4018/978-1-6684-6894-4.ch008

Boopathi, S. (2023b). Securing Healthcare Systems Integrated With IoT: Fundamentals, Applications, and Future Trends. In A. Suresh Kumar, U. Kose, S. Sharma, & S. Jerald Nirmal Kumar (Eds.), (pp. 186–209). Advances in Healthcare Information Systems and Administration. IGI Global. doi:10.4018/978-1-6684-6894-4.ch010

Boopathi, S. (2024a). Advancements in Machine Learning and AI for Intelligent Systems in Drone Applications for Smart City Developments. In *Futuristic e-Governance Security With Deep Learning Applications* (pp. 15–45). IGI Global.

Boopathi, S. (2024b). Energy Cascade Conversion System and Energy-Efficient Infrastructure. In Sustainable Development in AI, Blockchain, and E-Governance Applications (pp. 47–71). IGI Global.

Boopathi, S. (2024c). Sustainable Development Using IoT and AI Techniques for Water Utilization in Agriculture. In Sustainable Development in AI, Blockchain, and E-Governance Applications (pp. 204–228). IGI Global.

Boopathi, S., & Khang, A. (2023). AI-Integrated Technology for a Secure and Ethical Healthcare Ecosystem. In *AI and IoT-Based Technologies for Precision Medicine* (pp. 36–59). IGI Global.

Boopathi, S., Kumar, P. K. S., Meena, R. S., Sudhakar, M., & Associates. (2023a). Sustainable Developments of Modern Soil-Less Agro-Cultivation Systems: Aquaponic Culture. In Human Agro-Energy Optimization for Business and Industry (pp. 69–87). IGI Global.

Boopathi, S., Kumar, P. K. S., Meena, R. S., Sudhakar, M., & Associates. (2023b). Sustainable Developments of Modern Soil-Less Agro-Cultivation Systems: Aquaponic Culture. In Human Agro-Energy Optimization for Business and Industry (pp. 69–87). IGI Global.

Das, S., Lekhya, G., Shreya, K., Shekinah, K. L., Babu, K. K., & Boopathi, S. (2024). Fostering Sustainability Education Through Cross-Disciplinary Collaborations and Research Partnerships: Interdisciplinary Synergy. In Facilitating Global Collaboration and Knowledge Sharing in Higher Education With Generative AI (pp. 60–88). IGI Global.

Dhanya, D., Kumar, S. S., Thilagavathy, A., Prasad, D., & Boopathi, S. (2023). Data Analytics and Artificial Intelligence in the Circular Economy: Case Studies. In Intelligent Engineering Applications and Applied Sciences for Sustainability (pp. 40–58). IGI Global.

Galaz, V., Centeno, M. A., Callahan, P. W., Causevic, A., Patterson, T., Brass, I., Baum, S., Farber, D., Fischer, J., Garcia, D., & ... (2021). Artificial intelligence, systemic risks, and sustainability. *Technology in Society*, *67*, 101741.

Gnanaprakasam, C., Vankara, J., Sastry, A. S., Prajval, V., Gireesh, N., & Boopathi, S. (2023). Long-Range and Low-Power Automated Soil Irrigation System Using Internet of Things: An Experimental Study. In Contemporary Developments in Agricultural Cyber-Physical Systems (pp. 87–104). IGI Global.

Ingle, R. B., Swathi, S., Mahendran, G., Senthil, T., Muralidharan, N., & Boopathi, S. (2023). Sustainability and Optimization of Green and Lean Manufacturing Processes Using Machine Learning Techniques. In *Circular Economy Implementation for Sustainability in the Built Environment* (pp. 261–285). IGI Global.

Javaid, M., Haleem, A., Singh, R. P., Suman, R., & Gonzalez, E. S. (2022). Understanding the adoption of Industry 4.0 technologies in improving environmental sustainability. *Sustainable Operations and Computers*, *3*, 203–217.

Jeevanantham, Y. A., Saravanan, A., Vanitha, V., Boopathi, S., & Kumar, D. P. (2022). Implementation of Internet-of Things (IoT) in Soil Irrigation System. *IEEE Explore*, (pp. 1–5). IEEE.

Koshariya, A. K., Kalaiyarasi, D., Jovith, A. A., Sivakami, T., Hasan, D. S., & Boopathi, S. (2023). AI-Enabled IoT and WSN-Integrated Smart Agriculture System. In R. K. Gupta, A. Jain, J. Wang, S. K. Bharti, & S. Patel (Eds.), (pp. 200–218). Practice, Progress, and Proficiency in Sustainability. IGI Global. doi:10.4018/978-1-6684-8516-3.ch011

Koshariya, A. K., Khatoon, S., Marathe, A. M., Suba, G. M., Baral, D., & Boopathi, S. (2023). Agricultural Waste Management Systems Using Artificial Intelligence Techniques. In S. Kautish, N. K. Chaubey, S. B. Goyal, & P. Whig (Eds.), (pp. 236–258). Advances in Computational Intelligence and Robotics. IGI Global. doi:10.4018/978-1-6684-8171-4.ch009

Kumar, M., Kumar, K., Sasikala, P., Sampath, B., Gopi, B., & Sundaram, S. (2023). Sustainable Green Energy Generation From Waste Water: IoT and ML Integration. In Sustainable Science and Intelligent Technologies for Societal Development (pp. 440–463). IGI Global.

Kumar, P. R., Meenakshi, S., Shalini, S., Devi, S. R., & Boopathi, S. (2023). Soil Quality Prediction in Context Learning Approaches Using Deep Learning and Blockchain for Smart Agriculture. In Effective AI, Blockchain, and E-Governance Applications for Knowledge Discovery and Management (pp. 1–26). IGI Global.

Kumara, V., Mohanaprakash, T. A., Fairooz, S., Jamal, K., Babu, T., & B., S. (2023). Experimental Study on a Reliable Smart Hydroponics System: In P. Vasant, R. Rodríguez-Aguilar, I. Litvinchev, & J. A. Marmolejo-Saucedo (Eds.), *Advances in Environmental Engineering and Green Technologies* (pp. 27–45). IGI Global. https://doi.org/ doi:10.4018/978-1-6684-4118-3.ch002

Malathi, J., Kusha, K., Isaac, S., Ramesh, A., Rajendiran, M., & Boopathi, S. (2024). IoT-Enabled Remote Patient Monitoring for Chronic Disease Management and Cost Savings: Transforming Healthcare. In Advances in Explainable AI Applications for Smart Cities (pp. 371–388). IGI Global.

Mishra, H., & Mishra, D. (2023). Artificial Intelligence and Machine Learning in Agriculture: Transforming Farming Systems. *Res. Trends Agric. Sci*, *1*, 1–16.

Mohanty, A., Venkateswaran, N., Ranjit, P., Tripathi, M. A., & Boopathi, S. (2023). Innovative Strategy for Profitable Automobile Industries: Working Capital Management. In Handbook of Research on Designing Sustainable Supply Chains to Achieve a Circular Economy (pp. 412–428). IGI Global.

Myilsamy, S., & Sampath, B. (2017). Grey Relational Optimization of Powder Mixed Near-Dry Wire Cut Electrical Discharge Machining of Inconel 718 Alloy. *Asian Journal of Research in Social Sciences and Humanities*, *7*(3), 18–25.

Nanda, A. K., Sharma, A., Augustine, P. J., Cyril, B. R., Kiran, V., & Sampath, B. (2024). Securing Cloud Infrastructure in IaaS and PaaS Environments. In Improving Security, Privacy, and Trust in Cloud Computing (pp. 1–33). IGI Global.

Naval, N., & Yusta, J. M. (2021). Virtual power plant models and electricity markets-A review. *Renewable & Sustainable Energy Reviews*, *149*, 111393.

Nishant, R., Kennedy, M., & Corbett, J. (2020). Artificial intelligence for sustainability: Challenges, opportunities, and a research agenda. *International Journal of Information Management*, *53*, 102104.

Nishanth, J., Deshmukh, M. A., Kushwah, R., Kushwaha, K. K., Balaji, S., & Sampath, B. (2023). Particle Swarm Optimization of Hybrid Renewable Energy Systems. In *Intelligent Engineering Applications and Applied Sciences for Sustainability* (pp. 291–308). IGI Global.

Pachiappan, K., Anitha, K., Pitchai, R., Sangeetha, S., Satyanarayana, T., & Boopathi, S. (2024a). Intelligent Machines, IoT, and AI in Revolutionizing Agriculture for Water Processing. In *Handbook of Research on AI and ML for Intelligent Machines and Systems* (pp. 374–399). IGI Global.

Pachiappan, K., Anitha, K., Pitchai, R., Sangeetha, S., Satyanarayana, T., & Boopathi, S. (2024b). Intelligent Machines, IoT, and AI in Revolutionizing Agriculture for Water Processing. In *Handbook of Research on AI and ML for Intelligent Machines and Systems* (pp. 374–399). IGI Global.

Pachiappan, K., Anitha, K., Pitchai, R., Sangeetha, S., Satyanarayana, T. V. V., & Boopathi, S. (2023). Intelligent Machines, IoT, and AI in Revolutionizing Agriculture for Water Processing. In B. B. Gupta & F. Colace (Eds.), (pp. 374–399). Advances in Computational Intelligence and Robotics. IGI Global., doi:10.4018/978-1-6684-9999-3.ch015

Pallathadka, H., Mustafa, M., Sanchez, D. T., Sajja, G. S., Gour, S., & Naved, M. (2023). Impact of machine learning on management, healthcare and agriculture. *Materials Today: Proceedings*, *80*, 2803–2806.

Paul, A., & Thilagham, K. KG, J.-, Reddy, P. R., Sathyamurthy, R., & Boopathi, S. (2024). Multi-criteria Optimization on Friction Stir Welding of Aluminum Composite (AA5052-H32/B4C) using Titanium Nitride Coated Tool. Engineering Research Express.

Ragazou, K., Garefalakis, A., Zafeiriou, E., & Passas, I. (2022). Agriculture 5.0: A New Strategic Management Mode for a Cut Cost and an Energy Efficient Agriculture Sector. [s Note: MDPI stays neutral with regard to jurisdictional claims in published....]. *Energies*, *2022*(15), 3113.

Rahamathunnisa, U., Sudhakar, K., Murugan, T. K., Thivaharan, S., Rajkumar, M., & Boopathi, S. (2023). Cloud Computing Principles for Optimizing Robot Task Offloading Processes. In *AI-Enabled Social Robotics in Human Care Services* (pp. 188–211). IGI Global.

Rahamathunnisa, U., Sudhakar, K., Padhi, S., Bhattacharya, S., Shashibhushan, G., & Boopathi, S. (2024). Sustainable Energy Generation From Waste Water: IoT Integrated Technologies. In Adoption and Use of Technology Tools and Services by Economically Disadvantaged Communities: Implications for Growth and Sustainability (pp. 225–256). IGI Global.

Rangel-Martinez, D., Nigam, K., & Ricardez-Sandoval, L. A. (2021). Machine learning on sustainable energy: A review and outlook on renewable energy systems, catalysis, smart grid and energy storage. *Chemical Engineering Research & Design*, *174*, 414–441.

Rebecca, B., Kumar, K. P. M., Padmini, S., Srivastava, B. K., Halder, S., & Boopathi, S. (2024a). Convergence of Data Science-AI-Green Chemistry-Affordable Medicine: Transforming Drug Discovery. In *Handbook of Research on AI and ML for Intelligent Machines and Systems* (pp. 348–373). IGI Global.

Rebecca, B., Kumar, K. P. M., Padmini, S., Srivastava, B. K., Halder, S., & Boopathi, S. (2024b). Convergence of Data Science-AI-Green Chemistry-Affordable Medicine: Transforming Drug Discovery. In *Handbook of Research on AI and ML for Intelligent Machines and Systems* (pp. 348–373). IGI Global.

Reddy, M. A., Reddy, B. M., Mukund, C., Venneti, K., Preethi, D., & Boopathi, S. (2023). Social Health Protection During the COVID-Pandemic Using IoT. In *The COVID-19 Pandemic and the Digitalization of Diplomacy* (pp. 204–235). IGI Global.

Revathi, S., Babu, M., Rajkumar, N., Meti, V. K. V., Kandavalli, S. R., & Boopathi, S. (2024a). Unleashing the Future Potential of 4D Printing: Exploring Applications in Wearable Technology, Robotics, Energy, Transportation, and Fashion. In Human-Centered Approaches in Industry 5.0: Human-Machine Interaction, Virtual Reality Training, and Customer Sentiment Analysis (pp. 131–153). IGI Global.

Revathi, S., Babu, M., Rajkumar, N., Meti, V. K. V., Kandavalli, S. R., & Boopathi, S. (2024b). Unleashing the Future Potential of 4D Printing: Exploring Applications in Wearable Technology, Robotics, Energy, Transportation, and Fashion. In Human-Centered Approaches in Industry 5.0: Human-Machine Interaction, Virtual Reality Training, and Customer Sentiment Analysis (pp. 131–153). IGI Global.

Saeed, M. A., Ahmed, Z., Yang, J., & Zhang, W. (2020). An optimal approach of wind power assessment using Chebyshev metric for determining the Weibull distribution parameters. *Sustainable Energy Technologies and Assessments*, *37*, 100612.

Sampath, B., Sureshkumar, T., Yuvaraj, M., & Velmurugan, D. (2021). Experimental Investigations on Eco-Friendly Helium-Mist Near-Dry Wire-Cut EDM of M2-HSS Material. *Materials Research Proceedings*, *19*, 175–180.

Sankar, K. M., Booba, B., & Boopathi, S. (2023). Smart Agriculture Irrigation Monitoring System Using Internet of Things. In G. S. Karthick (Ed.), (pp. 105–121). Advances in Environmental Engineering and Green Technologies. IGI Global. doi:10.4018/978-1-6684-7879-0.ch006

Saravanan, A., Venkatasubramanian, R., Khare, R., Surakasi, R., Boopathi, S., Ray, S., & Sudhakar, M. (2022). POLICY TRENDS OF RENEWABLE ENERGY AND NON. *Renewable Energy*.

Sathish, T., Sunagar, P., Singh, V., Boopathi, S., Al-Enizi, A. M., Pandit, B., Gupta, M., & Sehgal, S. S. (2023). Characteristics estimation of natural fibre reinforced plastic composites using deep multi-layer perceptron (MLP) technique. *Chemosphere*, *337*, 139346. PMID:37379988

Shaikh, T. A., Rasool, T., & Lone, F. R. (2022). Towards leveraging the role of machine learning and artificial intelligence in precision agriculture and smart farming. *Computers and Electronics in Agriculture*, *198*, 107119.

Sharma, A., Jain, A., Gupta, P., & Chowdary, V. (2020). Machine learning applications for precision agriculture: A comprehensive review. *IEEE Access : Practical Innovations, Open Solutions*, *9*, 4843–4873.

Sharma, G. D., Yadav, A., & Chopra, R. (2020). Artificial intelligence and effective governance: A review, critique and research agenda. *Sustainable Futures : An Applied Journal of Technology, Environment and Society*, 2, 100004.

Sharma, M., Sharma, M., Sharma, N., & Boopathi, S. (2024). Building Sustainable Smart Cities Through Cloud and Intelligent Parking System. In *Handbook of Research on AI and ML for Intelligent Machines and Systems* (pp. 195–222). IGI Global.

Spanaki, K., Sivarajah, U., Fakhimi, M., Despoudi, S., & Irani, Z. (2022). Disruptive technologies in agricultural operations: A systematic review of AI-driven AgriTech research. *Annals of Operations Research*, *308*(1–2), 491–524.

Sundaramoorthy, K., Singh, A., Sumathy, G., Maheshwari, A., Arunarani, A., & Boopathi, S. (2024). A Study on AI and Blockchain-Powered Smart Parking Models for Urban Mobility. In *Handbook of Research on AI and ML for Intelligent Machines and Systems* (pp. 223–250). IGI Global.

Tzachor, A., Devare, M., King, B., & Avin, S., & Ó hÉigeartaigh, S. (2022). Responsible artificial intelligence in agriculture requires systemic understanding of risks and externalities. *Nature Machine Intelligence*, *4*(2), 104–109.

Ugandar, R., Rahamathunnisa, U., Sajithra, S., Christiana, M. B. V., Palai, B. K., & Boopathi, S. (2023). Hospital Waste Management Using Internet of Things and Deep Learning: Enhanced Efficiency and Sustainability. In Applications of Synthetic Biology in Health, Energy, and Environment (pp. 317–343). IGI Global.

Vanitha, S. K. R., & Boopathi, S. (2023). Artificial Intelligence Techniques in Water Purification and Utilization. In P. Vasant, R. Rodríguez-Aguilar, I. Litvinchev, & J. A. Marmolejo-Saucedo (Eds.), (pp. 202–218). Advances in Environmental Engineering and Green Technologies. IGI Global. doi:10.4018/978-1-6684-4118-3.ch010

Verma, R., Christiana, M. B. V., Maheswari, M., Srinivasan, V., Patro, P., Dari, S. S., & Boopathi, S. (2024). Intelligent Physarum Solver for Profit Maximization in Oligopolistic Supply Chain Networks. In *AI and Machine Learning Impacts in Intelligent Supply Chain* (pp. 156–179). IGI Global.

Wu, C.-J., Raghavendra, R., Gupta, U., Acun, B., Ardalani, N., Maeng, K., Chang, G., Aga, F., Huang, J., Bai, C., & ... (2022). Sustainable ai: Environmental implications, challenges and opportunities. *Proceedings of Machine Learning and Systems*, *4*, 795–813.

Yang, Z., Hu, J., Ai, X., Wu, J., & Yang, G. (2020). Transactive energy supported economic operation for multi-energy complementary microgrids. *IEEE Transactions on Smart Grid, 12*(1), 4–17.

Chapter 2
AI-Integrated Biosensors and Bioelectronics for Agriculture

Pankaj Bhambri
https://orcid.org/0000-0003-4437-4103
Guru Nanak Dev Engineering College, Ludhiana, India

Alex Khang
https://orcid.org/0000-0001-8379-4659
Global Research Institute of Technology and Engineering, USA

ABSTRACT

The integration of artificial intelligence (AI) alongside biosensors and bioelectronics has introduced a new age in precision agriculture, enabling the enhancement of agricultural processes. This book chapter examines the utilization of AI, biosensors, & bioelectronics in the domains of agriculture as well as aquaculture. The chapter commences with an examination of the agricultural sector's preoccupations, such as constraints on resources, environmental sustainability, and productivity. Subsequently, the text examines how biosensors and bioelectronics might effectively tackle these challenges by continuously monitoring and collecting agricultural data in real-time. A significant portion of the chapter focuses on the integration of artificial intelligence with biosensors. Farmers can make well-informed judgments about crop management, disease detection, and yield optimization by utilizing effective data analysis, predictive modeling, and decision-making. AI-powered biosensors have enhanced agricultural efficiency in several case studies and real-world applications.

INTRODUCTION

AI-integrated biosensors and bioelectronics have emerged as transformative tools in the realm of agriculture, revolutionizing traditional farming practices. These advanced technologies utilize artificial intelligence to understand and evaluate data gathered by biosensors integrated into the agricultural system (Patel and Lee, 2020). Through the seamless integration of artificial intelligence algorithms and biosensors, farmers may obtain immediate and accurate information about the state of soil, crop conditions, &

DOI: 10.4018/979-8-3693-1794-5.ch002

environmental elements. This empowers them to make precise and informed decisions based on data. This synergy between AI and bioelectronics not only enhances the efficiency of resource utilization but also facilitates early detection of diseases, optimizing crop management strategies. The result is a more sustainable and productive agricultural system that not only meets the growing demands for food but also promotes environmental stewardship through intelligent and adaptive farming practices.

Overview of the Convergence of AI, Biosensors, and Bioelectronics in Agriculture

The convergence of AI, biosensors, and bioelectronics in agriculture marks a paradigm shift in modern farming practices. This transformative integration involves the amalgamation of advanced technologies to enhance agricultural efficiency, sustainability, and productivity. Biosensors, which have the ability to identify and quantify particular biological indicators, are strategically placed in the agricultural setting to gather up-to-date information on the quality of soil, the health of crops, and the surrounding environmental conditions. Artificial intelligence systems process and analyze this abundant information, empowering farmers to make well-informed and accurate decisions. AI's ability to interpret complex datasets facilitates optimized resource management, early detection of diseases, and the development of tailored strategies for crop cultivation (Gonzalez-Dominguez et al., 2019). The synergy between AI, biosensors, and bioelectronics not only empowers farmers with actionable insights but also contributes to the creation of smart and adaptive farming systems, paving the way for a more resilient and sustainable future in agriculture.

Significance and Potential Impact on Precision Agriculture

The convergence of AI, biosensors, and bioelectronics holds immense significance for precision agriculture, promising to revolutionize the way crops are cultivated and managed. The integration of these technologies offers unprecedented opportunities for precise and data-driven decision-making, significantly impacting various aspects of precision agriculture.

An important advantage is the improved surveillance and control of soil conditions. Soil-embedded biosensors can offer immediate information on nutrient concentrations, moisture levels, and other pertinent factors. AI systems can subsequently analyze this data to provide customized suggestions for efficient irrigation, fertilizing, and soil health management. Such a high degree of accuracy guarantees the efficient utilization of resources, hence avoiding both waste and environmental consequences.

Moreover, the integration of artificial intelligence (AI) and biosensors allows for the timely identification of illnesses and pest (Li et al., 2018). Through the ongoing surveillance of crops for indications of distress or contamination, farmers can employ focused treatments, such as accurate administration of insecticides or the utilization of resilient crop breeds. This not only enhances agricultural productivity but also diminishes the dependence on chemical inputs, thereby promoting sustainable and environmentally-friendly farming methods.

AI-driven analytics have the ability to optimize the use of energy, water, and fertilizers in resource management, thereby aligning farming methods with environmental conservation aims. Additionally, the real-time nature of the data provided by biosensors ensures that farmers can adapt quickly to changing conditions, mitigating risks associated with unpredictable weather patterns or other environmental factors.

The convergence of AI, biosensors, and bioelectronics in precision agriculture signifies a transformative leap towards a more sustainable, efficient, and resilient farming future. The ability to harness real-time data for precise decision-making not only improves crop yields and quality but also contributes to environmental stewardship by minimizing the ecological footprint of agricultural activities.

CHALLENGES IN AGRICULTURE

The agricultural sector faces a multitude of challenges that impede its ability to meet the growing global demand for food while ensuring sustainability (He et al., 2020). Climate change poses a significant threat, leading to erratic weather patterns, extreme events, and shifts in growing seasons, affecting crop yields and livestock productivity (Chen and Liu, 2020). Water scarcity exacerbates these challenges, as agriculture is a major consumer of water resources. Additionally, the depletion of fertile soil, often accelerated by unsustainable farming practices, jeopardizes long-term productivity. The globalization of trade and the interconnectedness of supply chains expose farmers to market volatility and economic uncertainties. Moreover, there's a pressing need to address issues of rural livelihoods, as small-scale farmers often lack access to modern technology, credit, and market opportunities. Combating pests and diseases, adopting precision farming techniques, and embracing sustainable practices are pivotal in overcoming these challenges and fostering a resilient and productive agricultural sector for the future.

Exploration of Key Challenges in the Agricultural Sector

The agricultural sector faces various challenges that impact its productivity, sustainability, and ability to meet the growing global demand for food (Duan and Xu, 2020). These challenges are complex and interconnected, and addressing them requires a multi-faceted approach. Some key challenges in the agricultural sector include:

- Climate Change: Changes in meteorological patterns, severe occurrences, and erratic climatic circumstances can exert a substantial influence on agricultural output and cattle farming. Climate change exacerbates the proliferation of pests and diseases, hence impacting agricultural productivity.
- Water Scarcity: Agriculture is a water-intensive activity, and water scarcity is a growing concern in many regions. Competition for water resources among various sectors, including agriculture, poses a challenge for sustainable water management.
- Land Degradation: Soil erosion, deforestation, and improper land use contribute to the degradation of arable land. This reduces the land's fertility and overall productivity, making sustainable land management crucial for long-term agricultural success.
- Biodiversity Loss: The loss of biodiversity in agricultural ecosystems can lead to decreased resilience to pests and diseases, affecting crop yields (Bhambri et al., 2020). Monoculture and the use of certain agricultural practices contribute to the decline in biodiversity.
- Globalization and Market Access: Small-scale farmers often struggle to access global markets due to trade barriers, lack of infrastructure, and limited resources (Wu and Gao, 2019). Globalization can also lead to unequal distribution of benefits, favoring larger agribusinesses.

- Technological Gap: There is often a technology divide between large-scale commercial farms and smallholder farmers. Access to modern agricultural technologies, such as precision farming and advanced machinery, can significantly improve productivity, but smaller farmers may lack the resources to adopt these technologies.
- Rural Poverty: Many farmers, particularly in developing countries, face economic challenges and live in poverty. Insufficient access to credit, education, and markets can perpetuate the cycle of poverty and hinder agricultural development.
- Food Security and Hunger: Despite global food production being sufficient to feed the population, issues such as unequal distribution, food waste, and lack of access contribute to food insecurity and hunger in many parts of the world.
- Policy and Governance: Inconsistent or inadequate agricultural policies and governance can hinder the sector's growth (Sharma et al., 2020). Lack of supportive policies, corruption, and insufficient infrastructure can impede agricultural development.
- Labor Shortages and Rural-Urban Migration: In some regions, there is a shortage of skilled labor in agriculture due to rural-to-urban migration. This can affect productivity and exacerbate the challenges faced by those remaining in rural areas.

Analysis of Constraints on Resources, Environmental Sustainability, and Productivity

- Resource Constraints: The agricultural sector relies heavily on water, and numerous areas are confronted with water scarcity as a result of climate change, excessive extraction of water in the ground, and ineffective water management methods. The availability of arable land is limited, and expansion of agricultural areas often leads to deforestation and loss of biodiversity. Modern agriculture is heavily reliant on energy for machinery, irrigation, and transportation, contributing to environmental degradation and resource depletion.
- Environmental Sustainability: Intensive farming practices can lead to soil erosion, depletion of nutrients, and loss of soil fertility, affecting long-term agricultural productivity (Kaur et al., 2020). The use of pesticides, herbicides, and fertilizers can lead to environmental pollution, soil contamination, and harm to non-target species, impacting overall ecosystem health. Monoculture and the use of certain agricultural practices can lead to a decline in biodiversity, affecting natural pollinators and other essential ecosystem services.
- Productivity Challenges: Climate variations, encompassing severe weather phenomena, prolonged dry spells, and inundations, have the potential to disturb agricultural output and result in crop deficiencies. Globalization and climate change contribute to the spread of pests and diseases, posing threats to crops and livestock, thereby affecting productivity. Some regions lack access to modern agricultural technologies, limiting their productivity potential compared to areas with better technological infrastructure (Li and Wang, 2019).
- Social and Economic Factors: Inequitable distribution of resources such as land, water, and technology can contribute to disparities in agricultural productivity among different communities and regions. Fluctuations in commodity prices, lack of market access, and inadequate infrastructure for storage and transportation can impact the economic viability of farming.

- Policy and Governance: Inadequate policies and governance may lead to unsustainable agricultural practices, including over-extraction of resources, deforestation, and the misuse of chemicals (Chlingaryan and Sukkarieh, 2018). Subsidies that encourage unsustainable practices or favor certain crops over others can contribute to environmental degradation and resource mismanagement.

ROLE OF BIOSENSORS AND BIOELECTRONICS

Biosensors and bioelectronics play a pivotal role in modernizing agriculture by offering innovative solutions for monitoring and enhancing various aspects of the farming process. These technologies utilize biological components to identify particular substances, offering immediate data on soil conditions, crop well-being, and environmental elements (Devadutta et al., 2020). Biosensors are utilized in agriculture to quantify characteristics such as nutrient concentrations, moisture levels, and the existence of insecticides or pathogens in crop. Bioelectronics, on the other hand, integrate these biosensors with electronic devices, enabling the collection and interpretation of data for informed decision-making. By facilitating precision farming, these technologies help optimize resource use, minimize environmental impact, and improve overall crop yield and quality. The application of biosensors and bioelectronics in agriculture represents a transformative step towards sustainable and efficient farming practices, aligning with the growing demand for precision agriculture in the face of global food security challenges.

Examination of How Biosensors and Bioelectronics Address Agricultural Challenges

Biosensors and bioelectronics offer valuable tools to address various agricultural challenges by providing real-time, accurate, and cost-effective monitoring solutions. Here's an examination of how these technologies address key challenges in agriculture:

- Precision Farming: Traditional farming often relies on uniform application of resources, leading to inefficient use of water, fertilizers, and pesticides. Biosensors enable precise measurement of soil nutrient levels, moisture content, and other parameters. By integrating this information with bioelectronics, farmers can implement precision farming techniques, tailoring resource application based on specific needs, thereby optimizing resource usage.
- Crop Health Monitoring: Detecting early signs of diseases, pests, or nutrient deficiencies in crops is crucial for preventing yield loss (Qureshi and Zia, 2019). Biosensors can detect the presence of pathogens or changes in plant physiology indicative of stress. Bioelectronics facilitate the continuous monitoring of crop health, allowing for timely interventions such as targeted pesticide application or adjustments to nutrient supplementation.
- Environmental Impact and Sustainability: Agricultural practices can contribute to environmental degradation through the overuse of chemicals and inefficient resource management. Biosensors help monitor the environmental impact of farming activities, such as detecting the levels of pesticides or nutrients in runoff water. This information aids in adopting sustainable practices, minimizing the ecological footprint of agriculture.
- Resource Management: Efficient use of water is critical, especially in regions facing water scarcity. Biosensors can measure soil moisture levels in real-time, allowing for precise irrigation

scheduling. This prevents over-irrigation, conserving water resources and reducing energy costs associated with pumping water.

- Data-Driven Decision-Making: Farmers often face challenges in making informed decisions based on accurate and timely data. Biosensors generate a wealth of data on various agricultural parameters. Bioelectronics facilitate the processing and interpretation of this data, providing farmers with actionable insights for better decision-making, crop planning, and resource allocation.
- Early Warning Systems: Climate change and unpredictable weather patterns pose risks to crop production. Biosensors can be integrated into early warning systems that monitor weather conditions and provide alerts to farmers, allowing them to take preventive measures against extreme weather events, thereby reducing the impact on crops.
- Yield Optimization: Maximizing crop yield while minimizing resource input is a central goal in agriculture. Biosensors and bioelectronics help optimize the growing conditions by continuously monitoring and adjusting factors such as soil quality, temperature, and light conditions. This leads to improved crop yields and resource-use efficiency.

Continuous Monitoring and Real-Time Data Collection for Informed Decision-Making

The advanced technologies employ biological sensing elements coupled with electronic components, allowing for the constant measurement of crucial parameters such as soil health, crop conditions, and environmental factors (Zhang and Sun, 2019). The integration of AI enhances the interpretation and analysis of the collected data, providing farmers with actionable insights in real-time. Artificial intelligence systems analyze extensive quantities of data, detecting correlations and trends that may prove difficult for people to perceive. This dynamic approach facilitates proactive decision-making, allowing farmers to promptly address issues such as nutrient deficiencies, pest infestations, or adverse weather conditions. By harnessing continuous monitoring and AI-driven analytics, agriculture becomes more adaptive, precise, and efficient, ultimately contributing to sustainable practices and improved productivity in the face of evolving environmental and economic challenges.

INTEGRATION OF ARTIFICIAL INTELLIGENCE

AI technologies, such as machine learning as well as data analytics, are utilized to examine extensive datasets derived from diverse sources, like as sensors, satellites, & farm machines. This allows for the use of predictive modeling to optimize decisions on crop management and resource allocation. Applications of artificial intelligence in agriculture span a range of features, from crop monitoring as well as yield prediction to identifying diseases and pest control. AI-powered smart farming systems offer immediate analysis of soil quality, meteorological trends, and crop status, enabling farmers to make well-informed choices. The incorporation of AI not only enhances production by maximizing resource use but also adds to ecological sustainability by decreasing the use of inputs like water, pesticides, and fertilizers (Jiang and Li, 2019). As agriculture faces increasing challenges, the synergy between AI and farming practices promises to revolutionize the industry, fostering innovation and resilience in the pursuit of global food security.

Introduction to the Integration of AI With Biosensors in Agriculture

Biosensors, capable of detecting biological elements with high specificity, are now being seamlessly integrated with AI systems to create intelligent platforms for continuous monitoring and data analysis in agriculture. This synergy enables real-time and precise tracking of crucial parameters such as soil health, crop conditions, and environmental factors. AI algorithms, adept at processing large volumes of diverse data, enhance the interpretation of biosensor-generated information, providing farmers with actionable insights for informed decision-making. This convergence not only facilitates early detection of diseases, pests, and nutrient deficiencies but also optimizes resource utilization, contributing to sustainable and efficient farming practices. The integration of AI with biosensors stands at the forefront of technological innovation in agriculture, promising to transform the industry by promoting resilience, productivity, and environmental stewardship.

Overview of Data Analysis, Predictive Modeling, and Decision-Making Processes

Data analysis involves the systematic examination of vast datasets generated by various sources such as sensors, satellites, and farm machinery. This process unveils patterns, trends, and correlations, providing valuable insights into soil health, weather conditions, and crop performance. Predictive modeling leverages this data to forecast future scenarios, from crop yields to potential pest outbreaks or adverse weather events. Artificial Intelligence (AI) and machine learning algorithms play a crucial role in this phase, making sense of complex data sets and enhancing the accuracy of predictions. Informed by data analysis and predictive modeling, decision-making in agriculture becomes more strategic and adaptive (Tao and Zhang, 2018). Farmers can optimize resource allocation, implement precision farming techniques, and respond promptly to emerging challenges, fostering sustainable practices and improving overall productivity. This integration of data-driven analysis, predictive modeling, and informed decision-making represents a pivotal shift in agriculture, contributing to resilience, efficiency, and the pursuit of global food security.

APPLICATIONS IN CROP MANAGEMENT

AI-integrated biosensors provide real-time insights into soil conditions, nutrient levels, and crop health, allowing for precise and proactive interventions (Zhen and Yan, 2018). In crop monitoring, AI analyzes data from biosensors to detect early signs of diseases, pest infestations, and nutrient deficiencies, enabling farmers to implement targeted and timely responses. Predictive modeling, powered by AI, anticipates potential yield variations and optimizes irrigation and fertilization strategies (Shafieian and Le, 2019). Additionally, the integration of biosensors with AI facilitates the automation of farming processes, enabling smart irrigation systems and autonomous machinery that respond dynamically to changing environmental conditions. This holistic strategy not only improves the productivity and excellence of crops but also encourages sustainable agricultural methods by reducing the utilization of resources & agrochemicals, representing a substantial advancement towards more effective and resilient crop administration in agriculture.

Case Studies on how AI-Integrated Biosensors Optimize Crop Management

AI-integrated biosensors play a crucial role in optimizing crop management by providing real-time data and insights. Here are a few case studies that highlight the impact of such technologies:

- Prospera Technologies (Precision Agriculture in Greenhouses): Prospera Technologies is a company that utilizes AI and computer vision to optimize crop management in greenhouses (Islam and Zhong, 2020). Their system includes cameras and sensors that continuously monitor plant health, growth, and environmental conditions. Prospera's AI algorithms utilize data gathered from sensors to assess and deliver insights on several parameters, including water usage, nutritional levels, and the presence of pests. The technology assists greenhouse operators in making informed decisions to enhance irrigation plans, fertilization, and pest management through the utilization of data. By implementing Prospera's technology, greenhouse operators have reported significant improvements in crop yield, resource efficiency, and overall farm profitability. The AI-driven insights enable precise adjustments, leading to better resource allocation and reduced environmental impact.
- John Deere (Autonomous Farming with AI): John Deere, a leading agricultural machinery manufacturer, has been incorporating AI into its equipment to enable autonomous farming practices. AI-integrated sensors on John Deere tractors and other farming machinery collect data on soil conditions, crop health, and weather patterns. The AI algorithms use this data to make real-time decisions, such as adjusting planting depth, optimizing seed spacing, and applying the right amount of fertilizers. Farmers using John Deere's AI-integrated equipment experience increased efficiency, reduced fuel consumption, and improved crop yields. The autonomous features also help address labor shortages in the agriculture industry.
- IBM Watson Decision Platform for Agriculture (Crop Management Insights): The IBM Watson Decision System for Agriculture utilizes artificial intelligence to assess diverse data sources, such as satellite imagery, forecasts for the weather, and IoT sensors, in order to offer precise information on crop management. The platform integrates with existing farm management systems and collects data to generate personalized recommendations for farmers. It offers insights on optimal planting times, disease risk assessments, and irrigation scheduling. Users of IBM's platform have indicated that it has enhanced their decision-making process, resulting in higher crop yields, decreased input expenses, and enhanced farm sustainability. The AI-powered suggestions assist farmers in adjusting to fluctuating circumstances and minimizing potential hazards.
- Gamaya (Hyperspectral Imaging for Precision Agriculture): Gamaya specializes in hyperspectral imaging technology for agriculture. Their AI-based solution analyzes the reflected light spectrum from crops to provide detailed information about plant health, nutrient levels, and potential diseases. Gamaya's sensors are mounted on drones or satellites, capturing high-resolution images of fields. The AI algorithms then process this data to create detailed maps, allowing farmers to identify specific areas that may require attention. Farmers using Gamaya's technology have seen improvements in precision farming, enabling targeted interventions. By applying inputs such as fertilizers and pesticides only where needed, farmers reduce costs and minimize environmental impact while optimizing crop yields.

Disease Detection, Yield Optimization, and Other Agricultural Practices

AI-integrated biosensors have the potential to revolutionize agriculture by enhancing disease detection, optimizing yield, and improving overall crop management. Here's a discussion on these aspects:

- Disease Detection: AI-integrated biosensors can detect diseases in crops at an early stage by analyzing subtle changes in plant physiology, such as variations in leaf color, size, and texture (Amato and Moscatelli, 2018). Early detection enables farmers to promptly implement measures to avoid the proliferation of illnesses. Biosensors can utilize machine learning techniques to analyze collected data and detect trends linked to certain diseases. This enables the development of predictive models, helping farmers anticipate potential outbreaks and take preventive measures (Smith and Brown, 2018). By precisely identifying the affected areas, farmers can apply targeted treatments such as pesticides or fungicides, reducing the overall use of chemicals and minimizing environmental impact.
- Yield Optimization: AI algorithms may utilize data from biosensors, imagery from satellites, and various other sources to conduct analysis and offer valuable insights into the aspects that impact crop productivity. These parameters include soil health, meteorological conditions, and insect infestations (Singh et al., 2021). Subsequently, farmers can make well-informed decisions to maximize crop production. AI can enhance resource utilization by offering real-time suggestions based on prevailing environmental conditions and crop requirements, thereby optimizing the usage of water, fertilizers, and pesticides. This helps prevent overuse and reduces environmental impact. Continuous monitoring of crops through biosensors allows farmers to track growth patterns and identify areas of improvement (Sharma and Khamparia, 2020). This data can be utilized to fine-tune agricultural methods and optimize overall efficiency.
- Other Agricultural Practices: AI-integrated biosensors can assess soil health by measuring nutrient levels, pH, and moisture content (Bhambri, 2021). This information helps farmers make decisions on fertilization and irrigation, contributing to sustainable and efficient farming practices. AI can analyze weather data in real-time and predict future climate conditions. This information is crucial for farmers to adapt their practices to changing weather patterns, ensuring resilience against climate-related challenges. AI can be integrated with robotic systems for tasks like planting, harvesting, and weeding. This automation not only reduces labor costs but also allows for more precise and efficient farming operations.
- Challenges and Considerations: The incorporation of artificial intelligence (AI) in agriculture entails the gathering and examination of confidential information (Sun et al., 2019). To maintain the confidence of farmers and stakeholders, it is crucial to tackle issues pertaining to data privacy and security. The implementation of AI technology in agriculture may encounter obstacles in terms of availability, especially for small-scale growers (Wang and Zhang, 2018). It is imperative to exert efforts in order to enhance the affordability and user-friendliness of these technologies. It is imperative to provide education and training to farmers in order to enable them to proficiently utilize AI-integrated biosensors. Extension assistance and education programs are essential for promoting the use of these advances in technology.

REAL-WORLD EFFICIENCY ENHANCEMENTS

AI-powered biosensors have demonstrated significant potential in enhancing agricultural efficiency, offering innovative solutions to longstanding challenges (Dhiman and Aulakh, 2020). The integration of artificial intelligence with biosensors has led to real-world efficiency enhancements across various aspects of agriculture. Here are some areas and success stories that highlight the impact of AI-powered biosensors in agriculture:

- Precision Agriculture (Smart Farming with AI): Precision agriculture utilizes AI-driven biosensors to gather and assess data pertaining to weather patterns, soil conditions, a& crop health (Gao et al., 2018). These sensors provide real-time insights, allowing farmers to optimize irrigation, fertilization, and pest control.
- Crop Monitoring and Disease Detection (Plantix): Plantix is an application for mobile devices that employs artificial intelligence to examine photographs of plant diseases. Agriculturalists have the ability to capture images of their agricultural products, and the application detects possible diseases or deficits in nutrients. This allows for early detection and targeted treatment, reducing the risk of crop loss.
- Optimized Resource Management (Blue River Technology): Acquired by John Deere, Blue River Technology utilizes computer vision and machine learning to enable robotic systems that precisely apply herbicides to crops. This targeted approach reduces the use of chemicals, saving costs and minimizing environmental impact.
- Automated Harvesting (Octinion's Rubion): Rubion is an AI-powered strawberry-picking robot that uses sensors and machine learning algorithms to identify ripe strawberries and pick them with precision (Akhavan and Doroodgar, 2020). This technology addresses labor shortages and increases harvesting efficiency.
- Livestock Monitoring (Connecterra): Connecterra's AI-powered biosensors are used for monitoring dairy cows. The sensors track the health and behavior of individual cows, providing insights into their well-being and optimizing feeding schedules. This leads to improved milk production and overall herd management.
- Weather Forecasting for Agriculture (IBM's The Weather Company): AI is being employed to analyze weather patterns and predict conditions that can impact agriculture. By integrating this information into decision-making processes, farmers can make more informed choices about planting, harvesting, and other critical activities.
- Supply Chain Optimization (AgShift): AgShift uses AI to assess the quality of harvested produce. By automating the grading process, it helps streamline the supply chain by ensuring that only high-quality products reach the market. This reduces waste and ensures that consumers receive top-quality produce.
- Water Management (Hortau): Hortau's soil monitoring system uses AI to analyze data from sensors measuring soil moisture levels. This allows farmers to optimize irrigation, conserve water, and prevent over-watering, ultimately improving crop yield and resource efficiency.

EMERGING TRENDS AND ADVANCEMENTS

The incorporation of artificial intelligence with biosensors for agricultural use shows significant potential for the implementation of sustainable farming methods. The current trends indicate a movement towards increased connectivity, reliance on data, and precision in agriculture. Potential breakthroughs have the potential to significantly transform the industry, enhancing its resilience, efficiency, and environmental sustainability.

Overview of Current Trends in AI-Integrated Biosensors

- Miniaturization and Wearable Biosensors: Biosensors are becoming smaller and more portable, enabling their integration into wearable devices (Wang et al., 2017). AI algorithms can analyze real-time data from these biosensors, providing instant feedback on various health and environmental parameters.
- Internet of Things (IoT) Integration: AI-integrated biosensors are often connected to IoT platforms, creating a network of devices that can share and analyze data. This connectivity allows for comprehensive monitoring and management of agricultural systems.
- Machine Learning for Data Analysis: Advanced machine learning algorithms are being employed to analyze large datasets generated by biosensors. These algorithms can identify patterns, predict trends, and optimize agricultural practices based on real-time data.
- Precision Agriculture: AI-integrated biosensors contribute to precision agriculture by providing detailed information on soil health, crop conditions, and environmental factors. This enables farmers to make data-driven decisions, optimizing resource usage and increasing crop yield.
- Remote Sensing and Imaging: AI algorithms can process data from remote sensing technologies and imaging devices (Bhambri, 2020). This allows for the detection of diseases, pests, and nutrient deficiencies in crops, facilitating timely interventions.
- Biological and Environmental Monitoring: Biosensors can monitor the presence of specific biological agents, pathogens, or contaminants in the environment. AI helps in rapid and accurate identification of these elements, enhancing the overall safety of agricultural products.

Prospective Advancements and Potential Impact on Sustainable Agriculture

- Predictive Analytics for Crop Management: Future advancements in AI could lead to more sophisticated predictive models for crop management. Farmers could receive proactive recommendations for planting times, irrigation schedules, and pest control measures based on predictive analytics.
- Automated Decision-Making: AI could evolve to a point where it can autonomously make decisions regarding irrigation, fertilization, and pest control. This would optimize resource usage, reduce waste, and improve overall efficiency in agricultural practices.
- Climate-Resilient Agriculture: AI-integrated biosensors could play a crucial role in developing climate-resilient agriculture (Hossain and Ahmed, 2019). Advanced algorithms could analyze

climate data to predict extreme weather events, allowing farmers to take preventive measures and adapt their strategies accordingly.

- Customized Nutrient Management: AI could provide personalized nutrient management plans for each crop based on real-time data. This would ensure that crops receive the specific nutrients they need, minimizing environmental impact and maximizing yield.
- Collaborative Farming Networks: AI could facilitate the creation of collaborative farming networks where data from multiple farms is analyzed collectively. This shared knowledge could lead to more effective regional strategies for sustainable agriculture.
- Biological Pest Control Optimization: AI could enhance the use of biological pest control methods by analyzing data on pest populations and their natural predators. This would promote environmentally friendly pest control practices.

DATA SECURITY AND ETHICAL CONSIDERATIONS

As the agricultural sector increasingly integrates advanced technologies, data is generated and utilized at unprecedented levels. This data encompasses a wide range of information, including crop health, soil conditions, weather patterns, and more. Here's a discussion on the importance of data security and ethical considerations in the context of agricultural technology:

Importance of Data Security in Agricultural Technology

- Farm Operation Efficiency (Data-Driven Decision Making): Agricultural technologies, such as precision farming and IoT devices, generate large amounts of data. Farmers rely on this data to make informed decisions about crop management, irrigation, and resource allocation. Ensuring the security of this data is essential to maintaining the efficiency of farm operations.
- Protection of Sensitive Information (Farm and Crop Data): The data collected in agricultural technology often includes sensitive information about the farm's layout, crop yields, and farming practices. Protecting this information is crucial to prevent unauthorized access and potential misuse, which could harm the competitiveness of the farm.
- Market Competitiveness (Intellectual Property): Agricultural technology companies invest heavily in research and development (Liakos and Busato, 2018). Protecting the intellectual property related to innovative farming solutions is critical to maintaining a competitive edge. Data security helps prevent unauthorized access to proprietary algorithms, crop models, and other valuable assets.
- Supply Chain Integrity (Traceability and Certification): With the increasing demand for transparency in the food supply chain, data security is vital to maintaining the integrity of traceability and certification systems. Ensuring the accuracy and reliability of data along the supply chain is essential for meeting regulatory requirements and consumer expectations.

Ethical Considerations in Advanced Agricultural Technologies

- Data Ownership and Consent (Farmer Consent): Farmers should have control over the data generated on their farms. Clear policies on data ownership and obtaining informed consent from farmers are crucial ethical considerations. Farmers should be aware of how their data will be used and have the option to opt out if they choose.
- Environmental Impact (Sustainable Practices): Ethical considerations extend to the environmental impact of advanced agricultural technologies (Liu and Guo, 2021). Ensuring that technology adoption aligns with sustainable farming practices and environmental conservation is essential. Avoiding practices that lead to soil degradation, water pollution, or excessive resource consumption is imperative.
- Accessibility and Equity (Digital Divide): There is a risk of creating a digital divide where only certain farmers or regions have access to advanced technologies. Ethical considerations include promoting equal access to technology, training programs, and support services, ensuring that technological benefits are distributed equitably among farming communities.
- Algorithmic Bias (Fairness in Decision-Making): Advanced technologies often rely on algorithms for decision-making. Ethical considerations include addressing algorithmic bias to prevent discrimination and ensure fair outcomes. This is particularly important when technology is used to determine resource allocation, pricing, or access to support services.
- Community and Social Impacts (Community Engagement): Engaging with local communities and considering the social impacts of technological adoption are ethical imperatives (Bhambri et al., 2020). This includes assessing how technology might affect traditional farming practices, local economies, and community well-being.

ISSUES, PROBLEMS, AND TRENDS WITH RELEVANT EVIDENCE AND SUPPORTING ARGUMENTS

While AI-integrated biosensors and bioelectronics offer tremendous potential for transforming agriculture, addressing privacy concerns, ensuring affordability, promoting interoperability, and upholding ethical considerations are crucial for their successful and responsible deployment. The ongoing trends in predictive analytics, IoT integration, sensor technologies, and collaborative platforms are positive indicators that the agricultural industry is moving towards a more data-driven and sustainable future. However, several issues, problems, and trends deserve careful analysis to ensure the successful implementation and widespread adoption of these technologies.

Issues and Problems

- Data Privacy and Security Concerns: The integration of AI in biosensors generates vast amounts of sensitive data related to crop health, soil conditions, and environmental factors. There is a need for robust data privacy regulations and secure infrastructure to prevent unauthorized access and misuse of agricultural data, as breaches could have severe consequences for farmers and the entire food supply chain.

- Affordability and Accessibility: High initial costs and the need for specialized knowledge for AI-integrated systems may limit their accessibility, particularly for small-scale and resource-limited farmers. For widespread adoption, there must be efforts to make these technologies more affordable through subsidies, incentives, or collaborative initiatives that share costs and knowledge among farmers.
- Interoperability and Standardization: The diversity of AI and sensor technologies in the agriculture sector often leads to interoperability challenges, hindering seamless integration and data sharing. The development and adoption of industry standards can facilitate interoperability, allowing different devices and systems to work together effectively and promoting a more cohesive agricultural ecosystem.
- Ethical Considerations in Decision-Making: AI algorithms make critical decisions based on data, and biases in the data or algorithms can lead to unfair outcomes. Efforts should be made to address biases and ensure transparency in decision-making processes, emphasizing ethical considerations to prevent unintended consequences and promote trust among farmers and consumers.

Trends

- Machine Learning for Predictive Analytics: AI algorithms can analyze historical data to predict crop diseases, optimize planting schedules, and recommend precise interventions. The trend towards using machine learning for predictive analytics can significantly improve resource efficiency, reduce waste, and increase overall agricultural productivity.
- IoT Integration: The integration of biosensors with IoT devices allows real-time monitoring of various parameters such as soil moisture, nutrient levels, and weather conditions. This trend enhances the granularity and accuracy of data, enabling farmers to make informed decisions promptly, leading to more efficient resource utilization.
- Advancements in Sensor Technologies: Ongoing developments in sensor technologies, including miniaturization and improved sensitivity, contribute to more precise and real-time data collection. Continued advancements in sensor technologies enhance the effectiveness of AI-integrated biosensors, enabling farmers to gain deeper insights into their crops and the surrounding environment.
- Collaborative Platforms and Knowledge Sharing: Platforms that facilitate collaboration and knowledge sharing among farmers, researchers, and technology developers are gaining traction. The trend towards collaborative platforms fosters a community-driven approach, enabling the collective sharing of insights, best practices, and solutions, ultimately accelerating the adoption of AI-integrated biosensors in agriculture.

SCALABILITY AND FUTURE PROSPECTS

The integration of artificial intelligence enhances the capabilities of biosensors by enabling real-time data analysis, predictive modeling, and adaptive decision-making (Bahceci, 2018). However, as the scale of agricultural operations varies widely, from small farms to large agribusinesses, ensuring the scalability of AI-integrated biosensors becomes essential. To address this, scalable architectures and technologies need to be employed, allowing seamless integration and communication across diverse sensor networks

(Wang et al., 2021). Cloud computing and edge computing solutions can play a pivotal role in managing and processing data efficiently, providing a foundation for scalable AI applications in precision agriculture.

Looking towards future prospects, the evolving landscape of precision agriculture is promising, with AI-integrated biosensors poised to revolutionize farming practices. The continued development of advanced machine learning algorithms and sensor technologies will likely lead to even more accurate and context-aware agricultural insights. Future systems may involve autonomous drones or robotics equipped with AI-driven biosensors, contributing to increased automation and efficiency in farming processes. Moreover, the incorporation of precision agriculture alongside other burgeoning technologies, like the IoT & 5G connectivity, has the capacity to establish a holistic and integrated agricultural ecosystem.

As precision agriculture evolves, considerations for environmental sustainability and resource management will become increasingly significant (Bhambri et al., 2021). AI-integrated biosensors can contribute to optimized resource utilization by providing insights into soil health, crop conditions, and water usage. The scalability of these technologies will enable their adoption across diverse agricultural landscapes, benefiting both small-scale farmers and large-scale agribusinesses. However, it is crucial to address challenges related to data privacy, standardization, and the ethical use of AI in agriculture to ensure a responsible and inclusive deployment of these technologies on a global scale. In conclusion, the future prospects of AI-integrated biosensors in precision agriculture are promising, but careful attention to scalability, ethical considerations, and sustainable practices will be essential for long-term success.

CONCLUSION

The key findings underscore the critical role that AI plays in enhancing the capabilities of biosensors, allowing for real-time data analysis, predictive modeling, and adaptive decision-making in precision agriculture. The scalability of these technologies is identified as a fundamental consideration, emphasizing the need for scalable architectures and technologies to accommodate the diverse scales of agricultural operations. The insights provided throughout the chapter highlight the promising future prospects of AI-integrated biosensors and bioelectronics in agriculture. The evolving landscape of precision agriculture is poised for significant advancements, with the potential incorporation of autonomous drones, robotics, and other emerging technologies. This integration is expected to contribute to increased efficiency, resource optimization, and sustainable agricultural practices. As precision agriculture continues to evolve, the careful consideration of ethical implications, data privacy concerns, and the need for global standardization emerges as crucial aspects to ensure responsible and inclusive deployment.

Emphasizing the transformative potential of AI, biosensors, and bioelectronics in agriculture, this chapter envisions a future where these technologies drive a paradigm shift in farming practices. From monitoring soil health and crop conditions to optimizing resource utilization, the amalgamation of AI and biosensors holds the promise of creating a smarter, more resilient agricultural ecosystem. The transformative potential lies not only in technological advancements but also in the potential to address global challenges such as food security, environmental sustainability, and the increasing demands on agricultural productivity. As we navigate this transformative journey, it becomes evident that AI-integrated biosensors and bioelectronics are not merely tools; they are catalysts for a sustainable and efficient future in agriculture.

REFERENCES

Akhavan, M., & Doroodgar, A. (2020). Smart Farming: An Overview of Precision Agriculture Using IoT, Robotics, and AI. *Journal of Ambient Intelligence and Humanized Computing*, *11*(12), 5497–5521.

Amato, G., & Moscatelli, A. (2018). Internet of Things in Precision Agriculture for Crop Monitoring. *Sensors (Basel)*, *18*(5), 1515. PMID:29751623

Bahceci, S. (2018). Sensing Technologies for Precision Agriculture and Farming. *Biosensors (Basel)*, *8*(3), 67. PMID:30011807

Bhambri, P. (2020). Green Compliance. In S. Agarwal (Ed.), *Introduction to Green Computing* (pp. 95–125). AGAR Saliha Publication.

Bhambri, P. (2021). Electronic Evidence. In Textbook of Cyber Heal (pp. 86-120). AGAR Saliha Publication, Tamil Nadu.

Bhambri, P., Kaur, H., Gupta, A., & Singh, J. (2020). Human Activity Recognition System. *Oriental Journal of Computer Science and Technology*, *13*(2-3), 91–96.

Bhambri, P., Kaur, H., Gupta, A., & Singh, J. (2020). Human Activity Recognition System. *Oriental Journal of Computer Science and Technology*, *13*(2-3), 91–96.

Bhambri, P., Singh, M., Jain, A., Dhanoa, I. S., Sinha, V. K., & Lal, S. (2021). Classification of the GENE Expression Data with the Aid of Optimized Feature Selection. *Turkish Journal of Physiotherapy and Rehabilitation*, *32*(3), 1158–1167.

Chen, S., & Liu, C. (2020). An Intelligent Agriculture System with Wireless Sensor Networks. *Computers and Electronics in Agriculture*, *171*, 105322.

Chlingaryan, A., & Sukkarieh, S. (2018). Field Robotics in Precision Agriculture: A Review. *Precision Agriculture*, *19*(2), 152–169.

Devadutta, K., Bhambri, P., Gountia, D., Mehta, V., Mangla, M., Patan, R., Kumar, A., Agarwal, P. K., Sharma, A., Singh, M., & Gadicha, A. B. (2020). *Method for Cyber Security in Email Communication among Networked Computing Devices* [Patent application number 202031002649].

Dhiman, G., & Aulakh, N. S. (2020). Artificial Intelligence in Agriculture: A Review. *Journal of Agricultural Engineering and Biotechnology*, *3*(2), 75–83.

Duan, J., & Xu, C. (2020). Advances in Precision Agriculture Technologies: A Review. *Sustainable Production and Consumption*, *23*, 1–12.

Gao, L., & Sun, D. W. (2018). Smart Agriculture: Current Status and Future Perspectives. *Journal of Agricultural Science and Technology*, *20*(4), 815–824.

Gonzalez-Dominguez, J., Smith, H., & Li, Z. (2019). Bioelectronics for Plant Monitoring: Opportunities and Challenges. *Sensors (Basel)*, *19*(21), 4695. PMID:31671733

He, Q., Wu, D., & Wang, Z. (2020). Advances in Sensors for Agriculture and Environmental Monitoring: A Review. *Sensors (Basel)*, *20*(18), 5157. PMID:32927710

Hossain, M. A., & Ahmed, K. (2019). Applications of Bioelectronics in Agriculture: A Comprehensive Review. *Frontiers in Plant Science*, *10*, 411.

Islam, M. M., & Zhong, R. Y. (2020). Smart Agriculture Technologies for Sustainable Crop Protection: A Comprehensive Review. *Computers and Electronics in Agriculture*, *179*, 105827.

Jiang, Y., & Li, C. (2019). Applications of Internet of Things in Agriculture: A Systematic Review. *Computers and Electronics in Agriculture*, *157*, 436–447.

Kaur, K., Dhanoa, I. S., Bhambri, P., & Singh, G. (2020). Energy Saving VM Migration Techniques. *Journal of Critical Reviews*, *7*(9), 2359–2365.

Li, L., Zhang, Q., & Huang, D. (2018). Development of Agricultural Internet of Things and Its Applications in China. *Computers and Electronics in Agriculture*, *154*, 61–68.

Li, Y., & Wang, N. (2019). Precision Agriculture Monitoring System Based on IoT. *IOP Conference Series. Earth and Environmental Science*, *310*, 022015.

Liakos, K., Busato, P., Moshou, D., Pearson, S., & Bochtis, D. (2018). Machine Learning in Agriculture: A Review. *Sensors (Basel)*, *18*(8), 2674. doi:10.3390/s18082674 PMID:30110960

Liu, M., & Guo, J. (2021). Integration of Biosensors and Artificial Intelligence for Precision Agriculture. *Frontiers in Plant Science*, *11*, 604524.

Patel, V., & Lee, W. M. (2020). Artificial Intelligence in Agriculture: A Comprehensive Review. *Computers and Electronics in Agriculture*, *176*, 105554.

Qureshi, S., & Zia, A. (2019). Big Data in Agriculture: A Survey. *Computers and Electronics in Agriculture*, *156*, 133–146.

Shafieian, H., & Le, D. (2019). The Application of IoT in Agriculture: A Systematic Literature Review. *Computers and Electronics in Agriculture*, *157*, 436–447.

Sharma, P., & Khamparia, A. (2020). Internet of Things (IoT) and Artificial Intelligence (AI) in Agriculture: A Comprehensive Review. *Journal of King Saud University. Computer and Information Sciences*.

Sharma, R., Bhambri, P., & Sohal, A. K. (2020). Mobile Adhoc Networks. *Journal of Composition Theory*, *13*(2), 982–985.

Singh, M., Bhambri, P., Dhanoa, I. S., Jain, A., & Kaur, K. (2021). Data Mining Model for Predicting Diabetes. *Annals of the Romanian Society for Cell Biology*, *25*(4), 6702–6712.

Smith, J. A., & Brown, R. W. (2018). Advances in Biosensors for Agricultural Applications. *Biosensors & Bioelectronics*, *102*, 292–305.

Sun, B., Zhang, L., & Chen, Z. (2019). Smart Farming: Agriculture's Future. *Precision Agriculture*, *20*(6), 1114–1124.

Tao, Y., & Zhang, Y. (2018). Research on the Application of IoT Technology in Agriculture. *IOP Conference Series. Earth and Environmental Science*, *170*, 012007.

Wang, D., & Zhang, Y. (2018). Precision Agriculture—An Opportunity for IoT Technologies in Agriculture: A Review. *Sustainable Production and Consumption*, *13*, 37–47.

Wang, J., Zhang, C., & Zhang, L. (2021). A Review on the Use of Artificial Intelligence in Agriculture. *Precision Agriculture*, *22*(4), 701–722.

Wang, Z., Li, P., & Zhang, H. (2017). Internet of Things and Artificial Intelligence in Agriculture: A Review. *Computers and Electronics in Agriculture*, *142*, 283–297.

Wu, L., & Gao, S. (2019). A Survey of IoT Applications in Agricultural Field. *Computers and Electronics in Agriculture*, *157*, 218–229.

Zhang, L., & Sun, B. (2019). An Overview of Smart Agriculture in China. *Agricultural Systems*, *172*, 1–8.

Zhen, W., & Yan, L. (2018). Development of a Wireless Biosensor System for Agricultural Applications. *IEEE Transactions on Instrumentation and Measurement*, *67*(10), 2402–2411.

Chapter 3
Machine Learning and Deep Learning Solutions for Green Computing

Nagendra Singh Yadav
https://orcid.org/0000-0002-9591-4491
Engineering College, Bikaner, India

Vishal Kumar Goar
Engineering College, Bikaner, India

ABSTRACT

In recent years, the field of green computing has gained significant attention due to the growing concerns about the environmental impact of information technology. As the demand for computing resources continues to rise, it becomes imperative to develop sustainable solutions that reduce energy consumption, minimize electronic waste, and lower carbon emissions. This proposed chapter aims to explore the application of machine learning and deep learning algorithms as innovative solutions for addressing these challenges within the domain of green computing.

INTRODUCTION TO GREEN COMPUTING

In an increasingly digital world, the rapid expansion of technology has brought with it a pressing concern: the environmental impact of our digital infrastructure. The realm of Green IT, alternatively referred to as Eco-conscious Computing or Sustainable Technology, has emerged as a response to this concern (Kurp, 2008).

Remarkable advancements can be seen in various sectors, including business, healthcare, education, and entertainment (Gai et al., 2016). The devices we use, the data centers that power them, and the infrastructure that supports our digital activities consume vast amounts of energy and resources (Mao et al., 2017).

DOI: 10.4018/979-8-3693-1794-5.ch003

Back in 1992, the United States Environmental Protection Agency launched the Energy Star program, a mandatory labeling initiative aimed at promoting and acknowledging energy efficiency in various technologies, including displays, temperature control devices, and more. This initiative spurred the widespread adoption of sleep mode in commercial gadgets as a proactive energy-saving measure. Simultaneously, in Sweden, the TCO Advancement organization took the lead by launching the TCO Certification program (Sarkar & Misra, 2016). Originally focused on reducing magnetic and electric emissions from CRT-based display screens, this program quickly expanded its scope to encompass energy efficiency, user comfort, and restrictions on the use of environmentally harmful construction materials. In response to these pioneering efforts, a multitude of stakeholders, including government authorities, businesses, and environmental advocacy groups, embarked on a series of initiatives to foster the concept of Green Computing. This multifaceted approach to Green Computing includes practices such as reusing hardware, minimizing electronic waste, embracing digitalization, harnessing the power of cloud computing, implementing energy-saving measures, and adopting eco-friendly production methods.

The Environmental Impact

- **Energy Consumption -** One of the most significant environmental concerns associated with computing is energy consumption. Computers, data centers, and other digital devices require substantial amounts of electricity to operate. As the demand for digital services continues to grow, so does the energy consumption of the information technology (IT) sector (Fuchs, 2006).
- **Electronic Waste -** The swift speed of innovation in technology leads to the frequent obsolescence of electronic devices. This, in turn, results in a massive amount of electronic waste being generated worldwide.

Objectives of Green Computing

- **Energy Efficiency -** Green Computing focuses on designing and operating computer systems and data centers in ways that minimize energy consumption. This includes the development of energy-efficient hardware and software solutions.
- **Resource Conservation -** It aims to reduce the adoption and utilization of non-renewable resources in the manufacturing of electronic devices and promote reuse and responsible discarding of e-waste.
- **Carbon Reduction -** Green Computing strives to reduce the carbon footprint of IT operations through the utilization of renewable energy resources, energy-efficient cooling systems, & efficient data center design (Bougdah & Sharples, 2009).

Environmental Challenges Posed by Traditional Computing Practices

- **Energy Consumption -** One of the notable environmental challenges posed by traditional computing practices is their immense consumption of energy. Data centers, which house the servers that power our digital world, are voracious consumers of electricity. These facilities operate 24/7 to ensure that our data and applications are always available, and they require enormous amounts of energy for cooling and maintenance.

- **E-Waste -** The discarding of e-waste, is another pressing concern stemming from traditional computing practices. As technology marches forward, a rapid obsolescence cycle ensues, resulting in the mass disposal of electronic devices.
- **The Hidden Water Footprint -** Beyond energy consumption and e-waste, traditional computing practices also have a hidden water footprint. The manufacturing of electronic components, including semiconductors and microchips, relies on water-intensive processes
- **The Carbon Conundrum -** The carbon footprint of traditional computing practices is a critical issue in the fight against climate change. The production of electronics, data center operations, and the transportation of goods all contribute to carbon emissions.

The Role of Technology in Addressing Environmental Concerns

In an era marked by rapid technological advancements and growing environmental challenges, Technology's significance in tackling environmental challenges has reached an unprecedented level of importance.

- **Clean Energy Technologies -** One of the most prominent ways technology is contributing to environmental sustainability is through the development and adoption of clean energy technologies. Solar panels, wind turbines, and advanced battery storage systems have seen remarkable progress, making renewable energy sources increasingly viable and cost-effective.
- **Environmental Monitoring and Big Data -** Advances in sensor technology and data analytics have assisted in a new age of environmental monitoring. Satellites, drones, & ground-based sensors provide comprehensive data on various aspects of the environment, including air quality, soil health, and deforestation rates. This real-time data collection allows more informed decision-making & faster responses to environmental threats.
- **Sustainable Agriculture and Food Systems -** Agriculture is a significant contributor to environmental challenges, from deforestation for crop expansion to excessive use of water and pesticides. However, technology is driving innovations in sustainable agriculture.

Review of Literature

Creating robust and compact networks is the primary long-term aim of engineers and researchers. Based on this goal, (Pedrycz, 2022) discussed Green machine learning by focusing on reducing the environmental impact of machine learning. This is done by making machine learning models more efficient. Some of the ways to do this include using less data, transferring knowledge between models, and using more interpretable models. Granular computing is a helpful tool for green machine learning. It can be used to make models more interpretable and efficient.

(Ojo et al., 2019) described a three-fold training method that will be helpful for the right mindset about green computing internally in an organization. (Anthony et al., 2018) suggested a life cycle method for green computing which utilizes metrics to organize data centers into measurable entities which can lead to further computing based on cost saving, energy efficiency, and CO2 emissions. (El-Kassar & Singh, 2019) suggested that working from home also helps green computing as it lowers the fuel emissions generated by vehicles and saves heavily on operation costs. A classic example is set by Google as they are committed to lowering the carbon emissions generated by computing resources to 0%. This is achieved by energy awareness related to "IT equipment energy" which is utilized for cooling & lighting.

(Prakash et al., 2019) utilized a mix of genetic algorithms & artificial neural networks to optimize the energy consumption in wireless networks to meet the objective of green computing. they aimed to lower the consumption of energy which is considered an objective method in genetic algorithm.

MACHINE LEARNING AND DEEP LEARNING FUNDAMENTALS

Overview of Machine Learning

Arthur Samuel, a visionary in the realm of AI and computer gaming, is credited with coining the term "Machine Learning." He elucidated the concept of machine learning as the "Science of enabling computers to learn autonomously, without explicit programming." To put it simply, Machine Learning (ML) can be likened to the process of empowering computers to enhance their performance through experience, all without human intervention. The journey commences by first ingesting impeccable data, paving the way for the training of machines through the creation of intricate machine learning models, leveraging both the data and an array of algorithms (Mathew et al., 2020). The choice of algorithms depends on both the data's characteristics and the particular automation goal one have in mind (Deng & Yu, 2014).

In the world of Traditional Machine Learning, the synergy between data and statistical tools takes center stage, culminating in the prophecy of outputs that subsequently furnish us with actionable insights (Lundervold & Lundervold, 2019).

It's noteworthy that machine learning models have transcended these domains, empowering autonomous vehicles, drones, and robots, thereby augmenting their intelligence and adaptability to ever-evolving surroundings (Maier et al., 2019).

Among the many applications, recommender systems are particularly prominent, harnessing historical data to offer personalized suggestions. In the context of platforms like Netflix, a blend of collaborative filtering and content-based filtering techniques is deployed to suggest movies and TV shows to users, leveraging their viewing history, ratings, and genre preferences (Denil et al., 2013).

Reinforcement learning presents an agent that acquires knowledge through its dynamic engagement with the surroundings, and making decisions based on feedback. This feedback loop can be cleverly harnessed to fine-tune the recommendations users are provided with. i.e. the system might monitor how frequently a user views recommended content & utilize this data to fine-tune future recommendations (Sejnowski, 2018).

Figure 1. Machine and deep learning

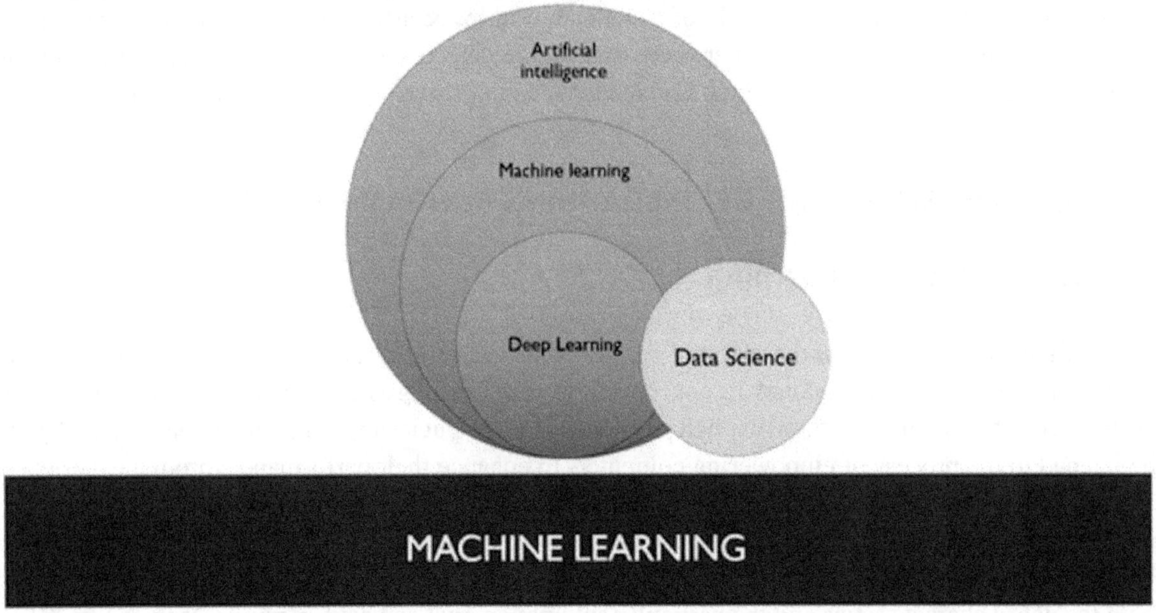

Demystifying the Inner Workings of Machine Learning Algorithms

- **Forward Pass -** The machine learning model leverages input data to produce an output, with the exact prediction computation adapting according to the algorithm employed within the model.
- **Loss Method –** It is often referred to as the error or cost method, serves as a crucial metric for assessing the fidelity of the model's projection. Essentially, this function quantifies how well the model's projection lines up with the real-world outcomes. It measures the disparity between what the model predicts and the actual observed results, and this measure of disparity is commonly called an error or loss.
- **Optimization Process of the model -** The model optimization journey unfolds as a continuous refinement of the model's inner workings, involving the fine-tuning of its internal parameters in pursuit of minimizing the error or loss function. This intricate task is achieved through the application of a specialized optimization algorithm, often exemplified by gradient descent.

After successfully training and fine-tuning the model with the training data, it becomes ready to perform projections on fresh, hidden data. To evaluate the model's projection quality, one can utilize various performance metrics, i.e. precision, recall, and the F1-score, among others.

Life Cycle of Machine Learning

The lifecycle of a machine learning endeavor embarks on a series of phases, encompassing the following:

- **Understand the Problem -** The initial stride in this journey is to delve into the intricacies of the issue at hand. This entails immersing oneself in a comprehensive comprehension of the business quandary and the formulation of clear-cut objectives for the model.
- **Collection of Data -** When faced with a clearly defined problem, we have the opportunity to gather the necessary data to build our model effectively. This data can be sourced from a variety of places, including databases, and APIs, or by employing web scraping techniques.

Figure 2. Life cycle of machine learning

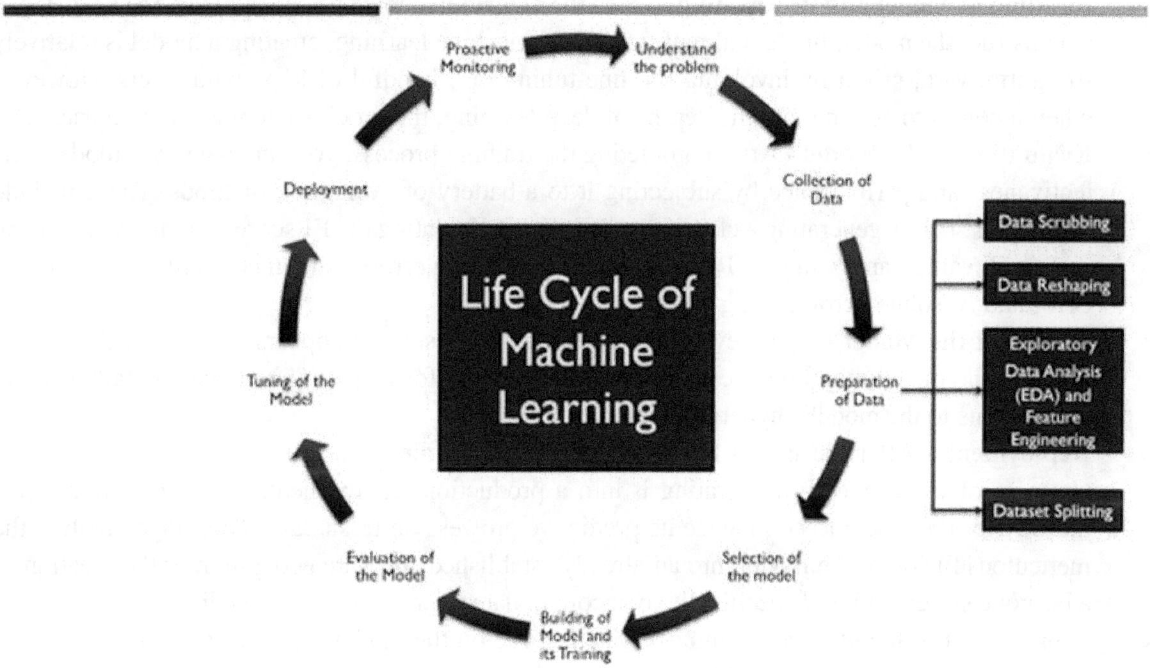

- **Preparation of Data -** Once we've gathered our problem-specific data, it's crucial to make sure that the data is thoroughly reviewed & formatted appropriately to enable the model to uncover concealed patterns. One can achieve this by following these guidelines:
    - **Data Scrubbing -** Start the data preparation process by meticulously inspecting the dataset, and diligently addressing any inaccuracies, discrepancies, or null entries in the data.
    - **Data Reshaping -** Next, reshape the data to ensure it's in the desired format for analysis. This may involve reorganizing columns, merging datasets, or aggregating data points as necessary.
    - **Exploratory Data Analysis (EDA) and Feature Engineering -** Perform an initial data exploration to uncover key patterns and attributes within the dataset. In this stage, you have the opportunity to innovate by crafting new features or reshaping existing ones to more accurately capture the inherent patterns. EDA helps in identifying outliers, trends, and potential relationships among variables.

- **Dataset Splitting -** Divide the dataset into training and testing subsets. The training data is employed for the machine learning model's training, and the testing data serves to assess the model's performance. Ensure that the splitting is done in a way that maintains the data's representativeness.

- **Selection of the model -** The subsequent phase involves the careful curation of a machine learning algorithm tailored to address our specific problem. This stage demands a profound understanding of the capabilities and limitations inherent in various algorithms. On occasion, one may employ multiple models, conducting a comprehensive comparison of their outcomes, and ultimately opting for the one that best aligns with our particular needs.
- **Building of Model and its Training -** Once the algorithm choice has been made, the next step is to construct the model. In the realm of traditional machine learning, creating a model is relatively straightforward, primarily involving the fine-tuning of a handful of hyperparameters. However, when it comes to delving into the depths of deep learning, the process becomes more intricate.
- **Evaluation of the Model -** After completing the training process, you can assess the model's effectiveness and performance by subjecting it to a battery of evaluation methods. These include techniques such as generating a classification report, calculating the F1 score, examining precision and recall values, analyzing the ROC Curve, and assessing errors using metrics like Mean Square Error and Absolute Error, among others.
- **Tuning of the Model -** Upon reviewing the evaluation results, it appears that fine-tuning or optimizing the model might be necessary to enhance its performance. This process entails making adjustments to the model's hyperparameters.
- **Deployment -** After the model has been meticulously trained and fine-tuned, the next crucial phase involves seamlessly integrating it into a production environment. This integration paves the way for the model to commence its predictive prowess on fresh data. This stage involves the meticulous fusion of the model into an already established software ecosystem or the creation of a bespoke system tailored specifically to accommodate the model's functionality.
- **Proactive Monitoring -** Maintaining a watchful eye on the model's performance in the production environment & promptly addressing any maintenance requirements is of utmost importance. This encompasses the continuous vigilance of data drift, retraining the model when necessary, & keeping the model up-to-date with the influx of fresh data.

Varieties of Machine Learning Approaches

- **Supervised Machine Learning -** It stands out as a prominent branch of machine learning, where the algorithm undergoes training using meticulously labeled datasets. It grasps the art of connecting input features with their respective desired outcomes, relying on the wealth of knowledge within the labeled training data.

Figure 3. Supervised learning

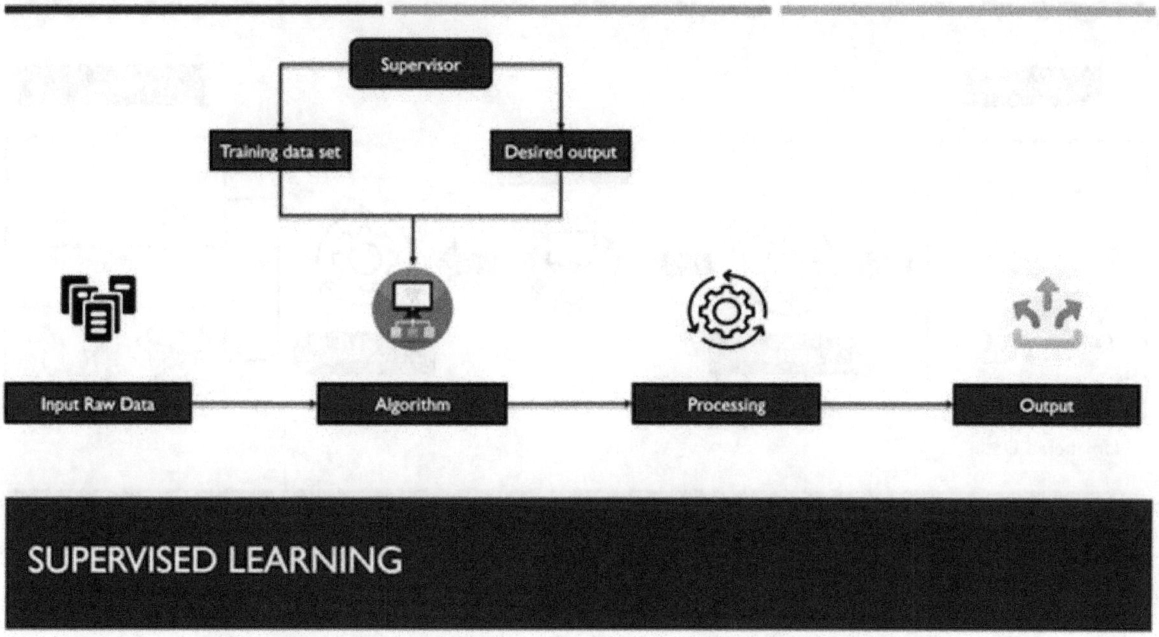

In this framework, The algorithm is equipped with input variables alongside their corresponding output labels, thus enabling it to distill essential patterns and relationships, ultimately empowering it to extrapolate and make predictions with accuracy when presented with novel, uncharted data. Two distinct categories exist of supervised machine learning –

- **Regression** - Regression, a cornerstone of supervised learning, empowers algorithms to forecast continuous outcomes by leveraging input features. Within regression tasks, the target labels span a continuous spectrum, encompassing real-world examples like stock prices and housing values. In the realm of machine learning, diverse regression techniques stand ready to tackle these challenges, including but not limited to i.e. Linear, Polynomial, Ridge, Decision Tree, Random Forest & Support Vector Regression and so on.
- **Classification -** Categorization, a fundamental concept within supervised learning, involves training algorithms to categorize input data into distinct classes or categories, relying on the provided input features. In classification, the output labels are discrete, representing the various classes or categories. Classification algorithms come in two primary forms: binary and multiclass. In binary classification, the end result is sorted into one of two probable classes, while multiclass classification entails assigning data to one of several potential classes. In the realm of machine learning, numerous classification algorithms exist, including but not limited to Logistic Regression, Naive Bayes, Decision Trees, SVM (Support Vector Machines), KNN (K-Nearest Neighbors), and many more.
- **Unsupervised Machine Learning -** Unsupervised learning stands as a facet of machine learning wherein algorithms autonomously unearth patterns within datasets without the reliance on labeled examples for training.

Figure 4. Unsupervised learning

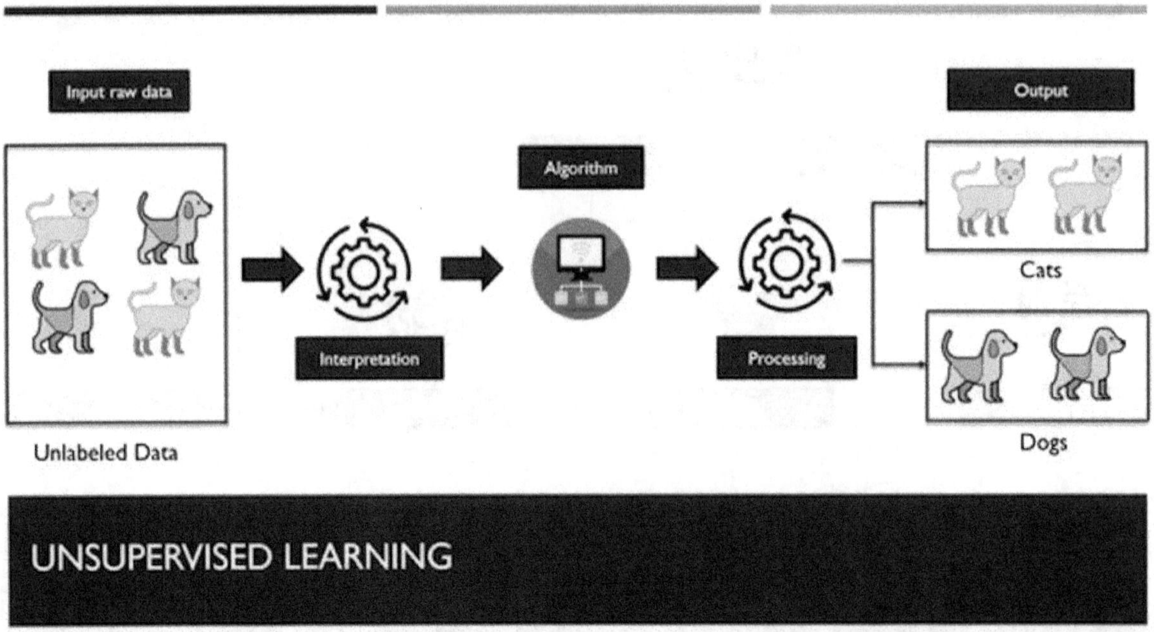

Its primary objective revolves around the revelation of latent structures or inherent distributions concealed within the data. There are two types of Unsupervised Machine Learning -

- **Clustering -** Clustering algorithms work by organizing data points into cohesive groups or clusters, emphasizing their shared characteristics and setting them apart from other data points. These algorithms aim to uncover underlying structures within datasets, revealing patterns and relationships. Widely recognized clustering techniques encompass K-means, Hierarchical clustering, and DBSCAN, among others.
- **Dimensionality reduction -** Dimensionality reduction techniques streamline datasets by reducing the multitude of input variables, all the while striving to retain the essence of the original information. These methods prove invaluable in simplifying dataset intricacies, ultimately facilitating enhanced visualization and analytical capabilities. Among the well-regarded dimensionality reduction algorithms are Principal Component Analysis (PCA), t-SNE, and Autoencoders, each offering distinct advantages for data analysis and visualization.
    - **Reinforcement Machine Learning -** Reinforcement learning stands as a distinctive facet within the realm of machine learning. In this paradigm, an intelligent agent embarks on a journey of interaction with its surroundings, engaging in actions that bring about rewards or penalties depending on their consequences.

Figure 5. Reinforcement learning

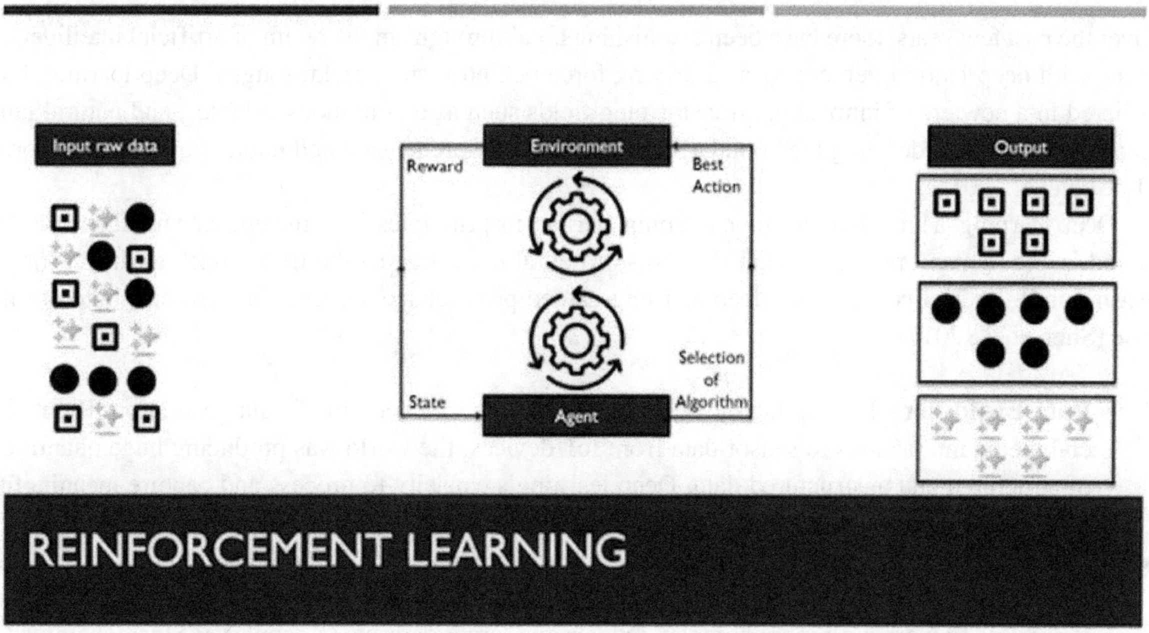

The overarching objective of reinforcement learning revolves around the acquisition of a policy—a set of guidelines that guide the agent in selecting actions corresponding to different states. The ultimate aim here is to fashion a policy that skillfully navigates the intricate landscape of the environment, ensuring the accrual of the highest anticipated cumulative reward throughout its exploration. Two categories exist for Reinforcement Machine Learning –

- **Model-based reinforcement learning -** In the realm of model-centric reinforcement learning, the agent embarks on a journey of comprehension by constructing an intricate model of the environment. This model encapsulates the enigmatic dance between states, revealing the mysterious transition probabilities that guide the agent's path, as well as the treasures hidden within each state-action pair, known as rewards. Armed with this newfound knowledge, the agent takes on the role of a master planner, meticulously devising its actions to orchestrate a symphony of rewards and maximize its expected returns. Among the stars in this constellation of model-centric reinforcement learning algorithms shine the luminous Value Iteration and the ingenious Policy Iteration.
- **Model-free reinforcement learning -** In the realm of model-free reinforcement learning, the agent eschews the conventional approach of constructing a detailed model of the environment. Instead, it embarks on a journey of experiential learning, directly from its interactions with the environment. The agent actively engages with its surroundings, making adjustments to its strategy solely on the basis of the rewards it garners. Among the prominent methodologies within the domain of model-free reinforcement learning, one encounters the likes of Q-Learning, SARSA, and the revolutionary field of Deep Reinforcement Learning.

Introduction to Deep Learning

Over the past few years, there have been astonishing breakthroughs in the realm of artificial intelligence (AI), with deep learning emerging as a driving force behind many breakthroughs. Deep learning has ushered in a new era of innovation, transforming fields such as autonomous vehicles, and natural language processing, redefining the boundaries of machine comprehension and interaction with our world (LeCun et al., 2015).

Deep learning, a branch of machine learning, derives its principles from the intricate neural networks found in the human brain. Although the roots of neural networks can be traced back to the 1940s, it wasn't until the 21st century that deep learning gained prominence. Several factors contributed to its rise (Shen et al., 2017):

- **Data Explosion -** The digital age ushered in an unprecedented era of data generation. From social media interactions to sensor data from IoT devices, the world was producing huge quantities of structured and unstructured data. Deep learning's capacity to process and capture meaningful insights from such data made it highly relevant (Gulshan et al., 2016).
- **Computational Power -** Advancements in hardware, particularly GPUs (Graphics Processing Units), given the required computational energy to train deep neural networks effectively. This made it possible to work with more complex models and larger datasets (Khanafer & Shirmohammadi, 2020).
- **Algorithmic Improvements -** Innovations in neural network architectures and training techniques, such as the introduction of CNNs (convolutional neural networks) & RNNs (recurrent neural networks), played a pivotal role in the resurgence of deep learning. These breakthroughs ushered in a new era of deep learning models, revolutionizing their capabilities across diverse domains (Lei et al., 2020).

Understanding Deep Learning

Deep learning models draw inspiration from the intricate workings of the human brain to inform their design and operation. Fundamentally, they comprise interconnected nodes, akin to artificial neurons, organized into layers, forming the heart of these systems (Goodfellow et al., 2016). These layers consist of the input layer, one or more invisible layers, and an end result layer.

- **Neurons and Layers -** Every artificial neuron accepts input signals, computes a weighted sum, introduces a bias factor, and then subjects the outcome to an activation method. This process mimics how biological neurons fire when stimulated. Layers of neurons work in tandem to process information hierarchically, with each layer capturing increasingly abstract features (Carletti et al., 2020).
- **Training and Learning -** Deep learning models gain understanding from data with the help of a process known as training. While in the training phase, the model fine-tunes its weights & biases to reduce the disparity between its projections and the true target values within the training dataset. This optimization is usually obtained using backpropagation and gradient descent algorithms.
- **Representation Learning -** Deep learning excels in the realm of automatic feature discovery from raw data, a fundamental technique referred to as representation learning. This terminates the

requirement for manual feature engineering & enables the model to adapt to different datasets and tasks.

Applications of Deep Learning

Across a multitude of fields, deep learning has left an indelible mark, encompassing, yet not confined to:

- **Computer Vision** - In computer vision, deep learning models have obtained human-level performance in tasks i.e. image classification, object detection, & facial recognition. CNNs, in particular, have been instrumental in these advancements (Shorten & Khoshgoftaar, 2019).
- **Natural Language Processing (NLP)** - NLP applications have been revolutionized by deep learning methods, including RNNs (recurrent neural networks) for sequence modeling & attention mechanisms for machine translation. Chatbots, sentiment analysis, and language generation are just a few examples of NLP applications (Guo et al., 2023).
- **Healthcare** - Deep learning models have reflected promising results in medical image analysis, disease diagnosis, drug discovery, and personalized treatment recommendations. They possess the capability to interpret medical imagery i.e. X-rays, MRI scans, and pathology slides with remarkable accuracy (Litjens et al., 2019).
- **Autonomous Systems** - At the core of autonomous vehicles, deep learning plays a pivotal role, enabling them to perceive their surroundings through sensors and create real-time decisions. These systems utilize deep neural networks to process data from cameras, LiDAR, & radar.

MACHINE LEARNING APPLICATIONS IN GREEN COMPUTING

Energy-Efficient Data Centers

In the digital age, data centers are the backbone of our connected world, powering everything from social media platforms to cloud services. Nevertheless, the exponential expansion of data centers has sparked apprehensions regarding their ecological footprint, particularly with regard to energy usage. Machine learning, with its ability to optimize and automate processes, plays a pivotal role in making data centers more energy-efficient (Ali et al., 2017).

Here are some key areas where machine learning is creating a notable effect in green computing for data centers:

- **Predictive Analytics** - Machine learning models can foresee future server workloads & temperature fluctuations based on historical data. By forecasting demand and performance, data centers can dynamically allocate resources, such as turning off underutilized servers during low demand periods, which reduces energy consumption.
- **Dynamic Resource Allocation** - Real-time monitoring of server and cooling system performance, coupled with machine learning algorithms, enables dynamic resource allocation. This means that servers can be powered up or down as needed, and cooling systems can adjust their settings in response to changing conditions, ultimately optimizing energy usage.

- **Anomaly Detection -** Machine learning algorithms can identify unusual behavior in server clusters or cooling systems, signaling potential problems. By taking this proactive stance, data center operators can preemptively tackle potential problems, minimizing system downtime and curbing energy inefficiencies.
- **Temperature Management -** Maintaining the right temperature within a data center is crucial for efficient operations. Machine learning can fine-tune cooling systems to achieve the desired temperature while reducing the consumption of energy. It can also predict hotspots and adjust cooling accordingly.
- **Renewable Energy Integration -** Machine learning models can optimize the use of renewable energy sources, i.e. solar or wind power, by predicting energy availability and coordinating data center operations accordingly. This diminishes the need for traditional energy sources and diminishes the environmental impact of data centers.
- **Workload Scheduling -** Machine learning can optimize workload scheduling to ensure that energy-intensive tasks are executed during periods of lower demand or when renewable energy sources are more abundant. This helps balance energy usage throughout the day.

Dynamic Resource Allocation

Dynamic resource allocation includes the efficient distribution of computing resources i.e. CPU, memory, storage, & network bandwidth to meet the changing demands of applications and workloads. In green computing, the primary goal is to allocate resources in a manner that minimizes energy consumption while ensuring satisfactory performance. This is achieved by continuously monitoring the resource usage and adapting the allocation in real-time (Xu et al., 2018).

Traditional approaches to resource allocation often involve static provisioning, which leads to inefficiencies as resources may remain underutilized during periods of low demand and result in increased energy consumption. Dynamic resource allocation addresses this issue by allowing resources to be allocated and deallocated based on current workload requirements.

Machine learning techniques have proven to be invaluable for optimizing dynamic resource allocation in green computing. Here are some key ways in which machine learning is applied:

- **Workload Prediction -** Machine learning models can be trained to project future workload patterns based on past data. These predictions enable proactive resource allocation adjustments, ensuring that resources are allocated in anticipation of demand spikes or lulls. By avoiding over-provisioning or under-provisioning, energy consumption is minimized.
- **Anomaly Detection -** Anomalies in resource utilization can indicate inefficient allocation or potential performance issues. Machine learning algorithms can identify such anomalies in real-time and trigger appropriate resource allocation changes to rectify the situation. This helps in maintaining energy efficiency while ensuring application performance.
- **QoS Optimization -** Quality of Service (QoS) is a crucial consideration in resource allocation. Machine learning can optimize QoS by dynamically adjusting resource allocations to meet service-level agreements (SLAs) while minimizing energy usage. This guarantees that essential applications obtain the required resources while non-critical ones are allocated fewer resources, further reducing energy consumption.

- **Load Balancing -** Load balancing is essential for evenly distributing workloads across computing resources. Machine learning algorithms can intelligently balance workloads by considering historical data, current usage, and predicted demand. This prevents resource bottlenecks and allows for efficient energy utilization.

While machine learning offers promising solutions for dynamic resource allocation in green computing, several challenges must be addressed:

- **Data Quality and Quantity -** Machine learning models require high-quality data for training and inference. Obtaining comprehensive and accurate data about resource utilization and workload patterns can be challenging. Additionally, the volume of data needed to train accurate models can be substantial.
- **Real-time Adaptation -** Green computing demands real-time adaptation to changing conditions. Machine learning models must make rapid decisions and adjustments to resource allocations, which can be challenging to achieve, especially for complex models.
- **Model Interpretability -** Interpreting machine learning models is crucial for understanding their resource allocation decisions. Black-box models may not be suitable in scenarios where transparency and accountability are required.
- **Security and Privacy -** Machine learning-based resource allocation systems must be resilient to security threats and protect sensitive data. Ensuring the confidentiality and integrity of resource allocation decisions is essential.

Predictive Maintenance for Hardware

As the demand for computational power and data processing continues to grow, the information technology industry faces the challenge of ensuring the sustainability of its infrastructure. Hardware components such as servers, data centers, and networking equipment are essential for modern computing, but they also consume significant amounts of energy and contribute to electronic waste when they fail. Predictive maintenance for hardware hinges on the essential role of machine learning. The process usually consists of the following steps:

- **Data Collection -** To predict hardware failures, relevant data is collected from several sources, i.e. sensors, logs, and performance metrics. This data can include temperature readings, voltage levels, fan speeds, and other hardware-specific parameters.
- **Data Preprocessing -** Before applying machine learning algorithms, the collected data should be cleaned and prepared. This includes handling missing values, normalizing data, and converting it into an appropriate format for analysis.
- **Feature Engineering -** Feature engineering includes the selection and creation of relevant features using the data. This step is crucial for building accurate predictive models. Domain knowledge is often essential for identifying meaningful features.
- **Model Development -** Machine learning models, i.e. regression, decision trees, random forests, or deep learning models, are trained on past data to predict hardware failures. These models can learn patterns & anomalies in the data that reflect probable issues.

- **Model Evaluation -** The performance of predictive maintenance models is assessed with the help of metrics i.e. accuracy, precision, recall, and F1-score. Cross-validation approaches are often used to evaluate model generalization.
- **Deployment -** Once a reliable predictive maintenance model is developed, it can be deployed in real-time environments. The model continuously monitors hardware conditions and triggers maintenance alerts when anomalies are detected.

While predictive maintenance offers significant benefits, several challenges must be overcome for successful implementation:

- **Data Quality -** Ensuring the quality and consistency of data from diverse hardware sources can be challenging.
- **Feature Engineering -** Identifying and engineering relevant features requires domain expertise and can be time-consuming.
- **Model Interpretability -** Some machine learning models are complex and lack interpretability, making it difficult to understand why a prediction was made.
- **Scalability -** Scaling predictive maintenance solutions to large data centers or extensive hardware ecosystems can be complex.
- **Cost of Implementation -** Developing and deploying predictive maintenance systems can be costly, especially for smaller organizations.

Optimizing Power Usage in IT Infrastructure

In recent years, the convergence of technology and environmental concerns has led to a growing emphasis on sustainability within the IT industry. With the proliferation of data centers, cloud computing, & the growing power demands of modern computing equipment, there is a pressing requirement to discover methods to minimize the environmental impact of information technology. Machine learning can address these challenges and help optimize power usage in IT infrastructure. Here are some key applications:

- **Predictive Maintenance -** By harnessing the power of machine learning, we can predict impending failures in IT equipment, enabling us to take preemptive maintenance actions. By preventing unexpected breakdowns, organizations can avoid the energy costs associated with emergency repairs and downtime.
- **Dynamic Resource Allocation -** Machine learning algorithms can analyze real-time data on server workloads and allocate resources accordingly. This ensures that servers operate at peak efficiency while minimizing energy waste during idle periods.
- **Temperature and Cooling Optimization -** Sensors and IoT devices can monitor temperature and humidity within data centers. Machine learning models can then adjust cooling systems to maintain optimal conditions, reducing energy consumption.
- **Renewable Energy Integration -** Machine learning can predict renewable energy availability (e.g., solar and wind) and align data center operations with periods of high green energy supply, reducing reliance on fossil fuels.
- **Data Center Design -** Machine learning can assist in designing energy-efficient data centers, optimizing layout, cooling infrastructure, and server placement to minimize energy consumption.

Carbon Footprint Reduction

As the world grapples with the consequences of climate change, the information technology (IT) industry has come under increased scrutiny for its significant carbon footprint. Green computing, a discipline aimed at reducing the environmental impact of computing technologies, has gained traction as a means to mitigate this issue. To effectively address this complex problem, machine learning techniques are being leveraged across various stages of the IT lifecycle.

- **Data Center Management -** Data centers are among the largest energy consumers in the IT domain. Machine learning algorithms are being deployed to optimize server utilization, cooling systems, and energy distribution within data centers. By predicting server workloads and dynamically allocating resources, ML can significantly reduce energy waste.
- **Energy-Efficient Hardware Design -** Machine learning is also influencing the design of energy-efficient hardware. Researchers and engineers are using ML to simulate and analyze the energy consumption of different hardware components. This enables the creation of more power-efficient devices, from servers and network equipment to consumer electronics.
- **Renewable Energy Integration -** Green computing aims to transition towards renewable energy sources. Machine learning models help in predicting energy generation from renewable sources i.e. solar and wind, facilitating better integration into computing infrastructure. ML can also optimize the use of energy storage systems to reduce reliance on fossil fuels.
- **Supply Chain Optimization -** The carbon footprint of computing extends beyond the data center to the entire supply chain, including the extraction of raw materials and manufacturing. ML models are used to optimize supply chains, reduce waste, and ensure that components are sourced from environmentally responsible suppliers.

Several machine learning techniques and approaches are applied to reduce the carbon footprint of IT operations:

- **Predictive Analytics -** Predictive models use past data to predict energy consumption, server utilization, and equipment failures. These predictions enable proactive measures to optimize energy use, leading to reduced carbon emissions.
- **Anomaly Detection -** Anomaly detection algorithms can identify irregular patterns in energy consumption, signaling potential inefficiencies or equipment malfunctions. By promptly addressing these anomalies, organizations can prevent energy waste and reduce their carbon footprint.
- **Reinforcement Learning for Energy Management -** Reinforcement learning is applied to optimize data center operations dynamically. RL agents learn to make decisions that minimize energy consumption while maintaining service quality and adapting to changing workloads and environmental conditions.
- **Energy-Efficient Routing -** Machine learning algorithms optimize network routing to minimize data transmission distances and reduce energy consumption. This is particularly relevant in cloud computing environments where data often travels across vast distances.
- **Lifecycle Assessment -** Machine learning can assist in conducting lifecycle assessments of IT products, considering their environmental impact from production to disposal. This information guides decisions about product design and end-of-life management.

DEEP LEARNING APPLICATIONS IN GREEN COMPUTING

Smart Grid Management

The integration of deep learning techniques into the field of green computing has revolutionized the way we manage and optimize smart grids. Smart grids are complex, dynamic systems that play a critical part in the efficient distribution & consumption of electrical energy. Utilizing cutting-edge communication and control systems, these smart grids are engineered to enhance the efficiency of electricity generation, transmission, and utilization.

- **Load Forecasting -** One of the primary challenges in smart grid management is predicting electricity demand accurately. Deep learning models i.e. LSTM (Long Short-Term Memory) networks, have shown exceptional performance in load forecasting. Through the examination of past data, these models can forecast future demand trends with remarkable accuracy.
- **Anomaly Detection -** Deep learning algorithms excel in anomaly detection, which is critical for identifying faults or cyberattacks on the smart grid. Convolutional Neural Networks (CNNs) can analyze data from various sensors & cameras to catch anomalies in real-time. Early detection allows for rapid response and minimizes the impact of disruptions.
- **Energy Theft Detection -** Deep learning algorithms can analyze consumption patterns & detect irregularities that might indicate energy theft. By flagging suspicious behavior, utilities can take action to prevent revenue loss and maintain the integrity of the grid.
- **Renewable Energy Integration -** Deep reinforcement learning (DRL) is employed to enhance the seamless integration of solar and wind energy sources into the grid. DRL algorithms can make real-time decisions on when and how to use renewable energy based on weather forecasts, grid conditions, and cost considerations.
- **Demand Response -** Deep learning models can facilitate demand response programs by predicting when and where peak demand will occur. This empowers utility companies to encourage customers to lower their energy usage during peak periods, thereby reducing stress on the grid & the requirement for additional power production.
- **Energy Storage Management -** Deep learning techniques are also applied to optimize power reserve solutions, i.e. batteries. By analyzing historical data and grid conditions, these models can determine the most efficient times to charge and discharge energy storage systems, thus maximizing their lifespan and utility.

Renewable Energy Optimization

In an era marked by increasing concerns about environmental sustainability and the depletion of fossil fuels, renewable energy sources have gained prominence as a viable alternative to meet the world's growing energy demands. Green computing, which focuses on environmentally responsible and energy-efficient computing practices, plays a pivotal role in achieving sustainability goals.

- **Solar Energy Prediction -** One of the significant challenges in solar energy utilization is predicting energy generation based on weather conditions, time of day, and other variables. Advanced deep learning models like Convolutional Neural Networks (CNNs) and Recurrent Neural Networks (RNNs) have the capability to leverage past data for precise solar energy predictions.

- **Wind Energy Forecasting -** Wind energy, like solar energy, is highly dependent on meteorological conditions. Deep learning models can process vast datasets of wind speed, direction, and atmospheric pressure to provide short-term and long-term wind energy forecasts.
- **Energy Storage Optimization -** Renewable energy sources are often characterized by fluctuations in supply. Utilizing deep reinforcement learning (DRL) algorithms, we can enhance the efficiency of energy storage systems i.e. batteries & pumped hydro storage, enabling them to efficiently capture surplus energy during periods of high generation and disburse it during peak demand, thereby upholding a reliable power grid.
- **Grid Management -** Deep learning plays a critical part in optimizing the distribution of renewable energy within the power grid. Smart grid systems use deep learning algorithms to balance energy production, storage, and consumption efficiently.
- **Energy-Efficient Data Centers -** Green computing also extends to data centers, which are significant energy consumers. Deep learning-based predictive maintenance models can anticipate equipment failures, optimize cooling systems, and allocate computing resources efficiently, reducing energy consumption and environmental impact.

Electronic Waste Management and Recycling

The breakneck pace of technological progress has resulted in an alarming surge of electronic waste on a global scale. In addition to harming the environment, the careless disposal of electronic devices squanders precious resources and exacerbates the problem of electronic waste. As the world becomes increasingly aware of the need for sustainable practices, the domain of green computing has obtained prominence. Improper disposal of these hazardous materials in landfills can result in the contamination of both soil and water, representing a significant hazard to the environment and human well-being. To mitigate this issue, effective e-waste management and recycling solutions are essential.

- **E-Waste Classification -** Deep learning models can be trained to acknowledge & classify dissimilar types of electronic waste. In the realm of this task, Convolutional Neural Networks (CNNs) truly excel, as they can process images and identify specific electronic components and materials. By automating the classification process, e-waste can be sorted more efficiently at recycling facilities.
- **Fault Detection and Diagnosis -** Electronic devices often end up as e-waste due to minor faults or defects that can be repaired. Deep learning algorithms can be employed to detect these faults by analyzing device sensor data or diagnostic images. This can extend the lifespan of electronic products & reduce the overall generation of e-waste.
- **Waste Routing and Logistics -** Efficiently managing the collection & transport of electronic waste is of paramount importance for reducing environmental impact. Deep learning can be used to develop predictive models that consider factors such as location, volume, and type of e-waste to optimize routing and logistics, ensuring that recycling centers receive the waste efficiently.
- **Materials Recovery -** Recycling electronic waste involves extracting precious materials i.e. gold, silver, & copper from discarded devices. Deep learning algorithms can help refine the efficiency

of the recovery process by identifying and separating valuable components from non-reusable materials.

Sustainable Manufacturing and Supply Chain Management

In today's rapidly evolving world, the importance of sustainable practices in manufacturing and supply chain management cannot be overstated. With the increasing global awareness of environmental issues, Companies are facing increasing demands to minimize their environmental impact, minimize waste, & operate in an ecologically responsible manner. Deep Learning Applications in Sustainable Manufacturing are listed below -

- **Predictive Maintenance -** One of the key challenges in manufacturing is ensuring the continuous operation of machinery while minimizing downtime and maintenance costs. Deep learning models can analyze sensor data from manufacturing equipment to project when maintenance is required. This proactive approach reduces unpredicted failure, prolongs the operational longevity of equipment, and minimizes resource wastage.
- **Energy Optimization -** Deep learning algorithms can optimize energy consumption in manufacturing procedures by analyzing past data & real-time sensor readings. By identifying patterns and anomalies, these models can suggest energy-saving strategies and help reduce the carbon footprint of manufacturing facilities.
- **Waste Reduction -** Manufacturers can use deep learning to optimize production processes and minimize waste generation. By analyzing data from sensors & cameras, these models can recognize inefficiencies & suggest process improvements that lead to reduced waste production.
  Deep Learning Applications in Sustainable Supply Chain Management are listed below -
- **Demand Forecasting -** Precise demand prediction is essential for minimizing excess inventory and transportation costs. Deep learning models can analyze past sales data, market trends, & external factors to give more correct demand projections. This aids in optimizing inventory levels and minimizing unnecessary resource consumption.
- **Route Optimization -** Efficient transportation is a key aspect of sustainable supply chain management. Utilizing deep learning algorithms, transportation efficiency can be enhanced by optimizing routes to reduce both fuel consumption and emissions. These models take into account variables i.e. traffic conditions, weather, and delivery schedules to determine the most eco-friendly routes.
- **Ethical Sourcing -** Deep learning can be used to ensure ethical sourcing practices by analyzing supplier data and monitoring compliance with ethical standards. This helps companies avoid suppliers with unethical practices and promotes responsible sourcing.

Environmental Monitoring and Impact Assessment

Environmentally conscious computing, often referred to as eco-smart technology, is an idea that has obtained substantial traction in recent years. It centers on developing and using computing technologies that are environmentally friendly, energy-efficient, and sustainable. As the world grapples with the escalating environmental crises, leveraging cutting-edge technologies i.e. Deep learning has become indispensable.

- **Air Quality Monitoring -** Deep learning models can analyze data from air quality sensors, satellite imagery, and weather patterns to predict air quality trends. For instance, convolutional neural networks (CNNs) can identify pollutants and forecast air quality levels, aiding in pollution control efforts.
- **Water Quality Assessment -** Recurrent neural networks (RNNs) can be employed to analyze data from water quality sensors, drones, and remote sensing platforms. They can detect anomalies, such as the presence of contaminants, in real-time, enabling rapid responses to water pollution incidents.
- **Wildlife Conservation -** Deep learning plays a crucial role in wildlife conservation by helping in the identification and tracking of endangered species. Object detection algorithms, such as YOLO (You Only Look Once), can identify animals in camera trap images, allowing researchers to monitor populations and assess their well-being.
- **Deforestation Detection -** Satellite imagery combined with deep learning techniques can monitor deforestation and illegal logging activities. Convolutional neural networks can detect changes in vegetation cover and identify areas at risk, facilitating timely intervention.

CASE STUDIES AND EXAMPLES

Real-world examples of machine learning in green computing - Lately, the convergence of machine learning and green computing has brought about transformative changes in the way we approach sustainability and environmental conservation. Harnessing the power of machine learning algorithms, capable of analyzing vast datasets and recognizing complex patterns, has opened up new avenues for streamlining power usage, reducing waste, & mitigating the environmental impact of technology.

- **Predictive Maintenance in the Smart Grid -** One of the key challenges in managing the energy grid is ensuring the reliability of power distribution infrastructure. Predictive maintenance heavily relies on machine learning for its success, where algorithms analyze data from sensors and past maintenance data to Forecast the probable timing of equipment failure. By proactively replacing or repairing components, energy companies can reduce downtime and avoid catastrophic failures, ultimately leading to significant energy savings. For instance, companies like General Electric and Siemens have implemented predictive maintenance solutions that use machine learning to monitor the condition of transformers, circuit breakers, and other critical grid components.
- **Google's DeepMind and Data Center Cooling -** Data centers are notorious energy hogs, consuming massive amounts of electricity to power servers and maintain optimal operating conditions. Google's DeepMind, in collaboration with their subsidiary, DeepMind Energy, has developed an AI-powered system to optimize data center cooling. DeepMind's machine learning algorithms analyze data from thousands of sensors, including temperature, humidity, and cooling system performance, to predict and optimize cooling requirements. By adjusting cooling settings in real-time based on predictive models, Google has reduced its data center energy consumption by up to 40%. This achievement not only saves energy but also demonstrates the potential of AI in managing large-scale, energy-intensive operations more efficiently.
- **Wind Farm Optimization -** Renewable energy sources i.e. wind and solar are inherently intermittent, making grid management complex. Machine learning is helping to address this challenge

by predicting energy generation from renewables and optimizing grid operations accordingly. In Denmark, the utility company Ørsted has implemented machine learning algorithms to predict wind turbine output based on weather forecasts and historical data. These predictions enable grid operators to better integrate wind energy into the grid, reducing the need for fossil fuel backup generation and improving energy efficiency.

- **IBM's AI-Powered E-Waste Recycling -** E-waste poses a notable environmental threat because of its harmful elements and inadequate waste management. IBM has developed an AI-powered recycling approach that leverages computer vision & machine learning to identify valuable components in discarded electronics. IBM's system can separate and categorize components like semiconductors, capacitors, and connectors, enabling more efficient recycling processes.
- **Uber's Machine Learning-Driven Route Optimization -** Ride-sharing companies like Uber are using machine learning to optimize driver routes, reducing carbon emissions and fuel consumption. By analyzing past traffic data, weather conditions, and real-time traffic updates, Uber's algorithms suggest the most efficient routes for drivers, minimizing idle time and fuel consumption.

Success Stories of Deep Learning Applications in Sustainability

Deep learning has revolutionized various industries by enabling computers to mimic the human brain's neural networks, making it possible to process huge volumes of data and capture meaningful insights. In the realm of sustainability, deep learning has appeared as a powerful tool to address complex environmental and social challenges.

- **Solar Power Forecasting -** One of the most prominent applications of deep learning in sustainability is in renewable energy generation, particularly in solar power forecasting. Deep neural networks have been employed to analyze historical weather data, satellite images, and solar panel characteristics to predict solar energy generation accurately.

Google's DeepMind developed a deep learning model called "DeepMind for Energy" that improved the accuracy of wind and solar power forecasting by up to 30%. This technology helps grid operators balance energy supply and demand efficiently.

- **Wind Turbine Optimization -** Deep learning also plays a crucial role in optimizing wind turbine operations. Neural networks can analyze data from sensors mounted on turbines to project maintenance needs, minimize downtime, and increase energy output. By doing so, they contribute to the overall efficiency of wind farms and reduce the environmental impact of energy production.

General Electric (GE) has implemented deep learning algorithms in their Digital Wind Farm platform, which uses real-time data to improve wind turbine performance and increase energy production by up to 20%.

- **Precision Farming -** Deep learning is transforming agriculture by enabling precision farming practices. Farmers can use drones equipped with deep learning algorithms to monitor crop health, detect pests and diseases, and optimize irrigation. This doesn't just boost agricultural productivity but also minimizes the need for chemical inputs, promoting sustainable farming practices.

Blue River Technology, a subsidiary of John Deere, developed the "See & Spray" system, which uses computer vision powered by deep learning to identify and precisely spray herbicides only where needed. This reduces herbicide usage by up to 90% and minimizes environmental impact.

FUTURE DIRECTIONS AND RESEARCH OPPORTUNITIES

Emerging Trends in Machine Learning for Green Computing

This explosion in data and computing resources has inevitably led to a substantial increase in power usage, contributing to environmental concerns and rising operational costs. Green computing, which aims to reduce the environmental effect of computing systems, has become more critical than ever before (Harmon & Auseklis, 2009). Machine learning, with its capacity to optimize and automate processes, is playing a pivotal role in driving green computing forward. Green computing involves designing, using, and managing computer systems in an environmentally responsible manner (Soomro & Sarwar, 2012). The ultimate goal is to minimize power usage, minimize e-waste, & decrease the carbon footprint associated with computing. Machine learning has appeared as a powerful tool to address these challenges, as it can help optimize hardware, improve resource allocation, and enhance energy efficiency in various computing environments (El-Kassar & Singh, 2019b).

Energy-Efficient Data Centers

- **Predictive Maintenance -** Machine learning models are being deployed to predict and prevent hardware failures in data centers. By analyzing sensor data and historical performance, these models can identify when equipment is prone to malfunction, allowing for proactive maintenance. This not only reduces downtime but also conserves energy by avoiding the operation of backup systems.
- **Dynamic Cooling Management -** Dynamic cooling management is another application of machine learning in data centers. By continuously monitoring temperature and airflow data, machine learning algorithms can optimize the cooling system to reduce energy consumption while ensuring that servers remain within safe operating temperatures.

Energy-Efficient Algorithms

- **Algorithm Selection -** Machine learning algorithms are being used to automatically select algorithms that consume less computational power. By analyzing the characteristics of a given task and the available hardware, these models can choose the most energy-efficient algorithm for execution.
- **Approximation Algorithms -** Approximation algorithms, which sacrifice a degree of accuracy for reduced computational complexity, are gaining traction in green computing. Machine learning models can be utilized to determine the acceptable level of approximation for a specific task, optimizing the trade-off between accuracy and energy consumption.

Energy-Efficient Edge Computing

- **Edge Device Optimization** - Machine learning models are being deployed on edge devices to optimize their energy usage. These models can adapt the device's behavior based on user patterns and requirements, ensuring that energy is conserved when the device is idle or underutilized.
- **Federated Learning** - Federated learning, a decentralized machine learning approach, is ideal for energy-efficient edge computing. Instead of sending data to a central server, edge devices collaborate locally to train machine learning models. This reduces the need for data transmission and central processing, ultimately saving energy.

Green AI

- **Model Pruning** - Machine learning models are becoming more energy-efficient through techniques like model pruning, where redundant or unnecessary parameters are removed. This not only reduces the model's computational requirements but also its energy consumption during inference.
- **Low-power Hardware Accelerators** - The creation of low-power hardware accelerators, specialized for machine learning workloads, is a significant trend in green AI. These accelerators, often based on neuromorphic computing or dedicated ASICs, enable energy-efficient model training and inference.

Environmental Monitoring

- **Environmental Impact Assessment** - Machine learning models are utilized to assess the environmental impact of computing operations. These models consider factors such as energy consumption, carbon emissions, and resource utilization to provide insights into the sustainability of computing systems (B. Guo et al., 2009).
- **E-waste Management** - Machine learning can help in the efficient recycling & management of electronic waste (e-waste) by automating the identification and sorting of electronic components for recycling.

The Role of Policymakers and Industry Stakeholders in Promoting Sustainable IT Practices

As the world becomes increasingly reliant on information technology (IT), the environmental and social impacts of this industry have come under scrutiny. Sustainable IT practices have emerged as a critical aspect of the broader sustainability agenda. Policymakers and industry stakeholders play pivotal roles in shaping the trajectory of IT sustainability. Here are some key ways in which policymakers contribute:

- **Regulation and Standards** - Policymakers can establish regulations and standards that mandate sustainable IT practices. This includes requirements for energy efficiency, responsible e-waste management, and reduction of hazardous materials in IT products. For example, the European

Union's Eco-design Directive sets energy efficiency requirements for various products, including IT equipment.

- **Incentives and Subsidies -** Governments have the ability to promote sustainable IT practices through the implementation of incentives and subsidies. Tax credits for green IT investments or research grants for sustainable technology development can motivate businesses to prioritize sustainability.
- **Eco-labeling and Certification -** Policymakers can promote eco-labeling and certification programs that help consumers make informed choices. These labels certify products that meet specific environmental and social criteria, making it easier for consumers to choose sustainable options.
- **Education and Awareness -** Governments can invest in public awareness campaigns to educate citizens and businesses about the importance of sustainable IT practices. Raising awareness can drive demand for eco-friendly products and practices.

Industry stakeholders, including IT companies, manufacturers, and trade associations, are pivotal in promoting sustainability within the sector. They have direct control over product design, supply chains, and manufacturing processes. Here are their primary roles:

- **Product Innovation -** IT companies can drive sustainability by designing energy-efficient products, using eco-friendly materials, and optimizing product lifecycles. For instance, Apple's commitment to using recycled materials in its products and reducing its carbon footprint demonstrates the industry's potential for innovation.
- **Supply Chain Management -** Sustainable IT practices extend beyond the final product. Manufacturers and suppliers must adopt eco-friendly practices in their supply chains. This involves the conscientious procurement of materials, reducing waste, and optimizing transportation logistics.
- **Transparency and Reporting -** Industry stakeholders should provide transparent information about their environmental and social impact. Sustainability reports, which disclose metrics related to energy consumption, carbon emissions, and waste reduction, are essential for accountability and benchmarking progress.
- **Collaboration and Advocacy -** Collaboration among industry players is essential to advance sustainability. Trade associations can facilitate knowledge sharing and best practices. Moreover, industry stakeholders can collectively advocate for policies that promote sustainable IT practices.

Policymakers and industry stakeholders can achieve the greatest impact by working together. Collaboration fosters synergy and ensures that policies are practical and aligned with industry capabilities. Here are some examples of collaborative efforts:

- **Public-Private Partnerships -** Governments and IT companies can form partnerships to jointly fund research and development of sustainable technologies. These collaborations can lead to breakthroughs in energy efficiency and waste reduction.
- **Multi-Stakeholder Initiatives -** Multi-stakeholder initiatives involving governments, businesses, non-profits, and academia can set industry-wide sustainability goals and standards. Examples include the Electronic Product Environmental Assessment Tool (EPEAT) and the Sustainable Electronics Initiative (SEI).

- **Advocacy and Lobbying -** Industry stakeholders can leverage their influence to advocate for policies that promote sustainability. They can engage with policymakers to shape legislation that aligns with their sustainability goals.

CONCLUSION

In conclusion, the importance of machine learning and deep learning in achieving green computing objectives cannot be overstated. As our world grapples with the urgent need to address environmental challenges and reduce our carbon footprint, these advanced technologies offer invaluable tools and solutions.

Machine learning algorithms have the capacity to optimize energy consumption, improve resource allocation, and enhance overall system efficiency. They can help businesses and individuals make data-driven decisions that minimize energy waste and reduce environmental impact. Additionally, deep learning, with its ability to process huge volumes of complex data, plays a pivotal role in areas such as climate modeling, weather prediction, and renewable energy optimization, contributing to a more sustainable future.

The deployment of machine learning and deep learning models in green computing not only enhances energy efficiency but also promotes innovation. It opens up new possibilities for the development of eco-friendly technologies and smart systems that can adapt and respond to changing environmental conditions.

Furthermore, these technologies empower us to make informed choices about energy consumption, conservation, and resource management. They enable the creation of intelligent, self-regulating systems that can optimize energy usage in real-time, reducing costs and environmental impact simultaneously.

In a world where the intersection of technology and environmental sustainability is increasingly crucial, machine learning and deep learning stand as powerful allies in the pursuit of green computing objectives. As we continue to harness the potential of these technologies and integrate them into our systems and processes, we move closer to a greener, more sustainable future where technology and environmental responsibility go hand in hand. It is imperative that we continue to invest in research, innovation, and the responsible application of machine learning and deep learning to ensure a harmonious coexistence between technology and the environment.

REFERENCES

Ali, H. M. M., El-Gorashi, T. E. H., Lawey, A. Q., & Elmirghani, J. M. H. (2017). Future energy efficient data centers with disaggregated servers. *Journal of Lightwave Technology*, *35*(24), 5361–5380. doi:10.1109/JLT.2017.2767574

Anthony, B., Majid, M. A., & Romli, A. (2018). A Descriptive Study towards Green Computing Practice Application for Data Centers in IT Based Industries. *MATEC Web of Conferences, 150*, 05048. 10.1051/matecconf/201815005048

Bougdah, H., & Sharples, S. (2009). Environment, Technology and Sustainability. Taylor & Francis eBooks. doi:10.4324/9780203878408

Carletti, V., Greco, A., Percannella, G., & Vento, M. (2020). Age from Faces in the Deep Learning Revolution. *IEEE Transactions on Pattern Analysis and Machine Intelligence*, *42*(9), 2113–2132. doi:10.1109/TPAMI.2019.2910522 PMID:30990174

Deng, L., & Yu, D. (2014). *Deep Learning: methods and applications*. IEEE. doi:10.1561/9781601988157

DenilM.ShakibiB.DinhL.RanzatoM.De FreitasN. (2013). Predicting parameters in deep learning. *Neural Information Processing Systems, 26*, 2148–2156. doi:10.14288/1.0165555

El-Kassar, A., & Singh, S. K. (2019a). Green innovation and organizational performance: The influence of big data and the moderating role of management commitment and HR practices. *Technological Forecasting and Social Change*, *144*, 483–498. doi:10.1016/j.techfore.2017.12.016

El-Kassar, A., & Singh, S. K. (2019b). Green innovation and organizational performance: The influence of big data and the moderating role of management commitment and HR practices. *Technological Forecasting and Social Change*, *144*, 483–498. doi:10.1016/j.techfore.2017.12.016

Fuchs, C. (2006). The implications of new information and communication technologies for sustainability. *Environment, Development and Sustainability*, *10*(3), 291–309. doi:10.1007/s10668-006-9065-0

Gai, K., Qiu, M., Zhao, H., Tao, L., & Zong, Z. (2016). Dynamic energy-aware cloudlet-based mobile cloud computing model for green computing. *Journal of Network and Computer Applications*, *59*, 46–54. doi:10.1016/j.jnca.2015.05.016

Goodfellow, I. J., Bengio, Y., & Courville, A. (2016). *Deep learning*. ACM. https://dl.acm.org/citation.cfm?id=3086952

Gulshan, V., Peng, L., Coram, M., Stumpe, M. C., Wu, D., Narayanaswamy, A., Venugopalan, S., Widner, K., Madams, T., Cuadros, J., Ramasamy, K., Raman, R., Nelson, P. C., Mega, J. L., & Webster, D. R. (2016). Development and validation of a deep learning algorithm for detection of diabetic retinopathy in retinal Fundus photographs. *Journal of the American Medical Association*, *316*(22), 2402. doi:10.1001/jama.2016.17216 PMID:27898976

Guo, B., Yan, S., & Shao, Z. (2009). The redefinition and some discussion of green computing. *Chinese Journal of Computers*, *32*(12), 2311–2319. doi:10.3724/sp.j.1016.2009.02311

Guo, Z., Liu, J., Wang, Y., Chen, M., Wang, D., Xu, D., & Cheng, J. (2023). Diffusion models in Bioinformatics: A new wave of deep learning revolution in action. *arXiv (Cornell University).* https://doi.org//arxiv.2302.10907 doi:10.48550

Harmon, R. R., & Auseklis, N. (2009). Sustainable IT Services: Assessing the impact of green Computing Practices. IEEE. doi:10.1109/PICMET.2009.5261969

Khanafer, M., & Shirmohammadi, S. (2020). Applied AI in instrumentation and measurement: The deep learning revolution. *IEEE Instrumentation & Measurement Magazine*, *23*(6), 10–17. doi:10.1109/MIM.2020.9200875

Kurp, P. (2008). Green computing. *Communications of the ACM*, *51*(10), 11–13. doi:10.1145/1400181.1400186

LeCun, Y., Bengio, Y., & Hinton, G. E. (2015). Deep learning. *Nature*, *521*(7553), 436–444. doi:10.1038/nature14539 PMID:26017442

Lei, Y., Yang, B., Jiang, X., Jia, F., Li, N., & Nandi, A. K. (2020). Applications of machine learning to machine fault diagnosis: A review and roadmap. *Mechanical Systems and Signal Processing*, *138*, 106587. doi:10.1016/j.ymssp.2019.106587

Litjens, G., Ciompi, F., Wolterink, J. M., De Vos, B. D., Leiner, T., Teuwen, J., & Išgum, I. (2019). State-of-the-Art deep learning in cardiovascular image analysis. *JACC: Cardiovascular Imaging*, *12*(8), 1549–1565. doi:10.1016/j.jcmg.2019.06.009 PMID:31395244

Lundervold, A. S., & Lundervold, A. (2019). An overview of deep learning in medical imaging focusing on MRI. *Zeitschrift für Medizinische Physik*, *29*(2), 102–127. doi:10.1016/j.zemedi.2018.11.002 PMID:30553609

Maier, A., Syben, C., Lasser, T., & Rieß, C. (2019). A gentle introduction to deep learning in medical image processing. *Zeitschrift für Medizinische Physik*, *29*(2), 86–101. doi:10.1016/j.zemedi.2018.12.003 PMID:30686613

Mao, Y., You, C., Zhang, J., Huang, K., & Letaief, K. B. (2017). A survey on Mobile Edge Computing: The Communication Perspective. *IEEE Communications Surveys and Tutorials*, *19*(4), 2322–2358. doi:10.1109/COMST.2017.2745201

Mathew, A., Amudha, P., & Sivakumari, S. (2020). Deep Learning Techniques: An Overview. In Advances in intelligent systems and computing (pp. 599–608). Springer. doi:10.1007/978-981-15-3383-9_54

Ojo, A. O., Raman, M., & Downe, A. G. (2019). Toward green computing practices: A Malaysian study of green belief and attitude among Information Technology professionals. *Journal of Cleaner Production*, *224*, 246–255. doi:10.1016/j.jclepro.2019.03.237

Pedrycz, W. (2022). Towards green machine learning: Challenges, opportunities, and developments. *Journal of Smart Environments and Green Computing*, *2*(4), 163–174. doi:10.20517/jsegc.2022.16

Prakash, B., Jayashri, S., & Karthik, T. (2019). A hybrid genetic artificial neural network (G-ANN) algorithm for optimization of energy component in a wireless mesh network toward green computing. *Soft Computing*, *23*(8), 2789–2798. doi:10.1007/s00500-019-03789-8

Sarkar, S., & Misra, S. (2016). Theoretical modelling of fog computing: A green computing paradigm to support IoT applications. *IET Networks*, *5*(2), 23–29. doi:10.1049/iet-net.2015.0034

Sejnowski, T. J. (2018). The deep learning revolution. The MIT Press eBooks. doi:10.7551/mitpress/11474.001.0001

Shen, D., Wu, G., & Suk, H. (2017). Deep learning in medical image analysis. *Annual Review of Biomedical Engineering*, *19*(1), 221–248. doi:10.1146/annurev-bioeng-071516-044442 PMID:28301734

Shorten, C., & Khoshgoftaar, T. M. (2019). A survey on Image Data Augmentation for Deep Learning. *Journal of Big Data*, *6*(1), 60. doi:10.1186/s40537-019-0197-0

Soomro, T. R., & Sarwar, M. (2012). Green Computing: From current to future trends. Zenodo *(CERN European Organization for Nuclear Research)*. doi:10.5281/zenodo.1062818

Xu, X., Fu, S., Cai, Q., Tian, W., Liu, W., Dou, W., Sun, X., & Liu, A. X. (2018). Dynamic resource allocation for load balancing in fog environment. *Wireless Communications and Mobile Computing*, *2018*, 1–15. doi:10.1155/2018/6421607

Chapter 4

Machine Learning-Integrated Sustainable Engineering and Energy Systems: Innovations at the Nexus

M. Sarat Chandra Prasad
Department of Computer Science and Engineering, Aditya College of Engineering, Surampalem, India

M. Dhanalakshmi
https://orcid.org/0000-0003-3920-6949
Department of Computer Science and Engineering, New Horizon College of Engineering, Bangalore, India

M. Mohan
Department of Computer Science and Engineering, KPR Institute of Engineering and Technology, Arasur, India

B. Somasundaram
School of Mechanical Engineering, REVA University, Bengaluru, India

R. Valarmathi
Department of S&H (Mathematics), RMK Engineering College, Kavaraipettai, India

Sampath Boopathi
https://orcid.org/0000-0002-2065-6539
Mechanical Engineering, Muthayammal Engineering College, Namakkal, India

ABSTRACT

Machine learning (ML) has revolutionized various fields, including engineering, energy systems, and sustainability. This abstract explores the synergies between ML and these domains, focusing on its optimization in predictive maintenance, energy consumption efficiency, and smart grids. ML's role in renewable energy forecasting, building energy management, and materials science is also explored. It also highlights its impact on supply chain optimization, environmental monitoring, and sustainability assessments. The holistic approach extends to smart city initiatives and infrastructure development, paving the way for intelligent urban planning. ML enhances decision-making processes, enabling more resilient, efficient, and sustainable practices in engineering and energy systems. This exploration serves as a beacon for researchers, practitioners, and policymakers seeking innovative solutions at the intersection of ML, engineering, and sustainability.

DOI: 10.4018/979-8-3693-1794-5.ch004

INTRODUCTION

Machine Learning (ML) is transforming sustainable engineering and energy systems by reshaping traditional paradigms and fostering a more sustainable future. Its ability to discern intricate patterns within vast datasets and adaptive learning mechanisms makes it a crucial enabler in addressing complex issues within these sectors. As the global community strives to reduce environmental impact, optimize resource usage, and enhance efficiency, ML is a key enabler in achieving these goals. The convergence of ML with sustainable engineering and energy systems represents a pivotal moment in our pursuit of innovative solutions(Rangel-Martinez et al., 2021).

This section delves into the fundamental principles and applications of machine learning (ML) in engineering, highlighting its role in enhancing decision-making processes, optimizing workflows, and predicting maintenance needs. It also discusses the potential of ML in laying the groundwork for sustainable practices by identifying patterns and making predictions(Zhou et al., 2020). The increasing global energy demand necessitates a shift towards sustainable energy systems, focusing on renewable energy sources, smart grids, and efficient consumption. Machine learning (ML) is crucial in this context, providing predictive capabilities for renewable energy outputs, optimizing energy distribution through smart grids, and enabling real-time adjustments to fluctuating energy demands, thus transforming the energy landscape towards sustainability(Willard et al., 2022).

Machine learning (ML) is a key tool in engineering, particularly in predictive maintenance. By analyzing historical data and real-time sensor inputs, ML algorithms can predict equipment failures, reducing downtime and environmental impact. This chapter explores the integration of ML in sustainable engineering and energy systems, focusing on optimizing energy consumption in buildings, forecasting renewable energy outputs, and revolutionizing materials science(Duchesne et al., 2020). The goal is to demonstrate ML's transformative potential in driving engineering practices towards sustainability. In the ever-evolving landscape of engineering and energy systems, the integration of Machine Learning (ML) stands as a transformative force, ushering in a new era of efficiency, resilience, and sustainability. This introduction sets the stage for a comprehensive exploration into the profound impacts of ML applications in sustainable engineering practices and energy systems(Yang et al., 2020).

In recent years, Machine Learning has emerged as a paradigm-shifting technology with the potential to revolutionize traditional engineering practices and reshape the landscape of energy systems. At its core, ML is a branch of artificial intelligence that empowers systems to learn and adapt from data, enabling them to make informed decisions without explicit programming. The integration of ML in engineering applications and energy systems has opened up unprecedented possibilities for optimization, predictive analysis, and data-driven decision-making(Ibrahim et al., 2020). In engineering, the traditional approach to system maintenance often involved reactive responses to equipment failures, leading to downtime and substantial costs. However, with ML algorithms, a paradigm shift has occurred towards predictive maintenance. These algorithms analyze vast datasets from sensors and historical records, predicting potential equipment failures before they occur. This proactive strategy not only minimizes downtime but also maximizes the lifespan of machinery, translating to significant cost savings and increased operational efficiency(Ibrahim et al., 2020).

The imperative for sustainable engineering practices and energy systems has never been more critical. As societies grapple with the challenges of climate change and resource depletion, the role of ML in navigating this complex landscape becomes paramount. Sustainable engineering aims to design and implement processes that minimize environmental impact, promote resource efficiency, and contribute to long-term ecological balance. ML acts as a catalyst in achieving these goals by optimizing existing systems and driving innovation across diverse domains(P. Sharma et al., 2022). The integration of machine learning (ML) in energy systems is revolutionizing energy generation, distribution, and consumption. Smart grids, powered by ML algorithms, are crucial for sustainable energy transition. These intelligent grids balance energy supply and demand, integrate renewable sources, and improve grid efficiency. ML also aids in forecasting renewable energy outputs, enabling efficient resource management and sustainable energy supply(Cao et al., 2020).

This chapter explores the role of machine learning (ML) in sustainable engineering and energy systems. It explores its applications in predictive maintenance, energy consumption optimization, renewable energy forecasting, building energy management, materials science, supply chain optimization, and environmental monitoring. The chapter aims to provide a nuanced understanding of how ML is reshaping sustainable engineering and energy systems, paving the way for a future where efficiency and sustainability are intertwined(Alimi et al., 2020). Machine Learning (ML) has emerged as a game-changer in the realm of energy systems, offering unprecedented opportunities to enhance efficiency, reliability, and sustainability. At the heart of its significance lies the ability of ML algorithms to analyze vast and complex datasets, derive patterns, and make predictions. This transformative capacity has far-reaching implications for energy systems across various domains(Zhang et al., 2021).

In the context of energy consumption, ML excels at optimizing and managing demand. Traditional energy systems often grapple with the challenge of balancing supply and demand dynamically. ML algorithms, however, can process real-time data on consumption patterns, weather conditions, and other variables to predict demand fluctuations. This predictive capability enables utilities to adjust energy production and distribution in advance, ensuring a more stable and efficient energy supply(Yang et al., 2020). Consequently, the integration of ML mitigates the risk of overloads or shortages, contributing to grid resilience. The advent of smart grids epitomizes the significance of ML in energy systems. Smart grids leverage ML algorithms to intelligently manage energy distribution, integrating renewable sources and optimizing energy flows. These systems continuously learn from historical data, adapting to changes in demand patterns and incorporating insights for efficient energy allocation. The result is a more adaptive, responsive, and sustainable energy infrastructure that can accommodate the growing complexities of modern energy networks(P. Sharma et al., 2022).

ML also plays a pivotal role in the forecasting of renewable energy outputs. As the world increasingly turns to renewable sources such as solar and wind, the intermittent nature of these resources poses challenges for grid stability. ML algorithms analyze weather data, historical energy production, and other variables to predict when and how much energy will be generated. Accurate forecasts enable better integration of renewable sources into the energy mix, reducing dependency on traditional fossil fuels and fostering a more sustainable energy landscape(Kumar B et al., 2024). Moreover, in energy efficiency and conservation efforts, ML can be employed to optimize building energy management systems. These systems learn from patterns of occupancy, external environmental factors, and historical energy usage to dynamically adjust heating, cooling, and lighting, thereby minimizing waste and reducing overall energy consumption(Ingle et al., 2023).

The role of machine learning (ML) in energy systems is crucial for optimizing energy consumption, improving grid management, integrating renewable sources, and promoting sustainability. As energy systems evolve towards a more intelligent and eco-friendly future, ML plays a pivotal role.

MACHINE LEARNING IN ENGINEERING

Machine Learning (ML) has become an integral part of engineering applications, revolutionizing traditional processes by leveraging advanced algorithms to extract insights from data. This section delves into the foundational aspects of ML in engineering, elucidating the basics of ML algorithms and their diverse applications, as well as exploring the seamless integration of ML into engineering processes. At its core, Machine Learning refers to the capability of systems to learn from data and improve their performance over time without being explicitly programmed. The foundation of ML lies in algorithms, mathematical models that enable computers to identify patterns, make predictions, and optimize decision-making processes. Supervised learning, unsupervised learning, and reinforcement learning are fundamental categories of ML algorithms.

In supervised learning, algorithms are trained on labeled datasets, where the model learns to map inputs to corresponding outputs. This is particularly valuable in predictive tasks, such as regression or classification problems. Unsupervised learning, on the other hand, deals with unlabeled data, aiming to discover inherent patterns or structures within the dataset. Clustering and dimensionality reduction are common applications of unsupervised learning. Reinforcement learning involves training algorithms through trial and error, where the system learns by receiving feedback on its actions in a dynamic environment(Ingle et al., 2023; Pachiappan et al., 2023; Ramudu et al., 2023). In engineering applications, these ML algorithms find diverse uses. In predictive maintenance, supervised learning models analyze historical data to predict equipment failures, enabling proactive maintenance strategies. Classification algorithms contribute to fault detection, while clustering algorithms group similar operational patterns for analysis. The adaptability of ML algorithms allows them to be tailored to specific engineering tasks, ranging from optimizing energy consumption in buildings to predicting material properties in materials science(Boopathi & Kanike, 2023; Maheswari et al., 2023).

The integration of ML into engineering processes marks a paradigm shift in how systems are designed, monitored, and optimized. One of the key areas where ML is making a substantial impact is in predictive maintenance. Traditional maintenance practices often rely on fixed schedules or reactive responses to equipment failures. ML algorithms, however, analyze sensor data and historical records to predict when equipment is likely to fail. This predictive approach minimizes downtime, extends the lifespan of machinery, and reduces operational costs. In the optimization of energy consumption, ML algorithms play a pivotal role in learning and adapting to dynamic conditions(Syamala et al., 2023; Zekrifa et al., 2023). Building energy management systems, for instance, utilize ML to analyze patterns of occupancy, external factors like weather conditions, and historical energy usage. By dynamically adjusting heating, cooling, and lighting based on these insights, ML contributes to significant energy savings and enhanced sustainability in engineering applications.

ML plays a crucial role in materials science and engineering by predicting properties and accelerating material discovery. It analyzes vast datasets, identifying optimal compositions for specific applications. ML integration promotes a data-driven, adaptive approach, continuously learning from new data. This integration is essential for modern engineering challenges, offering innovative solutions and new domains.

PREDICTIVE MAINTENANCE IN ENGINEERING

Predictive maintenance is a critical component of modern engineering practices, aiming to anticipate and mitigate equipment failures before they occur. Machine Learning (ML) has emerged as a powerful tool in this domain, revolutionizing the approach to maintenance strategies. Here are seven key points outlining the application of ML for predictive maintenance in engineering(Boopathi & Kanike, 2023; Zekrifa et al., 2023). The Figure 1 depicts the use of machine learning (ML) for predictive maintenance in engineering.

- **Data-Driven Decision Making:** ML algorithms thrive on data, and predictive maintenance leverages this by utilizing historical and real-time data from sensors and equipment. These algorithms learn patterns of normal operation and deviations that precede failures, enabling data-driven decision-making in maintenance planning.
- **Fault Detection and Classification:** ML models are adept at detecting anomalies and classifying faults based on patterns identified in data. These models can analyze sensor readings and performance metrics to pinpoint irregularities indicative of potential equipment failures. By categorizing faults, maintenance teams can prioritize and address issues efficiently.
- **Proactive Equipment Monitoring:** Traditional maintenance practices often rely on fixed schedules or reactive responses to equipment failures. ML transforms this paradigm by enabling proactive monitoring. Predictive maintenance systems continuously assess the health of equipment in real-time, allowing for timely interventions before critical failures occur.

Figure 1. Machine learning (ML) predictive maintenance in engineering

- **Condition-Based Monitoring:** ML facilitates condition-based monitoring, where the health of equipment is assessed based on its actual state rather than relying on predetermined maintenance schedules. Algorithms learn the normal operational conditions and deviations, allowing for personalized and adaptive maintenance strategies tailored to each piece of equipment.
- **Failure Prediction and Time-to-Failure Estimates:** ML algorithms excel at predicting when equipment is likely to fail. By analyzing patterns in historical data, these models can provide accurate estimates of the remaining useful life of machinery. This information allows maintenance teams to plan interventions strategically, minimizing downtime and optimizing resource allocation.
- **Integration of Multiple Data Sources:** Predictive maintenance with ML involves the integration of diverse data sources, including sensor data, maintenance records, environmental conditions, and even external factors like supply chain data. The holistic analysis of these data sets provides a comprehensive view of equipment health, enhancing the accuracy and reliability of predictions.
- **Cost Savings and Increased Operational Efficiency:** One of the primary advantages of applying ML in predictive maintenance is the potential for significant cost savings. By shifting from reactive to proactive maintenance strategies, organizations can reduce downtime, extend equipment lifespan, and optimize resource utilization. This not only lowers maintenance costs but also enhances overall operational efficiency.

Machine learning (ML) is revolutionizing asset management in engineering through predictive maintenance, utilizing data analytics and algorithms. This proactive, efficient, and cost-effective approach anticipates and addresses equipment failures before they impact operations, transforming the maintenance landscape.

Examples

Proactive equipment management in India involves the strategic use of technology and data-driven approaches to optimize the performance and longevity of machinery(Dhanya et al., 2023; Rebecca et al., 2023). Here are five examples of proactive equipment management practices implemented in various sectors across India:

- **Railway Predictive Maintenance:** Indian Railways, one of the world's largest railway networks, has implemented predictive maintenance practices using sensors and data analytics. Various components of trains, tracks, and signaling systems are monitored in real-time. Predictive algorithms analyze the data to identify potential issues, allowing for timely maintenance interventions. This approach helps reduce unplanned downtime, enhance safety, and improve overall operational efficiency.
- **Power Plant Asset Management:** In the power sector, thermal and renewable energy plants in India utilize proactive equipment management for optimizing the performance of turbines, generators, and other critical components. Sensors and IoT devices are deployed to monitor equipment health and performance parameters. Machine Learning algorithms analyze this data to predict potential failures, enabling power plant operators to schedule maintenance activities during planned shutdowns and minimize the impact on energy production.

- **Automotive Manufacturing Industry:** Automotive manufacturing plants in India leverage proactive equipment management to ensure the continuous operation of production lines. Predictive maintenance techniques are applied to robotic arms, conveyor belts, and other machinery on the assembly line. By monitoring equipment health in real-time, manufacturers can identify and address issues before they lead to production disruptions. This approach contributes to increased efficiency and reduced downtime in the automotive manufacturing process.
- **Telecommunication Infrastructure:** Telecommunication companies in India deploy proactive equipment management strategies to maintain the reliability of their network infrastructure. This includes monitoring the health of cell towers, antennas, and network equipment. ML algorithms analyze data from sensors to predict equipment failures, allowing for targeted maintenance activities. By addressing potential issues before they impact network performance, telecom companies enhance the reliability and availability of communication services.
- **Wind Energy Farms:** Wind energy farms across India utilize proactive equipment management to optimize the performance of wind turbines. Sensors installed in turbines monitor various parameters such as blade health, gearbox condition, and generator performance. ML algorithms analyze the data to predict component failures and optimize maintenance schedules. This proactive approach helps maximize the energy output of wind farms by minimizing downtime and ensuring the efficient operation of equipment.

Proactive equipment management in India is being utilized across various sectors to improve the reliability, efficiency, and lifespan of critical equipment. As technology advances, this practice is expected to play a crucial role in ensuring sustainable and cost-effective operation of industrial assets in India.

OPTIMIZING ENERGY CONSUMPTION WITH ML

Machine Learning (ML) models are being used for energy consumption optimization, providing a dynamic and adaptive approach to managing energy resources. These models use advanced algorithms to analyze complex datasets, enabling organizations to make informed decisions that reduce energy consumption and enhance sustainability. They focus on dynamic adaptation to varied conditions(Agrawal et al., 2024; Kumar B et al., 2024; Syamala et al., 2023). The figure 2 depicts the use of machine learning (ML) models to optimize energy consumption.

Figure 2. Machine learning (ML) models for optimizing energy consumption

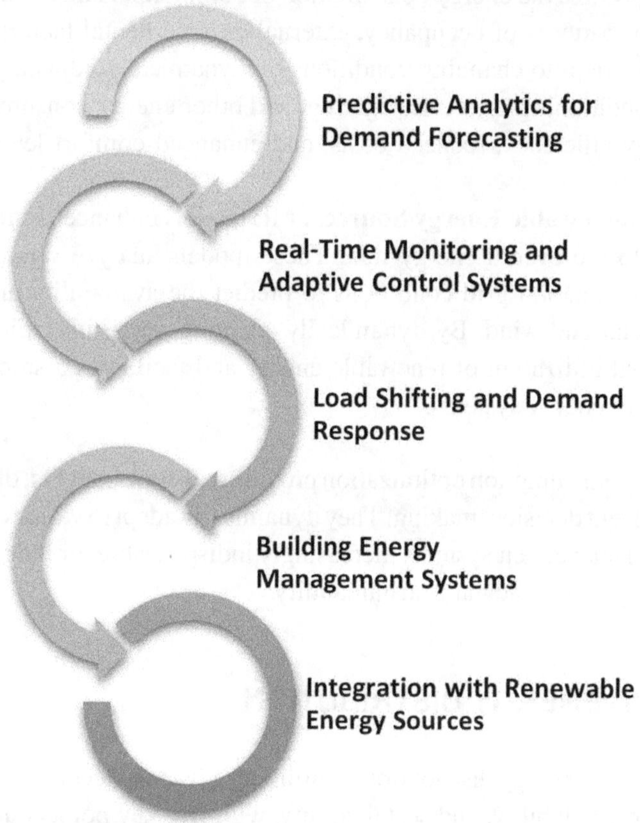

- **Predictive Analytics for Demand Forecasting:** ML models excel at predicting energy demand patterns by analyzing historical data, weather conditions, and other relevant factors. These predictive analytics enable organizations to anticipate peak demand periods and adjust their energy production and distribution accordingly. By forecasting demand with greater accuracy, companies can optimize energy generation, minimize waste, and avoid overloading the grid during peak periods(Rebecca et al., 2023).
- **Real-Time Monitoring and Adaptive Control Systems:** ML models integrated into real-time monitoring systems provide a continuous assessment of energy consumption patterns. These models can dynamically adapt to changing conditions by analyzing incoming data streams from sensors, smart meters, and other monitoring devices. Adaptive control systems use this information to optimize energy distribution, adjust equipment settings, and manage loads in real-time, ensuring efficient energy utilization under varying conditions(Boopathi, 2023b).
- **Load Shifting and Demand Response:** ML models play a crucial role in load shifting strategies, where energy consumption is redistributed to off-peak hours. By learning from historical data and external factors, these models can recommend optimal times for certain energy-intensive processes, reducing the strain on the grid during peak periods. Additionally, ML contributes to demand response programs by automatically adjusting energy consumption based on real-time pricing and grid conditions, promoting a more flexible and responsive energy ecosystem.

- **Building Energy Management Systems:** ML is extensively utilized in building energy management systems to optimize the energy consumption of commercial and residential structures. These systems learn from patterns of occupancy, external environmental factors, and historical energy usage. ML models adapt to changing conditions by dynamically adjusting HVAC (heating, ventilation, and air conditioning) systems, lighting, and other energy-consuming devices. The result is improved energy efficiency, reduced costs, and enhanced comfort levels within buildings(M. Sharma et al., 2023).
- **Integration with Renewable Energy Sources:** ML models enhance the integration of renewable energy sources into the overall energy mix. These models analyze weather forecasts, historical energy production data, and grid conditions to predict the availability and output of renewable sources such as solar and wind. By dynamically adapting to the intermittency of these sources, ML enables optimal utilization of renewable energy and facilitates a smoother transition to sustainable and greener energy systems.

ML models for energy consumption optimization provide real-time insights, optimize energy distribution, and facilitate intelligent decision-making. They dynamically adapt to varied conditions, making their integration into energy management systems increasingly indispensable for industries and organizations seeking to enhance energy efficiency and sustainability.

SMART GRIDS AND ENERGY DISTRIBUTION

Smart grids are transforming energy distribution by utilizing advanced technologies like Machine Learning to improve efficiency, reliability, and sustainability, with five key points outlining their application in energy distribution(Hema et al., 2023; Kavitha et al., 2023).

- **Grid Optimization and Load Balancing:** ML algorithms play a pivotal role in optimizing the operation of smart grids by dynamically balancing energy loads. These algorithms analyze real-time data from smart meters, sensors, and other grid components to predict demand patterns. With this information, the smart grid can intelligently distribute energy, allocate resources efficiently, and prevent overloads or disruptions. ML's ability to adapt to changing conditions in real-time contributes to a more resilient and responsive energy distribution system.
- **Predictive Maintenance for Grid Components:** Smart grids incorporate ML for predictive maintenance of grid components, such as transformers, switches, and sensors. By analyzing historical data and monitoring the performance of grid assets, ML models can predict when specific components are likely to fail. This proactive approach allows utilities to schedule maintenance activities strategically, minimizing downtime and optimizing the overall reliability of the grid.
- **Integration of Renewable Energy Sources:** ML facilitates the seamless integration of renewable energy sources, such as solar and wind, into smart grids. These sources are inherently variable, and ML algorithms analyze weather forecasts, historical energy production data, and grid conditions to predict the availability and output of renewable energy. By dynamically adapting to the intermittency of these sources, smart grids optimize energy distribution, balance supply and demand, and maximize the utilization of clean energy.

- **Demand Response and Consumer Engagement:** ML enables demand response programs within smart grids, allowing utilities to engage consumers in optimizing energy consumption. ML algorithms analyze consumption patterns, preferences, and real-time pricing to recommend optimal times for energy-intensive activities. This engagement not only empowers consumers to make informed decisions about their energy usage but also helps utilities manage peak demand more efficiently, contributing to overall grid stability.
- **Anomaly Detection and Grid Security:** ML models enhance grid security by detecting anomalies and potential cyber threats. These models continuously analyze data from various sources to identify abnormal patterns that may indicate security breaches or malfunctions. By monitoring for unusual behavior in real-time, smart grids equipped with ML-based security systems can respond swiftly to mitigate potential risks, ensuring the integrity and reliability of the energy distribution network.

Machine Learning (ML) is revolutionizing the energy sector by enhancing smart grid operations, integrating renewable energy more efficiently, and improving system reliability. Its adaptive capabilities, predictive analytics, and real-time decision-making contribute to the future of intelligent, sustainable, and resilient energy distribution, making its role increasingly critical in the evolving landscape of smart grids.

RENEWABLE ENERGY FORECASTING

Renewable energy forecasting is crucial for sustainable energy systems, and Machine Learning (ML) is vital for improving predictions. This section explores ML algorithms' application in renewable energy forecasting, focusing on energy output predictions and integrating weather data and historical trends(Nishanth et al., 2023). The figure 3 illustrates the various factors that influence the development of machine learning algorithms for energy forecasting.

ML Algorithms for Energy Output Predictions

- **Supervised Learning for Output Prediction:** Supervised learning algorithms, such as regression models and neural networks, are widely used for renewable energy output predictions. These algorithms learn from historical data, considering various input factors such as solar radiation, wind speed, temperature, and time of day. The models are trained to establish patterns between these inputs and the corresponding energy output. Once trained, the ML model can predict future energy production based on real-time or forecasted weather conditions.
- **Time Series Analysis for Temporal Patterns:** Time series analysis, a specialized ML technique, is employed to capture temporal patterns in renewable energy production. This is particularly crucial for solar and wind energy, which exhibit variations based on daily and seasonal cycles. ML algorithms analyze historical time series data to identify patterns, trends, and seasonality, allowing for more accurate predictions of energy output at different timescales.
- **Ensemble Learning for Robust Predictions:** Ensemble learning methods, such as random forests or ensemble of neural networks, are used to enhance the robustness of predictions. These algorithms combine the outputs of multiple individual models, leveraging diverse approaches to capture different aspects of the complex relationships between input variables and energy output.

Ensemble learning helps mitigate the impact of uncertainties and improves overall forecasting accuracy.

Figure 3. Factors affecting in ML algorithms for energy forecasting

Integration of Data

- **Weather Data Integration:** ML algorithms integrate diverse weather data sources to refine energy output predictions. This includes parameters like solar irradiance, wind speed, humidity, and temperature. Weather forecasts are utilized to provide input variables for the ML models, enabling them to adapt to changing weather conditions and improve the precision of energy output predictions(Hussain et al., 2023; Kumar B et al., 2024).
- **Feature Engineering for Relevant Variables:** ML models in renewable energy forecasting employ feature engineering techniques to select and optimize relevant input variables. By identifying which weather parameters and historical trends have the most significant impact on energy output, feature engineering enhances the model's ability to make accurate predictions. This process involves selecting the most informative features and transforming them to improve their relevance.
- **Real-Time Data Assimilation:** ML models are designed to assimilate real-time data seamlessly. As new weather information becomes available, these models can adapt and update predictions accordingly. Real-time data assimilation ensures that the forecasting system remains responsive to immediate changes in weather conditions, providing more accurate and timely predictions for energy production.

- **Integration with Remote Sensing Technologies:** ML-based renewable energy forecasting often integrates remote sensing technologies such as satellite imagery and ground-based sensors. Remote sensing data provide additional insights into localized weather patterns, cloud cover, and other environmental factors that influence renewable energy generation. ML models trained on this integrated dataset offer a more comprehensive understanding of the conditions impacting energy output(Boopathi, 2023b).

ML algorithms are revolutionizing renewable energy forecasting by utilizing supervised learning, time series analysis, ensemble learning, and diverse data sources. These algorithms improve prediction precision and reliability. By integrating weather data, historical trends, and real-time data assimilation, ML optimizes the integration of renewable energy into the energy landscape. As technology advances, the synergy between ML and renewable energy forecasting will be crucial for a sustainable and efficient energy future.

BUILDING ENERGY MANAGEMENT

This section explores the application of Machine Learning (ML) in Building Energy Management, focusing on enhancing energy efficiency and learning from occupancy patterns and external factors, as it is increasingly being used to optimize energy consumption in commercial and residential structures(M. Sharma et al., 2023).

ML for Building Energy Efficiency

- **Predictive Analytics for Energy Consumption:** ML algorithms are employed to predict and optimize building energy consumption by analyzing historical data, weather forecasts, and other relevant variables. These models learn intricate patterns within the building's energy usage, enabling accurate predictions of future consumption. Predictive analytics allow BEMS to proactively adjust heating, ventilation, air conditioning (HVAC) systems, lighting, and other energy-consuming components to meet anticipated demand, reducing energy wastage.
- **Fault Detection and Diagnostics:** ML models contribute to building energy efficiency by detecting faults and anomalies in the operation of energy-consuming systems. These faults may include HVAC malfunctions, inefficient lighting, or equipment degradation. By continuously monitoring building systems, ML algorithms can identify deviations from normal operation, triggering alerts for maintenance or repairs. This proactive approach minimizes energy wastage due to faulty equipment.
- **Dynamic Optimization of HVAC Systems:** HVAC systems are a major contributor to building energy consumption. ML algorithms analyze occupancy patterns, external weather conditions, and historical data to dynamically optimize HVAC settings. By learning from patterns of occupancy and adjusting the temperature and ventilation settings accordingly, ML contributes to significant energy savings while maintaining occupant comfort.
- **Personalized Energy Management:** ML enables personalized energy management within buildings by learning individual occupant preferences and adjusting environmental conditions accordingly. Occupant behavior, such as preferred temperature ranges and lighting preferences, can be

learned over time. ML algorithms use this information to optimize the building environment, contributing to energy efficiency while ensuring a comfortable and personalized experience for occupants.

Occupancy Patterns and External Factors

- **Occupancy Prediction and Scheduling:** ML models analyze historical occupancy patterns and external factors to predict future occupancy within a building. This information is valuable for optimizing energy usage, allowing BEMS to schedule heating, cooling, and lighting systems based on anticipated occupancy. By avoiding unnecessary energy consumption during unoccupied periods, buildings can significantly reduce their overall energy footprint.
- **Adaptive Lighting Systems:** ML contributes to energy efficiency in buildings by optimizing lighting systems based on real-time occupancy and external lighting conditions. Occupancy patterns and external factors, such as natural daylight availability, are considered by ML algorithms to dynamically adjust lighting levels. This adaptive approach minimizes unnecessary energy usage, especially in areas with varying occupancy and changing daylight conditions(Das et al., 2024).
- **Integration with Renewable Energy Sources:** ML algorithms facilitate the integration of renewable energy sources into building energy management. By considering external factors like solar radiation and weather conditions, ML models can predict the availability of renewable energy. BEMS can then optimize the utilization of renewable energy sources, such as solar panels, by adjusting energy consumption patterns based on the forecasted renewable energy production.
- **Continuous Adaptation to Changing Conditions:** ML models in BEMS are designed for continuous learning and adaptation. As occupancy patterns evolve and external factors change, these models dynamically adjust their predictions and recommendations. This adaptability ensures that building energy management remains responsive to real-time conditions, optimizing energy efficiency in a constantly changing environment.

Machine Learning (ML) is revolutionizing building energy management by predicting future energy needs and adapting to changing conditions. As sustainable building practices demand increase, ML's role is crucial for creating intelligent and energy-efficient built environments.

MATERIALS SCIENCE AND ENGINEERING APPLICATIONS

Materials Science and Engineering (MSE) applications benefit significantly from the integration of Machine Learning (ML), enhancing the ability to predict material properties and accelerating the materials discovery processes. This section explores how ML is employed in MSE, focusing on predicting material properties and expediting the discovery of novel materials(Boopathi et al., 2023; Boopathi & Davim, 2023; Senthil et al., 2023).

ML in Predicting Material Properties

- **Quantitative Structure-Property Relationships (QSPR):** ML techniques, particularly those under the umbrella of QSPR, are used to establish relationships between the structure of materials at a molecular or atomic level and their resulting properties. By leveraging large datasets that include information about material composition, crystal structures, and known properties, ML models can predict specific material characteristics. This allows researchers to estimate properties such as mechanical strength, thermal conductivity, or electrical conductivity without the need for extensive experimental testing.
- **Predictive Modeling for Complex Material Systems:** ML excels in handling the complexity of materials with intricate structures and compositions. Advanced algorithms, including neural networks and ensemble methods, are employed to create predictive models that capture the non-linear relationships between multiple parameters. ML models can handle high-dimensional data, making them effective in predicting material properties in complex systems, such as alloys, polymers, and composite materials.
- **Materials Informatics for Database Analysis:** ML is utilized in materials informatics to analyze vast databases of materials properties and characteristics. By applying algorithms to identify patterns and correlations within these datasets, researchers can uncover valuable insights into the relationships between material composition, processing conditions, and performance. This data-driven approach enhances the understanding of materials behavior and facilitates the prediction of properties for new compositions.

Accelerating Materials Discovery Processes

- **High-Throughput Screening with ML:** ML accelerates materials discovery by enabling high-throughput screening of potential candidates. Through the analysis of databases containing information on a myriad of materials and their properties, ML models can prioritize and identify promising materials for specific applications. This screening process significantly reduces the time and resources required for experimental testing, allowing researchers to focus on the most viable candidates.
- **Materials Design Optimization:** ML contributes to materials design optimization by guiding researchers towards the most promising directions. Algorithms can analyze the relationships between different material properties and suggest modifications to compositions or processing conditions that are likely to enhance specific characteristics. This iterative process, guided by ML, facilitates the targeted design of materials with improved performance and desired properties.
- **Virtual Materials Prototyping:** ML models enable virtual materials prototyping, allowing researchers to simulate the behavior of materials in silico before physical testing. This virtual experimentation reduces the need for extensive laboratory work and speeds up the materials development process. ML algorithms, trained on experimental data, can predict material responses to different conditions, providing valuable insights into the potential performance of new materials.
- **Adaptive Learning in Materials Discovery:** ML's adaptive learning capabilities allow it to continuously improve and refine its predictions as more data becomes available. In the context of materials discovery, this means that ML models can adapt to new experimental results and iteratively

enhance their predictive accuracy. This adaptive learning approach accelerates the feedback loop between computational predictions and experimental validations, fostering a more efficient and collaborative materials discovery process.

The integration of Machine Learning in Materials Science and Engineering is revolutionizing material properties prediction and discovering new materials. As ML techniques evolve and datasets grow, the synergy between computational models and experimental work will drive advancements in various industries and technologies.

SUPPLY CHAIN OPTIMIZATION

This section delves into the application of Machine Learning (ML) in Supply Chain Optimization, highlighting its crucial role in demand prediction, logistics optimization, waste reduction, and resource allocation, thereby enhancing efficiency and reducing operational costs for businesses(Boopathi, 2023a; Syamala et al., 2023).

ML for Demand Prediction and Logistics Optimization

- **Demand Forecasting with ML:** ML models analyze historical sales data, market trends, and external factors to predict future demand accurately. By identifying patterns and correlations within large datasets, these models can provide more precise demand forecasts compared to traditional methods. Accurate demand predictions enable businesses to optimize inventory levels, reduce stockouts, and enhance overall supply chain efficiency.
- **Dynamic Routing and Logistics Optimization:** ML algorithms optimize logistics and routing decisions by considering real-time factors such as traffic conditions, weather, and delivery constraints. These models dynamically adapt to changing circumstances, providing optimal routes for transportation and distribution. This dynamic optimization improves delivery timelines, reduces transportation costs, and enhances the overall responsiveness of the supply chain.
- **Predictive Maintenance for Fleet Management:** ML is applied in predictive maintenance models for fleet management, particularly in logistics and transportation. By analyzing data from sensors and historical maintenance records, ML algorithms predict when vehicles and equipment are likely to require maintenance. This proactive approach minimizes unplanned downtime, ensures the reliability of the transportation fleet, and contributes to efficient logistics operations.

Waste Reduction and Efficient Resource Allocation

The figure 4 depicts waste reduction and efficient resource allocation.

Figure 4. Waste reduction and efficient resource allocation

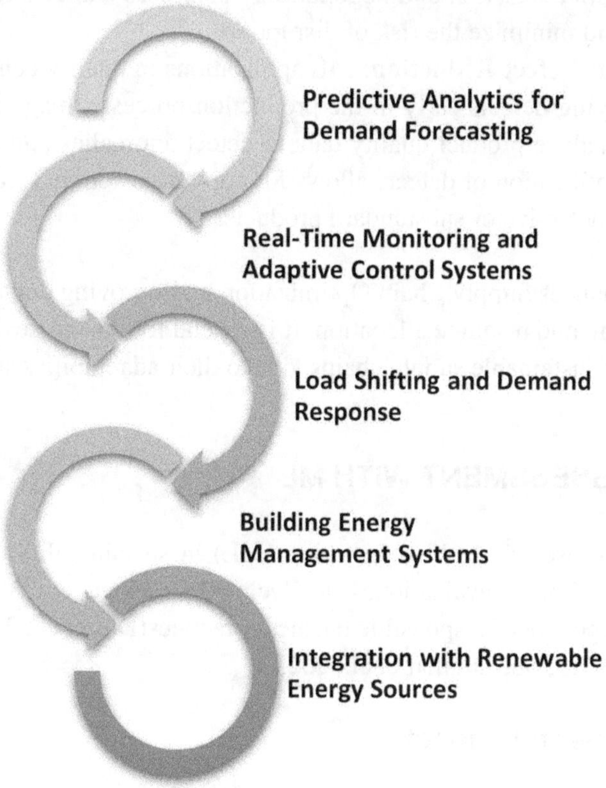

- **Optimizing Inventory Levels:** ML models contribute to waste reduction by optimizing inventory levels based on demand predictions. Overstocking or understocking can lead to excess waste and operational inefficiencies. ML algorithms continuously analyze demand patterns, supplier performance, and lead times to recommend optimal reorder points and quantities. This ensures that inventory is aligned with actual demand, reducing waste associated with surplus or expired goods(Reddy et al., 2023).
- **Dynamic Pricing Strategies:** ML-driven dynamic pricing strategies optimize resource allocation by adjusting prices based on real-time demand and market conditions. These algorithms analyze various factors, including competitor pricing, consumer behavior, and inventory levels, to dynamically set prices that maximize revenue and minimize waste. Dynamic pricing helps in selling excess inventory during peak demand periods and prevents overstocking by adjusting prices during slower periods.
- **Efficient Resource Allocation in Manufacturing:** ML is employed in manufacturing processes to optimize resource allocation and production scheduling. By analyzing historical production data, demand forecasts, and equipment performance metrics, ML models can recommend optimal production schedules, machine utilization, and resource allocation. This reduces the likelihood of production bottlenecks, minimizes idle time, and enhances overall manufacturing efficiency.
- **Supplier Relationship Management:** ML contributes to efficient resource allocation by optimizing supplier relationships. These models analyze supplier performance data, market trends, and

historical transaction data to identify reliable and cost-effective suppliers. By making informed decisions about supplier selection and negotiation, businesses can ensure a stable supply chain, reduce lead times, and minimize the risk of disruptions.

- **Quality Control and Defect Reduction:** ML applications in quality control contribute to waste reduction by identifying defects early in the production process. Image recognition and sensor-based ML models analyze product quality data to detect anomalies and deviations from quality standards. Early identification of defects allows for corrective actions, reducing waste associated with manufacturing defective or substandard products.

Machine Learning enhances Supply Chain Optimization by improving demand prediction, logistics efficiency, waste reduction, and resource allocation. It is crucial for businesses to adopt ML-driven solutions for responsive and sustainable supply chains, due to their adaptability and learning capabilities.

SUSTAINABILITY ASSESSMENT WITH ML

This section discusses the use of Machine Learning (ML) in sustainability assessment, specifically focusing on environmental impact evaluation, Life Cycle Assessments (LCA), and Carbon Footprint Analysis, as a crucial component of responsible business practices(Das et al., 2024; Hanumanthakari et al., 2023; Hussain et al., 2023; Selvakumar et al., 2023).

Evaluating Environmental Impact

- **Environmental Monitoring and Data Collection:** ML algorithms are applied to analyze large datasets from various sources, including remote sensing technologies, sensors, and environmental monitoring devices. These algorithms can identify patterns and trends in environmental data, such as air and water quality, biodiversity, and land use changes. ML aids in automating the collection and analysis of environmental data, providing a comprehensive understanding of the impact of human activities on ecosystems.
- **Predictive Modeling for Environmental Risk Assessment:** ML models enable predictive modeling for environmental risk assessment. By analyzing historical data on environmental incidents, pollution levels, and other relevant factors, these models can predict potential risks and assess the likelihood of environmental harm. This proactive approach helps businesses and policymakers make informed decisions to prevent or mitigate adverse environmental impacts.
- **Natural Language Processing (NLP) for Impact Assessment Reports:** ML, particularly Natural Language Processing (NLP), is applied to analyze impact assessment reports, environmental impact statements, and regulatory documents. NLP algorithms can extract valuable insights from unstructured text, identifying key environmental indicators, compliance issues, and mitigation measures. This facilitates a more efficient and comprehensive assessment of the environmental impact of projects or activities.

Life Cycle Assessments and Carbon Footprint Analysis

- **Life Cycle Assessment (LCA) Modeling:** ML plays a crucial role in modeling Life Cycle Assessments, which evaluate the environmental impact of a product or service throughout its entire life cycle. ML algorithms analyze complex relationships between inputs, processes, and outputs at each life cycle stage, considering factors such as raw material extraction, production, transportation, use, and end-of-life. This holistic approach provides a comprehensive view of the environmental impact of a product or service.
- **Carbon Footprint Prediction and Reduction Strategies:** ML models predict and assess carbon footprints by analyzing data on energy consumption, emissions, and other relevant factors. These models can identify areas of high carbon intensity in the supply chain or production processes and recommend strategies for reduction. ML-driven insights contribute to the development of sustainable practices, including energy-efficient operations, renewable energy adoption, and emissions reduction initiatives.
- **Supply Chain Sustainability:** ML is applied to assess and enhance the sustainability of supply chains. ML algorithms analyze vast amounts of data related to suppliers, transportation, and production processes to identify areas where sustainability improvements can be made. This includes optimizing transportation routes, reducing waste in the supply chain, and selecting environmentally responsible suppliers. ML contributes to the development of more sustainable and resilient supply chains.
- **Scenario Analysis for Sustainable Decision-Making:** ML enables scenario analysis for sustainable decision-making by assessing the environmental impact of different scenarios or strategies. Businesses can use ML models to simulate the outcomes of various decisions, helping them make informed choices that align with sustainability goals. Scenario analysis is particularly valuable in exploring the potential impact of different interventions or policy changes on the environment.
- **Monitoring Renewable Energy Integration:** ML algorithms monitor and optimize the integration of renewable energy sources into operations. By analyzing real-time data on energy consumption, production, and weather conditions, ML models can predict the availability and impact of renewable energy sources. This facilitates the efficient utilization of renewable energy, reducing reliance on fossil fuels and minimizing the overall carbon footprint.

Machine Learning (ML) is revolutionizing sustainability assessment by providing accurate predictions, proactive risk assessments, and comprehensive life cycle evaluations. As businesses and industries prioritize sustainability, ML-driven assessments become a powerful tool for informed decision-making and environmentally responsible practices, enhancing the overall sustainability assessment process.

SMART CITIES AND INFRASTRUCTURE DEVELOPMENT

This section explores the application of Machine Learning (ML) in Smart Cities and Infrastructure Development, focusing on its role in enhancing sustainability, urban planning, and project efficiency in urban planning and infrastructure projects. The figure 5 depicts the development of smart cities and infrastructure(Maguluri et al., 2023; Venkateswaran et al., 2023).

ML in Urban Planning for Sustainability

- **Traffic Management and Optimization:** ML is employed in traffic management systems to analyze real-time data from sensors, cameras, and other sources. These algorithms can predict traffic patterns, congestion, and optimal routes for vehicles. By optimizing traffic flow, ML contributes to reducing emissions, fuel consumption, and overall environmental impact. Smart traffic management enhances the sustainability of urban transportation systems.
- **Energy Consumption Prediction and Management:** ML models analyze historical data on energy consumption patterns in urban areas. These models predict future energy demands based on factors such as population growth, weather conditions, and economic activities. Urban planners can use these predictions to optimize energy distribution, implement energy-efficient technologies, and promote sustainable practices in buildings and infrastructure.
- **Waste Management and Recycling Optimization:** ML is applied in waste management systems to optimize collection routes, predict waste generation patterns, and enhance recycling efforts. By analyzing historical data and real-time information from waste sensors, ML models can recommend efficient waste collection schedules, reducing the environmental impact associated with transportation and improving recycling rates.
- **Green Space Planning and Biodiversity Preservation:** ML aids in urban planning for green spaces and biodiversity preservation. Algorithms analyze satellite imagery, climate data, and ecological information to identify optimal locations for green spaces, parks, and wildlife corridors. This approach helps balance urban development with environmental conservation, fostering sustainability and enhancing the quality of life for residents.

Figure 5. Smart cities and infrastructure development

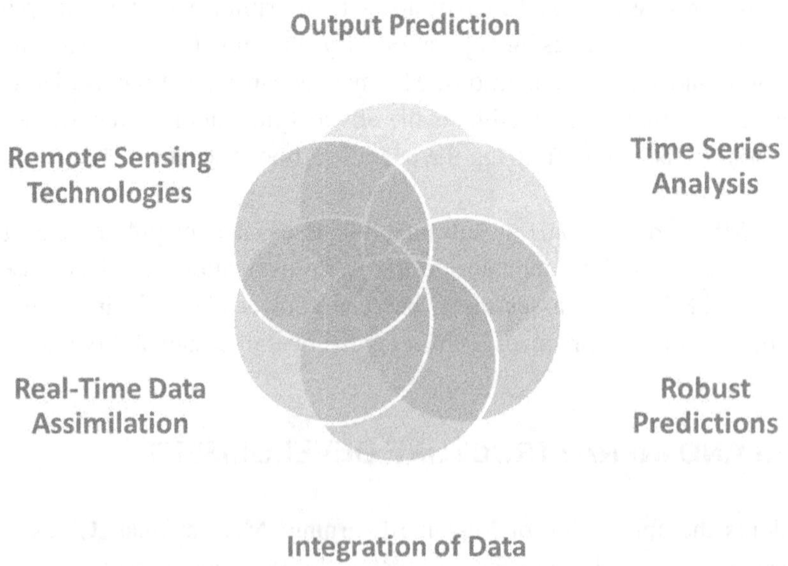

Enhancing Efficiency in Infrastructure Projects

- **Predictive Maintenance for Infrastructure:** ML is utilized for predictive maintenance in infrastructure projects, such as bridges, roads, and utilities. By analyzing sensor data, historical maintenance records, and environmental conditions, ML models predict when infrastructure components are likely to require maintenance or repairs. This proactive approach minimizes downtime, extends the lifespan of assets, and ensures the efficient operation of critical infrastructure(Hussain et al., 2023; Ugandar et al., 2023).
- **Project Scheduling and Resource Optimization:** ML algorithms optimize project scheduling and resource allocation in infrastructure development. These models consider various factors, including project timelines, resource availability, and potential risks. ML-driven project management systems can predict delays, optimize resource utilization, and recommend strategies to enhance overall project efficiency.
- **Cost Estimation and Budget Optimization:** ML contributes to cost estimation and budget optimization in infrastructure projects. By analyzing historical project data, market trends, and external factors, ML models can provide more accurate cost estimates. This helps prevent budget overruns, enhances financial planning, and ensures that infrastructure projects are completed within the allocated resources.
- **Risk Management and Decision Support:** ML is applied for risk management in infrastructure projects, helping identify potential risks and uncertainties. These models analyze historical project data, external market conditions, and project-specific factors to predict and assess risks. ML-driven decision support systems assist project managers in making informed decisions to mitigate risks and optimize project outcomes.
- **Quality Control and Defect Detection:** ML models are employed for quality control in infrastructure projects, particularly in construction. Image recognition and sensor-based algorithms analyze data to detect defects, deviations from design specifications, and potential construction errors. ML-driven quality control ensures that infrastructure projects meet the required standards, reducing the likelihood of rework and enhancing overall project efficiency.

Machine Learning is revolutionizing smart cities and infrastructure development by optimizing urban planning for sustainability and enhancing efficiency in infrastructure projects. As cities grow and infrastructure demands increase, ML-driven solutions become increasingly essential for sustainable and efficient urban development, creating more resilient, environmentally friendly, and technologically advanced urban landscapes.

CONCLUSION

Machine Learning (ML) has revolutionized engineering and sustainability by predicting material properties, optimizing energy consumption, and enhancing urban planning and infrastructure development. In engineering, ML has been instrumental in predictive maintenance, demand forecasting, and process optimization, offering a data-driven approach that improves efficiency, reduces costs, and advances sustainability.

Sustainability is a growing focus for machine learning (ML) in engineering and technology. It helps businesses assess environmental impact, optimize resource allocation, and make informed decisions. Future trends include increased automation, improved model interpretability, and sophisticated algorithms. The integration of ML with emerging technologies like IoT and edge computing is expected to further enhance its impact. Ethical considerations, transparency, and accountability are also being prioritized in ML applications to ensure responsible and equitable deployment.

Collaboration between machine learning (ML), engineering, and sustainability professionals is crucial for addressing complex challenges. As technology evolves, ML's role in shaping a sustainable future for engineering applications offers endless possibilities for innovation, optimization, and positive environmental impact. The intersection of ML, engineering, and sustainability holds immense potential for a resilient world.

ABBREVIATIONS

BEMS: Building Energy Management Systems
HVAC: Heating, Ventilation, and Air Conditioning
IoT: Internet of Things
LCA: Life Cycle Assessment
ML: Machine Learning
MSE: Materials Science and Engineering
NLP: Natural Language Processing
QSPR: Quantitative Structure-Property Relationships

REFERENCES

Agrawal, A. V., Shashibhushan, G., Pradeep, S., Padhi, S. N., Sugumar, D., & Boopathi, S. (2024). Synergizing Artificial Intelligence, 5G, and Cloud Computing for Efficient Energy Conversion Using Agricultural Waste. In B. K. Mishra (Ed.), (pp. 475–497). Practice, Progress, and Proficiency in Sustainability. IGI Global. doi:10.4018/979-8-3693-1186-8.ch026

Alimi, O. A., Ouahada, K., & Abu-Mahfouz, A. M. (2020). A review of machine learning approaches to power system security and stability. *IEEE Access : Practical Innovations, Open Solutions*, *8*, 113512–113531. doi:10.1109/ACCESS.2020.3003568

Boopathi, S. (2023a). An Investigation on Friction Stir Processing of Aluminum Alloy-Boron Carbide Surface Composite. In R. V. Vignesh, R. Padmanaban, & M. Govindaraju (Eds.), *Advances in Processing of Lightweight Metal Alloys and Composites* (pp. 249–257). Springer Nature Singapore. doi:10.1007/978-981-19-7146-4_14

Boopathi, S. (2023b). Internet of Things-Integrated Remote Patient Monitoring System: Healthcare Application. In A. Suresh Kumar, U. Kose, S. Sharma, & S. Jerald Nirmal Kumar (Eds.), (pp. 137–161). Advances in Healthcare Information Systems and Administration. IGI Global. doi:10.4018/978-1-6684-6894-4.ch008

Boopathi, S., & Davim, J. P. (2023). Applications of Nanoparticles in Various Manufacturing Processes. In S. Boopathi & J. P. Davim (Eds.), (pp. 1–31). Advances in Chemical and Materials Engineering. IGI Global. doi:10.4018/978-1-6684-9135-5.ch001

Boopathi, S., & Kanike, U. K. (2023). Applications of Artificial Intelligent and Machine Learning Techniques in Image Processing. In B. K. Pandey, D. Pandey, R. Anand, D. S. Mane, & V. K. Nassa (Eds.), (pp. 151–173). Advances in Computational Intelligence and Robotics. IGI Global. doi:10.4018/978-1-6684-8618-4.ch010

Boopathi, S., Umareddy, M., & Elangovan, M. (2023). Applications of Nano-Cutting Fluids in Advanced Machining Processes. In S. Boopathi & J. P. Davim (Eds.), (pp. 211–234). Advances in Chemical and Materials Engineering. IGI Global. doi:10.4018/978-1-6684-9135-5.ch009

Cao, D., Hu, W., Zhao, J., Zhang, G., Zhang, B., Liu, Z., Chen, Z., & Blaabjerg, F. (2020). Reinforcement learning and its applications in modern power and energy systems: A review. *Journal of Modern Power Systems and Clean Energy*, *8*(6), 1029–1042. doi:10.35833/MPCE.2020.000552

Das, S., Lekhya, G., Shreya, K., Lydia Shekinah, K., Babu, K. K., & Boopathi, S. (2024). Fostering Sustainability Education Through Cross-Disciplinary Collaborations and Research Partnerships: Interdisciplinary Synergy. In P. Yu, J. Mulli, Z. A. S. Syed, & L. Umme (Eds.), (pp. 60–88). Advances in Higher Education and Professional Development. IGI Global. doi:10.4018/979-8-3693-0487-7.ch003

Dhanya, D., Kumar, S. S., Thilagavathy, A., Prasad, D. V. S. S. S. V., & Boopathi, S. (2023). Data Analytics and Artificial Intelligence in the Circular Economy: Case Studies. In B. K. Mishra (Ed.), (pp. 40–58). Advances in Civil and Industrial Engineering. IGI Global. doi:10.4018/979-8-3693-0044-2.ch003

Duchesne, L., Karangelos, E., & Wehenkel, L. (2020). Recent developments in machine learning for energy systems reliability management. *Proceedings of the IEEE*, *108*(9), 1656–1676. doi:10.1109/JPROC.2020.2988715

Hanumanthakari, S., Gift, M. D. M., Kanimozhi, K. V., Bhavani, M. D., Bamane, K. D., & Boopathi, S. (2023). Biomining Method to Extract Metal Components Using Computer-Printed Circuit Board E-Waste. In P. Srivastava, D. Ramteke, A. K. Bedyal, M. Gupta, & J. K. Sandhu (Eds.), (pp. 123–141). Practice, Progress, and Proficiency in Sustainability. IGI Global. doi:10.4018/978-1-6684-8117-2.ch010

Hema, N., Krishnamoorthy, N., Chavan, S. M., Kumar, N. M. G., Sabarimuthu, M., & Boopathi, S. (2023). A Study on an Internet of Things (IoT)-Enabled Smart Solar Grid System. In P. Swarnalatha & S. Prabu (Eds.), (pp. 290–308). Advances in Computational Intelligence and Robotics. IGI Global. doi:10.4018/978-1-6684-8098-4.ch017

Hussain, Z., Babe, M., Saravanan, S., Srimathy, G., Roopa, H., & Boopathi, S. (2023). Optimizing Biomass-to-Biofuel Conversion: IoT and AI Integration for Enhanced Efficiency and Sustainability. In N. Cobîrzan, R. Muntean, & R.-A. Felseghi (Eds.), (pp. 191–214). Advances in Finance, Accounting, and Economics. IGI Global. doi:10.4018/978-1-6684-8238-4.ch009

Ibrahim, M. S., Dong, W., & Yang, Q. (2020). Machine learning driven smart electric power systems: Current trends and new perspectives. *Applied Energy*, *272*, 115237. doi:10.1016/j.apenergy.2020.115237

Ingle, R. B., Swathi, S., Mahendran, G., Senthil, T. S., Muralidharan, N., & Boopathi, S. (2023). Sustainability and Optimization of Green and Lean Manufacturing Processes Using Machine Learning Techniques. In N. Cobîrzan, R. Muntean, & R.-A. Felseghi (Eds.), (pp. 261–285). Advances in Finance, Accounting, and Economics. IGI Global. doi:10.4018/978-1-6684-8238-4.ch012

Kavitha, C. R., Varalatchoumy, M., Mithuna, H. R., Bharathi, K., Geethalakshmi, N. M., & Boopathi, S. (2023). Energy Monitoring and Control in the Smart Grid: Integrated Intelligent IoT and ANFIS. In M. Arshad (Ed.), (pp. 290–316). Advances in Bioinformatics and Biomedical Engineering. IGI Global. doi:10.4018/978-1-6684-6577-6.ch014

Kumar, B. M., Kumar, K. K., Sasikala, P., Sampath, B., Gopi, B., & Sundaram, S. (2024). Sustainable Green Energy Generation From Waste Water: IoT and ML Integration. In B. K. Mishra (Ed.), (pp. 440–463). Practice, Progress, and Proficiency in Sustainability. IGI Global. doi:10.4018/979-8-3693-1186-8.ch024

Maguluri, L. P., Arularasan, A. N., & Boopathi, S. (2023). Assessing Security Concerns for AI-Based Drones in Smart Cities. In R. Kumar, A. B. Abdul Hamid, & N. I. Binti Ya'akub (Eds.), (pp. 27–47). Advances in Computational Intelligence and Robotics. IGI Global. doi:10.4018/978-1-6684-9151-5.ch002

Maheswari, B. U., Imambi, S. S., Hasan, D., Meenakshi, S., Pratheep, V. G., & Boopathi, S. (2023). Internet of Things and Machine Learning-Integrated Smart Robotics. In M. K. Habib (Ed.), (pp. 240–258). Advances in Computational Intelligence and Robotics. IGI Global. doi:10.4018/978-1-6684-7791-5.ch010

Nishanth, J. R., Deshmukh, M. A., Kushwah, R., Kushwaha, K. K., Balaji, S., & Sampath, B. (2023). Particle Swarm Optimization of Hybrid Renewable Energy Systems. In B. K. Mishra (Ed.), (pp. 291–308). Advances in Civil and Industrial Engineering. IGI Global., doi:10.4018/979-8-3693-0044-2.ch016

Pachiappan, K., Anitha, K., Pitchai, R., Sangeetha, S., Satyanarayana, T. V. V., & Boopathi, S. (2023). Intelligent Machines, IoT, and AI in Revolutionizing Agriculture for Water Processing. In B. B. Gupta & F. Colace (Eds.), (pp. 374–399). Advances in Computational Intelligence and Robotics. IGI Global. doi:10.4018/978-1-6684-9999-3.ch015

Ramudu, K., Mohan, V. M., Jyothirmai, D., Prasad, D. V. S. S. S. V., Agrawal, R., & Boopathi, S. (2023). Machine Learning and Artificial Intelligence in Disease Prediction: Applications, Challenges, Limitations, Case Studies, and Future Directions. In G. S. Karthick & S. Karupusamy (Eds.), (pp. 297–318). Advances in Healthcare Information Systems and Administration. IGI Global. doi:10.4018/978-1-6684-8913-0.ch013

Rangel-Martinez, D., Nigam, K., & Ricardez-Sandoval, L. A. (2021). Machine learning on sustainable energy: A review and outlook on renewable energy systems, catalysis, smart grid and energy storage. *Chemical Engineering Research & Design*, *174*, 414–441. doi:10.1016/j.cherd.2021.08.013

Rebecca, B., Kumar, K. P. M., Padmini, S., Srivastava, B. K., Halder, S., & Boopathi, S. (2023). Convergence of Data Science-AI-Green Chemistry-Affordable Medicine: Transforming Drug Discovery. In B. B. Gupta & F. Colace (Eds.), (pp. 348–373). Advances in Computational Intelligence and Robotics. IGI Global. doi:10.4018/978-1-6684-9999-3.ch014

Reddy, M. A., Gaurav, A., Ushasukhanya, S., Rao, V. C. S., Bhattacharya, S., & Boopathi, S. (2023). Bio-Medical Wastes Handling Strategies During the COVID-19 Pandemic. In C. S. V. Negrão, I. G. P. Maia, & J. A. F. Brito (Eds.), (pp. 90–111). Advances in Logistics, Operations, and Management Science. IGI Global. doi:10.4018/978-1-7998-9213-7.ch006

Selvakumar, S., Shankar, R., Ranjit, P. S., Bhattacharya, S., Gopala Gupta, A. S. A. L. G., & Boopathi, S. (2023). E-Waste Recovery and Utilization Processes for Mobile Phone Waste. In P. Srivastava, D. Ramteke, A. K. Bedyal, M. Gupta, & J. K. Sandhu (Eds.), (pp. 222–240). Practice, Progress, and Proficiency in Sustainability. IGI Global. doi:10.4018/978-1-6684-8117-2.ch016

Senthil, T. S., Ohmsakthi Vel, R., Puviyarasan, M., Babu, S. R., Surakasi, R., & Sampath, B. (2023). Industrial Robot-Integrated Fused Deposition Modelling for the 3D Printing Process. In R. Keshavamurthy, V. Tambrallimath, & J. P. Davim (Eds.), (pp. 188–210). Advances in Chemical and Materials Engineering. IGI Global. doi:10.4018/978-1-6684-6009-2.ch011

Sharma, M., Sharma, M., Sharma, N., & Boopathi, S. (2023). Building Sustainable Smart Cities Through Cloud and Intelligent Parking System. In B. B. Gupta & F. Colace (Eds.), (pp. 195–222). Advances in Computational Intelligence and Robotics. IGI Global. doi:10.4018/978-1-6684-9999-3.ch009

Sharma, P., Said, Z., Kumar, A., Nizetic, S., Pandey, A., Hoang, A. T., Huang, Z., Afzal, A., Li, C., Le, A. T., Nguyen, X. P., & Tran, V. D. (2022). Recent advances in machine learning research for nanofluid-based heat transfer in renewable energy system. *Energy & Fuels*, *36*(13), 6626–6658. doi:10.1021/acs.energyfuels.2c01006

Syamala, M. C. R., K., Pramila, P. V., Dash, S., Meenakshi, S., & Boopathi, S. (2023). Machine Learning-Integrated IoT-Based Smart Home Energy Management System: In P. Swarnalatha & S. Prabu (Eds.), Advances in Computational Intelligence and Robotics (pp. 219–235). IGI Global. doi:10.4018/978-1-6684-8098-4.ch013

Ugandar, R. E., Rahamathunnisa, U., Sajithra, S., Christiana, M. B. V., Palai, B. K., & Boopathi, S. (2023). Hospital Waste Management Using Internet of Things and Deep Learning: Enhanced Efficiency and Sustainability. In M. Arshad (Ed.), (pp. 317–343). Advances in Bioinformatics and Biomedical Engineering. IGI Global. doi:10.4018/978-1-6684-6577-6.ch015

Venkateswaran, N., Vidhya, K., Ayyannan, M., Chavan, S. M., Sekar, K., & Boopathi, S. (2023). A Study on Smart Energy Management Framework Using Cloud Computing. In P. Ordóñez De Pablos & X. Zhang (Eds.), (pp. 189–212). Practice, Progress, and Proficiency in Sustainability. IGI Global. doi:10.4018/978-1-6684-8634-4.ch009

Willard, J., Jia, X., Xu, S., Steinbach, M., & Kumar, V. (2022). Integrating scientific knowledge with machine learning for engineering and environmental systems. *ACM Computing Surveys*, *55*(4), 1–37. doi:10.1145/3514228

Yang, T., Zhao, L., Li, W., & Zomaya, A. Y. (2020). Reinforcement learning in sustainable energy and electric systems: A survey. *Annual Reviews in Control*, *49*, 145–163. doi:10.1016/j.arcontrol.2020.03.001

Zekrifa, D. M. S., Kulkarni, M., Bhagyalakshmi, A., Devireddy, N., Gupta, S., & Boopathi, S. (2023). Integrating Machine Learning and AI for Improved Hydrological Modeling and Water Resource Management. In V. Shikuku (Ed.), (pp. 46–70). Advances in Environmental Engineering and Green Technologies. IGI Global., doi:10.4018/978-1-6684-6791-6.ch003

Zhang, L., Wen, J., Li, Y., Chen, J., Ye, Y., Fu, Y., & Livingood, W. (2021). A review of machine learning in building load prediction. *Applied Energy*, *285*, 116452. doi:10.1016/j.apenergy.2021.116452

Zhou, Y., Zheng, S., Liu, Z., Wen, T., Ding, Z., Yan, J., & Zhang, G. (2020). Passive and active phase change materials integrated building energy systems with advanced machine-learning based climate-adaptive designs, intelligent operations, uncertainty-based analysis and optimisations: A state-of-the-art review. *Renewable & Sustainable Energy Reviews*, *130*, 109889. doi:10.1016/j.rser.2020.109889

Chapter 5
The Impact of AI on Sustainability

Arshiya Begum
https://orcid.org/0000-0002-7820-5976
King Khalid University, Saudi Arabia

Arshi Naim
https://orcid.org/0000-0003-1325-6964
King Khalid University, Saudi Arabia

Asfia Sabahath
King Khalid University, Saudi Arabia

ABSTRACT

Artificial intelligence (AI) has been increasingly popular in various businesses in recent years. AI has proven to be an effective tool for enhancing efficiency and decision-making across industries, from healthcare to finance. However, sustainability is the area in which AI is having a huge impact is sustainability. AI has the potential to transform the way we think about sustainability. AI can find trends and make predictions based on massive amounts of data, allowing us to understand better and handle environmental challenges. For example, artificial intelligence (AI) can analyze weather patterns and predict natural disasters, allowing for improved disaster preparedness and response. It can also be utilized in industries like manufacturing and transportation to optimize energy usage and eliminate waste.

INTRODUCTION

With the most recent advancements in AI technology, numerous sectors, particularly sustainability, have reached new highs. These innovative technologies have shown amazing potential for improving sustainable practices in the fields of renewable energy and environmental health. The emergence of artificial intelligence (AI) is indeed expected to have a significant impact on various sectors, including global productivity, equality and inclusion, environmental outcomes, and more.

DOI: 10.4018/979-8-3693-1794-5.ch005

In AI research, the financial and environmental implications of training and tuning neural network models have been underscored, emphasizing the need for sustainable practices (Strubell et al., 2020). AI techniques, including deep reinforcement learning (DRL), have indeed played a significant role in advancing smart building energy management in the renewable energy sector (Yu et al., 2021). AI is also being utilized in green construction to address cost reduction, resource management, and safety enhancement (Chen & He, 2023).

Fault detection within renewable energy systems is another area where AI has proven effective. Deep learning-based systems can detect anomalies in photovoltaic images, ensuring the health of renewable energy systems (Pierdicca et al., 2020). Additionally, the use of AI for generating synthetic datasets and analyzing solar cell defects represents significant advancements in the field (Romero et al., 2023).

The recycling industry has also benefited from AI, particularly in waste management. Systems using AI and IoT technologies can classify recycled clothing, improving the efficiency of recycling processes (Noh, 2021). Similarly, AI and machine learning algorithms are employed to optimize the production of sustainable biofuels from renewable resources (Khanna et al., 2023).

RESEARCH CONTEXT AND RELATED WORK

In the following, we first introduce the main areas of this paper, namely AI and sustainability, to ensure a common understanding of these concepts. Afterward, we provide an overview of related studies to derive a series of needs for additional research.

Artificial Intelligence

The concept of Artificial Intelligence (AI) originated around since the 1950s. It involves building machines that can mimic human capabilities such as sensing, reasoning, and thinking. Over the years, AI has evolved significantly due to advancements in algorithms, larger datasets, and increased adoption of digital technologies. AI is now used in various products and services, including virtual assistants and autonomous vehicles. It is predicted to have a significant impact on many aspects of life.

AI encompasses different algorithmic approaches, with machine learning (ML)being one of the prevailing ones. ML involves the ability of a system to learn from training data iteratively and improve its performance without explicit programming. It utilizes statistical techniques and pattern recognition algorithms to perform cognitive tasks such as forecasting and object detection. ML techniques include regression models, support vector machines, decision trees, and k-nearest neighbor algorithms.

ML tasks can be categorized into three learning task (Figure 1): supervised learning, unsupervised learning, and reinforcement learning. Supervised learning involves finding the best function mapping inputs to outputs based on given input data and target values. Unsupervised learning identifies structural information within datasets without pre-structured inputs or labels. Reinforcement learning aims to find the best path to achieve a goal within a specified system state, goals, allowable actions, and environmental boundaries.

Figure 1. Artificial intelligence concepts

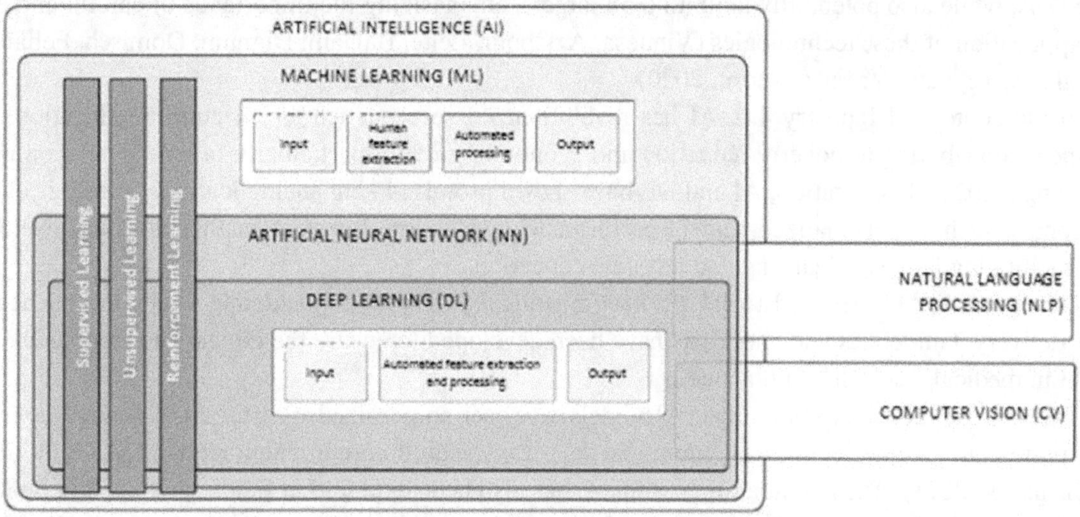

Artificial neural networks (ANNs) are a class of machine learning algorithms that have gained significant popularity in recent years. They are inspired by the information processing in biological systems and consist of interconnected processing units that transmit signals using weights and activation functions. Modern ANNs often have deep nested architectures with multiple layers, which enables them to process raw input data without the need for manual feature engineering. This capability is known as deep learning (DL).

DL networks, with their increased number of parameters to be trained, require a rich set of training examples. However, they have shown remarkable performance in many tasks involving large and high-dimensional data compared to traditional ML algorithms and statistical approaches. Specific DL architectures have emerged to handle different types of data. For example, convolutional neural networks (CNNs) are well-suited for processing image data, while recurrent neural networks (RNNs) are effective for sequential data processing.

The advancements in ANNs and DL have also contributed to progress in related fields of AI. Natural language processing (NLP) has benefited from DL techniques for tasks such as information retrieval, topic modeling, sentiment analysis, and content generation. Computer vision (CV) has also seen advancements in analyzing image data for tasks like object recognition, detection, and tracking.

These developments in ANNs, DL, NLP, and CV have significantly expanded the capabilities of AI systems in handling complex and diverse data types, opening up new possibilities for various applications.

AI, Sustainability, and Sustainable Development Goals

Artificial intelligence (AI) and deep learning (DL) have transformative potential in supporting the achievement of the Sustainable Development Goals (SusDevGs). These technologies have already made substantial impacts in various domains, including healthcare, industry, agriculture, education, and finance (Mhlanga, 2021).

A comprehensive analysis has identified that AI can support the achievement of 128 targets across all SDGs, while also potentially inhibiting 58 targets, emphasizing the importance of careful and ethical application of these technologies (Vinuesa, Azizpour, Leite, Balaam, Dignum, Domisch, Felländer, Langhans, Tegmark, & Fuso Nerini, 2020).

In the context of Industry 4.0, AI has revolutionized sectors such as agriculture, education, and finance, contributing to poverty reduction and economic growth, particularly in emerging economies (Mhlanga, 2021). In education, AI and DL have shown promise in enhancing learning experiences and outcomes, such as personalized student performance evaluation, boosting mathematics achievement, and facilitating language learning and physical education.

In healthcare, AI has played a vital role in combating the COVID-19 pandemic and improving healthcare delivery. Guidelines and checklists have been developed to ensure the ethical and responsible use of AI in medical research and practice.

Transparency and interpretability in AI models have been emphasized to ensure ethical and responsible use, particularly in developing AI systems and data-driven solutions compliant with the SDGs (Vinuesa & Sirmacek, 2021). The potential of AI and DL has also been explored in fields such as plant biology, sustainable infrastructure, and addressing ESG (environmental, social, governance) goals (Si, 2022). However, the unregulated roll-out of AI technologies poses risks, and regulatory oversight is necessary to minimize negative impacts on the SDGs.

Advancements in IoT and machine learning have also been highlighted in achieving the SDGs, including use cases in health, energy, and cities (Singh & Singh, 2023). Deep graph learning and knowledge graph-based frameworks have been proposed to address societal challenges and improve daily lives. While there have been rapid advancements in AI and DL, disparities exist in their implementation across different sectors and industries (Nahorniak & Hannouf, 2023). Some industries may face challenges in adopting and capitalizing on these technologies, indicating a gap between potential and implementation.

In agriculture, the integration of big data and AI presents opportunities to address challenges and optimize production. Similarly, in marine sciences, AI can contribute to fish behavior recognition and fishing gear selectivity, but comprehensive datasets are needed for the effective training of deep learning models (Franzo et al., 2023).

The innovative use of AI and DL can significantly contribute to the achievement of the SusDevGs (Table 1). However, comprehensive regulatory oversight is necessary to ensure the ethical and responsible application of these technologies. The future of sustainable development will be shaped by harnessing the potential of AI and DL in a manner that aligns with the SusDevGs.

Table 1. AI in sustainable development goals (SusDevGs)

SDG Goal	Application
No Poverty	Predicting poverty regions (Jean et al., 2016), optimizing social security payments (Cao, 2012), improving microfinance services (Gupta et al., 2015)
Zero Hunger	Crop yield prediction (Martos et al., 2021), Crop disease detection (Singh et al., 2022), precision agriculture (Singh & Singh, 2023)
Good Health and Well-being	Disease outbreak prediction (Habib et al., 2021; Leopord et al., 2016), telemedicine services (Lee & Yoon, 2021), AI-assisted diagnosis (Maghsudi et al., 2021; Tang et al., 2022)
Quality Education	Personalized education (Yang & Zhu, 2022), AI-assisted grading (Yang & Zhu, 2022)
Gender Equality	Analysis of gender bias data (Leavy et al., 2020)
Clean Water and Sanitation	Water quality monitoring (Dilmi & Ladjal, 2021; Solanki et al., 2015), water scarcity prediction (Hanoon et al., 2022; Jain et al., 2020), optimizing water distribution systems (Grbčić et al., 2021; Wu & Rahman, 2017)
Decent Work and Economic Growth	Enhancing productivity, job creation, forecasting economic trends (Tsai et al., 2017)
Industry, Innovation, and Infrastructure	Optimizing operations, reducing maintenance costs (Khan & Yairi, 2018), predictive maintenance (Serradilla et al., 2022)
Reduced Inequalities	Identifying and predicting social inequality, AI in policy-making (Milano et al., 2014), targeted interventions for inequality reduction
Climate Action	Creation and improvement of climate models, climate change prediction, carbon footprint tracking

RELATED STUDIES

The integration of AI and sustainability within the field of Information Systems (IS) research holds significant potential. Several studies have examined this intersection and provided valuable insights.

Nishant et al. (2020) conducted a review of journal articles published between 2015 and 2019 to outline a roadmap for future research on AI and sustainability. They focused on environmental themes such as biodiversity, water resources, energy conservation, smart cities, and climate change.

Vinuesa et al. (2020) examined the potential of AI in addressing the 17 Sustainable Development Goals (SDGs) defined by the United Nations. Through expert elicitation, they found that about two-thirds of the SDGs could be supported using AI.

Di Vaio et al. (2020) conducted a bibliometric analysis of publications from 1990 to 2019 to investigate the role of AI in sustainable business models. They provided a quantitative analysis but noted the need for qualitative in-depth conclusions. The analysis indicated a strong focus on the technology itself, with less emphasis on the relationship between AI and sustainability.

Holzinger et al. (2021) explored the risks associated with employing AI for sustainability. They highlighted potential issues such as bias, data privacy, data resilience, transparency of methods, and trust in results.

Leong et al. (2020) analyzed the links between IS research and the 17 SDGs. They found that the IS discipline mainly focused on goals related to healthy lives, sustainable growth, resilient infrastructure, and inclusive societies.

Other studies have examined specific fields where AI can contribute to sustainability, such as financial services, health, and public governance.

Collins et al. (2021) conducted a systematic review of how AI is used in IS literature from 2005 to 2020. They provided quantitative insights into the number of studies, publication channels, and research methods.

Enholm et al. (2021) reviewed and analyzed a sample of IS articles to extract business values resulting from the use of AI technologies. They highlighted sustainability performance as one of the values, particularly in terms of environmental AI for minimizing resources.

While valuable work has been done in exploring the intersection of AI, sustainability, and IS, there is still a need for an integrated perspective that considers sustainability from the Triple Bottom Line and SDGs standpoint. The IS discipline, with its socio-technical approach, can contribute to advancing knowledge and solutions that consider technical, social, and political perspectives. Design science research (DSR) can be utilized to design purposeful artifacts that solve practical problems and facilitate the implementation of AI technologies for decision-making and task execution.

TOWARDS SUSTAINABLE AI

AI in Renewable Energy

The integration of artificial intelligence (AI) and deep learning (DL) in the renewable energy sector has brought significant advancements and transformative potential. These technologies have been applied in various domains within the sector (Figure 2), including energy forecasting, system health assessment, solar power generation, and grid stability.

In the domain of solar power systems, AI and DL have been used for optimal power flow prediction. By combining deep learning and Lagrangian dual methods, researchers have improved the accuracy of optimal power flow models, enabling power grid operators to make more effective decisions (Fioretto et al., 2020). Deep reinforcement learning has also been employed to rapidly solve optimal AC power flow problems, providing real-time solutions.

Figure 2. Application of AI and DL in the field of renewable energy

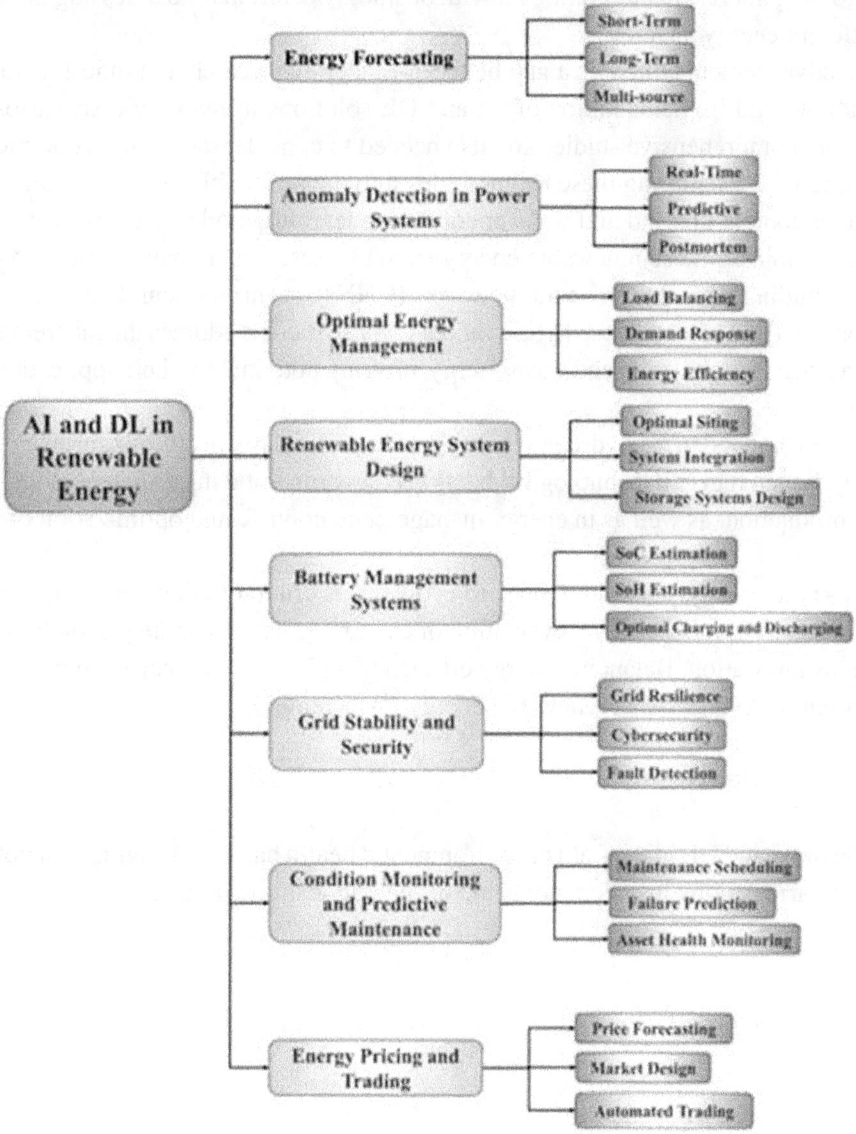

DL has proven invaluable in solar power generation, particularly in forecasting solar irradiance. By utilizing clustering techniques, convolutional neural networks (CNNs), long short-term memory (LSTM), and attention mechanisms, researchers have developed hybrid DL methods that overcome challenges associated with PV energy generation forecasting.

AI and DL have also made significant contributions to system health and fault detection in photovoltaic (PV) power plants. The Solar system, for example, integrates AI and DL with thermal infrared sensors on unmanned aerial vehicles (UAVs) to automate fault detection in PV power plants. This approach identifies and localizes anomalies in solar panels, enabling necessary maintenance actions and optimizing energy production and plant reliability. Research studies in the renewable energy sector have explored various methodologies and areas of focus. Some studies emphasize peer-to-peer energy trading, while others compare the performance of different algorithms for energy balance prediction. Additionally, there is growing interest in integrating AI with business operations and assessing the environmental impacts of different energy sources.

Despite the advancements, there is a gap between academic research and industry implementation. Practical scalability and implementation of AI and DL solutions in real-world scenarios need further exploration. More comprehensive studies are also needed to consider the socio-economic and environmental implications of integrating these technologies into the renewable energy sector.

In the broader context of wind and wave energy, deep learning models have shown effectiveness in managing and optimizing these renewable energy sources. Researchers have explored various models and methods, including convolution neural networks (CNNs), recurrent neural networks (RNNs), long short-term memory (LSTM), and deep hybrid models. These models address the inherent instability and complex data associated with wind and wave energy, offering potential for their application in renewable energy management.

AI has also been applied in the hydrogen energy sector, particularly in optimizing energy management strategies for hydrogen fuel cell technologies. It has been instrumental in predictive modeling for energy and hydrogen production, as well as in energy management, control, and optimization of hydrogen fuel cell vehicles.

As the renewable energy sector continues to evolve, it is crucial to develop energy-efficient methodologies for AI and DL models. The revaluation of existing research can help avoid redundancies and foster continuous innovation. Balancing energy efficiency with technological advancement is necessary for the integration of AI and DL in renewable energy management.

AI in Environmental Health

The integration of AI and DL in the field of environmental health has shown significant potential and has been applied in various domains. Here are some of the applications (Figure 3) of AI in Environmental Health.

Figure 3. AI in environmental health

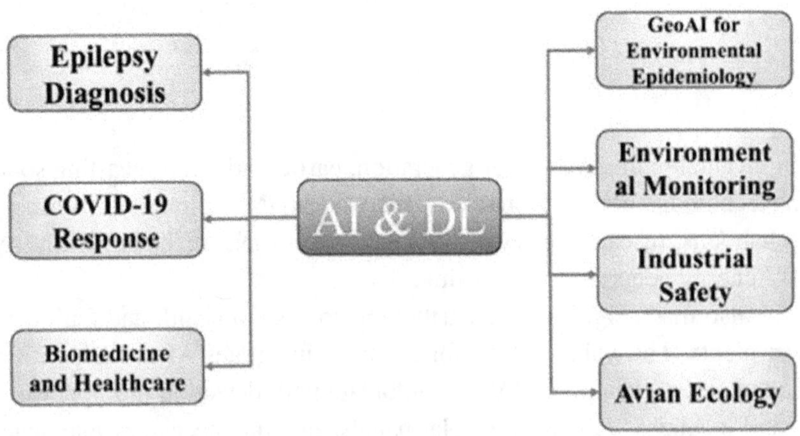

- **Epilepsy Diagnosis**: DL techniques have improved the efficiency and accuracy of epileptic seizure detection by automating the feature extraction process from EEG and MRI data (VoPham, 2018).
- **GeoAI in Environmental Epidemiology:** The combination of spatial science, AI, and DL in geoAI has revolutionized environmental epidemiology. It processes large-scale spatial data, models environmental exposures, and helps understand the relationships between environmental exposures and diseases.
- **AI and DL in Responding to Health Crises:** The COVID-19 pandemic highlighted the role of AI, ML, and DL in diagnosing, treating, and managing the disease. These technologies have been used for non-invasive detection, severity assessments, disease transmission modelling, data analysis, epidemic protection, and public monitoring efforts.
- **Prediction of Harmful Algal Blooms:** DL models have been effective in predicting harmful algal blooms, outperforming traditional regression analysis methods (Lee & Lee, 2018).
- **Biomedicine and Healthcare:** AI and DL have been applied in biomedicine and healthcare to address sex and gender gaps, optimize biomedical technologies for precision medicine, and improve global health outcomes (Cirillo et al., 2020).
- **Industrial Safety:** AI has been used in the development of a smart helmet prototype that monitors workers' environments and evaluates risks in real-time.
- **Avian Ecology:** AI has been employed to detect and monitor bird populations, offering the potential for computer-aided counting in real-world bird monitoring applications.

AI in Smart Building Energy Management

The integration of AI and DL in smart building energy management has shown significant progress and has led to improvements in prediction, optimization, and automation tasks. Several studies have highlighted the potential of AI and DL in this domain and have explored various applications and methodologies.

One significant application is the control of air conditioner operations to reduce energy consumption. Using algorithm such as YOLOv3, systems that can determine the number of people in a certain zone have been built by merging DL and the Internet of Things (IoT) (Elsisi et al., 2021). This data allows for the most efficient use of air conditioners in smart buildings. Deep reinforcement learning (DRL), a subset of AI, has also demonstrated promise in addressing smart building energy management issues. A review of DRL applications in this context has identified existing hurdles and suggested potential research directions (Yu et al., 2021).

Digital twin-based methods, which combine three-dimensional data models with the IoT, AI, and machine learning, have been instrumental in devising intelligent optimization and automation systems for residential district energy management. These methods have demonstrated their efficacy in enhancing renewable energy production.AI applications in smart buildings extend beyond energy management and encompass areas such as comfort, design, and maintenance. Building management systems (BMS) and demand response programs (DRPs) play a role in analysing and optimizing building performance in these domains (Farzaneh et al., 2021).

Efficient energy management in IoT networks can be achieved through robust DL frameworks. These frameworks utilize preprocessing techniques to handle the heterogeneity of electricity data and employ efficient decision-making algorithms for short-term forecasting (Han et al., 2020).

AI applications in smart buildings also extend to real-time production logistics in smart factories, where the interconnectedness between IoT-based systems and cyber-physical process monitoring is crucial for managing complexity and flexibility. Various studies have explored machine learning and DL models for energy forecasting tasks. These models, such as long short-term memory (LSTM) and support vector machines (SVM), have shown promise in accurately predicting electric load and electricity demand (Nishant et al., 2020).

While significant advancements have been made in AI and DL for smart building energy management, there is a gap between academic research and industry demands. Industry requires scalable and adaptable solutions that can cater to diverse building types, environments, and energy management challenges. Integrating real-time data, user preferences, and external factors into AI-driven energy management systems remains a challenge.

CONCLUSION

The recent advancements and applications of artificial intelligence (AI) and deep learning (DL) in driving sustainability in various sectors (Mohammed, 2023). AI has the potential to contribute to achieving sustainable development goals (SDGs), optimizing renewable energy systems, improving environmental health, and enhancing smart building energy management.

AI and DL technologies offer significant opportunities to promote sustainable practices and address societal challenges. They can contribute to many targets across all SDGs, making them valuable tools for advancing sustainability. However, the rapid development of AI and DL also raises concerns regarding transparency, safety, and ethical standards. Regulatory oversight is necessary to ensure responsible and ethical use of these technologies (Mohammed, 2023).

In the context of renewable energy, AI and DL have demonstrated their potential in optimizing energy management, detecting faults, and ensuring stability in power grids. They also contribute to waste management and predictive analysis in photovoltaic power plants, improving the efficiency and sustainability of renewable energy systems.

In the field of environmental health, the integration of AI and DL enables the analysis of complex spatial data, leading to improved exposure modeling and disease prediction. These technologies also enhance the accuracy and efficiency of diagnostic procedures and environmental monitoring.

Future research should focus on developing techniques that enhance the explainability and transparency of AI and DL models, scalable algorithms for processing large datasets, integration with next-generation wireless networks, and addressing ethical and privacy considerations. Improving the energy efficiency of AI and DL models is also crucial for their sustainable use.

By addressing these challenges and promoting responsible and ethical practices, AI and DL can significantly contribute to a more sustainable future. Their integration across sectors can optimize processes, improve decision-making, and help achieve global sustainability goals.

In conclusion, AI and DL are powerful tools that can shape a more sustainable future. By leveraging AI techniques, the renewable energy sector and green construction industry can benefit from improved sustainability, increased efficiency, and reduced environmental impact. These applications of AI contribute to the overall goal of creating a more sustainable and eco-friendlier built environment.

Continued research and development in these fields will be crucial for unlocking their full potential and driving sustainability in various sectors.

REFERENCES

Aggarwal, K., Mijwil, M. M., Al-Mistarehi, A. H., Alomari, S., Gök, M., Alaabdin, A. M. Z., & Abdulrhman, S. H. (2022). Has the future started? The current growth of artificial intelligence, machine learning, and deep learning. *Iraqi J. Comput. Sci. Math.*, *3*, 115–123.

Agostinelli, S., Cumo, F., Guidi, G., & Tomazzoli, C. (2021). Cyber-physical systems improving building energy management: Digital twin and artificial intelligence. *Energies*, *14*(8), 2338. doi:10.3390/en14082338

Akçay, H. G., Kabasakal, B., Aksu, D., Demir, N., Öz, M., & Erdoğan, A. (2020). Automated bird counting with deep learning for regional bird distribution mapping. *Animals (Basel)*, *10*(7), 1207. doi:10.3390/ani10071207 PMID:32708550

Alafif, T., Tehame, A. M., Bajaba, S., Barnawi, A., & Zia, S. (2021). Machine and deep learning towards COVID-19 diagnosis and treatment: Survey, challenges, and future directions. *International Journal of Environmental Research and Public Health*, *18*(3), 1117. doi:10.3390/ijerph18031117 PMID:33513984

Anthony, L. F. W., Kanding, B., & Selvan, R. (2020). Carbontracker: Tracking and predicting the carbon footprint of training deep learning models. *arXiv*, arXiv:2007.03051.

BeckT.PamukH.RamrattanR.UrasB. (2015). Mobile money, trade credit and economic development: Theory and evidence. CentER Discussion Paper Series No. *2015-023, European Banking Center Discussion Paper No. 2015-005.* https://ssrn.com/abstract=2588392

Blumenstock, J., Cadamuro, G., & On, R. (2015). Predicting poverty and wealth from mobile phone metadata. *Science*, *350*(6264), 1073–1076. doi:10.1126/science.aac4420 PMID:26612950

Boykin, L. M., Ghalab, A., Rossitto De Marchi, B., Savill, A., Wainaina, J. M., Kinene, T., Lamb, S., Rodrigues, M., Kehoe, M. A., & Ndunguru, J. (2018). Real time portable genome sequencing for global food security. *F1000 Research*, *7*, 1101. doi:10.12688/f1000research.15507.1

Campero-Jurado, I., Márquez-Sánchez, S., Quintanar-Gómez, J., Rodríguez, S., & Corchado, J. M. (2020). Smart helmet 5.0 for industrial internet of things using artificial intelligence. . *Sensors (Basel)*, *20*(21), 6241. doi:10.3390/s20216241 PMID:33139608

Cao, L. (2012). Social security and social welfare data mining: An overview. *IEEE Trans. Syst. Man. Cybern. Part C*, *42*, 837–853.

Chen, Z., & He, Y. (2023). Artificial Intelligence and Environmental Protection of Buildings. *Probl. Ekorozwoju/Probl. Sustainable Development (Bradford)*, *18*, 254–262.

Cirillo, D., Catuara-Solarz, S., Morey, C., Guney, E., Subirats, L., Mellino, S., Gigante, A., Valencia, A., Rementeria, M. J., Chadha, A. S., & Mavridis, N. (2020). Sex and gender differences and biases in artificial intelligence for biomedicine and healthcare. *NPJ Digital Medicine*, *3*(1), 81. doi:10.1038/s41746-020-0288-5 PMID:32529043

Collins, C., Dennehy, D., Conboy, K., & Mikalef, P. (2021). Artificial intelligence in information systems research: A systematic literature review and research agenda. *International Journal of Information Management, 60*, 102383. doi:10.1016/j.ijinfomgt.2021.102383

Di Vaio, A., Palladino, R., Hassan, R., & Escobar, O. (2020). Artificial intelligence and business models in the sustainable development goals perspective: A systematic literature review. *Journal of Business Research, 121*, 283–314. doi:10.1016/j.jbusres.2020.08.019

Dilmi, S., & Ladjal, M. (2021). A novel approach for water quality classification based on the integration of deep learning and feature extraction techniques. *Chemometrics and Intelligent Laboratory Systems, 214*, 104329. doi:10.1016/j.chemolab.2021.104329

Elsisi, M., Tran, M. Q., Mahmoud, K., Lehtonen, M., & Darwish, M. M. (2021). Deep learning-based industry 4.0 and internet of things towards effective energy management for smart buildings. . *Sensors (Basel), 21*(4), 1038. doi:10.3390/s21041038 PMID:33546436

Enholm, I. M., Papagiannidis, E., Mikalef, P., & Krogstie, J. (2021). Artificial intelligence and business value: A literature review. *Information Systems Frontiers.*

Farzaneh, H., Malehmirchegini, L., Bejan, A., Afolabi, T., Mulumba, A., & Daka, P. P. (2021). Artificial intelligence evolution in smart buildings for energy efficiency. *Applied Sciences (Basel, Switzerland), 11*(2), 763. doi:10.3390/app11020763

Fioretto, F., Mak, T. W., & Van Hentenryck, P. (2020). Predicting ac optimal power flows: Combining deep learning and lagrangian dual methods. In *Proceedings of the AAAI Conference on Artificial Intelligence.* Association for the Advancement of Artificial Intelligence. 10.1609/aaai.v34i01.5403

Franzo, G., Legnardi, M., Faustini, G., Tucciarone, C. M., & Cecchinato, M. (2023). When Everything Becomes Bigger: Big Data for Big Poultry Production. *Animals (Basel), 13*(11), 1804. doi:10.3390/ani13111804 PMID:37889739

Grbčić, L., Kranjčević, L., & Družeta, S. (2021). Machine learning and simulation-optimization coupling for water distribution network contamination source detection. *Sensors (Basel), 21*(4), 1157. doi:10.3390/s21041157 PMID:33562175

Gupta, J., Pouw, N. R., & Ros-Tonen, M. A. (2015). Towards an elaborated theory of inclusive development. *European Journal of Development Research, 27*(4), 541–559. doi:10.1057/ejdr.2015.30

Habib, M., Faris, M., Qaddoura, R., Alomari, M., Alomari, A., & Faris, H. (2021). Toward an automatic quality assessment of voice-based telemedicine consultations: A deep learning approach. *Sensors (Basel), 21*(9), 3279. doi:10.3390/s21093279 PMID:34068602

Han, T., Muhammad, K., Hussain, T., Lloret, J., & Baik, S. W. (2020). An efficient deep learning framework for intelligent energy management in IoT networks. *IEEE Internet of Things Journal, 8*(5), 3170–3179. doi:10.1109/JIOT.2020.3013306

Hanoon, S. K., Abdullah, A. F., Shafri, H. Z., & Wayayok, A. (2022). A Novel Approach Based on Machine Learning and Public Engagement to Predict Water-Scarcity Risk in Urban Areas. *ISPRS International Journal of Geo-Information, 11*(12), 606. doi:10.3390/ijgi11120606

Holzinger, A., Weippl, E., Tjoa, A. M., & Kieseberg, P. (2021). Digital transformation for sustainable development goals — A security, safety and privacy perspective on AI. In A. Holzinger, P. Kieseberg, A. M. Tjoa, & E. Weippl (Eds.), *Machine Learning and Knowledge Extraction* (pp. 1–20). Springer International Publishing. doi:10.1007/978-3-030-84060-0_1

Jain, S., Parida, A. K., & Sankaranarayanan, S. (2020). Water scarcity prediction for global region using machine learning. *International Journal of Water*, *14*(1), 69–88. doi:10.1504/IJW.2020.112746

Jean, N., Burke, M., Xie, M., Davis, W. M., Lobell, D. B., & Ermon, S. (2016). Combining satellite imagery and machine learning to predict poverty. *Science*, *353*(6301), 790–794. doi:10.1126/science.aaf7894 PMID:27540167

Khan, S., & Yairi, T. (2018). A review on the application of deep learning in system health management. *Mechanical Systems and Signal Processing*, *107*, 241–265. doi:10.1016/j.ymssp.2017.11.024

Khanna, A., Lamba, B. Y., Jain, S., Bolshev, V., Budnikov, D., Panchenko, V., & Smirnov, A. (2023). Biodiesel Production from Jatropha: A Computational Approach by Means of Artificial Intelligence and Genetic Algorithm. *Sustainability (Basel)*, *15*(12), 9785. doi:10.3390/su15129785

Leavy, S., Meaney, G., Wade, K., & Greene, D. (2020). Mitigating gender bias in machine learning data sets. In *Proceedings of the Bias and Social Aspects in Search and Recommendation: First International Workshop*. Springer. 10.1007/978-3-030-52485-2_2

Lee, D., & Yoon, S. N. (2021). Application of artificial intelligence-based technologies in the healthcare industry: Opportunities and challenges. *International Journal of Environmental Research and Public Health*, *18*(1), 271. doi:10.3390/ijerph18010271 PMID:33401373

Lee, S., & Lee, D. (2018). Improved prediction of harmful algal blooms in four Major South Korea's Rivers using deep learning models. *International Journal of Environmental Research and Public Health*, *15*(7), 1322. doi:10.3390/ijerph15071322 PMID:29937531

Leong, C., Tan, F., & Ahuja, M. (2020). IS for Good – 10 years to SDG: Where we have been and where we need to go? In *Proceedings of the International Conference on Information Systems*. IEEE.

Leopord, H., Cheruiyot, W. K., & Kimani, S. (2016). A survey and analysis on classification and regression data mining techniques for diseases outbreak prediction in datasets. *International Journal of Engineering Science*, *5*, 1–11.

Maghsudi, S., Lan, A., Xu, J., & van Der Schaar, M. (2021). Personalized education in the artificial intelligence era: What to expect next. *IEEE Signal Processing Magazine*, *38*(3), 37–50. doi:10.1109/MSP.2021.3055032

Martos, V., Ahmad, A., Cartujo, P., & Ordoñez, J. (2021). Ensuring agricultural sustainability through remote sensing in the era of agriculture 5.0. *Applied Sciences (Basel, Switzerland)*, *11*(13), 5911. doi:10.3390/app11135911

Mhlanga, D. (2021). Artificial intelligence in the industry 4.0, and its impact on poverty, innovation, infrastructure development, and the sustainable development goals: Lessons from emerging economies? *Sustainability (Basel)*, *13*(11), 5788. doi:10.3390/su13115788

Milano, M., O'Sullivan, B., & Gavanelli, M. (2014). Sustainable policy making: A strategic challenge for artificial intelligence. *AI Magazine, 35*(3), 22–35. doi:10.1609/aimag.v35i3.2534

Mohammed, A. B. (2023). *AS Syed - Global Applications of the Internet of Things in Digital*. Applications and Impact of Internet of Things in Digital Marketing.

Mohammed, A. (2023). *Fraud Prevention, Confidentiality, and Data Security for Implementation of Big Data in Modern Business: Fraud Detection and Prevention Techniques*. Academic Press.

Nahorniak, H., & Hannouf, K. (2023). Features of the challenges and opportunities of formation of intellectual capital of domestic enterprises in the engineering sector ukraine in the conditions of digitalization. *Probl. Transp. Telecommun., 24*, 1–15.

Nishant, R., Kennedy, M., & Corbett, J. (2020). Artificial intelligence for sustainability: Challenges, opportunities, and a research agenda. *International Journal of Information Management, 53*, 53. doi:10.1016/j.ijinfomgt.2020.102104

Noh, S. K. (2021). Recycled clothing classification system using intelligent IoT and deep learning with AlexNet. . *Computational Intelligence and Neuroscience, 2021*, 5544784. doi:10.1155/2021/5544784 PMID:33854541

Pierdicca, R., Paolanti, M., Felicetti, A., Piccinini, F., & Zingaretti, P. (2020). Automatic faults detection of photovoltaic farms: SolAIr, a deep learning-based system for thermal images. *Energies, 13*(24), 6496. doi:10.3390/en13246496

Romero, H. F. M., Hernández-Callejo, L., Rebollo, M. Á. G., Cardeñoso-Payo, V., Gómez, V. A., Bello, H. J., Moyo, R. T., & Aragonés, J. I. M. (2023). Synthetic Dataset of Electroluminescence Images of Photovoltaic Cells by Deep Convolutional Generative Adversarial Networks. *Sustainability (Basel), 15*(9), 7175. doi:10.3390/su15097175

Serradilla, O., Zugasti, E., Rodriguez, J., & Zurutuza, U. (2022). Deep learning models for predictive maintenance: A survey, comparison, challenges and prospects. *Applied Intelligence, 52*(10), 10934–10964. doi:10.1007/s10489-021-03004-y

Si, D. (2022). A Framework to analyze the Impacts of AI with the Sustainable Development Goals. *Highlights Sci. Eng. Technol., 17*, 313–323. doi:10.54097/hset.v17i.2621

Singh, R., & Singh, R. (2023). Applications of sentiment analysis and machine learning techniques in disease outbreak prediction–A review. *Materials Today: Proceedings, 81*, 1006–1011. doi:10.1016/j.matpr.2021.04.356

Singh, R., Singh, R., Gehlot, A., Akram, S. V., Priyadarshi, N., & Twala, B. (2022). Horticulture 4.0: Adoption of Industry 4.0 Technologies in Horticulture for Meeting Sustainable Farming. *Applied Sciences (Basel, Switzerland), 12*(24), 12557. doi:10.3390/app122412557

Solanki, A., Agrawal, H., & Khare, K. (2015). Predictive analysis of water quality parameters using deep learning. *Int. J. Comput. Appl., 125*.

Strubell, E., Ganesh, A., & McCallum, A. (2020). Energy and policy considerations for modern deep learning research. *Proceedings of the AAAI Conference on Artificial Intelligence*. 10.1609/aaai.v34i09.7123

Tang, H., Huang, H., Liu, J., Zhu, J., Gou, F., & Wu, J. (2022). AI-Assisted Diagnosis and Decision-Making Method in Developing Countries for Osteosarcoma. *Health Care*, *10*, 2313. PMID:36421636

Tsai, S. B., Xue, Y., Zhang, J., Chen, Q., Liu, Y., Zhou, J., & Dong, W. (2017). Models for forecasting growth trends in renewable energy. *Renewable & Sustainable Energy Reviews*, *77*, 1169–1178. doi:10.1016/j.rser.2016.06.001

Vinuesa, R., Azizpour, H., Leite, I., Balaam, M., Dignum, V., Domisch, S., Felländer, A., Langhans, S. D., Tegmark, M., & Fuso Nerini, F. (2020). The role of artificial intelligence in achieving the Sustainable Development Goals. . *Nature Communications*, *11*(1), 233. doi:10.1038/s41467-019-14108-y PMID:31932590

Vinuesa, R., & Sirmacek, B. (2021). Interpretable deep-learning models to help achieve the Sustainable Development Goals. *Nature Machine Intelligence*, *3*(11), 926. doi:10.1038/s42256-021-00414-y

VoPham, T. (2018). Emerging trends in geospatial artificial intelligence (geoAI): Potential applications for environmental epidemiology. *Environmental Health*, *17*, 1–6.

Wu, Z. Y., & Rahman, A. (2017). Optimized deep learning framework for water distribution data-driven modeling. *Procedia Engineering*, *186*, 261–268. doi:10.1016/j.proeng.2017.03.240

Yang, H., & Zhu, K. (2022). Research on AI-assisted grading of math questions based on deep learning. In *International Conference on Internet of Things and Machine Learning (IoTML 2021)*. SPIE. 10.1117/12.2628649

Yu, L., Qin, S., Zhang, M., Shen, C., Jiang, T., & Guan, X. (2021). A review of deep reinforcement learning for smart building energy management. *IEEE Internet of Things Journal*, *8*(15), 12046–12063. doi:10.1109/JIOT.2021.3078462

Chapter 6
Applications of Blockchain in Building Marketing Framework:
Approach to Environmental Sustainability

Arshi Naim
https://orcid.org/0000-0003-1325-6964
King Khalid University, Saudi Arabia

ABSTRACT

Firms apply various technological applications to achieve their businesses objectives. Developing marketing framework (MF) has been a challenge for firms in all types of products and services. This is because firms are not sure of which MF can best suit their business practices and attain environmental sustainability. Another important reason is that there are some marketing models (MM) which are confused with the MF that leads to misinterpretation of applications. This research chapter is qualitative in nature and has two parts. The first part presents the definitions of MM and MF and presents a comparative analysis between them. The second part presents the application of blockchain for developing successful marketing framework for environmental sustainability. Results are presented by using observation methods. Analysis and are based on the firm's experience in the use of blockchain in building MF for environmental sustainability.

INTRODUCTION

Blockchain is essentially a digital chain of blocks that keeps a public ledger of transactions. It's a decentralized technology, meaning that no single person or entity owns or runs it. Because of this, there's no middle man and the technology is much more secure (Rathore 2019). However, blockchain holds a number of cryptocurrencies, or digital money and because no one is managing the blockchain instead, it's a public ledger kept on a network the amount of crypto one has should always be correct, and can always be trusted (Saini and Tarkar 2022).

DOI: 10.4018/979-8-3693-1794-5.ch006

While blockchain technology is the perfect place for creating and housing cryptocurrencies like Bitcoin and Ethereum, there are more applications than that including when it comes to marketing (Saini and Tarkar 2022).

The term "blockchain marketing," it's likely referring to the ability to use blockchain technology within a digital marketing strategy (Hermann 2022). While blockchain is currently used almost exclusively for buying, selling, and trading cryptocurrency (Hermann 2022).

Blockchain in marketing can create a more secure relationship between brands and their customers, allowing them to design more targeted advertisements without needing to share so much data with so many companies (Rathore 2019).

Pros and Cons of Blockchain Marketing in Environmental Sustainability

There are a few pros of blockchain adoption in the marketing industry such as blockchain is trustworthy, data can never be stolen or sold, it prevents fraud, it eliminates the need for a middleman in both transactions and advertising and it strengthens content monetization (Saini and Tarkar 2022).

The Cons of blockchain marketing are; data is expensive, blockchain requires a lot of energy which prevents in achieving environmental sustainability (Dash et al. 2023). Some of the pros and cons are shown in Figure 1.

Figure 1. Pros and cons in blockchain marketing in environmental sustainability
(Hermann 2022)

PROS	CONS
Blockchain is trustworthy	Data can sometimes still be controlled
Data can never be stolen or sold	It's expensive
It prevents fraud	It requires a lot of energy
It eliminates the need for a middleman	
It strengthens content monetization	

Blockchain has become a helpful tool for advertising and marketing. Its decentralized ledger technology helps marketing and advertising teams better manage data, gain deeper insights into audience interactions with ad campaigns and cultivate meaningful customer relationships (Dash et al. 2023).

Blockchain is a shared, immutable ledger that facilitates the process of recording transactions and tracking assets in a business network. An asset can be tangible (a house, car, cash, land) or intangible (intellectual property, patents, copyrights, branding) (Dash et al. 2023).

A blockchain is a decentralized, distributed and public digital ledger that is used to record transactions across many computers so that the record cannot be altered retroactively without the alteration of all subsequent blocks and the consensus of the network." (Hermann 2022).

A marketing framework is a template containing instructions for the execution of your marketing plan (Hemanand et al. 2022). This framework ensures that you're delivering the right content to the right audience, through the right channels, at the right time to achieve your core marketing objectives. To use an analogy, the marketing framework is to the marketing plan as DNA is to all life (Hemanand et al. 2022). Marketing is the activity, set of institutions, and processes for creating, communicating, delivering, and exchanging offerings that have value for customers, clients, partners, and society at large (Dash et al. 2023). Marketing research is the function that links the consumer, customer, and public to the marketer through information and information is used to identify and define marketing opportunities and problems; generate, refine, and evaluate marketing actions; monitor marketing performance; and improve understanding of marketing as a process (Skobelev 2018). Marketing research specifies the information required to address these issues, designs the method for collecting information, manages and implements the data collection process, analyzes the results, and communicates the findings and their implication (Dash et al. 2023). A brand is a name, term, design, symbol, or any other feature that identifies one seller's goods or service as distinct from those of other sellers. ISO brand standards add that a brand "is an intangible asset" that is intended to create "distinctive images and associations in the minds of stakeholders, thereby generating economic benefit/values." There are following types of marketing (Skobelev 2018).

There are 4 Ps of Marketing, Product which is defined as a bundle of attributes (features, functions, benefits, and uses) capable of exchange or use, usually a mix of tangible and intangible forms, Price is the amount a customer must pay to acquire a product, Place is represented by place or placement, and Promotion that includes tactics that encourage short-term purchase, influence trial and quantity of purchase, and are very measurable in volume, share and profit (Gheorghe et al. 2023).

There are two major categories of marketing models: top-down and bottom-up. Top-down models are focused on audience demographics and expectations. They divide the market into segments to determine how different audience groups and segments make purchases. Instead of using specified groups to predict demand, revenue depends on the sales that a single product or base unit generates. Many variations and models exist within these two categories (Hemanand et al. 2022). Here are some popular marketing models that can be used to predict customer behavior, company growth and revenue expectations and few of them are explained for their importance in the discussion and results section.

SWOT and TOWS Analysis

SWOT and TOWS are both acronyms for strengths, weaknesses, opportunities and threats. While both use the same basic ideas in their analysis, TOWS emphasizes the external environment whereas SWOT focuses on the internal environment (Gheorghe et al. 2023). These models help you visualize strategic options and pivot the strengths and minimize the weaknesses to avoid threats and maximize opportunities (Skobelev 2018). Use the y-axis of the matrix to list the strength and weaknesses and the x-axis for your opportunities and threats (see figure 2).

Figure 2. Comparison between SWOT and TOWS Analysis
(Gheorghe et al. 2023)

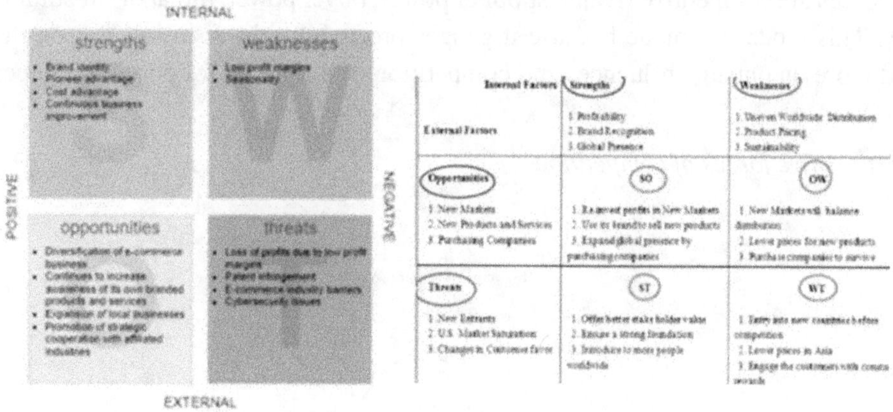

The result should be four quadrants: strengths and opportunities, strengths and threats, weaknesses and opportunities and weaknesses and threats. These categories can be used to create defensive strategies to combat challenges and illustrate potential avenues for success (Gheorghe et al. 2023).

STP Marketing Model

STP stands for segmentation, targeting and positioning. It's a popular model that uses the top-down approach by focusing on how a company addresses customers (Suryani and Luthfiyyah 2023). STP uses a four step process to deliver relevant, personalized messages to targeted audiences. Top-down models like the STP marketing model have gained attractiveness over the years as companies pivot to delivering catered content to their target audiences via social media (see figure 3 for STP marketing model).

Figure 3. STP marketing model
(Suryani and Luthfiyyah 2023)

The first step is market segmentation, where marketers determine important characteristics for each group within the market (Nghiêm and Suter 2022).

Porter's Five Forces

Porter's five forces are competitive rivalry, supplier power, buyer power, threat of substitution and threat of new entry. This model is unique because it gauges profitability by focusing less on the product or audience and more on outside influences and competition (see figure 4 for porter's 5 forces)

Figure 4. Porter's five forces of competition
(Skobelev 2018)

Using this analysis can be a simple but powerful way to understand the competitiveness within your business environment (Suryani and Luthfiyyah 2023).

AIDA

The AIDA marketing model focuses almost entirely on the customer (Nghiêm and Suter 2022). The acronym stands for awareness, interest, desire and action. These are the four stages a buyer goes through during the process of purchasing a service or product. Some models include an additional stage, retention, which addresses a buyer's choice to make return purchases and build brand loyalty (see figure 5 for details on AIDA model).

Figure 5. AIDA model
(Nghiêm and Suter 2022)

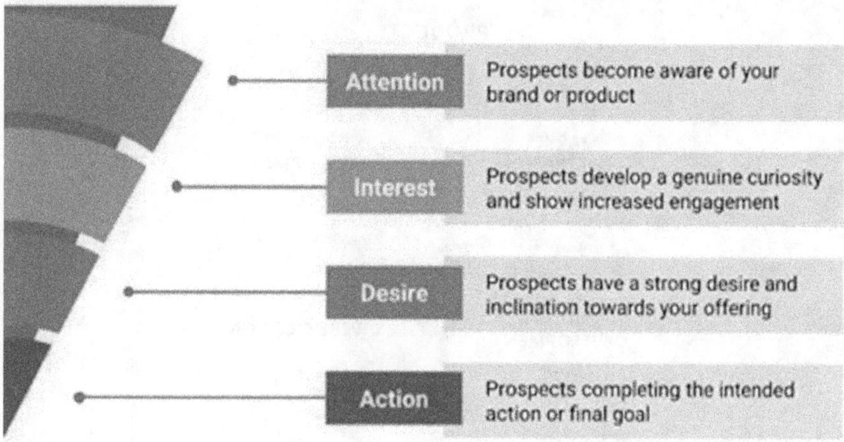

This model is unique because it acknowledges the influence social media has on buyer-seller relationships and incorporates that into selling strategies.

Ansoff Matrix

Also called the product or market expansion grid, the Ansoff matrix is a 2x2 grid that outlines four strategies a firm can use to grow its business and analyze potential risks (Suryani and Luthfiyyah 2023). Ansoff grids have markets on the y-axis. The lower end of the axis represents new markets, and the upper end represents existing markets. Products and services are on the x-axis. One side shows existing products and services, and the other represents those that are new.

The lower left quadrant shows an existing product entering an existing market. The lower right shows a new product and existing market (Suryani and Luthfiyyah 2023). The upper left shows a new product in an existing market, and the upper right shows new products and new markets. In this matrix, risk increases as firms move horizontally or vertically into a new quadrant (see figure 6 for Ansoff matrix for business growth and potential risks).

Figure 6. Ansoff model (product or market expansion grid)
(Nghiêm and Suter 2022)

The safest of the four options is market penetration, the lower left quadrant (Naim and Hassan 2022). Market penetration occurs when firms expand sales for the existing product in the existing market. The next safest is product development, which involves adding a new product to an existing market and occurs in the lower right quadrant. Market development, which occurs by introducing an existing product into a new market, has slightly more risk. According to the Ansoff matrix, the riskiest option is the upper right quadrant: diversification. Diversification means introducing an untested product into a new market you may not understand (Khatri et al. 2023).

Growth-Share Matrix

The growth-share matrix uses four quadrants to help businesses decide how to prioritize their different ventures. In this marketing model, the y-axis shows low to high growth and the x-axis displays high and low market shares. The matrix represents each of the four quadrants with the following symbols:

Stars: The growth-share matrix identifies opportunities that have both high growth and a high market share with a star symbol (Naim and Khan 2022). Often depicted in the top right quadrant, stars indicate the smartest investment opportunities with a high opportunity for success and stability (see figure 7 for BCG model or growth- share matrix).

Figure 7. Growth share matrix
(Khatri et al. 2023)

Cows: The cow symbol represents the bottom right quadrant, showing opportunities with low growth and a high market share. Cows typically generate a large return on income initially, but may not be ideal for further development (Malik et al.2022).

Question marks: The upper right quadrant uses a question mark symbol and shows low market share but high growth. The question mark symbol indicates that the opportunity has unpredictable potential in the current market (Malik et al.2022).

Pets: The bottom right quadrant shows opportunities that have both low growth and low market share, also known as pets. You can use an animal symbol, typically a dog, to indicate opportunities that the company should consider discarding or repositioning (Malik et al.2022).

SOSTAC

The SOSTAC model is a versatile planning model used to create marketing strategies (Chowdhury et al. 2023). SOSTAC stands for situation, objectives, strategy, tactics, action and control (Fu et al. 2020). It can be a suitable tool to review your process and discover areas of weakness (see figure 8 for SOSTAC model for marketing strategies).

Figure 8. SOSTAC model
(Chowdhury et al. 2023)

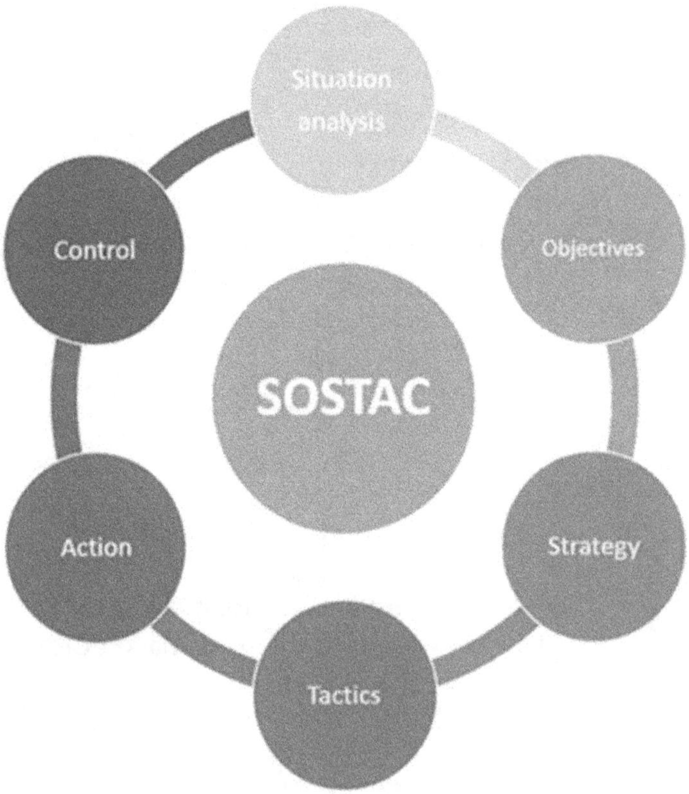

Each step in SOSTAC represents an important part of the development process (Naim 2022): Identifying the current conditions, defining your goals, crafting your strategy, outlining how you plan to execute your strategy, working your plan and reviewing these steps to ensure you're meeting your goals. Using this outline can be beneficial to finding potential holes in the marketing plan.

McKinsey 7-S model

The McKinsey 7-S model outlines seven major elements that need to work in harmony for a business to be successful. A watershed diagram with seven circles is the most common representation of the McKinsey 7-S model (see Figure 9).

Figure 9. McKinsey 7s model
(Fu et al. 2020)

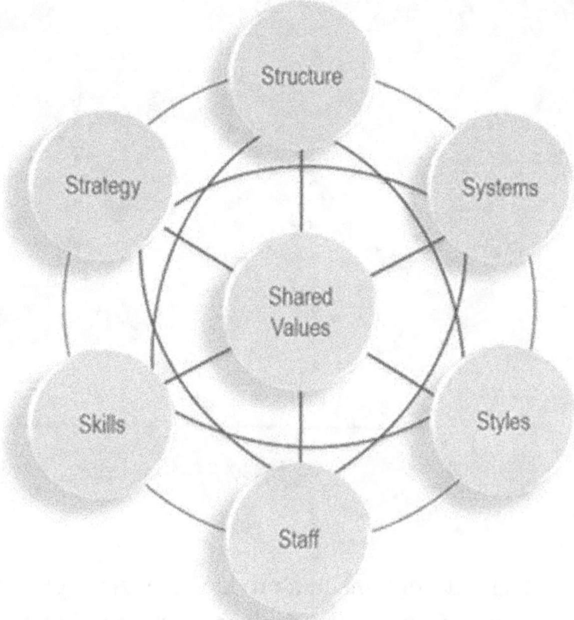

There are seven circles: strategy, structure, systems, styles, staff, skills and shared values. The shared values circle in the center connects the other circles to show that each of the elements is important to ensuring the success and adaptability of an organization (Naim and Alqahtani 2022). When working with this model, firms consider how their marketing efforts in each category could impact the others (Fu et al. 2020).

Product Life Cycle

The product life cycle model can help you develop new products, refine existing products and recognize when it is time to discontinue a product (Naim and Kautish 2022). It has four stages that can guide your marketing efforts throughout product development as explained in figure 10.

Figure 10. Phases of product life cycle
(Alahmari et al. 2023)

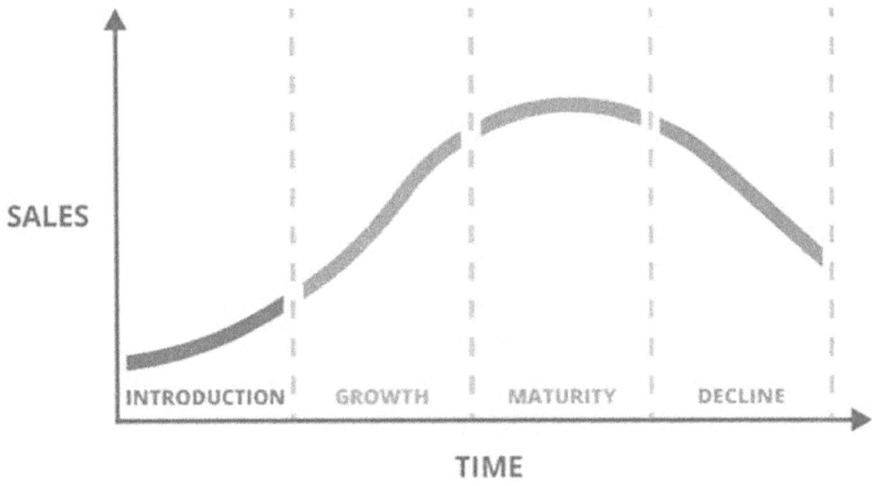

Introduction: After research and development, a product goes through the introduction stage where you first introduce it to consumers. This typically involves intense marketing and promotional efforts to develop public awareness of the new release (Chou et al. 2022).

Growth: As the product gains popularity and the company expands to support distribution, it enters the growth phase. This is when the product gains popularity, develops a dedicated customer base and increases market share among competitors (Chowdhury et al. 2023).

Maturity: Mature products often have many other competitors in a saturated market. During maturity, growth slows down and you may need to adjust your marketing strategy to find new audiences or applications of your product.

Decline: During decline, sales decrease and marketing efforts have less of an impact. When a product enters decline, you may shift your efforts to new product development instead of marketing existing ones.

LITERATURE REVIEW

The study of the history of marketing, as a discipline, is meaningful because it helps to define the baselines upon which change can be recognized and understand how the discipline evolves in response to those changes (Naim et al. 2023). The practice of marketing has been known for millennia, but the term "marketing" used to describe commercial activities buying and selling products or services came into popular use in the late nineteenth century (Fatima et al. 2022). The study of the history of marketing as an academic field emerged in the early twentieth century (Chou et al. 2022).

Marketers tend to distinguish between the history of marketing practice and the history of marketing thought: the history of marketing practice refers to an investigation into the ways that marketing has been practiced; and how those practices have evolved over time as they respond to changing socio-economic conditions and the history of marketing thought refers to an examination of the ways that marketing has been studied and taught (Fatima et al. 2022).

Although the history of marketing thought and the history of marketing practice are distinct fields of study, they intersect at different junctures (Fatima et al. 2022). Marketing practitioners engage in innovative practices that capture the attention of marketing scholars who codify and disseminate such practices. At the same time, marketing academics often develop new research methods or theories that are subsequently adopted by practitioners (Hoda and Naim 2023). Thus developments in marketing theory inform marketing practice and vice versa (Naim et al. 2021). The history of marketing will remain incomplete if one disassociates academia from practitioners.

Table 1. Marketing philosophies or orientation in popular texts (Hoda and Naim 2023)

Dibb and Simkin	Lancaster and Reynolds	Blythe	Drummond and Ensor	Morgan

Several scholars have attempted to describe the evolution of marketing thought chronologically and to connect it with broader intellectual and academic trends (Chou et al. 2022). Bartels (1965) provided a brief account of marketing's formative periods, and Shah and Gardner (1982) briefly considered the development of the six dominant schools in contemporary marketing. However, these initial attempts have been criticized as overly descriptive. One of the first theorists to consider the stages in the development of marketing thought was Robert Bartels, who in The History of Marketing Thought, (1965) used a periodisation approach. He categorized the development of marketing theory decade by decade from the beginning of the 20th century: (Naim et al. 2022)

1900s: discovery of basic concepts and their exploration

1910s: conceptualization, classification and definition of terms

1920s: integration on the basis of principles

1930s: development of specialization and variation in theory

1940s: reappraisal in the light of new demands and a more scientific approach (Hoda and Naim 2023)

1950s: re-conceptualization in the light of managerial social development and quantitative approaches

1960s: differentiation on bases such as managerial, holism, environmentalism, systems, and internationalism (Alqahtani and Naim 2022).

1970s: socialization; the adaptation of marketing to social change

Bartels was the first historian to provide a "long view of marketing's past and wide sweep of its sub disciplines" and in so doing, he nurtured an interest in the history of marketing thought (Hoda and Naim 2023).

Key innovations that influenced marketing practice are given below:

1450: Gutenberg's metal movable type, leading eventually to mass-production of flyers and brochures

1600s: Paid advertising in Italy

1600s: Use of handbills and posters is common practice in Elizabethan England

1605: World's first newspaper published in Germany

1600s and 1700s: Posters and handbills used for promotion in England

1665: The Oxford Gazette first published in England (later renamed the London Gazette and still published today)

1700s: Widespread emergence of newspapers and magazines in England and France; by the 1730s daily newspapers in London devoted more than half the available space to advertising (Naim and Malik)

1719: The Daily Post first published; early instance of a periodical dedicated to business, science and innovation

1836: Paid advertising in a newspaper (in France)

1839: Posters on private property banned in England

1864: Earliest recorded use of the telegraph for mass unsolicited spam

1867: Earliest recorded billboard rentals

1876: Films produced by French film-makers, Auguste and Louis Lumiere, made at the request of a representative of Lever Brothers in France and feature Sunlight soap, are thought to be the first recorded instance of paid product placement.

1880s: Early examples of trademarks as branding

1902: The first marketing course, taught by Edward David Jones, was offered at the University of Michigan

1905: The University of Pennsylvania offered a course in "The Marketing of Products"

1908: Harvard Business School opens - Harvard was an early influence on marketing thought

1920: The magazine, Variety, reports that 50% of cinemas show advertising program

1920s: Radio advertising commences

1940s: Electronic computers developed

1941: First recorded use of television advertising

1955: Television viewing exceeded radio listening for the first time in Britain

1950s: Systematization of telemarketing

1957: Three key scholarly texts published Wroe Alderson's Marketing Behavior and Executive Action; Howard's Marketing Management and Lazer's Managerial Marketing: Perspectives and Viewpoints

1960 E. Jerome McCarthy published his now classic, Basic Marketing: A Managerial Approach (1960).

1970s: E-commerce developed

1980s: Development of database marketing as precursor to CRM

1980s: Emergence of relationship marketing

1980s: Emergence of computer-oriented spam

1984: Introduction of guerrilla tactics

1985: Desktop publishing democratizes the production of print-advertising (precursor to consumer-generated media and content)

1991: IMC gains academic status

Mid 1990s: Modern search engines started appearing in the mid-1990s, with Google making its debut in 1998

1990s CRM and IMC (in various guises and names) gain dominance in promotions and marketing planning,

1996: Identification of viral marketing

2000s: Integrated marketing gains widespread acceptance with its first dedicated academic research centre opened in 2002

2003 -2006: Emergence of social media. MySpace and LinkedIn emerged in 2003, Facebook in 2004 and Twitter in 2006 (Saini and Tarkar 2022).

DISCUSSION/ RESULTS

In the discussion part, the paper presents the scope of Marketing Framework for business practices. A marketing framework is essentially how firms plan to execute your marketing strategy (Whig et al. 2022). This means that PEST and SWOT analysis are not part of the marketing framework. "PEST" comes at the earliest stage of the marketing plan. It sets the scene for your marketing efforts, explained in figure 11.

Figure 11. PEST analysis before marketing plan
(Chou et al. 2022)

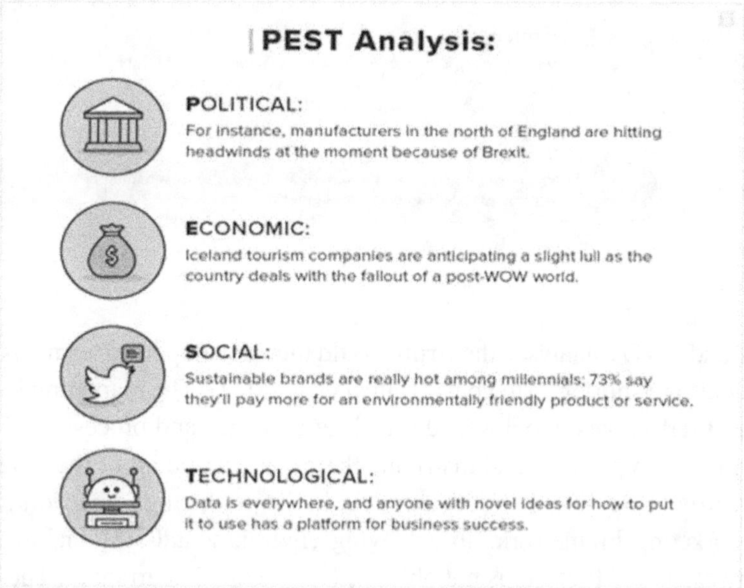

"SWOT" is also a pre- marketing framework activity. It's like a character assessment of the brand that helps firms to understand how equipped the business is to perform in the current market environment. Figure 12 gives the details of SWOT analysis and the scope of its application.

Figure 12. SWOT analysis description
(Bhardwaj 2022)

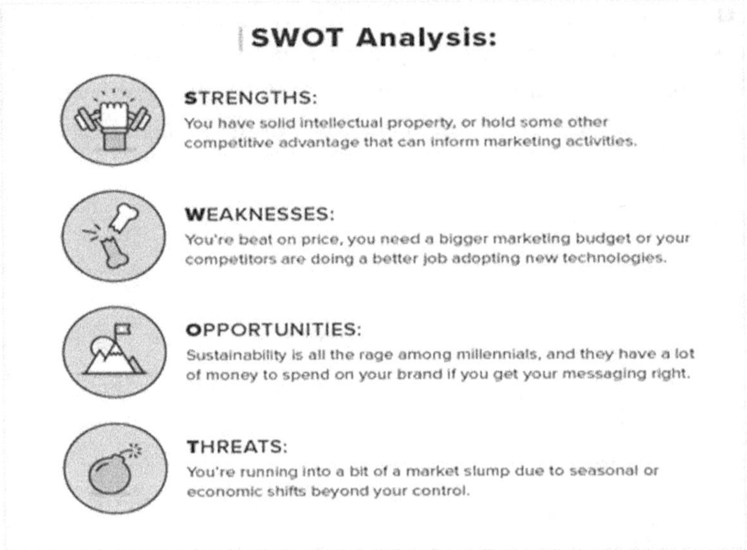

Based on PEST and SWOT analysis, the firms would then develop long-term business goals. Nested within those core goals are shorter-term goals (Kamal et al. 2022). The marketing framework organizes those long- and short-term goals into a well-defined set of stages and processes. Content strategy, the social channels firms use to promote and distribute their content and all of the other marketing tactics that support the specific goals should be accounted for in the marketing framework. The paper explains the examples of Marketing Frameworks in achieving environmental sustainability. Many marketing frameworks already exist, and the firm is not obligated to use one in particular. The organization might even switch among frameworks at various stages of its development. And often, the best marketing frameworks are tailor-made for the organization based on the current long-term marketing goals. Figure 3 shows the most commonly used marketing frameworks (Bag et al. 2022).

The Inbound Marketing Sales Funnel (Cabanillas 2016).

Often referred to as the content marketing framework, the inbound marketing sales funnel creates a systematic flow of inbound leads (Cabanillas 2016).

It uses content to engage users at every stage of the customer journey from brand awareness all the way through purchasing and choosing the echo friendly products (see Figure 13).

Figure 13. Inbound marketing sales funnel
(Gunasekaran and Boopathi 2023)

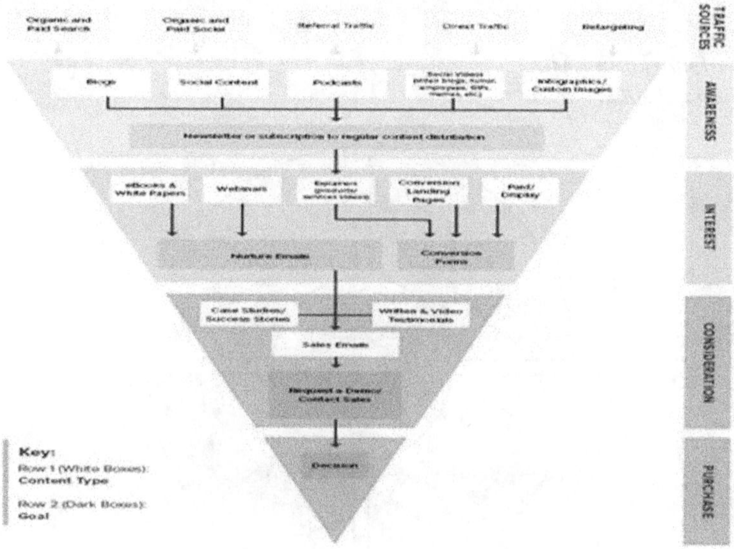

The overarching goal of inbound marketing is to automate the marketing funnel. As evidenced from the diagram above, there are multiple short-term goals involved at each stage of the funnel. All of them are priorities, but within a given company, some may be in dire need of TLC than others. Every modern organization relies on inbound marketing, also known as pull marketing, to some extent (Gunasekaran and Boopathi 2023). If the business has a website, then it's technically practicing inbound marketing. If that website has a blog, resource center or some other value-add content, then firms are also practicing content marketing.

Again, there is no single "best" marketing framework and even the representation of the inbound marketing and sales funnel is merely a template for guidance, not an exhaustive, step-by-step guide for the organization. Still, this is the most precise marketing framework when it comes to executing the digital marketing component of the marketing plan (Boudi et al. 2021).

It provides a roadmap for generating traffic with paid and organic content, lead nurturing, boosting conversion rates and ultimately turning potential customers into brand advocates through various forms of digital engagement.

The 7 PS of the Marketing Mix (Pillarisetty and Mishra, 2022)

For years, this was known as the 4 Ps of marketing. The original Ps was Product, Price, Promotion and Place, which are explained in the figure 14. However, three more Ps have been added, which are People, Process and Physical evidence. The 7 Ps can be exceptionally useful for helping answer some of the more open-ended questions associated with product marketing and your brand, and for that reason, it's a worthwhile exercise. But it's not quite as precise as the inbound-marketing model. The best way to think of the 7 Ps is as a product marketing framework since it also covers pricing, product features and other qualities that typically precede a content marketing strategy (Priya et al. 2023).

Figure 14. Four Ps of marketing M
mix (Pillarisetty and Mishra, P 2022)

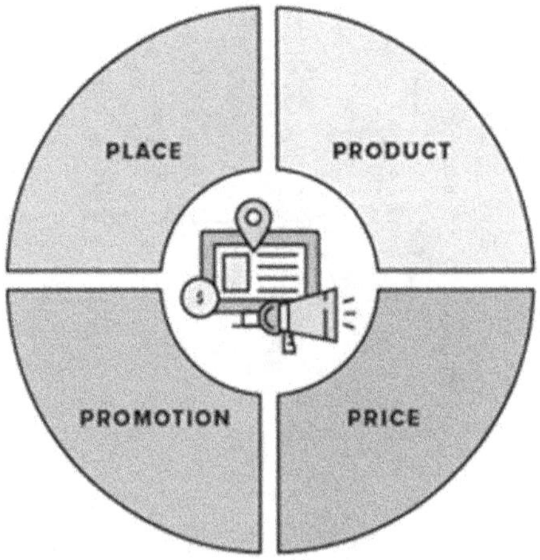

It covers some elements of the execution of your strategy, but it attempts to be a little more all-encompassing.

The Pragmatic Marketing Framework

This is the Pragmatic Marketing Framework created by Pragmatic Institute. It is also more of a product marketing framework than just a marketing framework (Zehir et al. 2020). It's perhaps even more all-encompassing than the 7 Ps in the sense that it also covers some of the same ground as PEST and SWOT analysis. Figure 15 shows the framework focuses on strategy development at the left side in the figure 15, but to the right, the framework becomes more about execution of that strategy (Zehir et al. 2020). For example, buyer personas, awareness, nurturing and advocacy, for instance, are important terms in content marketing.

Figure 15. Pragmatic marketing framework
(Nozari et al. 2021)

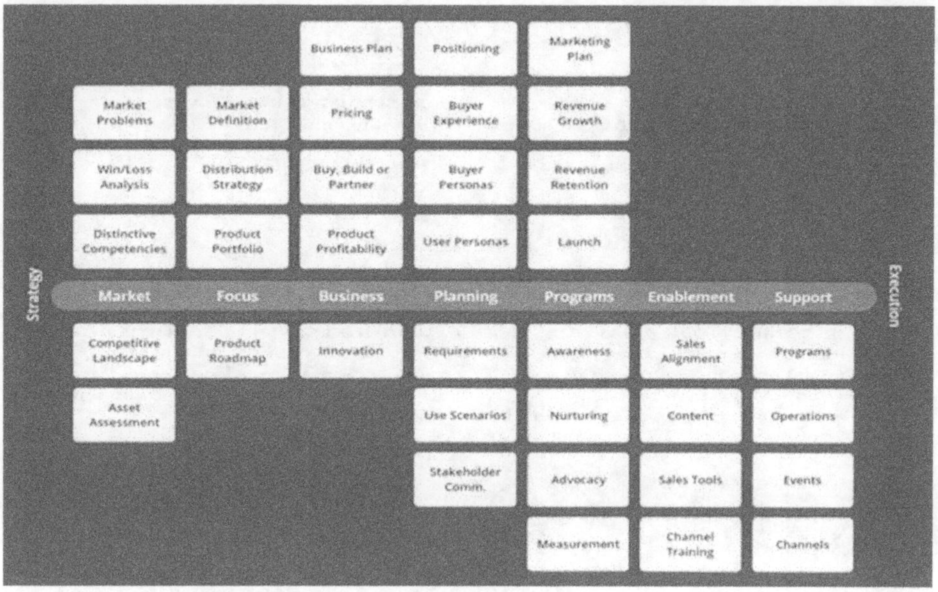

Buyer personas (or customer personas) help the firms to understand their target audience before the firm begins creating content, awareness is the highest level of the inbound marketing sales funnel, nurturing is what the firms do in the middle to boost interest and intent to buy, advocacy comes after purchase (Rathore 2019). It refers to user-generated content that praises the brand on social media, influencer marketing, case studies and testimonials, word-of-mouth and other forms of advocacy from satisfied customers and industry influencers. Clearly, the inbound marketing sales funnel has some common ground with the Pragmatic Marketing Framework (Nghiêm and Suter 2022).

Brafton's Marketing Framework Template

All marketing frameworks are basically just trying to visualize the firm's marketing efforts. Some dip more into the product roadmap and business side of things than others, but all of them are trying to do the same thing, which is to aid the firm in structuring the marketing campaigns (Suryani and Luthfiyyah 2023). Again, some of these pre-made templates may be more or less applicable to the brand, and the vision for marketing within the organization. Figure 16 shows the method of building Templates for the Marketing Framework for any firm.

Instead of giving another template to pore over, this method provides a set of steps that can help the firms to create their own template for the marketing framework (see figure 16).

Figure 16. Marketing framework template
(Samarasinghe and Medis 2020)

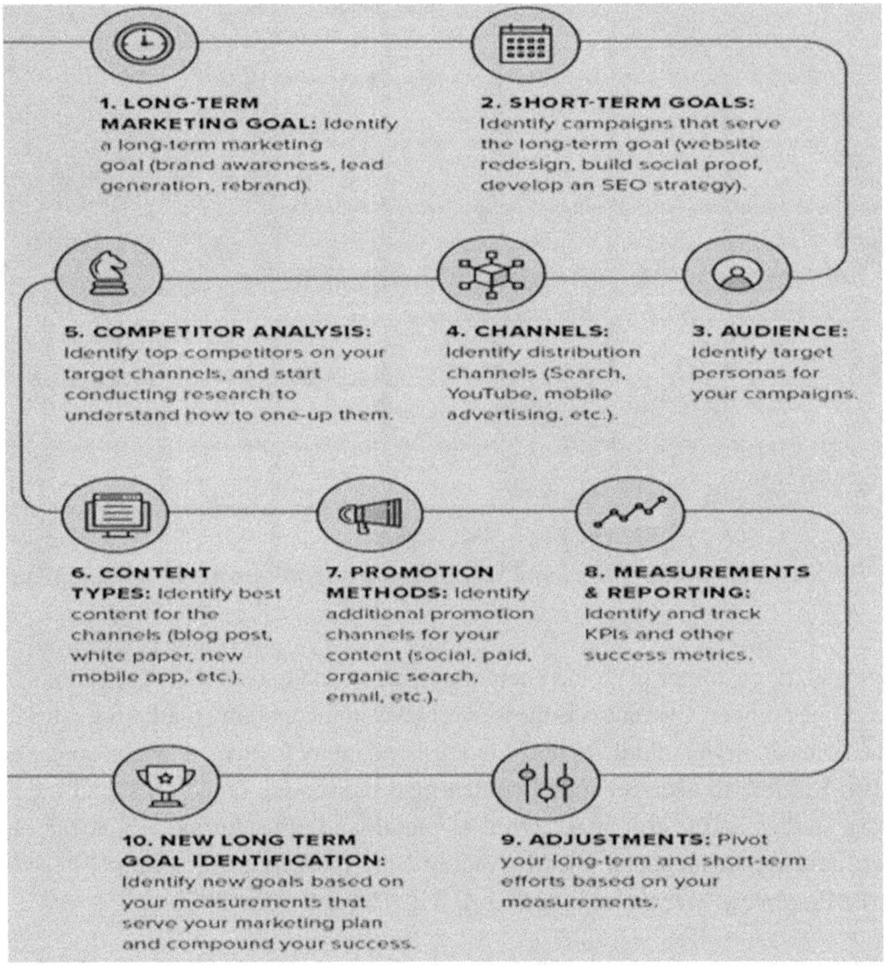

A firm may need a marketing template to help the business plan or the market problems. Instead, firms might just want to use the right side of the Pragmatic Marketing Framework given in the figure 17. Firms have many options to choose the framework as per their requirements and objectives and none of the framework can be said to be the perfect for all firms.

Figure 17. Right side of the pragmatic marketing framework
(Shen et al. 2021)

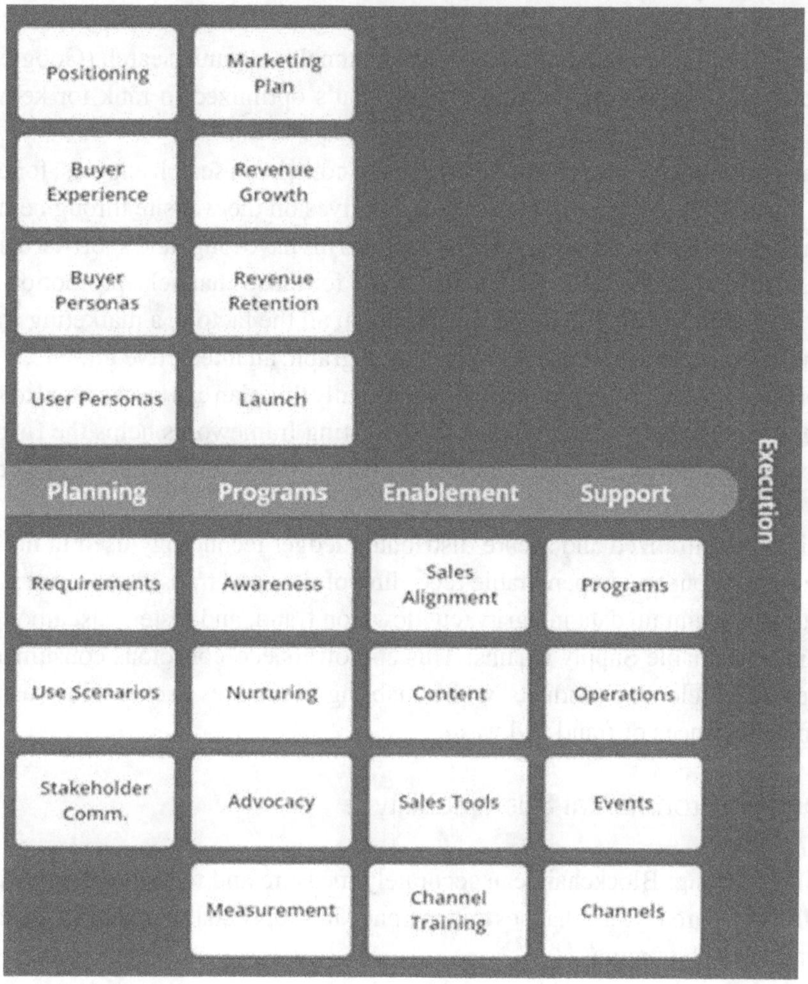

The best, most helpful marketing framework for the business depends on factors like your organizational structure, resources, market, whether its e-commerce or traditional commerce. Decision on choosing and applying marketing framework depends on following factors (Yeğin and Ikram 2022):

Identify Long- and Short-Term Marketing Goals: The long-term goal is to sell magic beans. The short-term goals within that might be to build awareness of the magic beans, or to increase the number of people who request a demonstration. Make sure to identify clear metrics to use as KPIs and to measure ROI.

Pinpoint Your Buyers and Influencers: Buyers are decision-makers. Influencers influence decision-makers. The head of accounting may be the one who signs off on the magic beans, but the team members might be the ones who have the most to gain from them (Gheorghe et al. 2023).

Create Detailed Personas for Those Buyers and Influencers: Personas will help you figure out more about who these individuals are, what they value, the technology they use and the types of messaging that resonates with them (Naim 2021).

Content distribution channels should be selected based on audience personas, but also on the firm's goals. Pick the marketing channels that will get the right content in front of the right audience at the right time.

Content should be tailored to that channel. If the channel is organic search (Google) the firms need blog posts, landing pages and other written content that's optimized to rank for keywords related to magic beans (Naim 2021).

Search engine optimization is one way to promote the content on search engines, for example. Google Ads is another. A third is to cross-promote content that lives on the website through email, and through the social media channels that your target personas use. Firms have long- and short-term marketing goals, target audience, distribution channels, and content types for those channels, promotion method, desired action and metrics (Vrontis et al. 2022). Finally, based on all the factors, a marketing framework can be substituted in the template to use a table, a flow chart, a graph, an interactive graphic, a story board, etc. that visually represents the marketing framework. Eventually this plan can be executed to sell magic beans.

The description and understanding of various marketing frameworks helps the firms to choose and apply the particular framework or the combination of framework for their business practices (Shen et al. 2021).

Blockchain is a decentralized and secure distributed ledger technology used in network marketing. It allows for the transparent and impenetrable recording of data and transactions over a network of connected nodes. It helps maintain data integrity, cut down on fraud, and foster trust among network users.

Blockchain in Sustainable Supply Chains: This empowers eco-conscious consumers to make well-informed choices when selecting products, while enabling businesses to champion sustainability objectives and minimize instances of fraud and waste.

Blockchain and Environmental Sustainability

Carbon Footprint Tracking: Blockchain can accurately measure and record carbon emissions data.

Supply Chain Transparency: It enhances transparency in supply chains, enabling sustainable practices like fair trade and ethical sourcing.

Blockchain can be used in renewable energy tracking by recording the production and consumption of renewable energy on a blockchain network. This promotes the use of renewable energy sources and reduces the carbon footprint of businesses (Yeğin and Ikram 2022).

Blockchain is not just another technology innovation or a general-purpose technology; it is a new form of institutional technology that could result in a radical change in organizations and subsequently drive them towards decentralized management. In this sense, blockchain offers a new model of economic coordination and governance in the pursuit of each stakeholder's interests and well-being (Vrontis et al. 2022). For this reason, there is huge potential for applying blockchain to sustainability strategy, which will transform from a firm's transient competitive advantage to sustainable competitive advantages, as well as enable a firm's decentralized sustainable management. In this regard, a transparent and real-time form of blockchain governance could be orchestrated to attract multistakeholder to commit to decentralized management to gain more benefits from a sustainability strategy (Yeğin and Ikram 2022).

Blockchain as well fundamentally changes the mechanisms enabling trust in exchanges. Remarkably, blockchain itself represents a guardian of trust by changing how we trust in buyer–seller relationships as compared to a traditional exchange (Whig et al. 2022). While performing business transactions on the blockchain-enabled exchanges, we put our trust in blockchain by relying on mathematical and cryptography-driven governance and network, information transparency, immutable data, and digital escrow for verifying ownership of valuable goods.

Corporate Sustainability Strategy and Blockchain

In a recent survey, more than 50% of investors have demanded public listed firms submit sustainability reports for improved transparency on firms' social and environmental impacts, rather than in demands for actions affecting sustainability decisions or operations (Whig et al. 2022). In the same vein, sustainability reports are a legitimized tool that corporations utilize to communicate sustainability efforts and their impact on the environment, ethical, philanthropic, and economic. Thus, sustainability reporting is mainly relevant to supporting investors' risk management purposes.

CONCLUSION

This research paper explains various marketing models and they aid in developing marketing framework to achieve environmental sustainability. The second part of the paper explains the blockchain marketing and how it uses the marketing models to plan and develop marketing framework to produce echo friendly product and contribute in environmental sustainability.

REFERENCES

Alahmari, F., Naim, A., & Alqahtani, H. (2023). E-Learning Modeling Technique and Convolution Neural Networks in Online Education. In IoT-enabled Convolutional Neural Networks: Techniques and Applications (pp. 261-295). River Publishers.

Alqahtani, H., & Naim, A. (2022). Critical Success Factors for Transforming CRM to SCRM for building E-CRM. In A. Naim & S. Kautish (Eds.), *Building a Brand Image Through Electronic Customer Relationship Management* (pp. 139–168). IGI Global. doi:10.4018/978-1-6684-5386-5.ch007

Anayat, S. (2023). Human Resources Management After Industry 4.0: Blending AI and HRM. In Strategic Human Resource Management in the Hospitality Industry: A Digitalized Economic Paradigm (pp. 97-115). IGI Global.

Chou, S. F., Horng, J. S., Liu, C. H., Yu, T. Y., & Kuo, Y. T. (2022). Identifying the critical factors for sustainable marketing in the catering: The influence of big data applications, marketing innovation, and technology acceptance model factors. *Journal of Hospitality and Tourism Management*, *51*, 11–21. doi:10.1016/j.jhtm.2022.02.010

Dash, G., Sharma, C., & Sharma, S. (2023). Sustainable marketing and the role of social media: An experimental study using natural language processing (NLP). *Sustainability (Basel)*, *15*(6), 5443. doi:10.3390/su15065443

Fatima, S., Alqahtani, H., Naim, A., & Alma'alwi, F. (2022). E-CRM Through Social Media Marketing Activities for Brand Awareness, Brand Image, and Brand Loyalty. In A. Naim & S. Kautish (Eds.), *Building a Brand Image Through Electronic Customer Relationship Management* (pp. 109–138). IGI Global. doi:10.4018/978-1-6684-5386-5.ch006

Fu, S., Yang, F., & Xiao, Y. (2020). AI inspired intelligent resource management in future wireless network. *IEEE Access : Practical Innovations, Open Solutions*, *8*, 22425–22433. doi:10.1109/ACCESS.2020.2968554

Gheorghe, G., Tudorache, P., & Roşca, I. M. (2023). The Contribution of Green Marketing in the Development of a Sustainable Destination through Advanced Clustering Methods. *Sustainability (Basel)*, *15*(18), 13691. doi:10.3390/su151813691

Hemanand, D., Mishra, N., Premalatha, G., Mavaluru, D., Vajpayee, A., Kushwaha, S., & Sahile, K. (2022). Applications of intelligent model to analyze the green finance for environmental development in the context of artificial intelligence. *Computational Intelligence and Neuroscience*, *2022*, 2022. doi:10.1155/2022/2977824 PMID:35845917

Hermann, E. (2022). Leveraging artificial intelligence in marketing for social good—An ethical perspective. *Journal of Business Ethics*, *179*(1), 43–61. doi:10.1007/s10551-021-04843-y PMID:34054170

Hermann, E. (2022). Leveraging artificial intelligence in marketing for social good—An ethical perspective. *Journal of Business Ethics*, *179*(1), 43–61. doi:10.1007/s10551-021-04843-y PMID:34054170

Hoda, N., & Naim, A. (Eds.). (2023). *Social Capital in the Age of Online Networking: Genesis, Manifestations, and Implications: Genesis, Manifestations, and Implications*. IGI Global.

Kamal, S., Naim, A., Magd, H., Khan, S. A., & Khan, F. M. (2022). The Relationship Between E-Service Quality, Ease of Use, and E-CRM Performance Referred by Brand Image. In A. Naim & S. Kautish (Eds.), *Building a Brand Image Through Electronic Customer Relationship Management* (pp. 84–108). IGI Global., doi:10.4018/978-1-6684-5386-5.ch005

Khatri, B., Shrimali, H., Khan, S. A., & Naim, A. (2023). Role of HR Analytics in Ensuring Psychological Wellbeing and Job Security: Learnings from COVID-19. In HR Analytics in an Era of Rapid Automation (pp. 36-53). IGI Global.

Malik, P. K., Naim, A., & Singh, R. (Eds.). (2022). *Printed Antennas: Design and Challenges*. CRC Press. doi:10.1201/9781003347057

Naim, A. (2021). Green Information Technologies in Business Operations. Periodica Journal of Modern Philosophy. *Social Sciences and Humanities*, *1*, 36–49.

Naim, A. (2021). Applications of MIS in building Electronic Relationship with customers: A case-based study. Periodica Journal of Modern Philosophy. *Social Sciences and Humanities*, *1*, 1–8.

Naim, A. (2021). New Trends in Business Process Management: Applications of Green Information Technologies. *British Journal of Environmental Studies*, *1*(1), 12–23.

A Naim. (2022). Applications of E-Learning Tools for Achieving Students Learning Outcomes for Business Programs. *Competitive Trends and Technologies in Business Management*, 125. https://. doi:10.52305/VIXO9830

Naim, A. (2022). Successful E-CRM by the Application of Management Information Systems, Business Issues. *Competition and Entrepreneurship*, *19*. doi:10.52305/VIXO9830

Naim, A. (2023). Application of HPC techniques to identify the business financial risks. *Academy of Marketing Studies Journal*, *27*(S5), 1–11.

Naim, A., Alahmari, F., & Rahim, A. (2021). Role of Artificial Intelligence in Market Development and Vehicular Communication. Smart Antennas. *Recent Trends in Design and Applications*, *2*, 28–39. doi: 10.2174/9781681088594121020006

Naim, A., & Alqahtani, H. (2023). Role of IoT in Industrial Transformation and Marketing Management. In *Global Applications of the Internet of Things in Digital Marketing* (pp. 67–93). IGI Global. doi:10.4018/978-1-6684-8166-0.ch005

Naim, A., Alqahtani, H., Fatima, S., & Khan, M. F. (2023). Applications of Artificial Intelligence of Things in Green Supply Chain Management: Challenges and Future Directions. In Blockchain, IoT, and AI Technologies for Supply Chain Management (pp. 99-126). CRC Press.

Naim, A., Alqahtani, H., Muniasamy, A., Bilfaqih, S. M., Mahveen, R., & Mahjabeen, R. (2023). Applications of Information Systems and Data Security in Marketing Management. In A. Naim, P. Malik, & F. Zaidi (Eds.), *Fraud Prevention, Confidentiality, and Data Security for Modern Businesses* (pp. 57–83). IGI Global. doi:10.4018/978-1-6684-6581-3.ch003

Naim, A., & Alqahtani, K. (2021). Role of Information Systems in Customer Relationship Management. *Pulse*, *2*(2).

Naim, A., & Alqahtani, K. (2022). Applications of Decision Support Systems in CRM, Business Issues. *Competition and Entrepreneurship*, *1*. Advance online publication. doi:10.52305/VIXO9830

Naim, A., Alshawaf, S. M., Kumar Malik, P., & Singh, R. (2023). Effective E-Learning Practices by HPC and Artificial Intelligence. *2023 International Conference on Artificial Intelligence and Smart Communication (AISC)*, Greater Noida, India. 10.1109/AISC56616.2023.10085391

Naim, A., & Hassan, S. Z. (2022). Economies of Scale for Security Applications in Business, Business Issues. *Competition and Entrepreneurship*, *73*. doi:10.52305/VIXO9830

Naim, A., & Kautish, S. K. (Eds.). (2022). *Building a Brand Image Through Electronic Customer Relationship Management*. IGI Global., doi:10.4018/978-1-6684-5386-5

Naim, A., & Khan, S. A. (2023). Impact and Assessment of Electronic Commerce on Consumer Buying Behaviour. In Global Applications of the Internet of Things in Digital Marketing (pp. 264-289). IGI Global. doi:10.4018/978-1-6684-8166-0.ch014

Naim, A., Malik, P. K., & Zaidi, F. A. (Eds.). (2023). *Fraud Prevention, Confidentiality, and Data Security for Modern Businesses*. IGI Global. doi:10.4018/978-1-6684-6581-3

Naim, A., Muniasamy, A., & Alqahtani, H. (2022). Application of Internet of Things in Marketing Management. *Ambient Intelligence and Internet of Things: Convergent Technologies*, 273-299.

Naim, A., Muniasamy, A., Clementking, A., & Rajkumar, R. (2022). Relevance of Green Manufacturing and IoT in Industrial Transformation and Marketing Management. In M. Lahby, A. Al-Fuqaha, & Y. Maleh (Eds.), *Computational Intelligence Techniques for Green Smart Cities. Green Energy and Technology*. Springer. doi:10.1007/978-3-030-96429-0_19

Naim, A., Syed, A., Bilfaqih, S. M., & Muniasamy, A. (2023). Impact of Individual and Organizational Social Capital of Marketing in a Global Network. In Social Capital in the Age of Online Networking: Genesis, Manifestations, and Implications (pp. 1-17). IGI Global. doi:10.4018/978-1-6684-8953-6.ch001

Nghiêm-Phú, B., & Suter, J. R. (2022). The Application of Anthropomorphism in Marketing: Implications for Green Economy in a digital world. In *Handbook of Research on Building Greener Economics and Adopting Digital Tools in the Era of Climate Change* (pp. 13–42). IGI Global. doi:10.4018/978-1-6684-4610-2.ch002

Nozari, H., Szmelter-Jarosz, A., & Ghahremani-Nahr, J. (2021). The Ideas of Sustainable and Green Marketing Based on the Internet of Everything—The Case of the Dairy Industry. *Future Internet*, *13*(10), 266. doi:10.3390/fi13100266

Rathore, B. (2019). From Trendy to Green: Exploring AI's Role in Sustainable Fashion Marketing. *International Journal of New Media Studies: International Peer Reviewed Scholarly Indexed Journal*, *6*(2), 12–22. doi:10.58972/eiprmj.v6i2y19.120

Saini, H., & Tarkar, P. (2022). Artificial intelligence in human resource practices with challenges and future directions. In *Handbook of Research on Innovative Management Using AI in Industry 5.0* (pp. 222–230). IGI Global. doi:10.4018/978-1-7998-8497-2.ch015

Suryani, W., & Luthfiyyah, A. (2023). Artificial Intelligence (AI) Opportunities and Challenges of Green Marketing Tourism. 5G. *Artificial Intelligence, and Next Generation Internet of Things: Digital Innovation for Green and Sustainable Economies*, 178-188.

Yeğin, T., & Ikram, M. (2022). Performance evaluation of green furniture brands in the marketing 4.0 period: An integrated MCDM approach. *Sustainability (Basel)*, *14*(17), 10644. doi:10.3390/su141710644

Chapter 7
Internet of Things Framework for Sustainable Agriculture in Smart Farming

Pratima Bais
Dr. C.V. Raman University, India

Preeti Singh Bahadur
Amity University, India

ABSTRACT

The global human population is constantly increasing whereas the agricultural land resources are declining for reasons of urban settlements and industrial growth. This has certainly strained the agricultural sector for increased annual yields, especially in the changing climate. To accomplish the rising demand of consumers for food, feed, and fiber, up to 70% upsurge in crop production is farsighted by 2050. The agricultural industry is getting more data-centric and requires precise, more advanced data and technologies than before, despite being familiar with agricultural processes. The agriculture industry is being advanced by various information and advanced communication technologies, such as the internet of things (IoT) and cloud computing are anticipated to inspire growth and initiate the use of robots and artificial intelligence in farming. The rapid emergence of these advanced technologies has restructured almost all other industries, as well as advanced agriculture, which has shifted the industry from a statistical approach to a quantitative one.

INTRODUCTION

It is estimated that by 2050, there will be 9.5 billion people on the planet. Given this steady population growth, it will be necessary to raise agricultural productivity by up to 70% in order to meet the need for food, feed, and fibre (Ehrlich et al., 2015). Over 35% of the world's population is currently experiencing chronic hunger and malnutrition. Reduced annual yields are the result of both the ever-changing environment and the stress conditions that crop plants are under. Global crop productivity is being negatively

DOI: 10.4018/979-8-3693-1794-5.ch007

impacted by ongoing climate change mostly due to rising greenhouse gas emissions, pollution releases into water and soil, and overuse of natural resources. Urbanisation and the expansion of industry are the main causes of the ongoing degradation of the land and water resources used for agriculture. Extreme heat waves, erratic rainfall patterns, and the current drought are mostly caused by the extraordinary rise in global temperatures. World yields of pulses and grains have been observed to decrease by up to 7% in regions with a unit increase in average world temperature (Zhao et al., 2017). Global food security is at risk due to the extensive impacts of climate change, which are reportedly disrupting agricultural productivity (Arora, 2019). The Food and Agriculture Organisation (www.fao.org) projects that, at the current rate of climate change, worldwide production of the main cereal crops will drop by up to 45% by the turn of the century. Therefore, a large increase in the rate of agricultural production is required to feed the world's hungry. Because of this, modern agriculture is severely limited in its ability to produce bioenergy, support accompanying commerce, and provide readily available resources to aid in climate change adaptation. Innovative technologies are undoubtedly helping global agriculture to meet consumer demands.

The resilience and sustainability of food grains produced in an environmentally friendly way are measured by sustainable agriculture (Srisruthi et al., 2016). Encouragement of farming methods and techniques that support the sustainability of farmers and resources is made possible by sustainable agriculture. In addition to being economically viable, it guarantees a natural and healthy environment, preserves soil quality, lessens soil degradation, conserves water, and increases land biodiversity. Reducing greenhouse gas emissions, stopping the loss of biodiversity, and protecting natural resources are all made possible by sustainable agriculture (Obaisi et al., 2022).

Sustainable agriculture farming is a way to increase farming productivity while protecting the environment and the basic needs of future generations. Crop rotation, preventing nutrient deficiencies in crops, managing pests and diseases, recycling, and water harvesting are the fundamental achievements of smart farming in terms of sustainable agriculture, all of which contribute to a safer environment. Because of pollution from waste emissions, pesticide and fertiliser use, decaying dead plants, and other factors, living things rely on biodiversity. Figure 1 illustrates the need for a better environment for living things due to the impact of greenhouse gas emissions on humans, animals, plants, and the environment.

Figure 1. Factors of sustainable agriculture

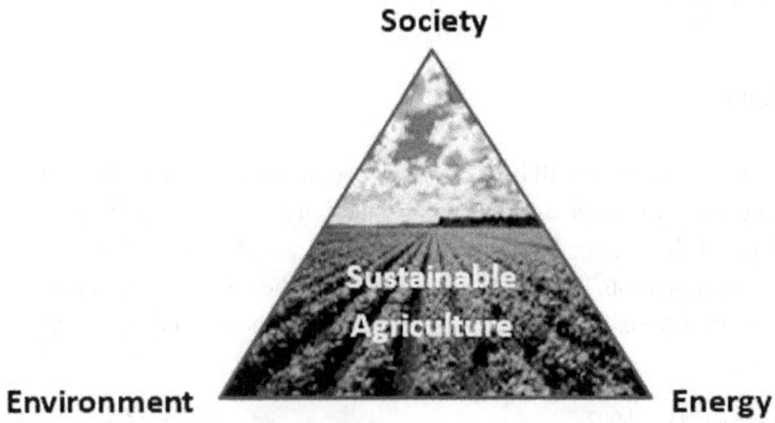

With an 18% GDP contribution and 57% of the population living in rural areas, agriculture is India's largest industry. The number of growers in India decreased from 71.9% in 1951 to 45.1% in 2011, despite an increase in the country's overall agronomic output over time (Reddy et al., 2018, Bais, P et al., 2023, Ekka et.al., 2023, Bais 2018). According to the 2018 Economic Survey, in 2050, the percentage of workers in agriculture will make up just 25.7% of the workforce overall. The next generation of farmers in rural areas gradually disappears from farming families due to factors such as increased cultivation costs, low productivity per capita, poor soil management, and migration to non-farming or higher paying occupations. Since the world is about to experience a digital revolution, now is the right moment to introduce and facilitate digital connectivity with farmers by using wireless technology to connect the agricultural landform.

Unfortunately, a number of factors, including soil quality, topography, temperature, climate, and the fact that most relevant cultivable areas are not homogenous, mean that not all areas of the Earth's surface are suitable for agriculture (Coulter 2004). The availability of arable land is continuously under pressure due to the fragmentation of current farming land caused by political, fiscal, and urbanisation factors, as illustrated in Figure 2. The total amount of agricultural land used to produce food has decreased recently. Additionally, each crop field has unique critical features, including the type of soil, irrigation flow, nutrient availability, and pest resistance, all of which are measured independently in terms of both quantity and quality with respect to a particular crop. Crop rotation and an annual crop growth development cycle require both spatial and temporal differences to maximise crop production in the same field (Hernández-Ochoa et al., 2022).

Figure 2. Key issues of technology in the agriculture industry

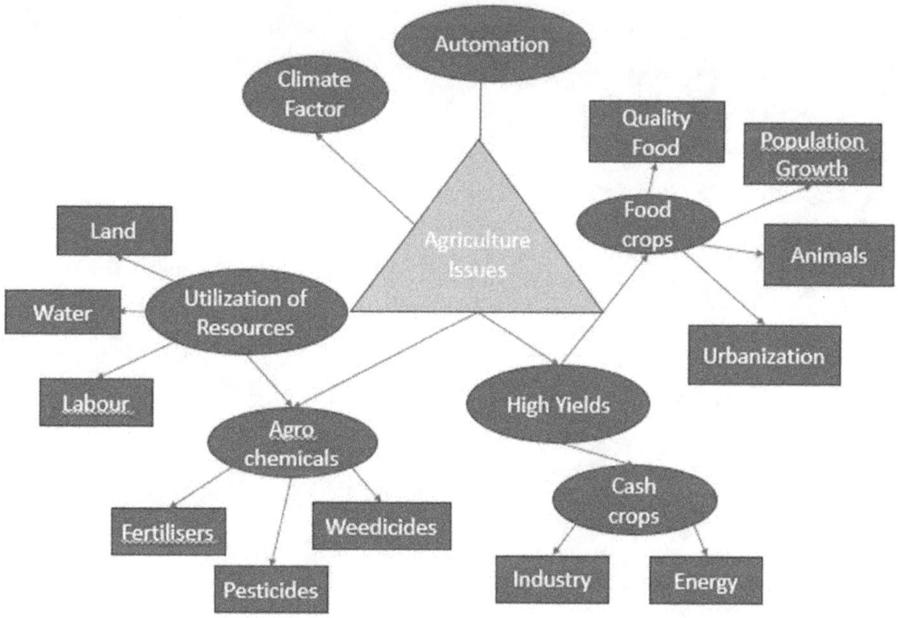

Most of the time, a single crop will have different characteristics from one another, or the same crop may be grown throughout the farm, necessitating site-specific analyses to produce the highest yields.

To solve these different problems and produce more on less land, new technology-based methods are required. In conventional farming methods, farmers visit their fields regularly during the crop's life to better understand the crop's conditions (Navuluret al., 2017). Present-day sensor and communication technologies provide farmers with an accurate field view, enabling them to identify ongoing field operations without physically being in the field. From the first sowing to the harvest of crops, wireless sensors monitor the crops more accurately and identify problems early on, which frequently makes the use of smart tools easier (Ayaz et al., 2018).

Because of accurate monitoring brought about by the timely use of sensors, farming as a whole has become smart and economical. The numerous robotic weed eaters, drones, and autonomous harvesters are equipped with sensors that gather data at brief intervals. But given the size of agriculture, technological solutions for sustainability with the least amount of negative ecological effects must meet extremely high demands. Wireless communication enables farmers to use sensor technology to remotely monitor crops' needs and requirements without having to physically be present in the fields. This allows farmers to take appropriate action from a distance (Lin et al., 2017).

SMART FARMING

The traditional agricultural era 1.0 refers to historical agricultural practises that were centred around the cultivation of cultivated lands to produce food for human survival and animal reproduction. Animals and human labour were mostly used in this. Simple implements like sickles and shovels were employed in farming operations. As a result, productivity remained low because the majority of work was done by hand (Figure 3).

Figure 3. Agricultural decision support system framework

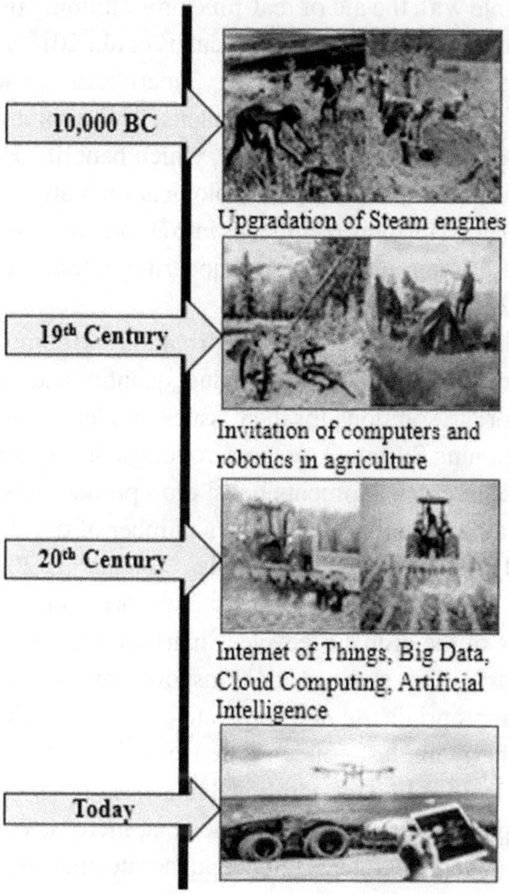

Steam engines were among the new machinery types that emerged in the agricultural industries during the 19th century. The agricultural era 2.0 was ushered in by farmers' widespread use of agricultural machinery and an abundance of chemicals, which also increased farms' and farmers' overall productivity and efficacy. But at the same time, extremely detrimental effects like chemical pollution, environmental destruction, resource waste, and overuse of energy emerged.

The 20th century saw the emergence of the agricultural era 3.0 as a result of the quick development of electronics and computation. Agricultural machinery with programming, robotics, and other technologies improved agricultural processes effectively. Agricultural era 2.0 problems were resolved, and policies were updated for agricultural era 3.0 with the help of effective pest control methods, site-specific nutrient application, precise irrigation, decreased chemical use, and labour distribution.

The agricultural era 4.0, which involves the use of modern technologies like the Internet of Things, big data analysis, artificial intelligence, cloud computing, and remote sensing, is the next iteration of agriculture. By creating low-cost sensor and network platforms with the goal of optimising production efficiency and minimising the use of energy and water resources while minimising negative environmental effects, the adoption of new technologies has greatly enhanced agricultural activities (Ferrández-Pastor et al., 2016). Extrapolative overviews of current agricultural conditions are made possible by big data in smart farming, which empowers farmers to make wise decisions (Wolfert et al., 2017). Farmers can make the best decisions possible with the aid of real-time programming that is embedded in IoT devices and developed using artificial intelligence concepts (Liakos et al., 2018).

With the use of cutting-edge, sophisticated technology, smart farming encourages precision agriculture and gives farmers the ability to remotely monitor the plants. The automation of sensors and machinery has increased the efficiency of the farming workforce, which benefits agricultural processes like crop yields and harvesting (O'Grady et al., 2017). A technological revolution in agriculture results from the technologies' conversion of manual farming techniques into automated machinery. The Internet of Things has revolutionised traditional farming methods, and modern agricultural technology has changed how farming is done (Quy et al., 2022).

Smart farming, a new modern technique utilising information and communication technologies (ICT), optimises farm labour requirements and increases product quantity and quality. To determine the needs of farmers and choose appropriate solutions for their issues, modern ICT technologies like the Internet of Things, GPS (Global Positioning Systems), sensors, robotics, drones, precision equipment, actuators, and data analytics are used. These developments boost crop productivity and decision-making timeliness and accuracy. To boost agricultural productivity, a number of developing nations and multilateral organisations have suggested smart farming technologies (Goel et al., 2021).

With increased accuracy, sensors are continuously monitoring crops to identify any unfavourable conditions in the early stages of the crop's lifecycle. Smart tools are used in farming today for everything from crop sowing to harvesting, storing, and transportation. Because of the accurate monitoring capabilities provided by the appropriate use of a wide range of sensors, the entire operation is now more profitable and efficient. Furthermore, fast-collecting sensors offer crop and site-specific agriculture for each site and are readily available online for additional analysis.

In order to improve the spatial management practises that increase crop production and prevent the overuse of fertilisers and pesticides, smart agriculture and monitoring are used to address the many issues related to crop production. This includes changes in soil characteristics, climate factors, soil moisture, etc (El Nahry et al., 2010). In smart irrigation water management (SIWM), artificial neural networks (ANNs) control irrigation scheduling support systems (DSS) and provide real-time data on irrigation efficiency, water productivity index, and irrigation water supply and demand. Climate-smart agriculture (CSA) is a relatively new technology that has the potential to reduce greenhouse gas emissions, increase farm system resilience, and improve food security, particularly in developing nations. IoT-based "smart agriculture" technology offers numerous benefits for all real-time agricultural processes and practises, such as fertilisation, irrigation, plant protection, enhanced product quality, disease prediction, and more (Adamides et al., 2020). The advantages of smart agriculture include the ability to gather data on crops in real time, evaluate soil and crops precisely, monitor crops remotely, manage water and other natural resources, and enhance livestock and agricultural output. As a result, smart agriculture is defined as precision agriculture advanced through modernization and clever ways to gather different farm activity data that are then remotely managed and supported by appropriate real-time farm maintenance alternatives.

INTERNET OF THINGS

A new technology called the Internet of Things (IoT) enables remote device connections for smart farming (Patil et al., 2016). In order to improve efficiency and performance across all markets, the

Internet of Things has started to have an impact on a wide range of industries, including health, trade, communications, energy, and agriculture (Sisinni et al., 2018, Shi et al., 2019, Elijah et al., 2018). Applications currently in use offer insights into the impact of IoT and its as-yet-unobserved practises. But as technology advances, it becomes clear that Internet of Things (IoT) technologies play a significant role in many farming activities. These include using IoT technologies for data acquisition, data acquisition, smart objects, sensors, mobile devices, cloud-based intelligent information, decision-making, and the automation of agricultural operations, as illustrated in Figure 4.

Figure 4. Barriers in the implementation of smart agriculture technology

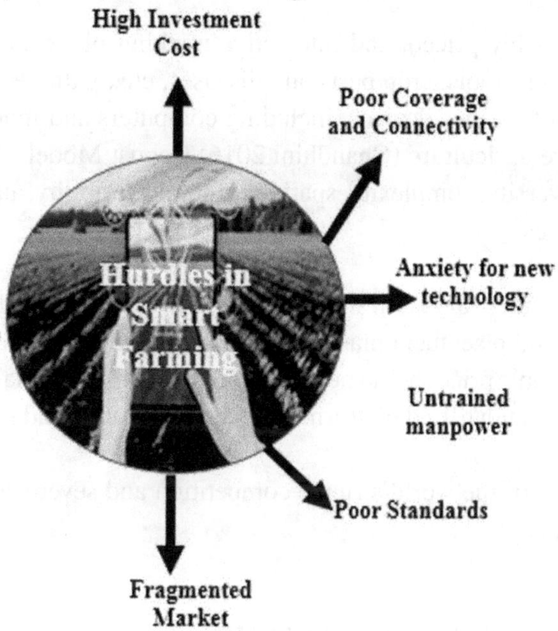

IoT technology remotely retrieves data from mobile phones and other devices and monitors plants and animals. Farmers are able to predict production levels and evaluate the weather thanks to sensors and instruments. More than ever, the Internet of Things contributes to water harvesting, flow amount monitoring and control, crop water requirements assessment, supply timing, and water conservation (Yong et al., 2018). According to the needs of the soil and plants, sensors and cloud connectivity via the gateway can remotely monitor the condition and water supply (Mekala et al., 2017). Farmers are unable to manually monitor and observe every plant to address nutrient deficiencies, pests, and diseases; however, IoT technology is still advantageous and has helped farmers reach a new milestone in modern agriculture (Mittal et al., 2007).

The farming industry has recently benefited greatly from the development of IoT technologies, especially in terms of communication infrastructure. Connecting smart objects, gathering data remotely, utilising cars and sensors via mobile devices and the internet, intelligent analysis on the cloud, interacting, making decisions, and automating agricultural processes are a few examples of this. The agriculture industry has undergone a revolution thanks to these proficiencies, which have improved crop yields, controlled the effects of climate change, and optimised resource use.

Based on a variety of crop and field types, researchers have proposed a variety of techniques, architectures, and equipment to monitor and convey crop information at different growth stages. To gather and then distribute data, numerous manufacturers offer communication devices, several sensors, robots, large machinery, and drones. The development of guidelines and policies for regulating the use of technologies to preserve food and environmental safety is done by food and agriculture organisations in collaboration with other government organisations (King et al., 2017).

Fundamentals of IoT Applications in Agriculture

The easily navigable, reasonably priced, and interactive tracking platform offers aggregated data on conventional farming practises, tools, crop pests and diseases, etc., gathered from multiple sources for sustainable farming. Users of various devices, including computers and mobile phones, can easily access data through interactive agriculture (Chandhini 2016).Robust Models: The distinctive features of the agriculture sector are diversity, complexity, spatio-temporal variability, and uncertainties of the right types of harvests and facilities.

1. Scalability: Since farm sizes vary from small to large, the outcomes ought to be scalable. Gradually increasing the testing and placement planning should be done at a lower cost.
2. Affordability: A reasonable price should be offered along with substantial support, as this is essential to successful farming. Standardised platforms, goods, equipment, and spaces could be acquired for an affordable price.
3. Sustainability: Because of the world's fierce competition and severe economic pressure, sustainability is a critical issue.

TECHNOLOGIES USED IN SMART FARMING

Global Positioning System (GPS)

Latitude, longitude, and elevation data are precisely recorded by GPS (Lang 1992). Satellites of the Global Positioning System send out signals that enable GPS receivers to calculate their current location in real time and provide continuous positions even when they are moving. Farmers can find the precise location of field data, including pest occurrence, soil type, weeds, and other obstacles, thanks to the precise location information. In order to apply the required inputs (seed, fertiliser, herbicide, pesticide, and water) to a specific field, the system makes it easier to identify different field locations (Batte 1999).

Sensor Technologies

Estimates of soil texture and structure, nutrient level, vegetation, humidity, vapour, air, temperature, and other parameters are made using techniques like photoelectricity, electromagnetics, conductivity, and ultrasonography. According to Chen et al. (2004), remote sensing data can identify different crop types, classify weeds and pests, identify stress in soil and plant conditions, and track drought.

Numerous elements, including soil moisture, nutrient availability, light exposure, humidity, rainfall totals, leaf colour, and more, all have an impact on the health of plants. By keeping the plants at the ideal temperature and light level and using micro-irrigation to save water and energy, the plants are kept under observation. Numerous parameters are detected using various sensors. The sensor detects changes when they go beyond a threshold and sends the information to the microcontroller, which then takes the necessary steps to bring the parameter back to its ideal level.

Sensing technologies are typically used with temperature, humidity, soil pattern monitoring, airflow sensor, location, CO2, pressure, light, and moisture sensors. According to Farooq et al. (2020), prominent sensor characteristics that make them suitable for agriculture include reliability, memory, portability, durability, coverage, and computational efficiency. The wireless sensors that are currently on the market are essential for gathering information and monitoring crop conditions. These sensors are stand-alone types that, depending on the needs of the application, can be integrated with heavy machinery and sophisticated agricultural tools

Variable-Rate of Technology (VRT) and Grid Soil Sampling

In farming, variable-rate technologies (VRT) are used to forecast the rate of input delivery based on a pre-made map that is extrapolated from a geographic information system (GIS) to place inputs at variable amounts at the appropriate location and time. Soil is collected from a systematic grid to create a map for each parameter, which is known as grid soil sampling. These maps are loaded into a variable-rate application and serve as the foundation for VRT. Based on map features, the GPS receiver and computer direct and control changes in fertiliser product delivery amounts (Berntsen et al., 2006, Ferguson et al., 2002).

Variable rate technology and related techniques (grid soil sampling) have the potential to enhance soil fertility management and determine the spatial distribution of yields and nutrients. By overlaying the grid lines onto the field, a field is divided into small areas, or cells, from which samples are taken for the grid sampling technique. At the intersections of grid lines, composite samples (grid-point sampling) encompass the whole area of each much smaller area. By applying interpolation techniques from non-measured locations between sampled points, soil-test values from grid sampling are mapped. Field-specific variations in phosphorus and potassium levels necessitate customised fertilisation strategies to enhance nutrient management techniques. This can be achieved through consistent application of fertilisers and manure, leading to more precise farming practises (Mallarino et al., 1997).

Geographic Information System (GIS)

In order to create maps and analyse characters and geography for statistics and spatial methods, the GIS consists of hardware and software that are designed to provide compilation, storage, retrieval, attributes analysis, and location data (Ehlers et al., 2008). The GIS database establishes the relationship between elements that affect a crop on a specific farming field and provides information on field soil types, nutrient status, topography, irrigation, surface and subsurface drainage, quantity of chemical applications, and crop production (Ojo et al., 2018). In addition to storing and displaying data, the GIS is used to evaluate management strategies, both current and alternative, by combining and modifying data layers to support decision-making.

Crop Management

Data on soil variations and crop performance impacted by field topography can be obtained from satellite imagery. Consequently, farmers are able to precisely monitor the production factors—seeds, fertiliser, and pesticides—that lead to increased yields and efficiency.

At the regional level, the information is provided almost instantly thanks to the satellite images' temporal revisit frequency and spatial coverage. Particularly in red and near-infrared combinations (vegetation indices) to monitor green foliage, the spectral reflectance properties of vegetation predict the relationship between the spectral properties of crops and their biomass/yield experiments (TUCKER et al., 1980). The closely related leaf area index (LAI) and photosynthetic activity of green vegetation make the normalised difference vegetation index (NDVI) the most widely used indicator among the various indices to evaluate crop productivity and vegetation health. Crop monitoring techniques rely on interpreting indicators derived from remote sensing and comparing the current crop status to normal or previous seasons (Muthumanickam et al., 2011). In certain times leading up to harvest, the correlation between biomass and vegetation indices allows for an early estimate of crop yield (Rembold et al., 2013). The more complex functionalities made possible by automated field management include automated data acquisition, processing, monitoring, decision-making, and management of farm operations (Chowdhury et al., 2020). These include the fundamental tasks of crop production (yields), profits and losses, farm weather prediction, field mapping, and tracking soil nutrients.

Soil and Plant Sensors

Water status, fertility, and soil properties are all provided by sensor technology, a key component of precision agriculture. Thus, in order to differentiate themselves from the sensors that are currently on the market, new sensors have been developed based on desirable features (Adamchuk et al., 2004).

Plant wearables and soil sensors track physical and chemical signals in the soil in real time, including temperature, pH, moisture content, and pollutants. The data they provide helps farmers improve crop growth conditions, combat biotic and abiotic stresses, and boost yields. The most crucial nutrients for crop production are nitrogen (N), potassium (K), phosphorus (P), and soil organic matter (SOMs). The spatial variation of soil nitrogen, both subsurface and surface, is measured by NIR reflectance-based sensors (Sudduth et al., 1993). By measuring the soil's spectral reflectance in the infrared and visible wavelength regions, SOM is estimated based on ideal wavelengths (Daniel et al., 2004). NIR spectrophotometry technology is used to predict the amount of phosphorus and nitrogen in the soil (Kuang et al., 2013; Maleki et al., 2006; Lvova et al., 2017). Since soil apparent electrical conductivity (ECa) is sensitive to variations in salinity and texture, data is continuously gathered on the field surface by the soil ECa sensors. Optoelectronic, acoustic, impedance, and nanostructured biosensors are used to identify soil insects and pests (Potamitis et al., 2019).

Rate Controllers

By tracking the speed of vehicles travelling across the field and adjusting the material flow rate in real-time to reach the desired rate, rate controllers are intended to regulate the delivery rate of inputs. As standalone systems, rate controllers are frequently employed (Adamchuk et al., 2004).

Precision Irrigation in Pressurized Systems

In order to achieve higher water utilisation efficiency by crop, recent advancements in irrigation systems have brought in irrigation machines that are dedicated to motion control, GPS-based controllers, sensor technologies, and wireless communication. These machines monitor soil and climate conditions in addition to evaluating irrigation parameters, such as flow and pressure. These technologies have a lot of promise, but more development is needed before they can be made widely accessible (Sudduth et al., 1993).

Yield Monitor

The set of sensors and other parts that manage integration and interaction components—such as a computer, data storage device, and user interface—are called yield monitors. The force of mass or volume of grain flow is evaluated by the sensor to continuously measure yield. The idea behind the mass flow sensor was to measure the energy that rebounds after being struck by microwave energy beams. GPS receivers in yield monitors use location yield data to generate yield maps (Daniel et al., 2011).

There is a yield monitor for the harvester installed, linked to a mobile app that displays harvest data in real time and automatically uploads it to the internet. The app allows farmers to export more farm management data for analysis, and agronomists can generate and receive high-quality yield maps. When it comes to accurately predicting the quantity and quality of product produced, fruit growth is considered to be one of the most significant factors in the crop progressing phase in horticulture crops (Kuang et al., 2013). Colour photos are used to monitor fruit conditions in order to determine when to harvest, target the appropriate market, and estimate fruit maturity (Maleki et al., 2006). One method for tracking crop yields in real time over large areas is the use of satellite images; Sentinel-1A images, for instance, are used to map crop intensity and rice yield in Myanmar.

In order to design the crop yield estimation system, hardware and software components were used. Crop yield is estimated mathematically through a mobile application, based on an Android application for a Bluetooth terminal and a yield estimator software programme (Lvova et al., 2017). Based on spectral signatures, satellite-based crop yield predictions show that estimated yields are just as dependable as actual yields. Using machine learning and data assimilation from satellites into crop models, maize yield predictions were successfully executed in a variety of environments (Potamitis et al., 2019).

Software

Numerous tasks are performed by the software, including data processing, analysis, interpretation, mapping, display controller interface, and more. Software is most frequently used to create maps of soil characteristics and nutrient status, yield maps, inputs with variable rate applications maps, and overlaying various map types with sophisticated geostatistical features (Ferrández et al., 2018).

ROLE OF IOT IN ADVANCED FARMING PRACTICES

The yield of crops was higher when the new, sensor- and IoT-based agricultural techniques were implemented than when traditional agricultural methods were used. Improving the quantity and quality of produce is largely dependent on the use of new, highly developed sensor based technologies in controlled environments.

Greenhouse Farming and Protected Cultivation

One of the first applications of smart farming was popularised in the 19th century: growing plants in a controlled environment. In nations experiencing extreme weather, these practises intensified even more during the 20th century (Theopoulos et al., 2018). Indoor crops are less susceptible to environmental influences. Thus, through the use of sensors and communication devices, crops that were once grown traditionally under favourable conditions are now raised whenever and wherever they are needed. Numerous elements, including shed structures, materials for mitigating wind effects, aeration systems, the precision of monitoring parameters, decision support systems, etc., are necessary for the success of crop production in a controlled environment (Shamshiri et al., 2018). Precisely monitoring environmental parameters is one of the biggest challenges in greenhouses; therefore, multiple measurement points are needed to predict the various parameters for regulating and guaranteeing the local climate. Sensors are used in an Internet of Things (IoT)-based greenhouse to measure and track internal parameters like light, pressure, temperature, humidity, and light (Akkaş et al., 2017).

In addition to shielding the plants from pest and insect attacks, hailstorms, wind, and UV radiation, the smart greenhouse has assisted farmers in conducting agricultural tasks automatically and eliminating the need for manual inspection. Using sensors for air humidity, temperature, and light, hibiscus plants are grown at night with the necessary wavelength. A study found that the amount of water needed could be reduced by 70–80%, and the Internet of Things allows for direct communication between farmers and consumers, increasing farming's efficiency and profitability (Kodali et al., 2016). By monitoring and controlling multiple parameters, including humidity, mist, CO2 level, UV light intensity, pH and EC value, water nutrients solution level, temperature and pesticide amount, through sensors for more effective detection and diagnosis, the IoT-enabled automated system increased the productivity of rose plants grown in a greenhouse (Tripathy et al., 2021).

Hydroponics

Growing plants without soil is known as hydroponics, a branch of hydroculture that enhances the advantages of greenhouse farming. A balanced rate of application of dissolved nutrients in water to crop roots is made possible by hydroponic-based irrigation systems. As of right now, a variety of parameters can be detected by the systems and sensors that are currently in use (Sambo et al., 2019), and data analysis is done on a scheduled basis. For plants to grow, the precise measurement and monitoring of the nutrient content in solution is essential, taking its requirements into account. According to Rouphae et al. (2018), the wireless-sensor-based prototype (Yang et al., 2020) has provided a real-time solution for soilless cultivation by measuring the concentration of multiple nutrients and water levels.

Three main parts make up an automated smart hydroponics system that is integrated with the Internet of Things: input data, cloud server, and output data. These use real-time analysis of parameters like pH level, water, nutrient-rich water-based solution, room temperature, and humidity to monitor lettuce cultivation from any location via the internet (Lakshmanan et al., 2020). Plants are grown using the hydroponic system of the deep flow technique, which involves putting roots in layers of deep water and making sure that the plant nutrient solution is constantly circulated. Sensors built into Raspberry Pis collect data on plant growth elements like pH, temperature, humidity, and water level in the hydroponic reservoir. The data is processed and monitored automatically in real-time to guarantee appropriate water circulation (Usman et al., 2018).

Vertical Farming

Soil quality is being harmed by industrial farming practises more quickly than nature can restore it. Arable land has decreased and the burden on current water reservoirs has increased due to the concerning rate of erosion and the use of fresh water for agriculture (Pimentel et al., 2013). With vertical farming (VF), it is possible to maintain plants in a carefully regulated environment, thereby decreasing resource consumption and increasing production on demand. Depending on the number of stacks, only a portion of the ground surface is required. Comparing VF to traditional farming, it is also very effective in increasing yields and using less water (Benke et al., 2017). Since the measurement of carbon dioxide is the most important factor, nondispersive infrared (NDIR) CO2 sensors are essential for monitoring and managing the environment in vertical farms.

Phenotyping

A new method of crop engineering that connects plant genomics to agronomy and ecophysiology is called phenotyping. While the development of genetic and molecular tools is important for crop breeding, the lack of efficient technologies and techniques makes quantitative analysis of crop behaviors—like pathogen resistance, grain weight, etc.—inadequate. According to Tripodi et al. (2018), plant phenotyping is very helpful in this situation when examining the quantitative traits in charge of growth, resistance to different stresses, yield quality, and quantity. According to Paul et al. (2019), biostimulant screening and an understanding of their mode of action are described by sensing technologies and image-based phenotyping. IoT-based phenotyping aims to monitor the crop and associated trait measurements, providing resources for crop breeding and digital agriculture (Zhou et al., 2017). Genetic variations, phenotypes, and growth conditions are related to each other, as determined by trait analysis algorithms and modelling support.

THE ROLE OF THE ENGINEER IN SMART FARMING

When farmers use IoT-based agriculture, they encounter numerous challenges. Engineers must therefore create solutions for particular issues pertaining to smart farming methods. Smart farming is a creative way to mechanise agricultural engineering through methods other than conventional mechanisation. An engineering role involves the application and use of innovative technologies and methods for precision agricultural machinery (Bochtis et al., 2019). Different technology domains, including agricultural mechanisation, mechatronics, instrumentation, control systems, and knowledge in artificial and computational intelligence, are the sources of the concepts and synergy-based information (Terence et al., 2020). Precision agriculture has undergone a revolution thanks to big data, satellite, and aerial images. These new technologies balance productivity and environmental protection, increasing production efficiency. Engineering combines technical expertise with potent business acumen to function as a system integrator in both the public and private sectors (James et al., 2019).

Engineering simultaneously takes advantage of the benefits of digital transformation across the whole agri-food chain, from daily farming operations to assisting with sales, logistics, and farm asset maintenance. For instance, familiarity with the Cloud, RPA, blockchain, IoT, AI, mobile, precision farming, remote sensing, and advanced analytics are required (Bacco et al., 2019). Responses pertaining to cereals, viticulture, fruit, and vegetables, as well as soil and monitoring, are produced by the data gathered from the various types of machinery through sensors and other devices (Ahmed et al., 2020).

Reducing manual human intervention is another benefit of automating production processes with digital technologies and control systems. From the field to the finished product, the production process is managed through the planning, organising, and analysis of machine data. Through traceable systems that rely on radio-frequency signals, the collected data are kept in historical archives and correlated with one another to extract relevant information for products (Adebayo et al., 2018).

Purpose

The ultimate requirement of the user determines the purpose, which affects crop monitoring during the growth season. The Internet of Things uses sensors to solve their issues. For instance, water level and moisture monitoring sensors are included to prevent water waste because the end-user is a maize farmer who faces challenges primarily related to water usage and making sure a crop receives enough water.

Technology

Since sensors gather data and transmit it to the server, distance is a crucial factor in the technology selection process. Consequently, different distances cannot be served by the same technology. For instance, data could be sent over hundreds or thousands of metres using low power, wide area network (LPWAN) and radio frequency identification (RFID) or near field communication (NFC) technologies.

Power Requirements

Developing low-power applications is preferable because most IoT solutions are dispersed throughout a sizable farm. However, greater data transmission comes with a higher cost and power consumption, so designers must think about creating affordable IoT solutions for farmers. Typically, engineers create apps for less frequent data transmission and save money with tailored IoT-based farming solutions.

Data Frequency

The needs of the final user play a crucial role in determining how many sensors and data packets are needed. While a farmer may not always need information, developers create Internet of Things applications that operate continuously in real-time with high data frequencies.

Placement of Sensors

Sensors are placed in such a way that they provide optimal performance, even if the farm has all the essential sensors with proper placement.

BARRIERS TO IMPLEMENTING SMART FARMING TECHNOLOGIES

Technology adoption is a method with a certain level of heterogeneity factors that are affective (Foster et. al., 2010). Technology implemented in farming systems has provided accuracy, efficiency, and eased time pressures. Although smart farming increases the productivity of crops, there are still problems in adopting these technologies

Cost of Technology

Current technologies reduce the need for labour and accomplish tasks very quickly and accurately. As such, it is expected that in the not too distant future, machines will most likely take the place of human labour. It is not feasible, though, as poverty has been experienced in many nations where the labour force was the primary driver of the agricultural industry. The cost of implementing devices and technologies is very high, so when farmers explore for alternatives to conventional tools, they encounter financial challenges.

Lack of Financial Resources

Financial supporters could provide adequate loans to farmers if farmers did not get the anticipated yield, perhaps because unexpected calamities like drought, flood, pests, and diseases impacted the crops.

Literacy Status of Farmers

One of the biggest obstacles to the adoption of new technologies in developing nations is the educational attainment of farmers. To operate the tools, one must possess both technical and educational knowledge. A farmer's ability to process information and make decisions using smart farming technologies increases with education level (Feder et al., 1985), making it easier for farmers to use computers (Alvarez et al., 2006). Due to a lack of interest in learning new things or awareness of emerging technologies, the majority of farmers in developing countries are illiterate and unskilled (Kimiti et al., 2009). According to Khan et al., 2007, Khan et al., 2022), this is one of the reasons farmers prefer traditional farming to smart farming. Because they use generic icons based on conventional understanding, farmers have complained that usage is too complicated and occasionally that they are unable to recognise the icons used in a mobile application. To further emphasise the benefits of smart farming technologies, farmers must be digitally literate. At the same time, agri-tech companies should make sure farmers can readily comprehend the technology's limitations.

Lack of Integration Between the Systems

One area where smart farming technologies need to advance further is integration across systems, which can be achieved by incorporating tools for decision-making, production, and property management. The divide between information science and agriculture must be bridged via interdisciplinary group communication and academic exchange. Increasing user effectiveness has received more attention when developing information systems (Abdul et al., 2016). Data integration is necessary to produce information and knowledge since timely access to high-quality data is the cornerstone of better decision-making.

Telecommunications Infrastructure

Most farming is done in rural areas, where arable land works better for farming than contaminated land. However, data transmission is unreliable due to inadequate telecommunications infrastructure, particularly when using tablets and mobile phones. A real-time internet connection is necessary for smart farming in order to utilise information. Furthermore, in order for different operation control systems to function, like those for seed volume, fertiliser, and pesticides, a strong internet connection is necessary. Rural producers now have access to mobile internet thanks to the proliferation of mobile phones; however, input speed and signal quality are constrained.

Data Management

Farmers are having trouble managing and organising the sensor-derived data. Farmers do not know how to use the data that the weather stations are producing or how to transform it into a format that is more readily available. Its intricate systems cause inaccurate calculations in addition to acceptability and usability problems. More accessibility to data and information in productive systems is required from farmers, consultants, and other production process participants.

CURRENT CHALLENGES AND FUTURE EXPECTATIONS

The international community and the United Nations set the 2030 Agenda for Sustainable Development as their goal to eradicate hunger by that year. According to the World Health Organisation, there are currently food shortages affecting over 800 million people worldwide (Frón et al., 2019). Furthermore, as the world's population grows, so does the demand for nutritious food; as a result, food and cash crops may enhance crop productivity as a whole.

There are three main issues: (1) feeding 10 billion people; (2) restricting land expansion; and (3) lowering greenhouse gas emissions. These difficulties force us to reevaluate our approaches to a number of issues, including climate change, rural labour, declining arable land, and water scarcity. In addition to causing communities to shrink and populations to age, the decline in rural populations brought about by urbanisation necessitates that younger growers take on greater responsibility. The population imbalance and generational shift have additional effects on the labour force and productivity.

The geographic and ecological conditions are responsible for the continued reduction of arable land and the suitability of specific crops in particular regions. Unpredictable weather patterns exacerbate environmental problems like soil erosion, groundwater depletion, and drought, which have an impact on crop productivity. Furthermore, conventional farming practises have traditionally used pesticides and fertilisers to meet food demands; however, this only partially boosts food production, and careless chemical use degrades the environment. In developing countries, various problems facing the agricultural sector include no suitable crop selection, soil testing, efficient irrigation systems, weather forecasting, animal husbandry, etc. Technological advancements have proved beneficial in developed countries, both quantitively and qualitatively, but, in developing countries, 50% of the population is already engaged in agriculture.

It is anticipated that big data services and artificial intelligence will be integrated into agriculture in the future. From seeding to production forecasting, farm machinery and management will begin as a single unit as a result of the systems merging. A small number of important techniques and technologies are geared towards attaining sustainable agriculture in the future.

Communication

The ability of devices to communicate with one another is crucial for the implementation of IoT in agriculture (Tzounisa et al., 2017). While they offer connectivity services, the majority of telecom providers only make up a small portion of smart farming overall. Particularly in rural areas, cellular operators target growers and improve market facilities by providing new services. The viability of cellular technology lies in its service providers' assurance of its advantages, which include cost-effective, flexible, and luxurious two-way communication. Smartphone technology and mobile services give farmers in developing nations hope for increasing crop yields in the future. The low power wide area technology (LPWA) is anticipated to play a major role in smart farming agriculture, due to its improved facilities, efficient coverage, low power consumption, and cost economics. The cellular operators with robust IoT create significant returns by offering smart agriculture facilities in collaboration with LPWA technology.

Wireless Sensors and IoT

Placing wireless sensors around the field provides timely information on a real-time basis to farmers in order to make decisions and act in order to obtain higher crop yields. Wireless sensor networks (WSNs) with GPS technology update all information on crop growth and terrain features. Recently, digital images and signal processing offer additional competencies to WSN, and precisely ascertain crop quality and health. By identifying crop requirements at every stage to maximise their effectiveness, IoT technology can predictably streamline tasks. IoT will eventually be upgraded to fifth-generation (5G) cellular mobile communication technologies, enabling farmers to access real-time data anywhere, at any time. Based on this accomplishment, it is anticipated that by the end of 2022, about 29 billion IoT-based components will be in use in the agriculture industry. Moreover, by 2050, farms are predicted to produce 4.1 million data points every day (Henriksen et al., 2019).

Drones and Unarmed Vehicles

Drones are frequently used by farmers to monitor crop growth, spray water and nutrient solutions, as well as pesticides in challenging terrain and at varying crop heights. In comparison to conventional machinery, drones have demonstrated their value for spraying speed, area coverage, and precision. Technology has advanced to the point where drones can now carry a variety of sensors, and farmers can manage their land more effectively thanks to 3D cameras. Farmers now face a number of challenges with the use of UAVs in agriculture, chief among them being the integration of new technologies and their use in inclement weather. Other than drones, robotics in agriculture has also enhanced productivity due to higher yields achieved by spraying and weeding without human intervention. The seeding, transplanting, and fruit harvest/picking robots have recently added a new efficiency level to traditional methods.

The UAV technology in smart agriculture provides information on fertilization, irrigation, use of pesticides, plant growth monitoring, weed management, crop disease management, and field-level phenotyping to enhance cultivation practices. Using UAVs to monitor crop growth parameters, a novel 3D modelling technique was utilised to measure the height of maize and sorghum plants in the field. The average root mean square error (RMSE) of sorghum height with hand sampling field data was 0.33 m. In order to extract the leaf area index (LAI) from soybean plants, the UAV and 3D models were also restored. Destructive LAI measurements (R2 = 0.89) were correlated with the measured LAI predicted accuracy corresponding to the handheld device (R2 = 0.92).

Weed detection and management were assessed by integrating low-resolution multi-spectral high-resolution RGB images (Chang et. al., 2017) using the Random Forest (RF) technique in field-grown rice and sugar beet crops (Barrero et. al., 2018). Multi-spectral digital images obtained by UAVs are used for evaluating vegetation indices (VIs) and multi-temporal VIs to predict grain yield in wheat (Lottes et. al., 2017). In wheat crops, the indices—which include the spectral vegetation index (SVI), green area index (GAI), and normalised difference vegetation index (NDVI)—are assessed in order to quantify plant density (Su et al., 2018), monitor breeding (Hassan et al., 2019), detect plant stress brought on by yellow rust disease (Duan et al., 2017), and predict grain yield (Stroppiana et al., 2015) [156]. An algorithm to self-adjust UAV routes during chemical spraying in a crop field has been developed and evaluated in an effort to reduce the waste of pesticides and fertilisers. The use of pesticides in agriculture is critical for crop yields and the environment (Jin et al., 2017).The usage of pesticides in agriculture is crucial for crop yields and the environment, and efforts have been made to develop and evaluate an algorithm to self-adjust UAV routes during chemical spraying in a crop field to reduce the waste of pesticides and fertilizers (Jin et. al., 2017)..

Vertical Farming and Hydroponics

The shrinking of arable land and rapid urbanization results in greater pressure on the present resources (Al-Kodmany et. al., 2018), which causes hardships for food production with current agriculture practices. Vertical farming (VF) navigates land and water shortage challenges, and is highly suitable for adoption in nearby cities. Hydroponics plays a key role in lowering water requirements. Hydroponics, along with VF, increases available arable land without distressing forests and other natural habitats. The presence of advanced technologies, especially the IoT, makes the agriculture industry highly remunerative with a reduction in labor requirements and other resources, in addition to minimizing environmental impact.

Performance Analysis Using Machine Learning

Data analytics and machine learning concepts are applied to analyze the real-time data. In crop production, identifying the best genes is an important process that can be conducted using machine learning techniques. In agriculture, machine learning is used to envisage the best genes suited for crop production, especially for selecting seed varieties that are highly suitable to specific climate conditions and locations. Machine learning algorithms identify high demand products and currently unavailable products. Recent developments in machine learning and analytics allow farmers to correctly categorize their harvests before it is processes and delivered to customers.

The problems associated with farmers' decision-making, crops, animal research, land, food availability and security, weather and climate change, and weeds are resolved by machine learning (ML) in big data systems (Cravero et al., 2022). Numerous agricultural activities are supported by ML-based applications, including yield prediction for maize based on a deep memory model (Kaneko et al., 2019), rainfall intensity assessment using a binary classification model and logistic regression technique (Oswa 2019), and soil water content prediction using a short-term memory model with higher R2 than artificial neural networks (Zhang et al., 2018) using data parameters of rainfall, temperature, water diversion, evaporation, and time for the next 1, 2, and 7 days. As a result, the agricultural sector is increasing farmers' incomes, and so communities are further integrated into the agricultural value chain to reduce poverty and provide access to health care, education, and nutritious food for their families (Ouyang et. al., 2019).

Renewable Energy, Microgrids, and Smart Grids

The high energy consumption of smart farming can be attributed to data transmission, GPS use, and long-term sensor placement. In remote places, long-term power problems are typically resolved by utilising renewable energy sources. Distributed energy sources (DERs) interact with smart grids and microgrids. Energy is stored and used by combining heat systems and electricity in new developments in storage devices.

Globally, energy security is ensured by smart grid technology, which facilitates a seamless switch from conventional to smart energy systems. Power-strengthening systems combined with renewable energy sources have improved the transport sector and raised the use of bioenergy in the power sector in developing nations. These systems were able to identify numerous renewable energy sources through the use of smart technologies like energy storage devices, smart appliances, computational intelligence, and the Internet of Things. As an illustration, the smart grid offers Nepal a wide range of opportunities for power sector reform. By establishing smart microgrids and eventually connecting to the national grid, it helps to alleviate the problem of rural electricity (Bhattarai et al., 2022). The Dayalbagh renewable energy smart microgrid in India is a small-scale electricity system comprising distributed loads and renewable energy resources, acting as a single controllable entity in the grid. The smart microgrids are integrated into renewable resources and form building blocks of smart grids, especially for the dairy plant to produce various dairy products (Kedri 2014).

The mixed integer linear programme (MILP) systematically and efficiently managed energy consumption and subsequently lowered the cost, especially in residential areas, by scheduling the use of smart appliances and charging/discharging electric vehicles (EVs). The model generates its own energy from a microgrid containing solar panels and wind turbines, and forecasts wind speed and solar radiation for effective energy management. MILP-based energy planning sustains the effectiveness and productiveness of energy-efficient techniques (Aslam et. al., 2020).

CONCLUSION

In order to meet the demands of an expanding global population for food and the problems of declining arable land, more intelligent and effective crop production techniques are required. It is imperative that everyone understands food security in relation to sustainable agriculture. the development of new technologies to boost crop yield and promote farming as a respectable career choice for creative youth. This paper focused on how various farming technologies, especially the Internet of Things, can make agriculture more efficient and intelligent in order to meet future demands. Scholars and engineers are said to be guided by the industry's present problems as well as its prospects for the future. Therefore, it is crucial to manage every square inch of land with sustainable Internet of Things (IoT)-based sensors and communication technologies in order to maximise crop productivity.

REFERENCES

Adamchuk, V. I., Hummel, J. W., Morgan, M. T., & Upadhyaya, S. K. (2004). On-the-go soil sensors for precision agriculture. *Computers and Electronics in Agriculture*, *44*(1), 71–91. doi:10.1016/j.compag.2004.03.002

Adamchuk, V. I., Hummel, J. W., Morgan, M. T., & Upadhyaya, S. K. (2004). On-the-go soil sensors for precision agriculture. [Google Scholar] [CrossRef] [Green Version]. *Computers and Electronics in Agriculture*, *44*(1), 71–91. doi:10.1016/j.compag.2004.03.002

Adamides, G., Kalatzis, N., Stylianou, A., Marianos, N., Chatzipapadopoulos, F., Giannakopoulou, M., Papadavid, G., Vassiliou, V., & Neocleous, D. (2020). Smart Farming Techniques for Climate Change Adaptation in Cyprus. *Atmosphere (Basel)*, *11*(6), 557. doi:10.3390/atmos11060557

Adebayo, S., Ogunti, E. O., Akingbade, F. K., & Oladimeji, O. (2018). A review of decision support system using mobile applications in the provision of day-to-day information about farm status for improved crop yield. *Period. Eng. Nat. Sci.*, *6*(2), 89–99. doi:10.21533/pen.v6i2.183

Ahmed, A. T., El Gohary, F., Tzanakakis, V. A., & Angelakis, A. N. (2020). Egyptian and Greek Water Cultures and Hydro-Technologies in Ancient Times. *Sustainability (Basel)*, *12*(22), 9760. doi:10.3390/su12229760

Akkaş, M. A., & Sokullu, R. (2017). An IoT-based greenhouse monitoring system with Micaz motes. *Procedia Computer Science*, *113*, 603–608. doi:10.1016/j.procs.2017.08.300

Al-Kodmany, K. (2018). The Vertical Farm: A Review of Developments and Implications for the Vertical City. *Buildings*, *8*(2), 24. doi:10.3390/buildings8020024

Alvarez, J., & Peter, N. (2006). Adoption of computer based information systems: The case of dairy farmers in Canterbury, NZ, and Florida, Uruguay. *Computers and Electronics in Agriculture*, *50*(1), 48–60. doi:10.1016/j.compag.2005.08.013

Arora, N. K. (2019). Impact of climate change on agriculture production and its sustainable solutions. *Environmental Sustainability*, *2*(2), 95–96. doi:10.1007/s42398-019-00078-w

Aslam, S., Khalid, A., & Javaid, N. (2020). Towards Efficient Energy Management in Smart Grids Considering Microgrids with Day-ahead Energy Forecasting. *Electric Power Systems Research*, *182*, 106232. doi:10.1016/j.epsr.2020.106232

Ayaz, M., Ammad-uddin, M., Baig, I., & Aggoune, H. M. (2018). Wireless Sensor's Civil Applications, Prototypes, and Future Integration Possibilities: A Review. *IEEE Sensors Journal*, *18*(1), 4–30. doi:10.1109/JSEN.2017.2766364

Bacco, M., Barsocchi, P., Ferro, E., Gotta, A., & Ruggeri, M. (2019). The digitization of agriculture: A survey of research activities on smart farming. *Array (New York, N.Y.)*, *3–4*, 100009. doi:10.1016/j.array.2019.100009

Bais, P. (2019). Study of workers in small scale industries. *Reports on Economics and Finance*, *5*(1), 49–54. doi:10.12988/ref.2019.911

Bais, P., & Bahadur, P. S. (2023). Agriculture in Indian Economy and Contribution of Science and Technology. [AJAST]. *Asian Journal of Applied Science and Technology*, *7*(1), 167–176. doi:10.38177/ajast.2023.7116

Barrero, O., & Perdomo, S. A. (2018). RGB and multispectral UAV image fusion for Gramineae weed detection in rice fields. *Precision Agriculture*, *19*(5), 809–822. doi:10.1007/s11119-017-9558-x

Batte, M. T., & VanBuren, F. N. (1999). *Precision farming—Factor influencing productivity*. In Proceedings of the Northern Ohio Crops Day Meeting, Wood County, OH, USA.

Benke, K., & Tomkins, B. (2017). Future food-production systems: Vertical farming and controlled-environment agriculture. *Sustainability*, *13*(1), 13–26. doi:10.1080/15487733.2017.1394054

Berntsen, J., Thomsen, A., Schelde, K., Hansen, O. M., Knudsen, L., Broge, N., Hougaard, H., & Hørfarter, R. (2006). Algorithms for sensor-based redistribution of nitrogen fertilizer in winter wheat. *Precision Agriculture*, *7*(2), 65–83. doi:10.1007/s11119-006-9000-2

Bhattarai, T. N., Ghimire, S., Mainali, B., Gorjian, S., Treiche, H., & Paudel, S. R. (2022). Applications of smart grid technology in Nepal: Status, challenges, and opportunities. *Environmental Science and Pollution Research International*, *30*(10), 1–25. doi:10.1007/s11356-022-19084-3 PMID:35138545

Bochtis, D., Sørensen, C. A. G., & Kateris, D. (2019). *Operations Management in Agriculture*. Elsevier. [Google Scholar] [CrossRef]

Chandhini, K. (2016). A Literature Study on Agricultural Production System Using IoT as Inclusive Technology. [Google Scholar]. *Int. J. Innov. Technol. Res.*, *4*, 2727–2731.

Chang, A., Jung, J., Maeda, M., & Landivar, J. (2017). Crop height monitoring with digital imagery from unmanned aerial system (UAV). *Computers and Electronics in Agriculture*, *141*, 232–237. doi:10.1016/j.compag.2017.07.008

Chen, F., Kissel, D. E., West, L. T., Adkins, W., Clark, R., Rickman, D., & Luvall, J. C. (2004). Field Scale Mapping of Surface Soil Clay Concentration. *Precision Agriculture*, *5*(1), 7–26. doi:10.1023/B:PRAG.0000013617.96272.9e

Chowdhury, M. E. H., Khandakar, A., Ahmed, S., Al-Khuzaei, F., Hamdalla, J., Haque, F., Reaz, M. B. I., Al Shafei, A., & Al-Emadi, N. (2020). Design, Construction and Testing of IoT Based Automated Indoor Vertical Hydroponics Farming Test-Bed in Qatar. *Sensors (Basel)*, *20*(19), 5637. doi:10.3390/s20195637 PMID:33023097

Coulter, J. K. (2004). World Agriculture: Towards 2015/2030. An FAO Perspective. Experimental Agriculture, 40(2), 269–269. doi:10.1017/S0014479704211796

Cravero, A., Pardo, S., Sepúlveda, S., & Muñoz, L. (2022). Challenges to Use Machine Learning in Agricultural Big Data: A Systematic Literature Review. *Agronomy (Basel)*, *12*(3), 748. doi:10.3390/agronomy12030748

Daniel, K., Tripathi, N. K., Honda, K., & Apisit, E. (2011). Analysis of spectral reflectance and absorption patterns of soil organic matter. In *Proceedings of the 22nd Asian Conference on Remote Sensing*.

Daniel, K. W., Tripathi, N. K., Honda, K., & Apisit, E. (2004). Analysis of VNIR (400–1100 nm) spectral signatures for estimation of soil organic matter in tropical soils of Thailand. *International Journal of Remote Sensing*, *25*(3), 643–652. doi:10.1080/0143116031000139944

Duan, T., Chapman, S., Guo, Y., & Zheng, B. (2017). Dynamic monitoring of NDVI in wheat agronomy and breeding trials using an unmanned aerial vehicle. *Field Crops Research*, *210*, 71–80. doi:10.1016/j.fcr.2017.05.025

Ehlers, M. (2008). Geoinformatics and digital earth initiatives: A German perspective. *International Journal of Digital Earth*, *1*(1), 17–30. doi:10.1080/17538940701781975

Ehrlich, P. R., & Harte, J. (2015). To feed the world in 2050 will require a global revolution. *Proceedings of the National Academy of Sciences of the United States of America*, *112*(48), 14743–14744. doi:10.1073/pnas.1519841112 PMID:26627228

Ekka, V., & Bais, P. (2023). Contribution of Sericulture in the Socio-economic Transformation of the Workers. *Asian Journal of Applied Science and Technology*, *7*(2), 217–224. doi:10.38177/ajast.2023.7223

El Nahry, A. H., & Mohamed, E. S. (2010). Potentiality of land and water resources in African Sahara: A case study of south Egypt. *Environmental Earth Sciences*, *63*(6), 1263–1275. doi:10.1007/s12665-010-0799-5

Elijah, O., Rahman, T. A., Orikumhi, I., Leow, C. Y., & Hindia, M. H. D. N. (2018). An Overview of Internet of Things (IoT) and Data Analytics in Agriculture: Benefits and Challenges. *IEEE Internet of Things Journal*, *5*(5), 3758–3773. doi:10.1109/JIOT.2018.2844296

Farooq, M. S., Riaz, S., Abid, A., Umer, T., & Zikria, Y. B. (2020). Role of IoT Technology in Agriculture: A Systematic Literature Review. *Electronics (Basel)*, *9*(2), 319. doi:10.3390/electronics9020319

Feder, G., Just, R. E., & Zilberman, D. (1985). Adoption of agricultural innovations in developing countries: A survey. *Economic Development and Cultural Change*, *33*(2), 255–298. doi:10.1086/451461

Ferguson, R. B., Hergert, G. W., Schepers, J. S., Gotway, C. A., Cahoon, J. E., & Peterson, T. A. (2002). Site-Specific Nitrogen Management of Irrigated Maize. *Soil Science Society of America Journal, 66*(2), 544–553. . doi:10.2136/sssaj2002.5440

Ferrández-Pastor, F., García-Chamizo, J., Nieto-Hidalgo, M., Mora-Pascual, J., & Mora-Martínez, J. (2016). Developing Ubiquitous Sensor Network Platform Using Internet of Things: Application in Precision Agriculture. *Sensors (Basel)*, *16*(7), 1141. doi:10.3390/s16071141 PMID:27455265

Ferrández-Pastor, F. J., García-Chamizo, J. M., Nieto-Hidalgo, M., & Mora-Martínez, J. (2018). Precision Agriculture Design Method Using a Distributed Computing Architecture on Internet of Things Context. *Sensors (Basel)*, *18*(6), 1731. doi:10.3390/s18061731 PMID:29843386

Foster, A. D., & Mark, R. R. (2010). Microeconomics of technology adoption. *Annual Review of Economics*, *2*(1), 395–424. doi:10.1146/annurev.economics.102308.124433 PMID:24386501

Fróna, D., Szenderák, J., & Rákos, M. H. (2019). The Challenge of Feeding the World. *Sustainability (Basel)*, *11*(20), 5816. doi:10.3390/su11205816

Goel, R. K., Yadav, C. S., Vishnoi, S., & Rastogi, R. (2021). Smart agriculture – Urgent need of the day in developing countries. *Sustainable Computing : Informatics and Systems*, *30*, 100512. doi:10.1016/j.suscom.2021.100512

Hassan, M. A., Yang, M., Rasheed, A., Yang, G., Reynolds, M., Xia, X., Xiao, Y., & He, Z. (2019). A rapid monitoring of NDVI across the wheat growth cycle for grain yield prediction using a multi-spectral UAV platform. *Plant Science*, *282*, 95–103. doi:10.1016/j.plantsci.2018.10.022 PMID:31003615

Henriksen, A. V., Edwards, T. C. G., Pesonen, L. A., Green, O., & Sørensen, C. A. G. (2019). Internet of Things in arable farming: Implementation, applications, challenges and potential. *Biosystems Engineering*, *191*, 60–84. doi:10.1016/j.biosystemseng.2019.12.013

Hernández-Ochoa, I. M., Gaiser, T., Kersebaum, K.-C., Webber, H., Seidel, S. J., Grahmann, K., & Ewert, F. (2022). Model-based design of crop diversification through new field arrangements in spatially heterogeneous landscapes. A review. *Agronomy for Sustainable Development*, *42*(4), 74. doi:10.1007/s13593-022-00805-4

James, A., Saji, A., Nair, A., & Joseph, D. (2019). CropSense–A Smart Agricultural System using IoT. [Google Scholar]. *J. Electron. Des. Eng.*, *5*, 1–7.

Jin, X., Liu, S., Baret, F., Hemerl, M., & Comar, A. (2017). Estimates of plant density of wheat crops at emergence from very low altitude UAV imagery. *Remote Sensing of Environment*, *198*, 105–114. doi:10.1016/j.rse.2017.06.007

Kaneko, A., Kennedy, T. W., Mei, L., Sintek, C., Burke, M., Ermon, S., & Lobell, D. B. (2019). *Deep Learning for Crop Yield Prediction in Africa*. In *Proceedings of the International Conference on Machine Learning AI for Social Good Workshop*, LongBeach, CA, USA.

Kedri, J. (2014). *Simulation and Validation of SPV Micro Grid Comprising 518.2 KWp Distributed Solar Power Plants at Dayalbagh Educational Institute*. [Ph.D. Thesis, Dayalbagh Educational Institute, Dayalbagh, Agra, India].

Khan, A. R., Dubey, M. K., Bisen, P. K., & Saxena, K. K. (2007). Constraints faced by farmers of Narsing Kheda village of Sihore district. [Google Scholar]. *Young*, *8*, 16.

Khan, S. A., Magd, H., Al Shamsi, I. R., & Masoom, K. (2022). Social Entrepreneurship Through Innovations in Agriculture. In International Perspectives on Value Creation and Sustainability Through Social Entrepreneurship (pp. 209-222). IGI Global. doi:10.4018/978-1-6684-4666-9.ch010

Kimiti, J. M., Odee, D. W., & Vanlauwe, B. (2009). *Area under Grain Legumes Cultivation and Problems Faced by Smallholder Farmers in Legume Production in the Semi-Arid Eastern Kenya*. Academic Press. [Google Scholar]

King, T., Cole, M., Farber, J. M., Eisenbrand, G., Zabaras, D., Fox, E. M., & Hill, J. P. (2017). Food safety for food security: Relationship between global megatrends and developments in food safety. Trends in Food Science &amp. *Trends in Food Science & Technology*, *68*, 160–175. doi:10.1016/j.tifs.2017.08.014

Kodali, R. K., Jain, V., & Karagwal, S. (2016). *IoT based smart greenhouse*. In 2016 IEEE Region 10 Humanitarian Technology Conference (R10-HTC), Agra, India. 10.1109/R10-HTC.2016.7906846

Kuang, B., & Mouazen, A. M. (2013). Non-biased prediction of soil organic carbon and total nitrogen with vis-NIR spectroscopy, as affected by soil moisture content and texture. *Biosystems Engineering*, *114*(3), 249–258. doi:10.1016/j.biosystemseng.2013.01.005

Kuang, B., & Mouazen, A. M. (2013). Non-biased prediction of soil organic carbon and total nitrogen with vis-NIR spectroscopy, as affected by soil moisture content and texture. *Biosystems Engineering*, *114*(3), 249–258. doi:10.1016/j.biosystemseng.2013.01.005

Lakshmanan, R., Djama, M., Selvaperumal, S., & Abdulla, R. (2020). Automated smart hydroponics system using internet of things. *Iranian Journal of Electrical and Computer Engineering*, *10*(6), 6389–6398. doi:10.11591/ijece.v10i6.pp6389-6398

Lang, L. (1992). GPS + GIS + remote sensing: An overview. [Google Scholar]. *Earth Observation Magazine*, *1*, 23–26.

Liakos, K., Busato, P., Moshou, D., Pearson, S., & Bochtis, D. (2018). Machine Learning in Agriculture: A Review. *Sensors (Basel)*, *18*(8), 2674. doi:10.3390/s18082674 PMID:30110960

Lin, J., Yu, W., Zhang, N., Yang, X., Zhang, H., & Zhao, W. (2017). A Survey on Internet of Things: Architecture, Enabling Technologies, Security and Privacy, and Applications. *IEEE Internet of Things Journal*, *4*(5), 1125–1142. doi:10.1109/JIOT.2017.2683200

Lottes, P., Khanna, R., Pfeifer, J., Siegwart, R., & Stachniss, C. UAV-based crop and weed classification for smart farming. In *Proceedings of the 2017 IEEE International Conference on Robotics and Automation (ICRA)*, (pp. 3024–3031). IEEE. 10.1109/ICRA.2017.7989347

Lvova, L., & Nadporozhskaya, M. (2017). Chemical sensors for soil analysis: Principles and applications. In A. M. Grumezescu (Ed.), *Series Nanotechnology in the Agri-Food Industry; New Pesticides and Soil Sensors* (Vol. 10, pp. 637–678). Elsevier. doi:10.1016/B978-0-12-804299-1.00018-7

Lvova, L., & Nadporozhskaya, M. (2017). Chemical sensors for soil analysis: Principles and applications. In A. M. Grumezescu (Ed.), *Series Nanotechnology in the Agri-Food Industry; New Pesticides and Soil Sensors* (Vol. 10, pp. 637–678). Elsevier. doi:10.1016/B978-0-12-804299-1.00018-7

Maleki, M. R., Van Holm, L., Ramon, H., Merckx, R., De Baerdemaeker, J., & Mouazen, A. M. (2006). Phosphorus Sensing for Fresh Soils using Visible and Near Infrared Spectroscopy. *Biosystems Engineering*, *95*(3), 425–436. doi:10.1016/j.biosystemseng.2006.07.015

Maleki, M. R., Van Holm, L., Ramon, H., Merckx, R., De Baerdemaeker, J., & Mouazen, A. M. (2006). Phosphorus Sensing for Fresh Soils using Visible and Near Infrared Spectroscopy. *Biosystems Engineering*, *95*(3), 425–436. doi:10.1016/j.biosystemseng.2006.07.015

Mallarino, A., & Wittry, D. (1997). Use of DGPs, Yield Monitors, Soil Testing, and Variable Rate Technology to Improve Phosphorus and Potassium Management. *Proceedings of the Integrated Crop Management Conference*. IEEE. 10.31274/icm-180809-589

Mekala, M. S., & Viswanathan, P. (2017). A Survey: Smart agriculture IoT with cloud computing. *2017 International Conference on Microelectronic Devices, Circuits and Systems (ICMDCS)*. IEEE. 10.1109/ICMDCS.2017.8211551

Mittal, A., & Singh, A. (2007). Microcontroller based pest management system. *Second International Conference on Systems (ICONS'07)*. IEEE. 10.1109/ICONS.2007.35

Muthumanickam, D., Kannan, P., Kumaraperumal, R., Natarajan, S., Sivasamy, R., & Poongodi, C. (2011). Drought assessment and monitoring through remote sensing and GIS in western tracts of Tamil Nadu, India. *International Journal of Remote Sensing*, *32*(18), 5157–5176. doi:10.1080/01431161.2010.494642

Navulur, S., & Sastry, S. C. S., A., & N. Giri Prasad, M. (. (2017). Agricultural Management through Wireless Sensors and Internet of Things. [IJECE]. *Iranian Journal of Electrical and Computer Engineering*, *7*(6), 3492. doi:10.11591/ijece.v7i6.pp3492-3499

O'Grady, M. J., & O'Hare, G. M. P. (2017). Modelling the smart farm. *Information Processing in Agriculture*, *4*(3), 179–187. doi:10.1016/j.inpa.2017.05.001

Obaisi, A. I., Adegbeye, M. J., Elghandour, M. M. M. Y., Barbabosa-Pliego, A., & Salem, A. Z. M. (2022). Natural Resource Management and Sustainable Agriculture. Handbook of Climate Change Mitigation and Adaptation, (pp. 2577–2613). Springer. doi:10.1007/978-3-030-72579-2_133

Ojo, O. I., & Ilunga, F. (2018). *Geospatial Analysis for Irrigated Land Assessment, Modeling and Mapping*. Multi-Purposeful Application of Geospatial Data. doi:10.5772/intechopen.73314

Ouyang, H., Wei, X., & Wu, Q. (2019). Agricultural commodity futures prices prediction via long- and short-term time series network. *Journal of Applied Econometrics*, *22*, 468–483.

Patil, K. A., & Kale, N. R. (2016). A model for smart agriculture using IoT. *2016 International Conference on Global Trends in Signal Processing, Information Computing and Communication (ICGTSPICC)*. IEEE. 10.1109/ICGTSPICC.2016.7955360

Paul, K., Sorrentino, M., Lucini, L., Rouphael, Y., Cardarelli, M., Bonini, P., Reynaud, H., Canaguier, R., Trtílek, M., Panzarová, K., & Colla, G. (2019). Understanding the Biostimulant Action of Vegetal-Derived Protein Hydrolysates by High-Throughput Plant Phenotyping and Metabolomics: A Case Study on Tomato. *Frontiers in Plant Science*, *10*, 47. doi:10.3389/fpls.2019.00047 PMID:30800134

Pimentel, D., & Burgess, M. (2013). Soil erosion threatens food production. *Agriculture*, *3*(3), 443–463. doi:10.3390/agriculture3030443

Potamitis, I., Rigakis, I., Tatlas, N. A., & Potirakis, S. (2019). In-Vivo Vibroacoustic Surveillance of Trees in the Context of the IoT. *Sensors (Basel)*, *19*(6), 1366. doi:10.3390/s19061366 PMID:30893798

Potamitis, I., Rigakis, I., Tatlas, N. A., & Potirakis, S. (2019). In-Vivo Vibroacoustic Surveillance of Trees in the Context of the IoT. *Sensors (Basel)*, *19*(6), 1366. doi:10.3390/s19061366 PMID:30893798

Quy, V. K., Hau, N. V., Anh, D. V., Quy, N. M., Ban, N. T., Lanza, S., Randazzo, G., & Muzirafuti, A. (2022). IoT-Enabled Smart Agriculture: Architecture, Applications, and Challenges. *Applied Sciences (Basel, Switzerland)*, *12*(7), 3396. doi:10.3390/app12073396

Reddy, T. K., & Dutta, M. (2018). Impact of Agricultural Inputs on Agricultural GDP in Indian Economy. *Theoretical Economics Letters*, *08*(10), 1840–1853. doi:10.4236/tel.2018.810121

Rembold, F., Atzberger, C., Savin, I., & Rojas, O. (2013). Using Low Resolution Satellite Imagery for Yield Prediction and Yield Anomaly Detection. *Remote Sensing (Basel)*, *5*(4), 1704–1733. doi:10.3390/rs5041704

Roth, L., Aasen, H., Walter, A., & Liebisch, F. (2018). Extracting leaf area index using viewing geometry effects new perspective on high-resolution unmanned aerial system photography. *ISPRS Journal of Photogrammetry and Remote Sensing*, *141*, 161–175. doi:10.1016/j.isprsjprs.2018.04.012

Rouphael, Y., Spíchal, L., Panzarová, K., Casa, R., & Colla, G. (2018). High-Throughput Plant Phenotyping for Developing Novel Biostimulants: From Lab to Field or From Field to Lab? *Frontiers in Plant Science*, *9*, 1197. doi:10.3389/fpls.2018.01197 PMID:30154818

Sambo, P., Nicoletto, C., Giro, A., Pii, Y., Valentinuzzi, F., Mimmo, T., Lugli, P., Orzes, G., Mazzetto, F., Astolfi, S., Terzano, R., & Cesco, S. (2019). Hydroponic Solutions for Soilless Production Systems: Issues and Opportunities in a Smart Agriculture Perspective. *Frontiers in Plant Science*, *10*, 923. doi:10.3389/fpls.2019.00923 PMID:31396245

Shamshiri, R. R., Kalantari, F., Ting, K. C., Thorp, K. R., Hameed, I. A., Weltzien, C., Ahmad, D., & Shad, Z. (2018). Advances in greenhouse automation and controlled environment agriculture: A transition to plant factories and urban agriculture. *International Journal of Agricultural and Biological Engineering*, *11*, 1–22. doi:10.25165/j.ijabe.20181101.3210

Shi, X., An, X., Zhao, Q., Liu, H., Xia, L., Sun, X., & Guo, Y. (2019). State-of-the-Art Internet of Things in Protected Agriculture. *Sensors (Basel)*, *19*(8), 1833. doi:10.3390/s19081833 PMID:30999637

Sisinni, E., Saifullah, A., Han, S., Jennehag, U., & Gidlund, M. (2018). Industrial Internet of Things: Challenges, Opportunities, and Directions. *IEEE Transactions on Industrial Informatics*, *14*(11), 4724–4734. doi:10.1109/TII.2018.2852491

Srisruthi, S., Swarna, N., Ros, G. M. S., & Elizabeth, E. (2016). Sustainable agriculture using eco-friendly and energy efficient sensor technology. *2016 IEEE International Conference on Recent Trends in Electronics, Information & Communication Technology (RTEICT)*. IEEE.

Stroppiana, D., Migliazzi, M., Chiarabini, V., Crema, A., Musanti, M., Franchino, C., & Villa, P. Rice yield estimation using multispectral data from UAV: A preliminary experiment in northern Italy. In *Proceedings of the 2015 IEEE International Geoscience and Remote Sensing Symposium (IGARSS)*, Milan, Italy. 10.1109/IGARSS.2015.7326869

Su, J., Liu, C., Coombes, M., Hu, X., Wang, C., Xu, Z., Li, Q., Guo, L., & Chen, W. H. (2018). Wheat yellow rust monitoring by learning from multispectral UAV aerial imagery. *Computers and Electronics in Agriculture*, *155*, 157–166. doi:10.1016/j.compag.2018.10.017

Terence, S., & Purushothaman, G. (2020). Systematic review of Internet of Things in smart farming. *Transactions on Emerging Telecommunications Technologies*, *31*(6), e3958. doi:10.1002/ett.3958

Theopoulos, A., Boursianis, A., Koukounaras, A., & Samaras, T. (2018). *Prototype wireless sensor network for real-time measurements in hydroponics cultivation*. In *Proceedings of the 7th International Conference on Modern Circuits and Systems Technologies (MOCAST)*, Thessaloniki, Greece. 10.1109/MOCAST.2018.8376576

Tripathy, P. K., Tripathy, A. K., Agarwal, A., & Mohanty, S. P. (2021). MyGreen: An IoT-Enabled Smart Greenhouse for Sustainable Agriculture. *IEEE Consumer Electronics Magazine*, *10*(4), 57–62. doi:10.1109/MCE.2021.3055930

Tripodi, P., Massa, D., Venezia, A., & Cardi, T. (2018). Sensing Technologies for Precision Phenotyping in Vegetable Crops: Current Status and Future Challenges. *Agronomy (Basel)*, *8*(4), 57. doi:10.3390/agronomy8040057

Tucker, C. J., Elgin, J. H., & McMurtrey, J. E. (1980). Relationship of crop radiance to alfalfa agronomic values. *International Journal of Remote Sensing*, *1*(1), 69–75. doi:10.1080/01431168008948211

Tzounisa, A., Katsoulasa, N., Bartzanasb, T., & Kittas, C. (2017). Internet of Things in agriculture, recent advances and future challenges. *Biosystems Engineering*, *164*, 31–48. doi:10.1016/j.biosystemseng.2017.09.007

Usman, N., Arief, P., Gilang, L., Erfan, R., & Hendra, P. (2018). Implementation IoT in System Monitoring Hydroponic Plant Water Circulation and Control. [Google Scholar]. *Int. J. Eng. Technol.*, *7*, 122–126.

Wolfert, S., Ge, L., Verdouw, C., & Bogaardt, M.-J. (2017). Big Data in Smart Farming – A review. *Agricultural Systems*, *153*, 69–80. doi:10.1016/j.agsy.2017.01.023

Yang, W., Feng, H., Zhang, X., Zhang, J., Doonan, J. H., Batchelor, W. D., Xiong, L., & Yan, J. (2020). Crop Phenomics and High-Throughput Phenotyping: Past Decades, Current Challenges, and Future Perspectives. *Molecular Plant*, *13*(2), 187–214. doi:10.1016/j.molp.2020.01.008 PMID:31981735

Yong, W., Shuaishuai, L., Li, L., Minzan, L., Ming, L., Arvanitis, K. G., Georgieva, C., & Sigrimis, N. (2018). Smart Sensors from Ground to Cloud and Web Intelligence. *IFAC-PapersOnLine*, *51*(17), 31–38. doi:10.1016/j.ifacol.2018.08.057

Zhang, J., Zhu, Y., Zhang, X., Ye, M., & Yang, J. (2018). Developing a Long Short-Term Memory (LSTM) based model for predicting water table depth in agricultural areas. *Journal of Hydrology (Amsterdam)*, *561*, 918–929. doi:10.1016/j.jhydrol.2018.04.065

Zhao, C., Liu, B., Piao, S., Wang, X., Lobell, D. B., Huang, Y., Huang, M., Yao, Y., Bassu, S., Ciais, P., Durand, J.-L., Elliott, J., Ewert, F., Janssens, I. A., Li, T., Lin, E., Liu, Q., Martre, P., Müller, C., & Asseng, S. (2017). Temperature increase reduces global yields of major crops in four independent estimates. *Proceedings of the National Academy of Sciences of the United States of America*, *114*(35), 9326–9331. doi:10.1073/pnas.1701762114 PMID:28811375

ZhouJ.ReynoldsD.WebsdaleD.Le CornuT.Gonzalez NavarroO.ListerC.OrfordS.LaycockS.FinlaysonG. StittT. (2017). Cropquant: An automated and scalable field phenotyping platform for crop monitoring and trait measurements to facilitate breeding and digital agriculture. bioRxiv. doi:10.1101/161547

Chapter 8
Circular Economy Digital Practices for Ethical Dimensions and Policies for Digital Waste Management

V. Vijaya Lakshmi
https://orcid.org/0000-0001-9522-7897
Humanities and Mathematics Department, G. Narayanamma Institute of Technology and Science, Hyderabad, India

Manoj Mishra
https://orcid.org/0009-0008-5222-3990
Department of Basic Science and Humanities, Institute of Technology and Management, Gwalior, India

Jitendra Singh Kushwah
https://orcid.org/0000-0001-5102-2160
Department of Information Technology, Institute of Technology and Management, Gwalior, India

Umaya Salma Shajahan
Department of Management Studies, Sona College of Technology, Salem, India

M. Mohanasundari
Department of Management Studies, Kongu Engineering College, Erode, India

Sampath Boopathi
https://orcid.org/0000-0002-2065-6539
Mechanical Engineering, Muthayammal Engineering College, Namakkal, India

ABSTRACT

Digital technology has revolutionized our lives, but it also presents a growing challenge: digital waste. This issue involves electronic, data, and obsolete devices, causing significant environmental and social consequences. To tackle this, a multi-faceted approach is needed, including ethical considerations, policy interventions, technological innovations, and consumer behavior change. Ethical frameworks, strengthened regulatory frameworks, eco-design, producer responsibility, and circular economy principles are essential. Industry best practices and technological innovations can reduce digital waste, while consumer education and behavior change initiatives encourage informed digital habits. Sustainable digital practices require a concerted effort from all stakeholders.

DOI: 10.4018/979-8-3693-1794-5.ch008

INTRODUCTION

The digital age has led to a surge in digital waste, posing ethical, environmental, and social challenges. Sustainable digital practices are needed to mitigate these negative impacts while promoting long-term environmental and social well-being. Recognizing the finite nature of resources and the interconnectedness of environmental, social, and economic systems is crucial. Digital products and services often lack physical form, making it easy to overlook their environmental consequences. However, their environmental footprint is substantial, including energy consumption, resource extraction, electronic waste generation, and carbon emissions from data centers and digital infrastructure. Therefore, sustainable digital practices are essential for a sustainable future(Khan et al., 2021).

The increasing use of digital devices has worsened social equity and digital divides, as technology access is crucial for modern society. Digital waste, including discarded devices and obsolete technologies, further exacerbates these disparities. Sustainable digital practices aim to minimize environmental and social impacts of digital consumption, incorporating principles of eco-design, resource efficiency, and circular economy to reduce waste, extend product lifespan, and promote responsible end-of-life management(Mies & Gold, 2021).

Sustainable digital practices involve eco-design, product lifecycle management, responsible consumption, and end-of-life management. They include extended producer responsibility (EPR), eco-labeling, and waste management policies. Technological innovation like AI, IoT, and blockchain is used to develop solutions for digital waste reduction, resource efficiency, and sustainable consumption. These practices aim to minimize resource consumption, energy use, and waste generation throughout digital products and services(Roberts et al., 2022).

Sustainable digital practices aim to address the environmental, social, and ethical aspects of digital consumption. By integrating sustainability principles into digital innovation, production, and consumption, we can create a more resilient and equitable digital future for future generations. As society increasingly relies on digital technologies for communication, commerce, and social interaction, the environmental and social impacts of digital consumption are under scrutiny. Digital waste management is a central issue, presenting complex ethical dilemmas and requiring robust policy frameworks to address its multifaceted challenges(D'Adamo et al., 2022).

Sustainable digital practices involve responsible stewardship of digital resources throughout their lifecycle, from production to disposal and recycling. Digital technologies contribute significantly to resource depletion, energy consumption, and electronic waste generation. The digital infrastructure underpins modern society, exacerbating environmental degradation and climate change. Ethical considerations are central to sustainable digital practices, prompting stakeholders to question their consumption habits and societal implications(Lehtokunnas et al., 2022). Digital waste disproportionately affects marginalized communities, exacerbating social inequalities and widening the digital divide. The pursuit of technological innovation and convenience often comes at the expense of ethical values like fairness, transparency, and accountability(Dhanalakshmi et al., 2024a).

Policymakers, businesses, and consumers are addressing ethical challenges in digital consumption and waste management by developing comprehensive policy frameworks that integrate environmental sustainability, social equity, and ethical principles into digital technology governance. These frameworks may include measures for product durability, eco-design, resource efficiency, and responsible end-of-life management and recycling systems(Persis et al., 2021).

Extended producer responsibility (EPR) is an ethical principle in digital waste management that holds manufacturers accountable for their products' environmental impact throughout their lifecycle. EPR schemes encourage eco-friendly design choices and promote closed-loop production and consumption systems. Eco-labeling initiatives empower consumers to make informed choices about digital products' environmental performance, driving demand for sustainable alternatives(Medaglia et al., 2024).

Ethical frameworks for digital waste management must consider environmental sustainability, social justice, and human rights. Environmental justice ensures that marginalized communities don't bear the burden of digital waste disposal, while human rights protection ensures privacy and data security in electronic waste handling. Sustainable digital practices require a holistic approach that integrates environmental sustainability, social equity, and ethical principles into digital technology governance. This approach can lead to a more resilient, equitable, and ethically sound digital future, ensuring that sensitive information is not compromised during the recycling process(Gupta et al., 2021).

Digital waste management is crucial for sustainability due to its environmental, social, and economic implications. With the rise of digital technologies, the generation of electronic waste has increased, posing a pressing sustainability challenge. By managing digital waste, we can mitigate its environmental impact, conserve resources, and reduce pollution, thereby advancing sustainability goals(Nascimento et al., 2019).

Digital waste management involves ethical considerations beyond environmental concerns, as it raises questions about resource depletion, environmental degradation, and social justice. Digital products often contain hazardous materials like lead, mercury, and brominated flame retardants, posing risks to human health and the environment. The rapid turnover of digital devices encourages wasteful consumption, contributing to resource depletion and environmental degradation. Addressing digital waste requires ethical decision-making based on environmental stewardship, social responsibility, and intergenerational equity(Nascimento et al., 2019).

DIGITAL WASTES

Digital waste refers to the collection of obsolete digital products, components, and data, including physical electronic devices and intangible digital assets, which contribute to environmental and social issues in unique ways(Ruohomaa & Ivanova, 2019).

- **E-Waste (Electronic Waste):** E-waste refers to discarded electronic devices and equipment, such as smartphones, laptops, tablets, desktop computers, televisions, and household appliances. These devices contain valuable materials like metals, plastics, and glass, but they also contain hazardous substances like lead, mercury, and cadmium. When improperly disposed of, either through landfilling or informal recycling methods, e-waste can leach toxic chemicals into the environment, contaminating soil, water, and air, and posing risks to human health and wildlife.
- **Data Waste:** Data waste refers to the inefficient or unnecessary accumulation, storage, and processing of digital data. In an era of big data and ubiquitous connectivity, organizations often collect and retain vast amounts of data without a clear purpose or strategy for its utilization. This data redundancy and inefficiency contribute to unnecessary energy consumption, storage costs, and cybersecurity risks. Moreover, the proliferation of digital data increases the demand for data cen-

ters and cloud infrastructure, which consume significant amounts of energy and resources, further exacerbating environmental impacts(Rahamathunnisa et al., 2023).

- **Obsolete Digital Devices:** Beyond e-waste, digital waste also includes obsolete or outdated digital devices that are no longer functional or in use. This category encompasses everything from old cell phones and computers to peripherals like printers, scanners, and routers. While these devices may not pose the same environmental hazards as e-waste, their disposal still requires proper handling to prevent pollution and resource depletion. Additionally, the rapid turnover of digital devices contributes to a culture of consumerism and planned obsolescence, perpetuating unsustainable consumption patterns and resource depletion.
- **Software and Digital Content Waste:** Digital waste is not limited to physical devices but also includes software and digital content that are no longer needed or used. This may include outdated software applications, digital media files, and online content. Like physical devices, digital content contributes to resource consumption and energy use during production, distribution, and storage. Moreover, the proliferation of digital content raises questions about digital preservation and archiving, as well as the environmental impact of data centers and cloud infrastructure hosting this content(Rahamathunnisa et al., 2023).

Digital waste, consisting of discarded digital products, components, and data, poses significant environmental, social, and economic challenges. It involves electronic devices, data redundancy, and obsolete digital assets. Understanding these forms can help develop strategies to reduce waste generation, promote responsible consumption, and advance digital sustainability. Addressing these consequences is crucial for addressing the broader implications of digital consumption and waste management(Mondal et al., 2023; Rusch et al., 2023).

Environmental Consequences

- **Resource Depletion:** The production of electronic devices requires significant amounts of natural resources, including metals like gold, silver, copper, and rare earth elements. The extraction and processing of these materials contribute to habitat destruction, water pollution, and ecosystem degradation, exacerbating pressures on already stressed ecosystems.
- **Energy Consumption:** Digital devices, data centers, and cloud infrastructure consume vast amounts of energy during manufacturing, operation, and disposal phases. This energy consumption, largely reliant on fossil fuels, contributes to greenhouse gas emissions and climate change. Data centers alone are estimated to account for a significant portion of global electricity consumption(Agrawal, Magulur, et al., 2023; Dhanya et al., 2023a; Ramudu et al., 2023).
- **E-Waste Contamination:** Improper disposal of electronic devices, either through landfilling or informal recycling methods, can lead to the release of hazardous substances into the environment. Components like lead, mercury, cadmium, and brominated flame retardants can leach into soil, water, and air, posing risks to human health and wildlife. E-waste contamination can persist for decades, impacting ecosystems and communities long after disposal(Hanumanthakari et al., 2023a; Janardhana et al., 2023a; Selvakumar, Adithe, et al., 2023a).
- **Landfill Pressure:** The rapid turnover of digital devices contributes to the mounting pressure on landfills, where electronic waste occupies valuable space and poses risks of leachate contamina-

tion and methane emissions. Landfilling e-waste also represents a missed opportunity for resource recovery and recycling, as valuable materials remain trapped in discarded devices.

Social Consequences:

- **Digital Divide:** The unequal distribution of digital resources exacerbates social inequalities and widens the digital divide between affluent and marginalized communities. Access to digital devices and technology is essential for participation in modern society, including education, employment, healthcare, and civic engagement. The lack of access to technology perpetuates socioeconomic disparities and limits opportunities for social and economic advancement(Agrawal, Magulur, et al., 2023; Dhanya et al., 2023a).
- **Health Risks:** E-waste recycling and informal dismantling activities, often conducted in developing countries with lax regulations, expose workers and communities to hazardous substances and pollutants. Workers in electronic waste recycling facilities face risks of chemical exposure, respiratory illnesses, and occupational injuries. Moreover, nearby communities may experience environmental contamination and health impacts due to e-waste pollution(Boopathi & Khang, 2023; Karthik et al., 2023; Satav et al., 2023).
- **Digital Inclusion:** The rapid turnover of digital devices and the proliferation of digital content contribute to digital obsolescence, rendering older technologies and formats obsolete. This can have implications for digital inclusion and access to information, particularly for individuals and communities with limited resources or technological literacy. Bridging the digital divide requires addressing barriers to access, affordability, and digital skills development(Boopathi & Kumar, 2024a; Dhanalakshmi et al., 2024b).

Digital waste accumulation has significant environmental and social impacts, emphasizing the need for sustainable practices and responsible waste management. Addressing these challenges holistically promotes environmental stewardship, social equity, and human well-being in the digital age. Understanding the lifecycle of digital products and services is crucial for identifying waste generation points and implementing strategies to minimize environmental impact(Maiurova et al., 2022; Mondal et al., 2023).

Design and Development:

- **Materials Selection:** During the design phase, manufacturers select materials for electronic components based on performance, cost, and availability. However, the extraction and processing of these materials can generate waste and pollution, particularly in regions with lax environmental regulations(Boopathi, Umareddy, et al., 2023; Boopathi & Davim, 2023b, 2023a).
- **Manufacturing:** The manufacturing process involves assembling electronic components into finished products, such as smartphones, laptops, and consumer electronics. Waste generation can occur through manufacturing defects, excess packaging, and the disposal of unused materials or byproducts(Boopathi & Davim, 2023a; Ingle et al., 2023; Mohanty, Jothi, et al., 2023).

Distribution and Packaging:

- **Packaging:** Digital products are often packaged in cardboard boxes, plastic wraps, and protective materials for shipping and retail distribution. While packaging serves to protect products during transit, it also contributes to packaging waste and environmental pollution, particularly if materials are not recyclable or biodegradable.
- **Transportation:** The transportation of digital products from manufacturing facilities to distribution centers, retailers, and end-users requires energy and resources, leading to carbon emissions and air pollution. Inefficiencies in transportation logistics can exacerbate waste generation and environmental impact.

Use Phase:

- **Energy Consumption:** Digital devices consume electricity during their use phase, contributing to energy consumption and greenhouse gas emissions. Inefficient devices, excessive energy usage, and standby power consumption can result in unnecessary waste of resources and environmental harm(Boopathi, 2024d; Revathi et al., 2024).
- **Data Storage:** Digital services and platforms rely on data storage infrastructure, including servers, data centers, and cloud storage. The energy-intensive operation of data centers and server farms contributes to environmental impact, including electricity consumption, heat generation, and carbon emissions.

End-of-Life Management:

- **Disposal and Recycling:** The disposal of electronic devices at the end of their lifecycle presents significant waste management challenges. Improper disposal, such as landfilling or incineration, can lead to environmental pollution and health risks from hazardous substances. Recycling electronic waste requires specialized processes to recover valuable materials like metals and plastics while minimizing environmental impact(Hanumanthakari et al., 2023a; Harikaran et al., 2023a).
- **E-Waste Export:** In some cases, electronic waste is exported to developing countries for recycling or disposal, often under inadequate environmental and labor standards. This practice can lead to environmental pollution, health risks, and exploitation of vulnerable communities.

Obsolescence and Upgrades:

- **Planned Obsolescence:** Manufacturers may intentionally design products with limited lifespans or compatibility to encourage consumers to upgrade to newer models. Planned obsolescence contributes to electronic waste generation and consumer dissatisfaction, as older devices become obsolete or incompatible with newer technologies.
- **Software Updates and Support:** Software updates and upgrades can extend the lifespan of digital devices by improving performance, security, and functionality. However, discontinued support for older devices or operating systems may render them obsolete, leading to premature disposal and waste generation(Rahamathunnisa et al., 2023).

The lifecycle of digital products and services involves stages with potential waste generation and environmental impact. By implementing sustainable design principles, resource efficiency, and responsible end-of-life management practices, stakeholders can reduce waste and promote environmental sustainability.

Ethical Considerations in Digital Waste Management: This study explores ethical frameworks for managing digital waste, outlines responsibilities for businesses, consumers, and policymakers, and examines its impact on social equity, environmental justice, and human rights(Lehtokunnas et al., 2022; Persis et al., 2021).

POLICIES AND REGULATIONS FOR SUSTAINABLE DIGITAL PRACTICES

Survey existing policies and regulations

The study of global digital waste management policies reveals a variety of strategies aimed at reducing the environmental and social impacts of e-waste and promoting sustainable digital practices, with key themes and initiatives emerging at international, regional, and national levels. The figure 1 depicts policies and regulations for sustainable digital practices. Global policies and regulations aim to tackle environmental and social issues related to e-waste by promoting sustainable digital practices. Implementing comprehensive frameworks, fostering international cooperation, and promoting industry best practices can help achieve a more sustainable and circular economy in the digital age(Gupta et al., 2021; Rusch et al., 2023).

- **Basel Convention on the Control of Transboundary Movements of Hazardous Wastes and Their Disposal:** The Basel Convention, adopted in 1989 and ratified by 187 countries, seeks to regulate the transboundary movement of hazardous wastes, including e-waste. Parties to the convention are required to minimize the generation of hazardous wastes and ensure environmentally sound management of waste disposal and recycling. In 2019, an amendment to the convention, known as the "Ban Amendment," was adopted to prohibit the export of hazardous waste from developed to developing countries for disposal(Harikaran et al., 2023a; Maguluri, Ananth, et al., 2023; Selvakumar, Shankar, et al., 2023a).
- **European Union (EU) Waste Electrical and Electronic Equipment (WEEE) Directive:** The WEEE Directive, established in 2003 and revised in 2012, sets requirements for the collection, treatment, recycling, and recovery of e-waste within the EU. It mandates the implementation of extended producer responsibility (EPR) schemes, requiring manufacturers to finance the collection and recycling of their electronic products at the end of their lifecycle. The directive also aims to promote eco-design and reduce the use of hazardous substances in electronic products.
- **United States Environmental Protection Agency (EPA) Regulations:** In the United States, e-waste management is primarily regulated at the state level, with several states implementing EPR laws and e-waste recycling programs. At the federal level, the EPA oversees the management of hazardous waste under the Resource Conservation and Recovery Act (RCRA), which sets standards for the treatment, storage, and disposal of hazardous wastes, including certain types of e-waste(Selvakumar, Shankar, et al., 2023a).
- **Asia-Pacific Economic Cooperation (APEC) Initiative:** APEC member economies have collaborated on initiatives to address e-waste management and promote sustainable consumption

and production practices. These initiatives include capacity-building workshops, policy dialogues, and knowledge-sharing activities aimed at improving e-waste management practices and fostering cooperation among member economies(Rahamathunnisa et al., 2024a; Ravisankar et al., 2023).

- **National E-Waste Management Policies:** Many countries have adopted national e-waste management policies and regulations tailored to their specific contexts and challenges. These policies often include provisions for EPR, e-waste collection and recycling infrastructure, public awareness campaigns, and data reporting requirements. Countries such as Japan, South Korea, and Taiwan have implemented comprehensive e-waste management systems and achieved high rates of e-waste recycling(Saravanan et al., 2022).

Evaluate the effectiveness of current regulatory frameworks

The effectiveness of current regulatory frameworks in reducing digital waste involves assessing their impact on e-waste generation, promoting responsible disposal and recycling practices, and fostering sustainable digital practices, considering key factors varying by jurisdiction(Henneman et al., 2017).

- **Compliance and Enforcement:** The effectiveness of regulatory frameworks depends on the extent to which regulations are enforced and compliance is achieved by manufacturers, retailers, consumers, and waste management entities. Strong enforcement mechanisms, including penalties for non-compliance, can incentivize stakeholders to adhere to regulatory requirements and ensure proper e-waste management practices.
- **Extended Producer Responsibility (EPR):** EPR schemes, which place the responsibility for e-waste management on manufacturers, have been implemented in many regions worldwide. The effectiveness of EPR policies depends on their design, including the scope of products covered, financial contributions required from producers, and the transparency and accountability of funding mechanisms. Evaluating the effectiveness of EPR schemes involves assessing their impact on increasing e-waste collection rates, promoting product design for recyclability, and reducing the burden on public waste management systems.
- **Data Reporting and Monitoring:** Effective regulatory frameworks include provisions for data reporting and monitoring to track progress towards e-waste reduction goals and identify areas for improvement. Evaluating the effectiveness of regulatory frameworks involves analyzing data on e-waste generation, collection rates, recycling rates, and disposal practices to assess the impact of policy interventions and inform future decision-making(Boopathi, 2023; Sankar et al., 2023; Subha et al., 2023).
- **Innovation and Technology Adoption:** Regulatory frameworks should incentivize innovation and the adoption of sustainable technologies for e-waste management and recycling. Evaluating effectiveness involves assessing the extent to which regulations promote research and development of recycling technologies, support the use of environmentally friendly materials in product design, and encourage the adoption of circular economy principles in the electronics industry(Boopathi, 2024b; Kushwah et al., 2024).
- **International Cooperation and Harmonization:** Given the global nature of e-waste flows, international cooperation and harmonization of regulatory frameworks are essential for effective e-waste management. Evaluating effectiveness involves assessing the extent to which regulations

facilitate cross-border cooperation, harmonize standards and guidelines, and address challenges related to e-waste trade and illegal dumping.

Figure 1. Policies and regulations for sustainable digital practices

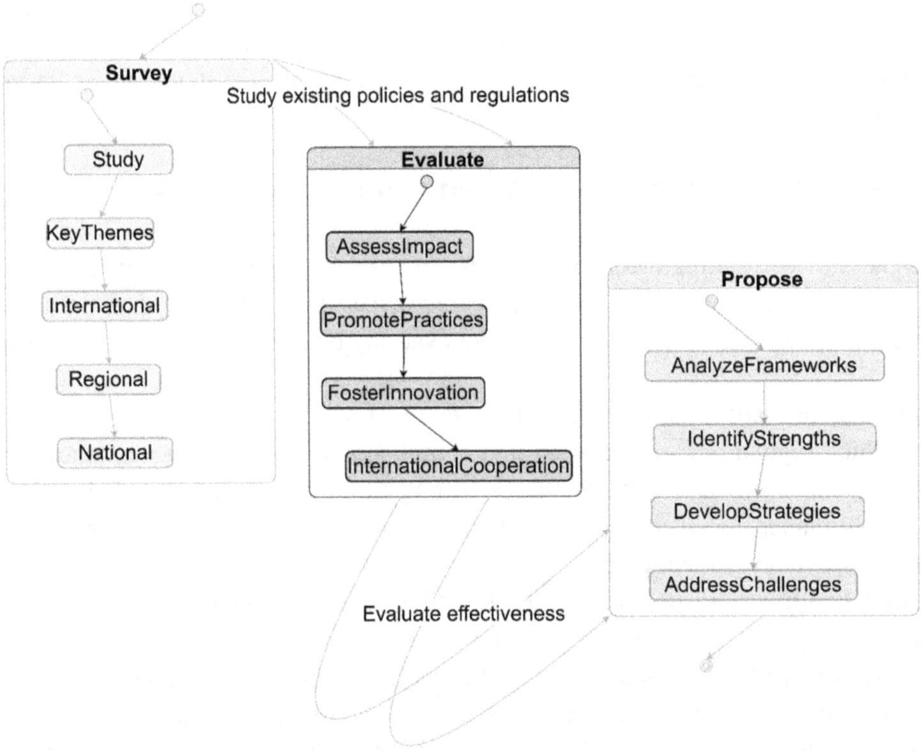

The effectiveness of current regulatory frameworks in mitigating digital waste requires a thorough analysis of their impact on reducing e-waste generation, promoting responsible disposal and recycling practices, fostering innovation, and facilitating international cooperation. By identifying strengths and weaknesses, policymakers can develop strategies to enhance effectiveness and achieve sustainable digital practices(Hanumanthakari et al., 2023b; Mohanty, Venkateswaran, et al., 2023; Selvakumar, Adithe, et al., 2023b).

- **Extended Producer Responsibility (EPR) Legislation:** Expand or strengthen EPR schemes to hold manufacturers accountable for the entire lifecycle of their products, including collection, recycling, and disposal of electronic devices. Require manufacturers to finance and participate in e-waste management programs, promote product design for longevity and recyclability, and incentivize the use of environmentally friendly materials.
- **Mandatory Product Labeling:** Implement mandatory labeling requirements to inform consumers about the environmental and social impacts of digital products, including their energy efficiency, recyclability, and use of hazardous substances. Labels could provide clear and standardized

information to help consumers make informed purchasing decisions and encourage manufacturers to improve product sustainability.

- **Consumer Awareness and Education Campaigns:** Launch public awareness campaigns to educate consumers about the environmental and social impacts of digital consumption and empower them to make sustainable choices. Provide information on e-waste recycling options, product repair and maintenance, and the importance of extending product lifespan through responsible use and disposal practices(Naveeenkumar et al., 2024).
- **Eco-Design Standards and Guidelines:** Develop and implement eco-design standards and guidelines for digital products to promote resource efficiency, recyclability, and durability. Encourage manufacturers to adopt eco-design principles in product development and provide support and incentives for compliance with sustainability standards.
- **Research and Innovation Funding:** Allocate funding for research and innovation in sustainable technologies and materials for digital products. Support research initiatives focused on improving e-waste recycling processes, developing alternative materials, and advancing circular economy models in the electronics industry.
- **International Collaboration and Harmonization:** Foster international cooperation and harmonization of policies and standards to address global challenges related to digital waste management. Collaborate with other countries and organizations to share best practices, exchange information, and coordinate efforts to reduce e-waste generation and promote sustainable digital practices on a global scale.

INDUSTRY BEST PRACTICES

Businesses Implementing Sustainable Digital Practices

Businesses worldwide are embracing sustainable digital practices to reduce environmental impact, enhance operational efficiency, and promote corporate social responsibility. Leading examples include companies that are implementing these practices(Jones & Wynn, 2021). The figure 2 depicts businesses implementing sustainable digital practices.

- **Apple Inc.:** Apple has made significant strides in promoting sustainability across its product lifecycle. The company has committed to using 100% renewable energy for its global operations, including manufacturing facilities and data centers. Apple also prioritizes product longevity and recyclability, offering trade-in programs for old devices and using recycled materials in new product designs.
- **Google:** Google has set ambitious sustainability goals, aiming to operate on 100% carbon-free energy by 2030 and achieve zero waste to landfill for all its operations. The company also invests in renewable energy projects and supports circular economy initiatives, such as the recycling of electronic waste and the development of sustainable packaging solutions.
- **Patagonia:** Patagonia, a leading outdoor apparel company, is known for its commitment to environmental sustainability and corporate social responsibility. The company invests in renewable energy, implements waste reduction measures, and promotes product repair and reuse through its

Worn Wear program. Patagonia also advocates for environmental protection and supports grassroots environmental organizations through its 1% for the Planet initiative.

- **IKEA:** IKEA, the Swedish furniture retailer, has adopted sustainable practices throughout its operations, from product design to supply chain management. The company focuses on resource efficiency, using renewable and recycled materials in its products and striving to minimize waste generation. IKEA also invests in renewable energy and promotes sustainable living through initiatives like energy-efficient home solutions and furniture recycling programs.

These businesses are leading the transition towards a more sustainable and resilient future by integrating sustainability into their operations, products, and supply chains, demonstrating the interconnectedness of environmental sustainability and business success.

Figure 2. Businesses implementing sustainable digital practices

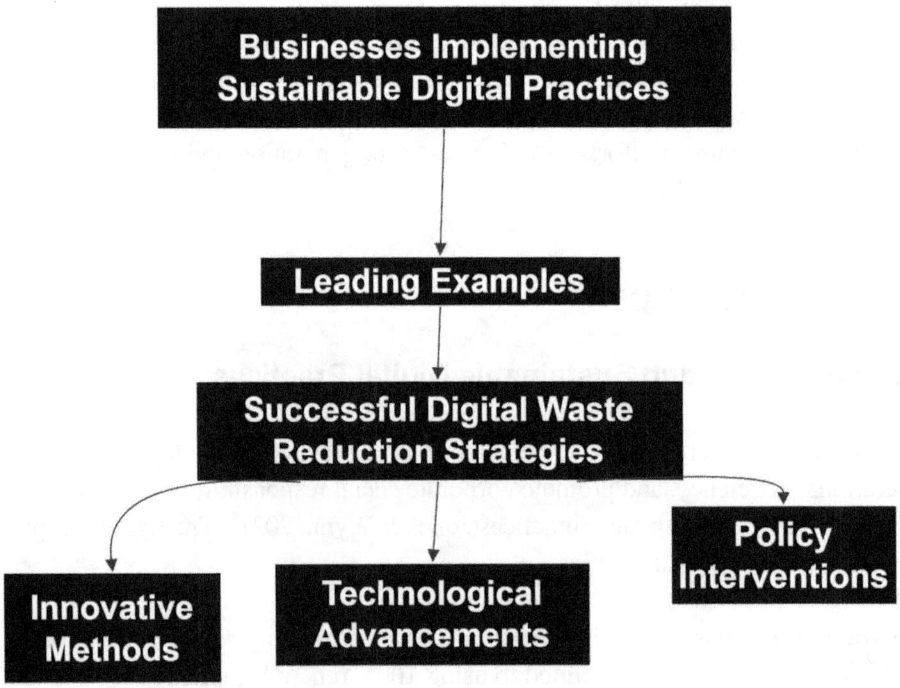

Successful Digital Waste Reduction Strategies

Successful digital waste reduction strategies involve innovative methods, technological advancements, and policy interventions to minimize waste generation, promote resource efficiency, and promote circular economy principles, with some effective strategies highlighted(Harikaran et al., 2023b; Janardhana et al., 2023b; Koshariya et al., 2023).

- **Consumer Education and Awareness:** Raising awareness among consumers about the environmental and social impacts of digital consumption and empowering them to make informed choices can drive demand for sustainable products and practices. Educational initiatives, public awareness campaigns, and consumer engagement programs can inform individuals about e-waste recycling options, product repair and maintenance, and the importance of responsible electronic waste disposal, thereby fostering a culture of sustainability and waste reduction.
- **Innovation in Materials and Technologies:** Investing in research and development of sustainable materials, alternative manufacturing processes, and eco-friendly technologies can spur innovation and drive advancements in digital waste reduction. Materials innovation, such as the development of biodegradable plastics, sustainable electronics, and closed-loop recycling systems, can help minimize the environmental impact of electronic products and promote a circular economy model(Boopathi et al., 2021; Sampath et al., 2021; Vignesh et al., 2018).
- Implementing digital waste reduction strategies collaboratively can significantly reduce electronic waste generation, promote resource efficiency, and advance environmental sustainability in the digital age(Janardhana et al., 2023a; Sampath, 2021).

TECHNOLOGIES AND INNOVATIONS FOR DIGITAL WASTE REDUCTION

Emerging Technologies for Extending the Lifecycle of Digital Products

Emerging technologies are enhancing the lifecycle of digital products, reducing electronic waste, and promoting a circular economy, with key innovations shaping the future of digital waste reduction(Agrawal, Shashibhushan, et al., 2023; Selvakumar, Shankar, et al., 2023b; Sengeni et al., 2023). The figure 3 depicts various technologies and innovations aimed at reducing digital waste.

- **Modular Design and Repairability:** Modular design principles enable the disassembly and replacement of individual components within digital products, such as smartphones and laptops, making them easier to repair and upgrade. Modular design allows for the replacement of faulty components, such as batteries, screens, and processors, without replacing the entire device, thereby extending its lifespan and reducing electronic waste. Companies like Fairphone and Google's Project Ara have pioneered modular smartphones that prioritize repairability and user customization.
- **3D Printing and Additive Manufacturing:** 3D printing, also known as additive manufacturing, enables the creation of complex shapes and custom-designed components using a variety of materials, including plastics, metals, and ceramics. 3D printing technology can be used to produce replacement parts and components for digital products, allowing for on-demand manufacturing and repair. By reducing the need for mass production and inventory storage, 3D printing can minimize waste and enable more sustainable supply chains(Boopathi, Khare, et al., 2023; Palaniappan et al., 2023; Senthil et al., 2023).
- **Blockchain Technology:** Blockchain technology offers decentralized and transparent systems for tracking and tracing digital products throughout their lifecycle, from manufacturing to disposal. Blockchain-based solutions can provide immutable records of product provenance, ownership, and repair history, enabling consumers to verify the authenticity and sustainability of digital prod-

ucts. By enhancing transparency and accountability in supply chains, blockchain technology can promote responsible manufacturing practices and reduce the risk of counterfeit products(Kumar et al., 2023; Maguluri, Arularasan, et al., 2023; Sundaramoorthy et al., 2024).

- **Internet of Things (IoT) and Product Lifecycle Management:** The Internet of Things (IoT) connects digital products and devices to the internet, enabling real-time monitoring, data collection, and remote management throughout their lifecycle. IoT technology can be used to track product usage, performance, and maintenance requirements, allowing manufacturers to proactively identify issues, schedule repairs, and optimize product lifespans. IoT-enabled product lifecycle management systems can improve efficiency, reduce downtime, and extend the usability of digital products, ultimately reducing electronic waste(Ali et al., 2024; Boopathi, 2024g; Rahamathunnisa et al., 2024a).
- **Artificial Intelligence (AI) for Predictive Maintenance:** Artificial intelligence (AI) algorithms can analyze large datasets and sensor data to predict equipment failures and maintenance needs in digital products and devices. AI-powered predictive maintenance systems can identify potential issues before they occur, enabling proactive maintenance and repair activities to prevent downtime and extend product lifespans. By optimizing maintenance schedules and reducing unplanned downtime, AI technology can improve the reliability and longevity of digital products, minimizing waste and resource consumption(Boopathi, 2024h, 2024a).
- **Remanufacturing and Refurbishment:** Remanufacturing and refurbishment processes involve restoring used or defective digital products to like-new condition through cleaning, repair, and replacement of components. Remanufactured products undergo rigorous quality testing and inspection to ensure they meet performance standards and warranty requirements. Remanufacturing and refurbishment programs can extend the lifespan of digital products, reduce electronic waste, and provide cost-effective alternatives to new purchases for consumers and businesses.

Figure 3. Technologies and innovations for digital waste reduction

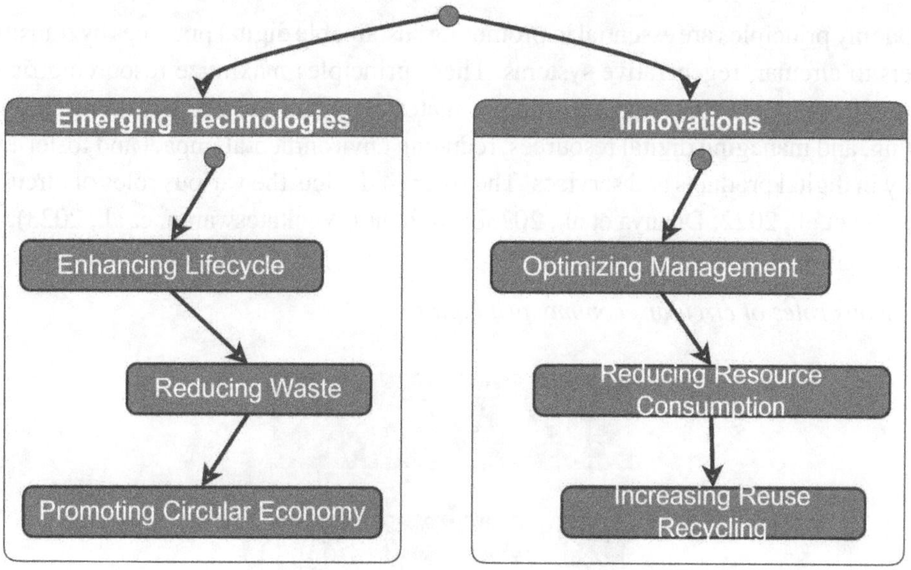

Innovations in Design, Manufacturing, and Recycling Aimed at Minimizing Digital Waste

Innovations in design, manufacturing, and recycling are crucial for reducing digital waste and promoting a circular economy in the digital age. These focus on optimizing product lifecycle management, reducing resource consumption, and increasing reuse and recycling of digital products(Rahamathunnisa et al., 2024b; Ugandar et al., 2023; Venkateswaran, Kumar, et al., 2023).

Recycling Innovations

- **Advanced Recycling Technologies:** Advanced recycling technologies, such as mechanical recycling, chemical recycling, and pyrolysis, enable the recovery of valuable materials from electronic waste. These technologies can process complex electronic products and separate materials into pure fractions for reuse in new manufacturing processes, reducing the need for virgin raw materials and minimizing waste(Naveeenkumar et al., 2024; Puranik et al., 2024).
- **Urban Mining:** Urban mining involves recovering valuable materials from electronic waste through processes such as shredding, sorting, and extraction. Urban mining techniques extract metals, plastics, and rare earth elements from discarded electronic products, providing a sustainable source of raw materials for manufacturing and reducing reliance on traditional mining practices.
- **Circuit Board Recycling:** Circuit boards, which contain valuable metals like gold, silver, and copper, can be recycled using specialized processes that recover and refine these materials for reuse. Innovations in circuit board recycling enable the recovery of precious metals from electronic waste, reducing the environmental impact of mining and promoting resource conservation.

Role of Circular Economy Principles

Circular economy principles are essential in promoting sustainable digital practices by transitioning from linear models to circular, regenerative systems. These principles maximize resource efficiency, minimize waste, and encourage reuse and recycling of materials. They provide a framework for designing, manufacturing, and managing digital resources, reducing environmental impact and fostering long-term sustainability in digital products and services. The figure 4 depicts the various roles of circular economy principles(Babu et al., 2022; Dhanya et al., 2023b; Mohanty, Venkateswaran, et al., 2023).

Figure 4. Various roles of circular economy principles

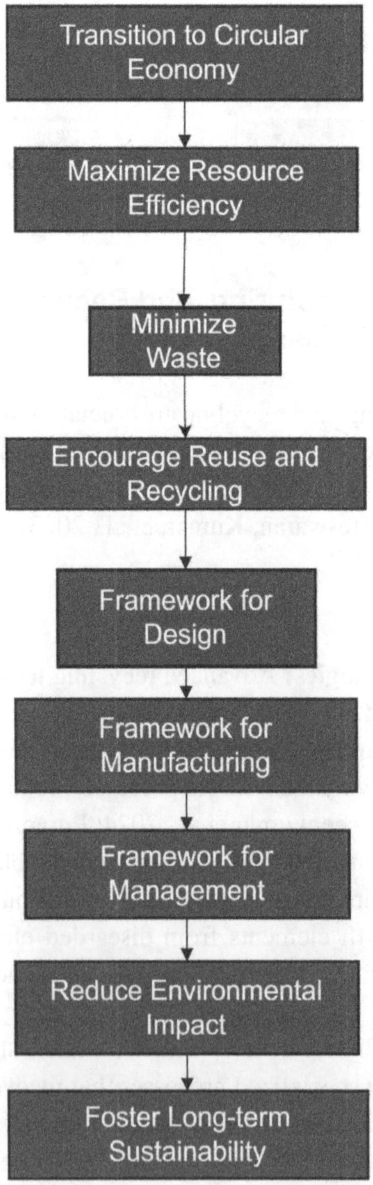

CONSUMER EDUCATION AND BEHAVIOR CHANGE

Consumer awareness and behavior significantly influence digital waste generation, as consumer choices drive demand for electronic products and services. Addressing these factors is crucial for reducing digital waste and promoting sustainable practices in the digital age. Strategies for educating consumers and encouraging sustainable digital behaviors are proposed(Boopathi, 2024e; Rebecca et al., 2024). Figure 5 illustrates the impact of consumer education on behavior change.

Figure 5. Illustration for consumer education and behavior change

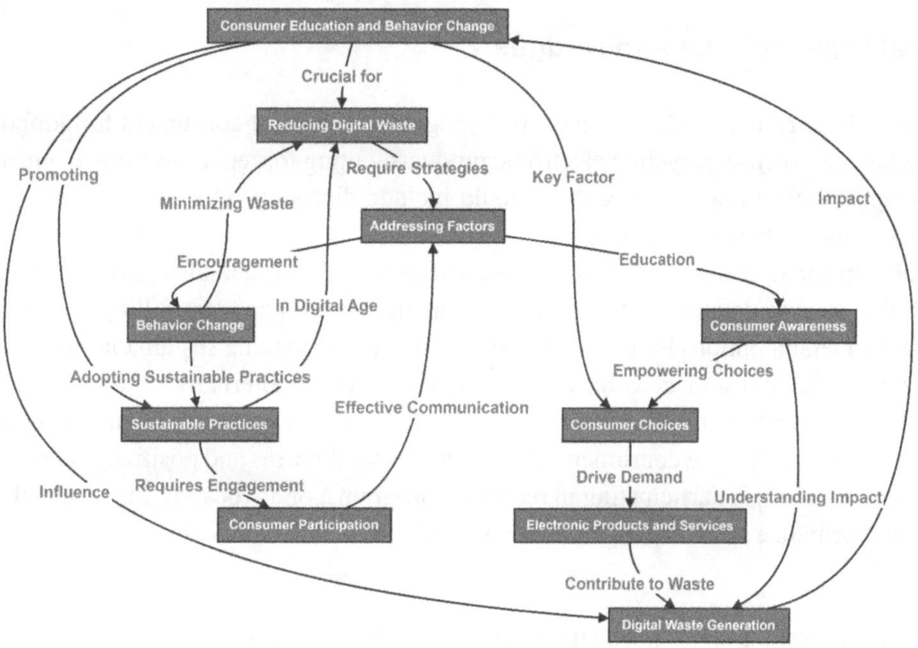

Incentive programs can be implemented to encourage sustainable behaviors, such as recycling electronic products or choosing energy-efficient devices. These programs can include discounts, rebates, or loyalty points. Digital platforms should also have default options that promote sustainability, such as energy-saving modes, paperless billing, or recycling programs. These options make sustainable choices the default choice, requiring consumers to actively opt out if they prefer alternative options. Social norms and peer influence can also be leveraged to encourage sustainable behaviors, such as sharing repair tips or participating in recycling programs(Boopathi, 2024c; Boopathi & Kumar, 2024b).

Strategies for Educating Consumers

- **Public Awareness Campaigns:** Launch public awareness campaigns to educate consumers about the environmental and social impacts of digital consumption, including the lifecycle impacts of electronic products, the importance of responsible disposal and recycling, and the benefits of sustainable consumption practices.

- **Product Labeling and Certification:** Implement product labeling and certification schemes to provide consumers with clear and transparent information about the environmental performance and sustainability attributes of digital products. Labels could include information on energy efficiency, recyclability, and environmental certifications to help consumers make informed purchasing decisions.
- **Consumer Education Programs:** Develop consumer education programs in collaboration with schools, universities, non-profit organizations, and community groups to raise awareness about sustainable digital practices, teach skills for product repair and maintenance, and promote responsible consumption behaviors.

Behavioral Interventions and Nudges

- **Incentive Programs:** Implement incentive programs to reward consumers for adopting sustainable behaviors, such as recycling electronic products, opting for repair over replacement, or choosing energy-efficient devices. Incentives could include discounts, rebates, loyalty points, or other rewards to motivate behavior change.
- **Default Options:** Design digital platforms and services with default options that promote sustainability, such as default settings for energy-saving modes, paperless billing, or opt-in recycling programs. Default options influence consumer behavior by making sustainable choices the default choice, requiring consumers to actively opt out if they prefer alternative options.
- **Social Norms and Peer Influence:** Leverage social norms and peer influence to encourage sustainable behaviors among consumers. Highlighting social norms and positive peer behaviors, such as sharing repair tips, participating in recycling programs, or advocating for sustainable products, can help normalize sustainable practices and motivate behavior change.

COLLABORATION AND PARTNERSHIPS FOR SUSTAINABLE SOLUTIONS

Collaboration among stakeholders is crucial for addressing digital waste and advancing sustainable solutions in the digital economy. Fostering partnerships between businesses, governments, NGOs, and academia allows them to leverage their expertise, resources, and networks to drive initiatives promoting sustainability and minimizing digital waste.

Stakeholder Collaboration

- **Shared Responsibility:** Digital waste management is a complex and multifaceted issue that requires coordinated action from multiple stakeholders, including manufacturers, retailers, policymakers, consumers, and waste management organizations. Collaborative approaches that involve all stakeholders in the value chain can ensure shared responsibility for reducing digital waste and promoting sustainable practices.
- **Knowledge Sharing:** Collaboration enables the sharing of knowledge, best practices, and lessons learned among stakeholders, facilitating collective learning and continuous improvement in digital waste management efforts. By exchanging insights and experiences, stakeholders can identify

common challenges, explore innovative solutions, and leverage each other's expertise to drive progress towards sustainability goals.
- **Resource Mobilization:** Collaborative initiatives can mobilize resources and funding from multiple sources to support sustainability projects and digital waste management efforts. By pooling financial resources, technical expertise, and infrastructure, stakeholders can achieve greater impact and scale in addressing digital waste challenges than they could individually.

Partnerships Between Stakeholders

- **Business-Government Partnerships:** Partnerships between businesses and governments can drive policy development, regulatory compliance, and industry standards for digital waste management. By working together, businesses and governments can develop effective regulatory frameworks, support EPR schemes, and promote sustainable practices throughout the product lifecycle.
- **Business-NGO Partnerships:** Collaboration between businesses and NGOs can enhance corporate social responsibility (CSR) initiatives, promote transparency and accountability, and engage consumers in sustainability efforts. NGOs can provide expertise in environmental advocacy, community engagement, and stakeholder consultation, while businesses can contribute resources, funding, and access to markets to support NGO-led initiatives.
- **Academic-Industry Partnerships:** Partnerships between academia and industry can drive research and innovation in sustainable technologies, materials science, and waste management practices. Academic institutions can conduct research, develop technologies, and provide technical expertise, while industry partners can offer funding, access to data, and real-world testing environments to validate research findings and bring innovations to market.

Synergies Between Sustainability Initiatives and Digital Waste Management

- **Circular Economy Initiatives:** Collaboration between stakeholders can promote circular economy principles in digital product design, manufacturing, and end-of-life management. By adopting circular economy strategies, such as eco-design, product take-back programs, and materials recycling, businesses can minimize waste, conserve resources, and promote sustainable consumption and production patterns(Dhanya et al., 2023a; Hussain et al., 2023).
- **Climate Action and Energy Efficiency:** Digital waste management efforts can align with broader sustainability initiatives, such as climate action and energy efficiency programs. By reducing energy consumption, optimizing resource use, and promoting renewable energy sources, stakeholders can mitigate environmental impacts and contribute to global efforts to address climate change and promote sustainable development.

Collaboration and partnerships are crucial for addressing digital waste and promoting sustainable solutions in the digital economy. They allow stakeholders to utilize their strengths, drive innovation, and promote responsible consumption and production practices, ultimately contributing to a more sustainable future.

FUTURE TRENDS AND CHALLENGES

Future digital consumption trends and waste generation require considering technological advancements, consumer behaviors, regulatory developments, and global sustainability challenges. While digital consumption offers convenience, connectivity, and efficiency, it also poses significant challenges in electronic waste generation and environmental impact. Future trends, challenges, and opportunities for sustainable digital practices are discussed(Boopathi & Kumar, 2024c; Durairaj et al., 2023a; Venkateswaran, Vidhya, Ayyannan, et al., 2023; Venkateswaran, Vidhya, Naik, et al., 2023).

Figure 6. Future digital consumption trends and waste generation

Future Trends in Digital Consumption

- **Rapid Technological Advancements:** Technological innovations, such as artificial intelligence, Internet of Things (IoT), and 5G connectivity, will continue to drive digital consumption trends and shape the way people interact with digital products and services.
- **Growth of E-commerce and Digital Services:** The continued growth of e-commerce platforms, streaming services, and digital entertainment content will lead to increased consumption of digital products and data, contributing to electronic waste generation and energy consumption(Rahamathunnisa et al., 2024a; Vijayakumar et al., 2024).
- **Emergence of New Technologies:** Emerging technologies, such as virtual reality (VR), augmented reality (AR), and wearable devices, will create new opportunities for digital consumption and interaction, but may also introduce new challenges in terms of waste management and resource consumption.

Challenges and Barriers to Achieving Sustainable Digital Practices

- **E-waste Management Challenges:** The rapid turnover of digital devices and the complexity of electronic products pose challenges for e-waste management and recycling. Limited infrastructure, inadequate recycling facilities, and lack of awareness about proper disposal practices contribute to electronic waste accumulation and environmental pollution(Boopathi, 2024f, 2024i; P. Das et al., 2024).

- **Global Supply Chain Impacts:** The global nature of the electronics supply chain, with components sourced from multiple countries, presents challenges in terms of environmental regulations, labor standards, and supply chain transparency. Ensuring ethical sourcing practices and responsible manufacturing processes across the supply chain remains a complex issue.

Opportunities for Innovation and Adaptation

- **Technological Innovations:** Investing in research and development of sustainable technologies, materials science, and waste management solutions can drive innovation and support the transition to a more sustainable digital economy. Advancements in recycling technologies, modular design, and materials recovery offer promising opportunities for reducing electronic waste and promoting resource efficiency(Boopathi et al., 2022; Gunasekaran et al., 2022; Yupapin et al., 2023).
- **Consumer Awareness and Education:** Educating consumers about the environmental and social impacts of digital consumption, raising awareness about proper disposal and recycling practices, and empowering consumers to make informed choices can drive demand for sustainable products and services and promote a culture of sustainability(S. Das et al., 2024; Durairaj et al., 2023b; Prabhuswamy et al., 2024; Sharma et al., 2024).

CONCLUSION

The chapter on sustainable digital practices and digital waste management emphasizes the need to address the environmental and social impacts of digital consumption in today's interconnected world. It explores ethical considerations, policy frameworks, technological innovations, and collaborative partnerships in sustainable digital practices. It also examines the role of circular economy principles in promoting resource efficiency, minimizing waste generation, and fostering a sustainability culture in the digital economy.

This chapter highlights the global challenge of digital waste management, emphasizing the need for collective action from various stakeholders. Circular economy principles provide a roadmap for designing, manufacturing, and managing digital products to maximize resource efficiency, minimize waste, and promote reuse and recycling. Collaboration and partnerships are crucial for sustainable solutions in digital waste management. Future trends in digital consumption, such as rapid technological advancements and e-commerce growth, present both opportunities and challenges for sustainable practices. Innovation and adaptation are essential for overcoming barriers to sustainable digital practices. Investing in research, technology development, and policy innovation can drive initiatives promoting sustainability and minimizing digital waste generation.

Sustainable digital practices necessitate a collective effort from all stakeholders, guided by environmental stewardship, social responsibility, and economic viability. By integrating sustainability into business strategies, policy frameworks, and consumer behaviors, we can foster a more resilient digital economy for future generations.

ABBREVIATIONS

- AI - Artificial Intelligence
- APEC - Asia-Pacific Economic Cooperation
- AR - Augmented Reality
- CSR - Corporate Social Responsibility
- EPA - Environmental Protection Agency
- EPR - Extended Producer Responsibility
- EU - European Union
- IEC - International Electrotechnical Commission
- IKEA - Ingvar Kamprad Elmtaryd Agunnaryd (Swedish furniture retailer)
- ISO - International Organization for Standardization
- NGO - Non-Governmental Organization
- RCRA - Resource Conservation and Recovery Act
- VR - Virtual Reality
- WEEE - Waste Electrical and Electronic Equipment

REFERENCES

Agrawal, A. V., Magulur, L. P., Priya, S. G., Kaur, A., Singh, G., & Boopathi, S. (2023). Smart Precision Agriculture Using IoT and WSN. In *Handbook of Research on Data Science and Cybersecurity Innovations in Industry 4.0 Technologies* (pp. 524–541). IGI Global. doi:10.4018/978-1-6684-8145-5.ch026

Agrawal, A. V., Shashibhushan, G., Pradeep, S., Padhi, S., Sugumar, D., & Boopathi, S. (2023). Synergizing Artificial Intelligence, 5G, and Cloud Computing for Efficient Energy Conversion Using Agricultural Waste. In Sustainable Science and Intelligent Technologies for Societal Development (pp. 475–497). IGI Global.

Ali, M. N., Senthil, T., Ilakkiya, T., Hasan, D. S., Ganapathy, N. B. S., & Boopathi, S. (2024). IoT's Role in Smart Manufacturing Transformation for Enhanced Household Product Quality. In *Advanced Applications in Osmotic Computing* (pp. 252–289). IGI Global. doi:10.4018/979-8-3693-1694-8.ch014

Babu, B. S., Kamalakannan, J., Meenatchi, N., Karthik, S., & Boopathi, S. (2022). Economic impacts and reliability evaluation of battery by adopting Electric Vehicle. *IEEE Explore*, (pp. 1–6). IEEE.

Boopathi, S., Hari Balaji, V., Mageswari, M., & Asif, M. M. (2022). Influences of Boron Carbide Particles on the Wear Rate and Tensile Strength of AA2014 Surface Composite Fabricated by Friction-Stir Processing. *Mater-Tehnol.Si. Materials Technology*, *56*(3), 263–270.

Boopathi, S. (2023). Internet of Things-Integrated Remote Patient Monitoring System: Healthcare Application. In *Dynamics of Swarm Intelligence Health Analysis for the Next Generation* (pp. 137–161). IGI Global. doi:10.4018/978-1-6684-6894-4.ch008

Boopathi, S. (2024a). Advancements in Machine Learning and AI for Intelligent Systems in Drone Applications for Smart City Developments. In *Futuristic e-Governance Security With Deep Learning Applications* (pp. 15–45). IGI Global. doi:10.4018/978-1-6684-9596-4.ch002

Boopathi, S. (2024b). Balancing Innovation and Security in the Cloud: Navigating the Risks and Rewards of the Digital Age. In Improving Security, Privacy, and Trust in Cloud Computing (pp. 164–193). IGI Global.

Boopathi, S. (2024c). Balancing Innovation and Security in the Cloud: Navigating the Risks and Rewards of the Digital Age. In Improving Security, Privacy, and Trust in Cloud Computing (pp. 164–193). IGI Global.

Boopathi, S. (2024d). Energy Cascade Conversion System and Energy-Efficient Infrastructure. In Sustainable Development in AI, Blockchain, and E-Governance Applications (pp. 47–71). IGI Global. doi:10.4018/979-8-3693-1722-8.ch004

Boopathi, S. (2024e). Energy Cascade Conversion System and Energy-Efficient Infrastructure. In Sustainable Development in AI, Blockchain, and E-Governance Applications (pp. 47–71). IGI Global. doi:10.4018/979-8-3693-1722-8.ch004

Boopathi, S. (2024f). Energy Cascade Conversion System and Energy-Efficient Infrastructure. In Sustainable Development in AI, Blockchain, and E-Governance Applications (pp. 47–71). IGI Global. doi:10.4018/979-8-3693-1722-8.ch004

Boopathi, S. (2024g). Sustainable Development Using IoT and AI Techniques for Water Utilization in Agriculture. In Sustainable Development in AI, Blockchain, and E-Governance Applications (pp. 204–228). IGI Global. doi:10.4018/979-8-3693-1722-8.ch012

Boopathi, S. (2024h). Sustainable Development Using IoT and AI Techniques for Water Utilization in Agriculture. In Sustainable Development in AI, Blockchain, and E-Governance Applications (pp. 204–228). IGI Global. doi:10.4018/979-8-3693-1722-8.ch012

Boopathi, S. (2024i). Sustainable Development Using IoT and AI Techniques for Water Utilization in Agriculture. In Sustainable Development in AI, Blockchain, and E-Governance Applications (pp. 204–228). IGI Global. doi:10.4018/979-8-3693-1722-8.ch012

Boopathi, S., Balasubramani, V., Kumar, R. S., & Singh, G. R. (2021). The influence of human hair on kenaf and Grewia fiber-based hybrid natural composite material: An experimental study. *Functional Composites and Structures*, *3*(4), 045011. doi:10.1088/2631-6331/ac3afc

Boopathi, S., & Davim, J. P. (2023a). Applications of Nanoparticles in Various Manufacturing Processes. In *Sustainable Utilization of Nanoparticles and Nanofluids in Engineering Applications* (pp. 1–31). IGI Global. doi:10.4018/978-1-6684-9135-5.ch001

Boopathi, S., & Davim, J. P. (2023b). *Sustainable Utilization of Nanoparticles and Nanofluids in Engineering Applications*. IGI Global. doi:10.4018/978-1-6684-9135-5

Boopathi, S., & Khang, A. (2023). AI-Integrated Technology for a Secure and Ethical Healthcare Ecosystem. In *AI and IoT-Based Technologies for Precision Medicine* (pp. 36–59). IGI Global. doi:10.4018/979-8-3693-0876-9.ch003

Boopathi, S., & Khare, R. KG, J. C., Muni, T. V., & Khare, S. (2023). Additive Manufacturing Developments in the Medical Engineering Field. In Development, Properties, and Industrial Applications of 3D Printed Polymer Composites (pp. 86–106). IGI Global.

Boopathi, S., & Kumar, P. (2024a). Advanced bioprinting processes using additive manufacturing technologies: Revolutionizing tissue engineering. *3D Printing Technologies: Digital Manufacturing, Artificial Intelligence, Industry 4.0*, 95.

Boopathi, S., Umareddy, M., & Elangovan, M. (2023). Applications of Nano-Cutting Fluids in Advanced Machining Processes. In *Sustainable Utilization of Nanoparticles and Nanofluids in Engineering Applications* (pp. 211–234). IGI Global. doi:10.4018/978-1-6684-9135-5.ch009

D'Adamo, I., Mazzanti, M., Morone, P., & Rosa, P. (2022). Assessing the relation between waste management policies and circular economy goals. *Waste Management (New York, N.Y.)*, *154*, 27–35. doi:10.1016/j.wasman.2022.09.031 PMID:36206656

Das, P., Ramapraba, P., Seethalakshmi, K., Mary, M. A., Karthick, S., & Sampath, B. (2024). Sustainable Advanced Techniques for Enhancing the Image Process. In *Fostering Cross-Industry Sustainability With Intelligent Technologies* (pp. 350–374). IGI Global. doi:10.4018/979-8-3693-1638-2.ch022

Das, S., Lekhya, G., Shreya, K., Shekinah, K. L., Babu, K. K., & Boopathi, S. (2024). Fostering Sustainability Education Through Cross-Disciplinary Collaborations and Research Partnerships: Interdisciplinary Synergy. In Facilitating Global Collaboration and Knowledge Sharing in Higher Education With Generative AI (pp. 60–88). IGI Global.

Dhanya, D., Kumar, S. S., Thilagavathy, A., Prasad, D., & Boopathi, S. (2023a). Data Analytics and Artificial Intelligence in the Circular Economy: Case Studies. In Intelligent Engineering Applications and Applied Sciences for Sustainability (pp. 40–58). IGI Global.

Durairaj, M., Jayakumar, S., Karpagavalli, V., Maheswari, B. U., Boopathi, S., & ... (2023a). Utilization of Digital Tools in the Indian Higher Education System During Health Crises. In *Multidisciplinary Approaches to Organizational Governance During Health Crises* (pp. 1–21). IGI Global. doi:10.4018/978-1-7998-9213-7.ch001

Gunasekaran, K., Boopathi, S., & Sureshkumar, M. (2022). Analysis of a Cryogenically Cooled Near-Dry WEDM Process using Different Dielectrics. *Mater-Tehnol.Si. Materials Technology*, *56*(2), 179–186.

Gupta, H., Kumar, A., & Wasan, P. (2021). Industry 4.0, cleaner production and circular economy: An integrative framework for evaluating ethical and sustainable business performance of manufacturing organizations. *Journal of Cleaner Production*, *295*, 126253. doi:10.1016/j.jclepro.2021.126253

Hanumanthakari, S., Gift, M. M., Kanimozhi, K., Bhavani, M. D., Bamane, K. D., & Boopathi, S. (2023a). Biomining Method to Extract Metal Components Using Computer-Printed Circuit Board E-Waste. In *Handbook of Research on Safe Disposal Methods of Municipal Solid Wastes for a Sustainable Environment* (pp. 123–141). IGI Global. doi:10.4018/978-1-6684-8117-2.ch010

Harikaran, M., Boopathi, S., Gokulakannan, S., & Poonguzhali, M. (2023a). Study on the Source of E-Waste Management and Disposal Methods. In *Sustainable Approaches and Strategies for E-Waste Management and Utilization* (pp. 39–60). IGI Global. doi:10.4018/978-1-6684-7573-7.ch003

Henneman, L. R., Liu, C., Mulholland, J. A., & Russell, A. G. (2017). Evaluating the effectiveness of air quality regulations: A review of accountability studies and frameworks. *Journal of the Air & Waste Management Association*, *67*(2), 144–172. doi:10.1080/10962247.2016.1242518 PMID:27715473

Hussain, Z., Babe, M., Saravanan, S., Srimathy, G., Roopa, H., & Boopathi, S. (2023). Optimizing Biomass-to-Biofuel Conversion: IoT and AI Integration for Enhanced Efficiency and Sustainability. In Circular Economy Implementation for Sustainability in the Built Environment (pp. 191–214). IGI Global.

Ingle, R. B., Swathi, S., Mahendran, G., Senthil, T., Muralidharan, N., & Boopathi, S. (2023). Sustainability and Optimization of Green and Lean Manufacturing Processes Using Machine Learning Techniques. In *Circular Economy Implementation for Sustainability in the Built Environment* (pp. 261–285). IGI Global. doi:10.4018/978-1-6684-8238-4.ch012

Janardhana, K., Singh, V., Singh, S. N., Babu, T. R., Bano, S., & Boopathi, S. (2023a). Utilization Process for Electronic Waste in Eco-Friendly Concrete: Experimental Study. In Sustainable Approaches and Strategies for E-Waste Management and Utilization (pp. 204–223). IGI Global.

Jones, P., & Wynn, M. (2021). The leading digital technology companies and their approach to sustainable development. *Sustainability (Basel)*, *13*(12), 6612. doi:10.3390/su13126612

Karthik, S., Hemalatha, R., Aruna, R., Deivakani, M., Reddy, R. V. K., & Boopathi, S. (2023). Study on Healthcare Security System-Integrated Internet of Things (IoT). In Perspectives and Considerations on the Evolution of Smart Systems (pp. 342–362). IGI Global.

Khan, S. A. R., Ponce, P., Thomas, G., Yu, Z., Al-Ahmadi, M. S., & Tanveer, M. (2021). Digital technologies, circular economy practices and environmental policies in the era of COVID-19. *Sustainability (Basel)*, *13*(22), 12790. doi:10.3390/su132212790

Koshariya, A. K., Khatoon, S., Marathe, A. M., Suba, G. M., Baral, D., & Boopathi, S. (2023). Agricultural Waste Management Systems Using Artificial Intelligence Techniques. In *AI-Enabled Social Robotics in Human Care Services* (pp. 236–258). IGI Global. doi:10.4018/978-1-6684-8171-4.ch009

Kumar, P. R., Meenakshi, S., Shalini, S., Devi, S. R., & Boopathi, S. (2023). Soil Quality Prediction in Context Learning Approaches Using Deep Learning and Blockchain for Smart Agriculture. In Effective AI, Blockchain, and E-Governance Applications for Knowledge Discovery and Management (pp. 1–26). IGI Global. doi:10.4018/978-1-6684-9151-5.ch001

Kushwah, J. S., Gupta, M., Shrivastava, S., Saxena, N., Saini, R., & Boopathi, S. (2024). Psychological Impacts, Prevention Strategies, and Intervention Approaches Across Age Groups: Unmasking Cyberbullying. In Change Dynamics in Healthcare, Technological Innovations, and Complex Scenarios (pp. 89–109). IGI Global.

Lehtokunnas, T., Mattila, M., Närvänen, E., & Mesiranta, N. (2022). Towards a circular economy in food consumption: Food waste reduction practices as ethical work. *Journal of Consumer Culture*, *22*(1), 227–245. doi:10.1177/1469540520926252

Maguluri, L. P., Ananth, J., Hariram, S., Geetha, C., Bhaskar, A., & Boopathi, S. (2023). Smart Vehicle-Emissions Monitoring System Using Internet of Things (IoT). In Handbook of Research on Safe Disposal Methods of Municipal Solid Wastes for a Sustainable Environment (pp. 191–211). IGI Global.

Maguluri, L. P., Arularasan, A., & Boopathi, S. (2023). Assessing Security Concerns for AI-Based Drones in Smart Cities. In Effective AI, Blockchain, and E-Governance Applications for Knowledge Discovery and Management (pp. 27–47). IGI Global. doi:10.4018/978-1-6684-9151-5.ch002

Maiurova, A., Kurniawan, T. A., Kustikova, M., Bykovskaia, E., Othman, M. H. D., Singh, D., & Goh, H. H. (2022). Promoting digital transformation in waste collection service and waste recycling in Moscow (Russia): Applying a circular economy paradigm to mitigate climate change impacts on the environment. *Journal of Cleaner Production*, *354*, 131604. doi:10.1016/j.jclepro.2022.131604

Medaglia, R., Rukanova, B., & Zhang, Z. (2024). Digital government and the circular economy transition: An analytical framework and a research agenda. *Government Information Quarterly*, *41*(1), 101904. doi:10.1016/j.giq.2023.101904

Mies, A., & Gold, S. (2021). Mapping the social dimension of the circular economy. *Journal of Cleaner Production*, *321*, 128960. doi:10.1016/j.jclepro.2021.128960

Mohanty, A., Jothi, B., Jeyasudha, J., Ranjit, P., Isaac, J. S., & Boopathi, S. (2023). Additive Manufacturing Using Robotic Programming. In *AI-Enabled Social Robotics in Human Care Services* (pp. 259–282). IGI Global. doi:10.4018/978-1-6684-8171-4.ch010

Mohanty, A., Venkateswaran, N., Ranjit, P., Tripathi, M. A., & Boopathi, S. (2023). Innovative Strategy for Profitable Automobile Industries: Working Capital Management. In Handbook of Research on Designing Sustainable Supply Chains to Achieve a Circular Economy (pp. 412–428). IGI Global.

Mondal, S., Singh, S., & Gupta, H. (2023). Green entrepreneurship and digitalization enabling the circular economy through sustainable waste management-An exploratory study of emerging economy. *Journal of Cleaner Production*, *422*, 138433. doi:10.1016/j.jclepro.2023.138433

Nascimento, D. L. M., Alencastro, V., Quelhas, O. L. G., Caiado, R. G. G., Garza-Reyes, J. A., Rocha-Lona, L., & Tortorella, G. (2019). Exploring Industry 4.0 technologies to enable circular economy practices in a manufacturing context: A business model proposal. *Journal of Manufacturing Technology Management*, *30*(3), 607–627. doi:10.1108/JMTM-03-2018-0071

Naveeenkumar, N., Rallapalli, S., Sasikala, K., Priya, P. V., Husain, J., & Boopathi, S. (2024). Enhancing Consumer Behavior and Experience Through AI-Driven Insights Optimization. In *AI Impacts in Digital Consumer Behavior* (pp. 1–35). IGI Global. doi:10.4018/979-8-3693-1918-5.ch001

Palaniappan, M., Tirlangi, S., Mohamed, M. J. S., Moorthy, R. S., Valeti, S. V., & Boopathi, S. (2023). Fused Deposition Modelling of Polylactic Acid (PLA)-Based Polymer Composites: A Case Study. In Development, Properties, and Industrial Applications of 3D Printed Polymer Composites (pp. 66–85). IGI Global.

Persis, D. J., Venkatesh, V., Sreedharan, V. R., Shi, Y., & Sankaranarayanan, B. (2021). Modelling and analysing the impact of Circular Economy; Internet of Things and ethical business practices in the VUCA world: Evidence from the food processing industry. *Journal of Cleaner Production*, *301*, 126871. doi:10.1016/j.jclepro.2021.126871

Prabhuswamy, M., Tripathi, R., Vijayakumar, M., Thulasimani, T., Sundharesalingam, P., & Sampath, B. (2024). A Study on the Complex Nature of Higher Education Leadership: An Innovative Approach. In *Challenges of Globalization and Inclusivity in Academic Research* (pp. 202–223). IGI Global. doi:10.4018/979-8-3693-1371-8.ch013

Puranik, T. A., Shaik, N., Vankudoth, R., Kolhe, M. R., Yadav, N., & Boopathi, S. (2024). Study on Harmonizing Human-Robot (Drone) Collaboration: Navigating Seamless Interactions in Collaborative Environments. In Cybersecurity Issues and Challenges in the Drone Industry (pp. 1–26). IGI Global.

Rahamathunnisa, U., Subhashini, P., Aancy, H. M., Meenakshi, S., Boopathi, S., & ... (2023). Solutions for Software Requirement Risks Using Artificial Intelligence Techniques. In *Handbook of Research on Data Science and Cybersecurity Innovations in Industry 4.0 Technologies* (pp. 45–64). IGI Global.

Rahamathunnisa, U., Sudhakar, K., Padhi, S., Bhattacharya, S., Shashibhushan, G., & Boopathi, S. (2024a). Sustainable Energy Generation From Waste Water: IoT Integrated Technologies. In Adoption and Use of Technology Tools and Services by Economically Disadvantaged Communities: Implications for Growth and Sustainability (pp. 225–256). IGI Global.

Ramudu, K., Mohan, V. M., Jyothirmai, D., Prasad, D., Agrawal, R., & Boopathi, S. (2023). Machine Learning and Artificial Intelligence in Disease Prediction: Applications, Challenges, Limitations, Case Studies, and Future Directions. In Contemporary Applications of Data Fusion for Advanced Healthcare Informatics (pp. 297–318). IGI Global.

Ravisankar, A., Sampath, B., & Asif, M. M. (2023). Economic Studies on Automobile Management: Working Capital and Investment Analysis. In Multidisciplinary Approaches to Organizational Governance During Health Crises (pp. 169–198). IGI Global.

Rebecca, B., Kumar, K. P. M., Padmini, S., Srivastava, B. K., Halder, S., & Boopathi, S. (2024). Convergence of Data Science-AI-Green Chemistry-Affordable Medicine: Transforming Drug Discovery. In *Handbook of Research on AI and ML for Intelligent Machines and Systems* (pp. 348–373). IGI Global.

Revathi, S., Babu, M., Rajkumar, N., Meti, V. K. V., Kandavalli, S. R., & Boopathi, S. (2024). Unleashing the Future Potential of 4D Printing: Exploring Applications in Wearable Technology, Robotics, Energy, Transportation, and Fashion. In Human-Centered Approaches in Industry 5.0: Human-Machine Interaction, Virtual Reality Training, and Customer Sentiment Analysis (pp. 131–153). IGI Global.

Roberts, H., Zhang, J., Bariach, B., Cowls, J., Gilburt, B., Juneja, P., Tsamados, A., Ziosi, M., Taddeo, M., & Floridi, L. (2022). Artificial intelligence in support of the circular economy: Ethical considerations and a path forward. *AI & Society*, 1–14.

Ruohomaa, H., & Ivanova, N. (2019). From solid waste management towards the circular economy and digital driven symbiosis. *IOP Conference Series. Earth and Environmental Science*, *337*(1), 012032. doi:10.1088/1755-1315/337/1/012032

Rusch, M., Schöggl, J.-P., & Baumgartner, R. J. (2023). Application of digital technologies for sustainable product management in a circular economy: A review. *Business Strategy and the Environment*, *32*(3), 1159–1174. doi:10.1002/bse.3099

Sampath, B. (2021). *Sustainable Eco-Friendly Wire-Cut Electrical Discharge Machining: Gas Emission Analysis*.

Sampath, B., Sureshkumar, T., Yuvaraj, M., & Velmurugan, D. (2021). Experimental Investigations on Eco-Friendly Helium-Mist Near-Dry Wire-Cut EDM of M2-HSS Material. *Materials Research Proceedings*, *19*, 175–180.

Sankar, K. M., Booba, B., & Boopathi, S. (2023). Smart Agriculture Irrigation Monitoring System Using Internet of Things. In *Contemporary Developments in Agricultural Cyber-Physical Systems* (pp. 105–121). IGI Global. doi:10.4018/978-1-6684-7879-0.ch006

Saravanan, A., Venkatasubramanian, R., Khare, R., Surakasi, R., Boopathi, S., Ray, S., & Sudhakar, M. (2022). POLICY TRENDS OF RENEWABLE ENERGY AND NON. *Renewable Energy*.

Satav, S. D., Hasan, D. S., Pitchai, R., Mohanaprakash, T., Sultanuddin, S., & Boopathi, S. (2023). Next generation of internet of things (ngiot) in healthcare systems. In *Sustainable Science and Intelligent Technologies for Societal Development* (pp. 307–330). IGI Global.

Selvakumar, S., Adithe, S., Isaac, J. S., Pradhan, R., Venkatesh, V., & Sampath, B. (2023a). A Study of the Printed Circuit Board (PCB) E-Waste Recycling Process. In Sustainable Approaches and Strategies for E-Waste Management and Utilization (pp. 159–184). IGI Global.

Selvakumar, S., Adithe, S., Isaac, J. S., Pradhan, R., Venkatesh, V., & Sampath, B. (2023b). A Study of the Printed Circuit Board (PCB) E-Waste Recycling Process. In Sustainable Approaches and Strategies for E-Waste Management and Utilization (pp. 159–184). IGI Global.

Selvakumar, S., Shankar, R., Ranjit, P., Bhattacharya, S., Gupta, A. S. G., & Boopathi, S. (2023a). E-Waste Recovery and Utilization Processes for Mobile Phone Waste. In *Handbook of Research on Safe Disposal Methods of Municipal Solid Wastes for a Sustainable Environment* (pp. 222–240). IGI Global. doi:10.4018/978-1-6684-8117-2.ch016

Sengeni, D., Padmapriya, G., Imambi, S. S., Suganthi, D., Suri, A., & Boopathi, S. (2023). Biomedical waste handling method using artificial intelligence techniques. In *Handbook of Research on Safe Disposal Methods of Municipal Solid Wastes for a Sustainable Environment* (pp. 306–323). IGI Global. doi:10.4018/978-1-6684-8117-2.ch022

Senthil, T., Puviyarasan, M., Babu, S. R., Surakasi, R., Sampath, B., & Associates. (2023). Industrial Robot-Integrated Fused Deposition Modelling for the 3D Printing Process. In Development, Properties, and Industrial Applications of 3D Printed Polymer Composites (pp. 188–210). IGI Global.

Sharma, D. M., Ramana, K. V., Jothilakshmi, R., Verma, R., Maheswari, B. U., & Boopathi, S. (2024). Integrating Generative AI Into K-12 Curriculums and Pedagogies in India: Opportunities and Challenges. *Facilitating Global Collaboration and Knowledge Sharing in Higher Education With Generative AI*, (pp. 133–161). IEEE.

Subha, S., Inbamalar, T., Komala, C., Suresh, L. R., Boopathi, S., & Alaskar, K. (2023). A Remote Health Care Monitoring system using internet of medical things (IoMT). *IEEE Explore*, (pp. 1–6). IEEE.

Sundaramoorthy, K., Singh, A., Sumathy, G., Maheshwari, A., Arunarani, A., & Boopathi, S. (2024). A Study on AI and Blockchain-Powered Smart Parking Models for Urban Mobility. In *Handbook of Research on AI and ML for Intelligent Machines and Systems* (pp. 223–250). IGI Global.

Ugandar, R., Rahamathunnisa, U., Sajithra, S., Christiana, M. B. V., Palai, B. K., & Boopathi, S. (2023). Hospital Waste Management Using Internet of Things and Deep Learning: Enhanced Efficiency and Sustainability. In Applications of Synthetic Biology in Health, Energy, and Environment (pp. 317–343). IGI Global.

Venkateswaran, N., Kumar, S. S., Diwakar, G., Gnanasangeetha, D., & Boopathi, S. (2023). Synthetic Biology for Waste Water to Energy Conversion: IoT and AI Approaches. *Applications of Synthetic Biology in Health. Energy & Environment*, 360–384.

Venkateswaran, N., Vidhya, K., Ayyannan, M., Chavan, S. M., Sekar, K., & Boopathi, S. (2023). A Study on Smart Energy Management Framework Using Cloud Computing. In 5G, Artificial Intelligence, and Next Generation Internet of Things: Digital Innovation for Green and Sustainable Economies (pp. 189–212). IGI Global. doi:10.4018/978-1-6684-8634-4.ch009

Venkateswaran, N., Vidhya, R., Naik, D. A., Raj, T. M., Munjal, N., & Boopathi, S. (2023). Study on Sentence and Question Formation Using Deep Learning Techniques. In *Digital Natives as a Disruptive Force in Asian Businesses and Societies* (pp. 252–273). IGI Global. doi:10.4018/978-1-6684-6782-4.ch015

Vignesh, S., Arulshri, K., SyedSajith, S., Kathiresan, S., Boopathi, S., & Dinesh Babu, P. (2018). Design and development of ornithopter and experimental analysis of flapping rate under various operating conditions. *Materials Today: Proceedings*, *5*(11), 25185–25194. doi:10.1016/j.matpr.2018.10.320

Vijayakumar, G. N. S., Domakonda, V. K., Farooq, S., Kumar, B. S., Pradeep, N., & Boopathi, S. (2024). Sustainable Developments in Nano-Fluid Synthesis for Various Industrial Applications. In Adoption and Use of Technology Tools and Services by Economically Disadvantaged Communities: Implications for Growth and Sustainability (pp. 48–81). IGI Global.

Yupapin, P., Trabelsi, Y., Nattappan, A., & Boopathi, S. (2023). Performance improvement of wire-cut electrical discharge machining process using cryogenically treated super-conductive state of Monel-K500 alloy. *Iranian Journal of Science and Technology. Transaction of Mechanical Engineering*, *47*(1), 267–283. doi:10.1007/s40997-022-00513-0

Chapter 9
Applications of Machine Learning Techniques for Achieving Financial Sustainability

Arshi Naim
https://orcid.org/0000-0003-1325-6964
King Khalid University, Saudi Arabia

ABSTRACT

In the contemporary business landscape, enterprises grapple with the pursuit of marginal profits, recognizing that the bedrock of their success lies in financial sustainability. Consequently, the mitigation or prevention of business financial risks (BFR) emerges as a pivotal factor in enhancing firms' profitability prospects. Various technological strategies are employed to address the challenges posed by financial risks, with machine learning (ML) standing out as a particularly pertinent and effective solution across diverse business sectors. In the current milieu, the significance and efficacy of ML are keenly acknowledged, with businesses leveraging this cutting-edge technology to tackle challenges associated with BFR.

INTRODUCTION

Machine Learning (ML) holds significant relevance in the realm of business, particularly in the domain of financial analysis, as highlighted by Naim (2022). The infusion of ML has ushered in a new and diverse dimension to finance, contributing to the feasibility of business operations (Lahmiri and Bekiros, 2019). Notably, the application of ML has proven instrumental in providing firms with a competitive edge, leading to substantial advantages in their business practices (Leo et al., 2019). The current business landscape witnesses a widespread integration of Artificial Intelligence (AI), ML, and Deep Learning (DL) across various operational facets, resulting in notable achievements and returns for enterprises. Figure 1, as presented by Wong et al. (2022), delineates the levels and definitions of AI, ML, and DL.

DOI: 10.4018/979-8-3693-1794-5.ch009

The year 2000 marked a pivotal moment with a severe crash affecting businesses across all sectors. Subsequently, Information Technologies began expanding their applications to mitigate risks and automate business processes extensively. Technologies such as Artificial Intelligence, Machine Learning, Deep Learning, and blockchain experienced rapid growth, impacting nearly every domain of business processes. During this period, ML and AI played crucial roles in assessing Business Financial Risk (BFR) and extending their applications to mitigate such risks.

It is imperative to note that Machine Learning (ML) is not a substitute for Artificial Intelligence (AI); rather, it is a subfield of AI, as elucidated by Naim (2022). In broad terms, AI is defined as the ability of a system to act, behave, and respond like a human. A paramount benefit of AI lies in its facilitation of complex decision-making processes, akin to the cognitive actions of humans (Leo et al., 2019). ML, on the other hand, can be conceptualized as a tool for transforming systems to think and respond in a manner reminiscent of human cognition. To achieve this objective, ML employs neural networks, employing sequences of algorithms that enable systems to emulate human responses and actions (Naim, 2022).

Figure 1. Definition of AI, ML and DL
(Wong et al., 2022)

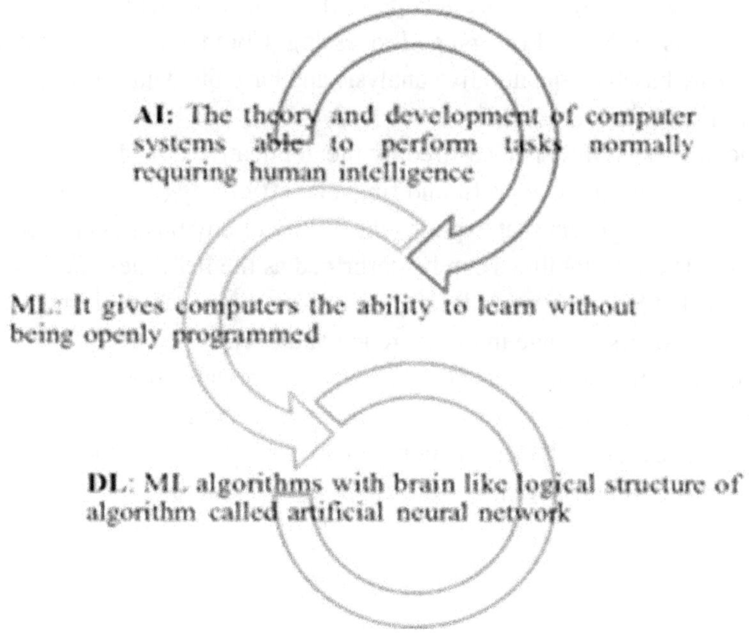

ML has four types and figure 2 shows the general depiction.

Figure 2. Types of ML
(Wong et al., 2022)

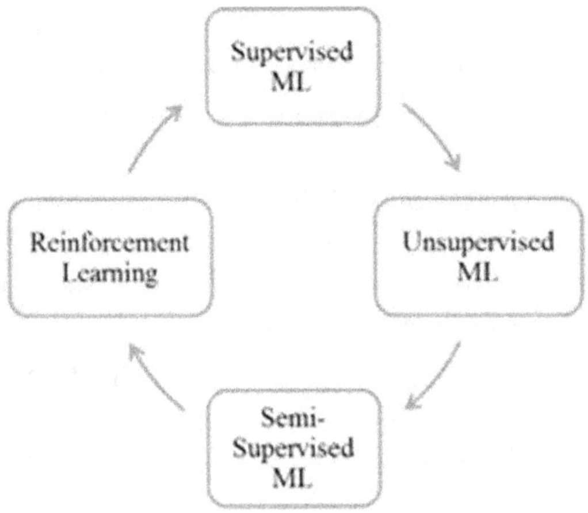

ML has received overwhelming acceptance from users from all disciplines such as financial analysts, social media analysts, advertising and media experts (Lahmiri and Bekiros 2019). This is because of ML's advancements in using Natural Language Processing, Computer Vision, and Robotics. The business practices which are based on quantitative analysis and have big data sets extensively use ML types for solutions (Lahmiri and Bekiros 2019).

Business Financial Risks (BFR) is measured by the applications of ML and financial analysts take decisions to reduce risks and improve profit and financial efficiency (Naim and Hassan, 2022).

Financial Risk (FR) is a concept that explains the ability of any business to take care and deal with financial instabilities. These instabilities can be described as the liabilities and financial commitments any firm has to pay back. FR mostly ascends because of loss in business and financial markets, changes in stock market and in process, change in the currency value, etc.

BFR is important not only for financial analysts, business organizations but also for the general economic scenario because it supports, synchronizes and devises commercial statistics and progressions. Also it offers clear policies for exploring financial openings and assesses the performance of any firm. Figure 3 gives the types of financial risks with further segmentation.

Figure 3. Types of financial risks
(Wong et al., 2022)

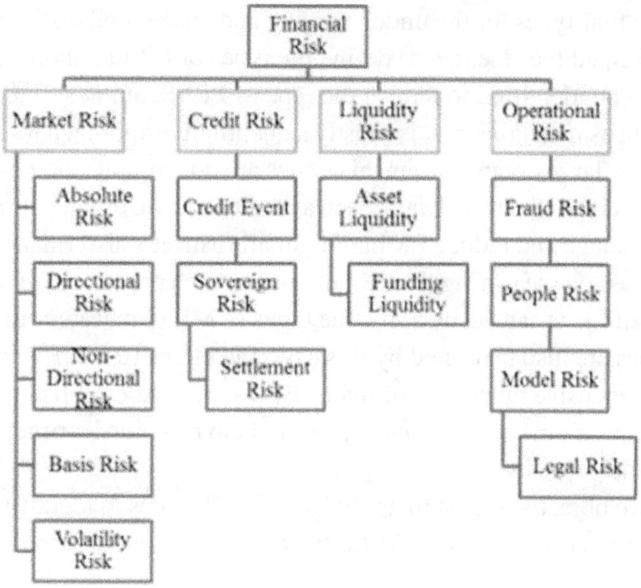

FR in financial institutions (FI) needs more attention because other firms are dependent on FI for financial stabilities and support. Data Scientists work to develop models, apply algorithms, use cases to automate the procedure to identify and measure BFR. In 2019 many FI applied ML and AI techniques to surpass human intelligence in measuring the BFR but it was not a simple or economic practice because many small scale and medium scale firms were not financially sound to invest in such consolidated technologies (Bhatore et al., 2020). Yet ML played a vital role in overcoming BFR and soon a survey conducted by Baker McKenzie presented 49% increase in profit and reduction in BFR after the applications of ML, AI and deep learning. The survey focused on several tasks facilitated by ML in FI such as for Fraud detection, International trade, Portfolio management, Mortgage risk, Trading and Loan and insurance underwriting, etc (Naim et al., 2022). FI and banking use ML for several financial services especially for risk detection such as PayPal uses ML to detect the risk for fraud during financial transactions; banks have established ML systems to benefit and succeed BFR and scan suspicious data. CO-OP applied ML to manage BFR risk for credit unions. The instrument worked to identify the fraud irrespective of frequency and volume of transactions.

ML has varied advantages but faces several challenges also for its applications and techniques in BFR and other business practices. Biggest challenge is Trusting ML techniques without human intervention and depending completely on algorithms for solving critical risks (Sanni et al., 2021). ML has prospective for reforming BFR but there are more challenges of ML like to keep follow up, preventing wrong alarms, etc.

BFR is always a demanding area of research and application of ML to meet the challenge is not a new approach, (Naim et al., 2021) however the application has used efforts and time in measuring, studying and identifying the FR clearly due to lack of conceptual knowledge in financial management in general and FR in specifically. In the current scenario taxonomy to correlate and discover ML types to measure the FR has not been streamlined effectively (Wong et al., 2022).

Much past research has used ML and explained the benefits of using ML in FR for specific firms but a general framework is yet to be explained that can be functional for any firm (Naim et al., 2022). FR is broadly segmented into four types for the understanding and conceptual distinction (Wong et al., 2022).

These distinctions helped the research to define the types of FR in a more specific manner so that a definite type of ML can be identified to reduce the type of FR (Sanni et al., 2021).

Reducing risk as well as managing risk is mostly a quantitative approach for the organizations (Naim and Alahmari, 2021). For large organizations this issues are not complex however for small scale firms, managing risks and reducing risk have always been a challenge. Big organizations can apply sophistical statistical methods to manage and reduce FR but for small firms it's also financial problems. In the past many studies have focused on measuring the FR but this research paper explains the correlation between ML and FR so that small firms can easily apply the types of ML to measure the FR (Naim and Kautish, 2022). Big organizations are also benefited by this easy and effective framework (Bhatore et al., 2020).

We present a comprehensive taxonomy of major BFR tasks and establish their connection with relevant machine learning problems. This taxonomy would help machine learning researchers navigate the complex domain of financial risks.

This research has two objectives, first to apply specific ML types to identify and measure BFR types and second objective to refer to cases of BFR for the application of ML types and verify the relevance for ML applications for BFR.

LITERATURE REVIEW

AI is facilitated by ML, as these tools help to leverage the objectives of technologies. Many times ML is synonymously used in place of AI due to the decision making capabilities until the 1970s (Sanni et al., 2021).

In later years ML grew as a separate branch and continues to become a popular tool for aiding various processes in business, medicine and other disciplines (Bhatore et al., 2020). ML is witnessed to be applied for business research and modern techniques in business (Sanni et al., 2021).

ML uses the algorithms and neural network models to help computer systems in cultivating and enhancing the performance of business at all levels (Bhatore et al., 2020). In the year 1952, Arthur Samuel developed the concept of ML when he marked a principal processor culturing codes. The codes were for checker's game and later the firm IBM enhanced the game by applying ML technologies (Hoepner et al., 2021).

In the late 1970s and early 1980s, AI research had focused on using logical, knowledge-based approaches rather than algorithms. Additionally, neural network research was abandoned by CS and AI researchers (Hoepner et al., 2021). Therefore, ML was considered to be a practicing program for AI and there was no clear demarcation between ML and AI until then (Lahmiri and Bekiros 2019). Presently ML is applied by many areas and disciplines for success and profit maximizations. Some of the breakthrough advancements occurred by the ML are self-driving vehicles, discovering the galaxy, and many more which are mentioned in figure 4.

Figure 4. Benefits and applications of ML in different areas
(Lahmiri and Bekiros 2019)

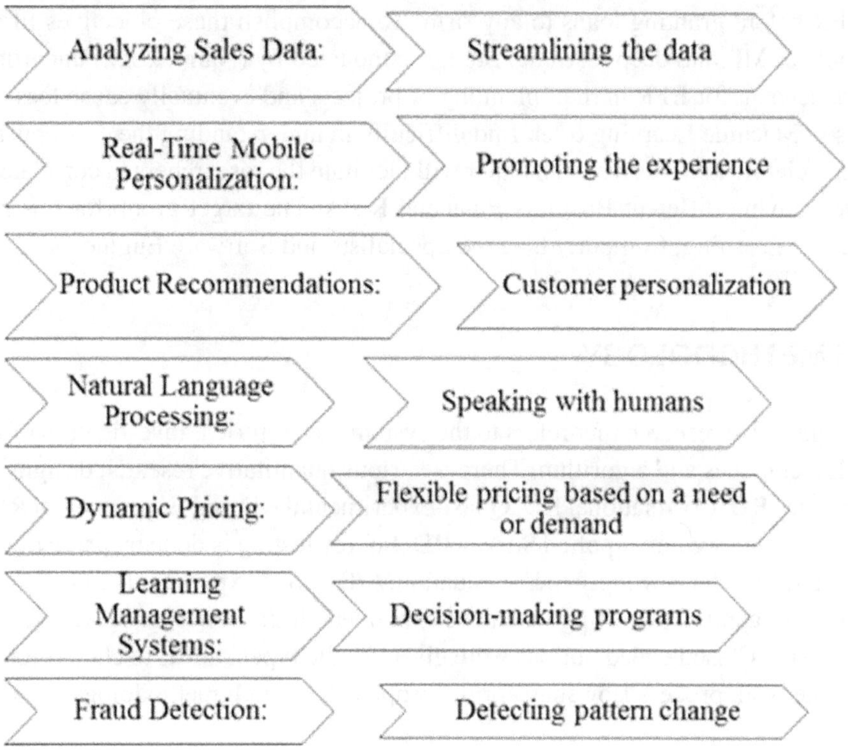

Another major contribution made by Stanford University while defining ML as "the discipline to make systems work without being programmed "(Lahmiri and Bekiros 2019). ML has encouraged a new collection of concepts and technologies such as chatbot, IOT, algorithms for robots, supervised and unsupervised learning, etc, (Lahmiri and Bekiros 2019).

The term finance is as old as the start of civilization. The first finance application was in about 3000 BC. Empires like Babylonians Empire was the first ones to give the concept of banking and centuries later in 2013 the Nobel laureate Eugene F. Fama, explained the term modern finance in economics. He is popularly known as the father of modern finance (Lahmiri and Bekiros 2019). There are three broad categories of finance and later two more were added, so currently 5 categories of finance exist. These categories are public finance, corporate finance, and personal finance, social finance and behavioral finance (Leo et al., 2019).

(Mashrur et al., 2020) explained ML for FR through the survey method but did not specify the target computer science and ML expert for the responses for identifying the ML techniques for FR. (Lin et al. 2011) illustrated the results of ML for Financial crisis and predicted the solutions through the survey but the research did not cover the FR through the types of ML.

To revitalize the economy globally in 2010, more financial investments are seen in Information technologies. Small and Medium size firms face more competition from the firms which were using ML and AI in their business practices and reducing BFR. Banking and FIs also explained advantages of ML in BFR and suggested more investments in using emerging technologies.

FI, Banks and other firms have been using ML techniques in the Fintech sector to identify the level of BFR and methods to reduce. Apart from that FI and banks have also introduced the column for calculating the BFR before granting loans to any firm. To accomplish these objectives FI and banks are applying the tools of ML and deep learning. Studies conducted by (Naim, 2022) underlined that many investors do not account for BFR in their planning of projects and eventually cause loss.

Researchers in Machine Learning often find difficulty in understanding the concepts and taxonomy of Business Financial Risks. This research paper will facilitate the researchers to correlate the ML types and methods for solving different Business Financial Risks. The target group for this research is the experts in Machine Learning, Computer Science Specialists and Software Engineers.

RESEARCH METHODOLOGY

This study is quantitative research that refers to the systematic empirical investigation of BFR concepts through the ML techniques and algorithm. There is various quantitative research designs (RD) like Descriptive RD, Survey RD, Correlational RD, Quasi-experimental RD and Experimental RD. To meet the objective of this research we have applied Survey RD. Survey testing is done by pretesting, pilot testing, data analysis, Cognitive Interviewing, expert evaluation, A/B testing experiments and observation studies. In this study we have applied pretesting where we explained all the concepts of BFR for CS specialists and Software experts. Closed ended options were given for the types of ML to choose for specific BFR. Figure 5 shows the work process flowchart for the proposed research methodology.

Figure 5. Work process flowchart for proposed research methodology

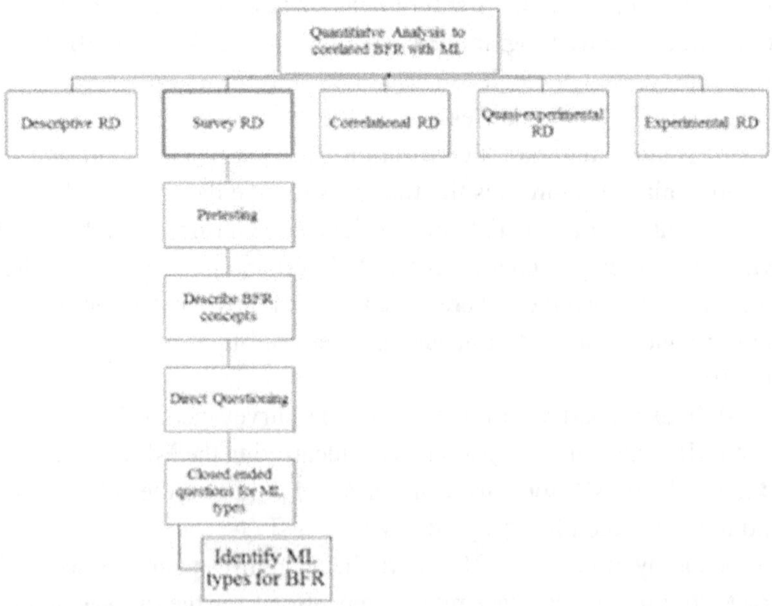

This research has a quantitative approach and results are derived from the responses submitted in the survey questions. Analysis for the results is based on the few case studies on similar research questions. Final inferences are prepared by comparing the results from the case studies results.

RQ1: The different types of BFR were explained in the questionnaire and respondents were asked to identify the ML types to assess the specific type of BFR.

RQ2: Refer to Case studies based on ML applications for reducing BFR and validate the results of RQ1 for the correlation of ML types for BFR.

The current study is based on the survey results from the computer science specialists and software engineers. The respondents were academicians from the southern universities in Saudi Arabia and professionals from Information Technology firms in Saudi Arabia. The survey questions have identified four types of ML types which are supervised, unsupervised, semi – supervised and reinforcement learning. Based on the responses the results show the methods of ML to measure the type of BFR.

DISCUSSION

Most of the financial issues based on quantitative analysis are solved by M, these issues are mostly related to profit maximization, predictive analysis, return on investments and FR. This research paper focuses on general understanding of BFR types and ML types and the taxonomy of Business FR to identify ML types for measuring specific types of FR (Song, et al.,2014).

There are four types of BFR which are given in figure 3 in the earlier part of the paper. Each type of FR is segmented into sub parts of FR based on applications and conceptual understanding. These four types are market risk, credit risk, liquidity risk and operational risk

Market Risk has five subparts namely Absolute Risk, Directional Risks, Non directional Risks, Basis risk and Volatile risk (Leo et al., 2019).

Absolute financial risk is measured by the total debt for certain or definite events. Directional risks explain the financial forfeiture ascends due to actual assets in the market. Non-Directional may occur because of any unexpected incident. Basis risk happens due to mismatch in a hedging situation. Volatility risk is caused during the change of price of a collection. This brings financial instability.

Credit risk has three sub parts such as credit event, sovereign risk and settlement risk. (Leo et al., 2019). A credit event is an adverse variation in a debtor's ability when he is unable to payback and this prompts for credit default swap (CDS) contract. Sovereign risk is a national risk where the government has an obligation to pay back the debt due to risk in making investment in other nations.

Settlement risk arises when one entity is unable to hold the cash for financial transactions and that eventually causes financial instability (Gotoh et al.,2014).

Liquidity risk has two subparts such as asset liquidity and funding liquidity (Leo et al., 2019). Asset Liquidity arises when the asset is unable to be converted to cash when needed for financial transactions or to pay debt. Funding liquidity risk is correlated to the financial institutions, where they are incompetent to settle debt or financial commitments. (Li et al., 2009)

Operational risk has four subparts namely, fraud risk, people risk, model risk and legal risk. Fraud risk depends on internal or external factors for illegal activity that can damage the image of the firm (Naim, 2022) People risk also depends on internal and external factors when there is lack of human resources or negative behavior or lack of motivation or contribution that cause loss or errors in the process. Model risk is due to errors in designing that cause financial damage (Kou et al., 2019). Legal risk is caused due to ethical issues or non-compliance regulations from the government.

The FR causes uncertainty to all types of firms and it is very important to reduce and implement a strategy. ML has successfully helped several firms for profit maximization and this research will focus on types of ML for measuring FR. Table 1 shows ML techniques and FR applications.

Table 1. ML techniques with FR applications (Leo et al., 2019)

Learning Methods	Learning Task	FR application
	Classification	Fraud Detection
		Portfolio optimization
Supervised Learning		Credit saving and bankruptcy policies
Unsupervised Learning		Volatile forecasting
		Claims modeling
Regression	Clustering	Loss Reserving
		Mortality Modeling
		Insurance Policy
		Sensitivity Analysis
		Credit saving and bankruptcy predictions

ML is a computational method which is generally dependent on improving the functions and reducing the errors and losses. ML can be classified into supervised learning, unsupervised learning, semi supervised learning and reinforcement learning (Leo et al., 2019).

Supervised Learning is a type of ML as well as AI that apply the labeled dataset to practice algorithms which aid in classification and eventually predicting the results for the discipline of areas where it is implemented (Gao et al., 2021)

Unsupervised Learning initiates ML algorithms without any human intervention. This ML sub category prompts to study and cluster unlabeled datasets and the algorithms applied here determine the concealed designs or data groups (Xuan, 2021)

Reinforcement learning is a ML practicing technique created on having required outcomes and removing irrelevant aspects. This subpart is very helpful in identifying the errors and it is focused on trial and error approach (Naim, 2021)

Semi-supervised is a type of ML which is a blend of supervised as well as unsupervised type of ML and AI. The combination is based on the application of small types of labeled data and large unlabeled dataset (Hoepner et al., 2021).

RESULTS

Types of ML can be applied in measuring FR and taxonomy can be developed to correlate ML types for FR. The results show the type of ML corresponding to the FR. Figure 6 shows the general scenario of Ml for FR with the correlated taxonomy.

Figure 6. Taxonomy of correlation between ML for FR
(Gotoh et al.,2014)

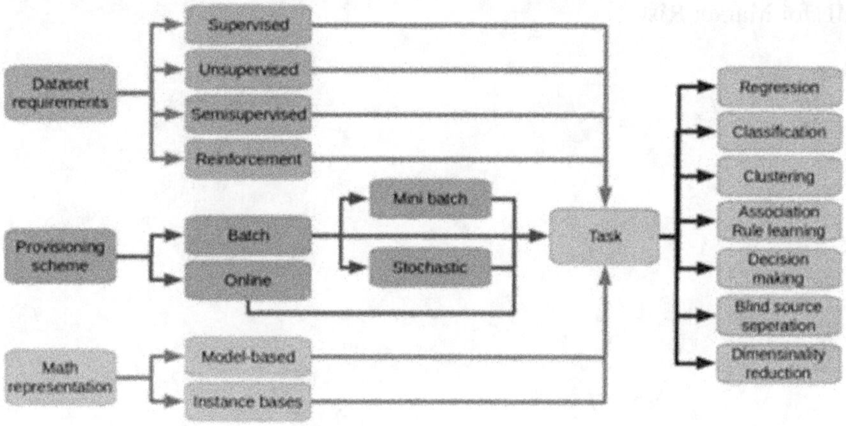

Further table 2 shows the segmented FR and types of ML for types and application. However, the results will show the specific ML type for measuring and reducing a definite FR type.

Table 2. General scenario of ML types for measuring FR

Financial Risk			
Market Risk	Credit Risk	Liquidity Risk	Operational Risk
• Portfolio Optimization	• Credit Scoring	• Claim Modeling	• Fraud Detection [Anomaly
• [Supervised Learning, Online Learning, Reinforcement Learning]	• [Multi Class Classification]	• [Supervised Learning, Clustering]	• Detection, Binary Classification, Text Mining]
	• Default or Bankruptcy Prediction	• Loss Reserving	
		• [Temporal Sequence Learning]	
• Sensitivity Analysis	• [Clustering]	• Mortality Forecasting	
• [Semi Supervised Learning]		• [Temporal Sequence Learning, Dimensionality Reduction]	
• Volatility Forecasting		• Insurance underwriting	
• [Temporal Sequence Learning]		• [Binary Classification, Clustering, Dimensionality Reduction]	
Machine Learning			

The closed ended survey was conducted for CS and ML experts, where the questionnaire explained the BFR concepts and subparts and was asked to identify the ML types to measure a particular type of FR.

RQI: The results will show four types of FR and which ML types are correlated to reduce and measure the type of BFR. As mentioned in the figure 3 above, four types of FR are further segmented into distinctive types. The four types included in the results are Market Risk, Credit Risk, Liquidity Risk and Operational Risk.

Types of ML for Market Risk

Figure 7. Absolute risk

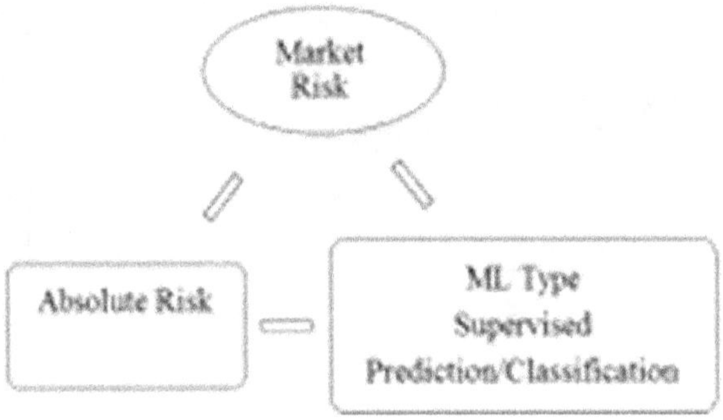

Figure 8. Directional risks

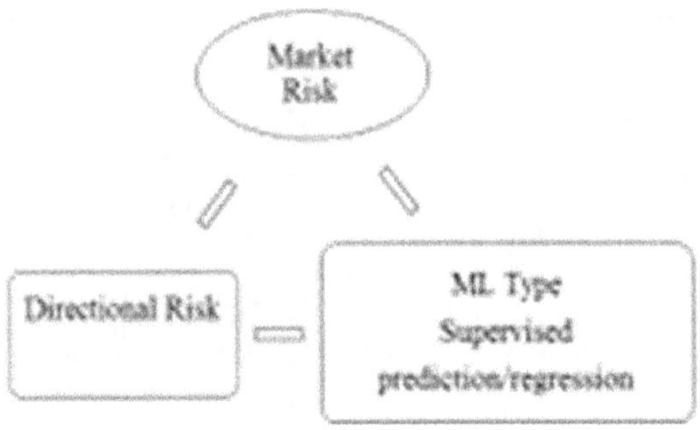

Figure 9. Non directional risks

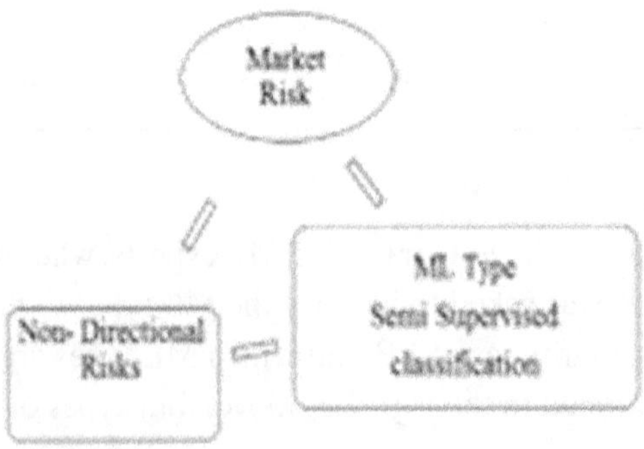

Figure 10. Basis risk

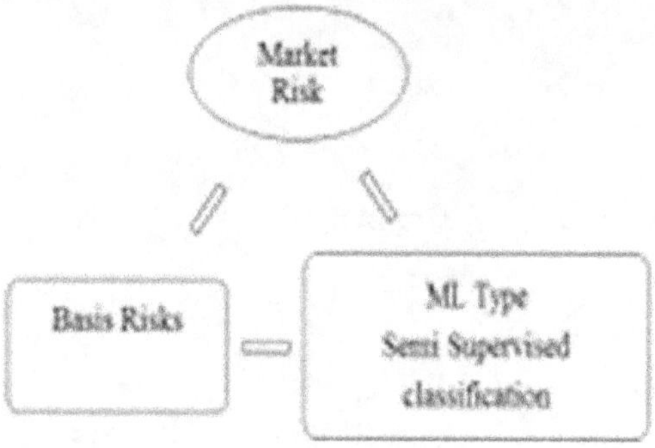

Figure 11. Volatile risk

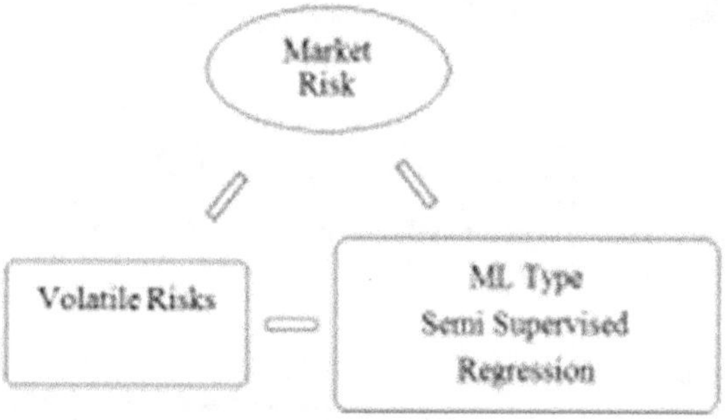

Figure 12. Credit event

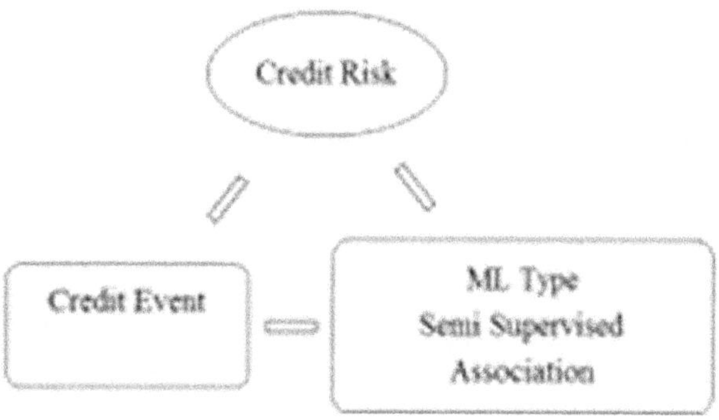

The results show that for absolute risk and directional risk, supervised learning can be applied for the FR measurement and for non-directional, basis risk and volatile risk; the best ML type should be semi-supervised learning.

Types of ML for Credit Risk

Figure 13. Sovereign risk

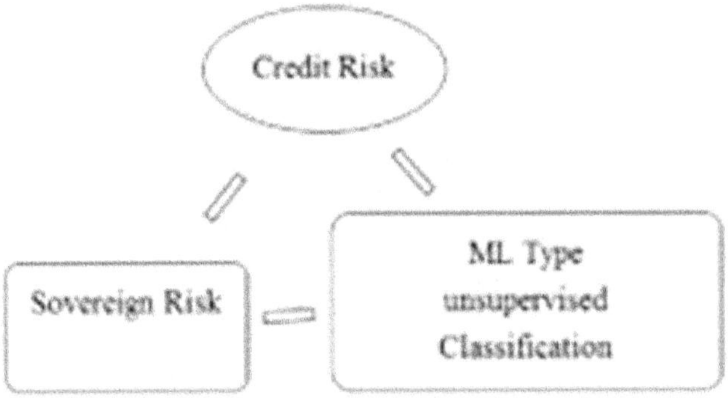

Figure 14. Settlement risk

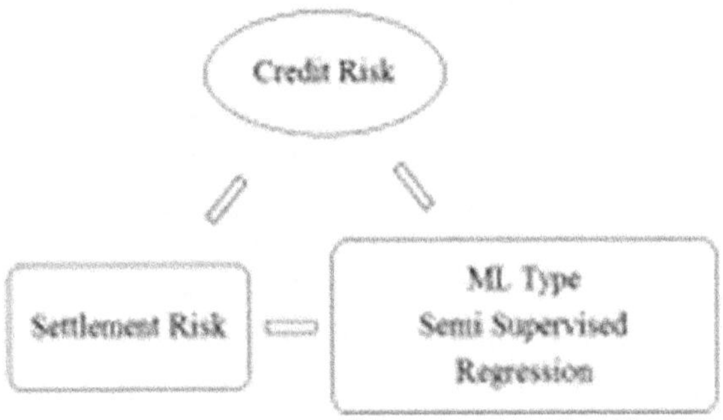

Figure 15. Asset liquidity

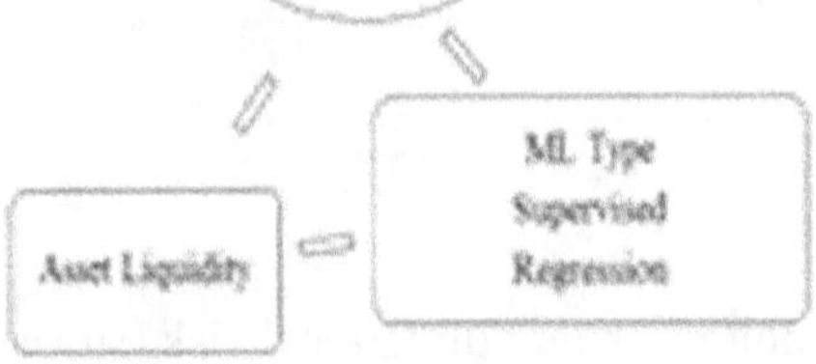

The results show for subparts of credit risk that for credit event and settlement risk, semi-supervised risk should be applied and for sovereign risk unsupervised ML type should be pragmatic for the measurement.

Types of ML for Liquidity Risk

Figure 16. Funding liquidity

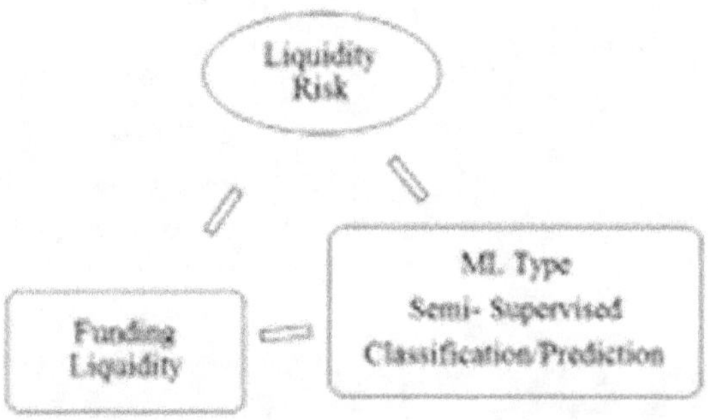

Figure 17. Fraud risk

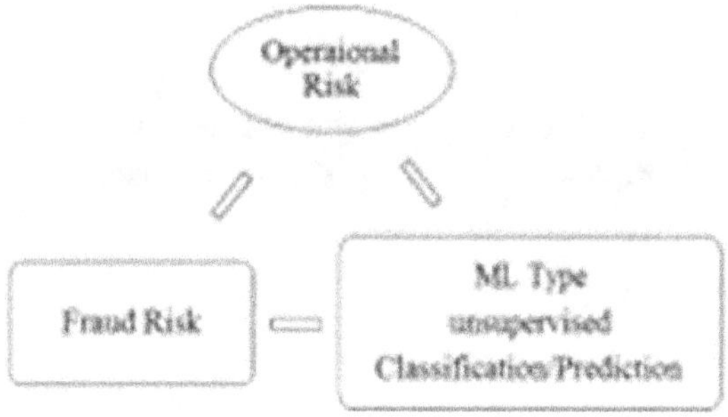

The results for subparts of liquidity risk, show that for asset liquidity supervised ML and for funding liquidity semi supervised ML type should be functional to measure the FR.

Types of ML for Operational Risk

Figure 18. People risk

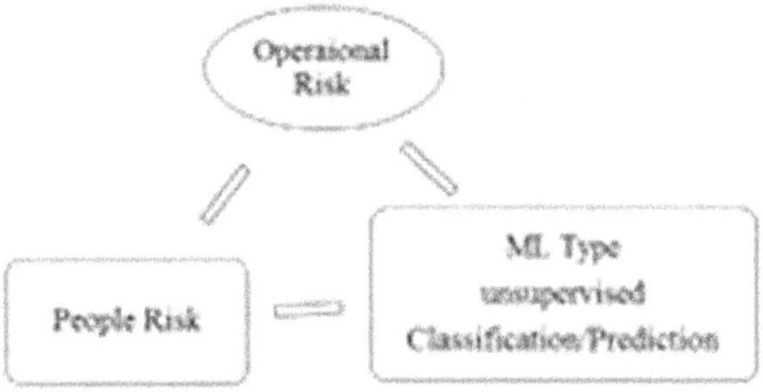

Figure 19. Model risk

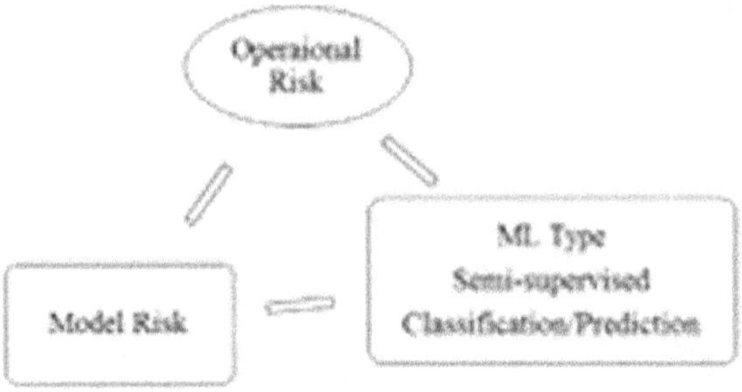

Figure 20. Legal risk

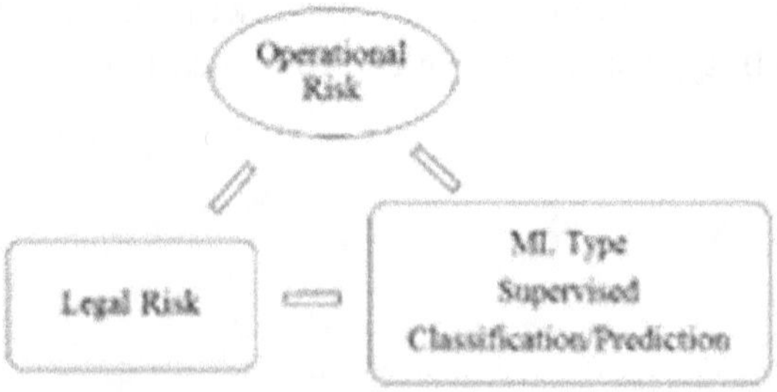

Operational risk is measured under four subparts and results show that for fraud risk and people risk, unsupervised risk is more practical type and for model risk, semi supervised and for legal risk, supervised

ML types are practical measurement tools.

ANALYSIS

BFR as explained is a point of improbability for return. The four major types of risks namely Market risk explains the improbability of impending returns because of market fluctuation, Credit risk can cause possible loss because of not competing the debt, Liquidity risk occurs when an organization is compelled to sell asset at less price to meet the financial need to pay the loan and Operational risk can happen any time due to discrepancies and mistakes in business processes or not abiding the laws at any levels. Modeling or automating BFR identification and measurement through ML types is presented in this study. Analysis shows that mostly pretesting methods help in developing models depending on correlation rather than causation.

Figure 21. Comprehensive result on correlation between BFR and ML types

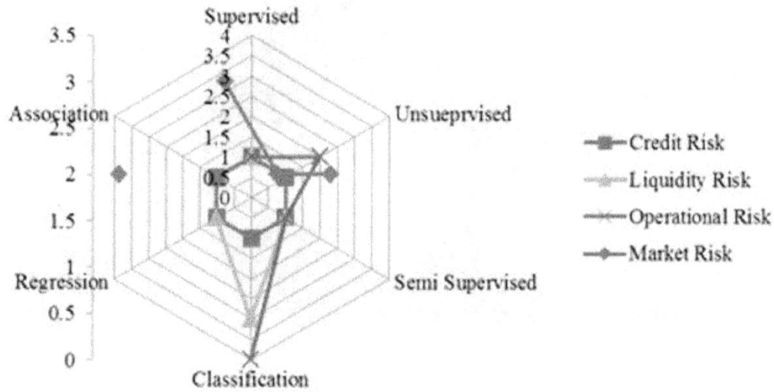

Goldman Sachs Risk Management Case Study shows 'Prediction to 'Anticipation of Risk' by ML techniques (Makhija and Chacko 2021). The case study explains the correlation methods for BFR and ML and gave the corollary for unsupervised ML for measuring all types of BFR. However, the approach of Goldman Sachs Risk Management Case Study was to present the application of blockchain technologies and ML tools for FR in large scale firms. Goldman Sachs suggested using Association, Regression and classification techniques for all types of BFR (Makhija and Chacko 2021).

Case study from European central bank applying ML to measure BFR in countries like China gave a clear inference on relevance of ML and automating the assessment techniques for BFR (Al-Qudah et al., 2022). The case study depicted the results by applying ML techniques to textual data. The analysis was based on predefined data collected from a popular newspaper (Velte, (2022). The next step was to select the important and applied keywords for BFR, later topical modeling based unsupervised ML techniques was applied for BFR (Velte, (2022).

Another example is banking regulation which gives a clear vision and protocol for measuring BFR for the growth of economy and business sector at the micro level (Luo, 2022). Asia–Pacific bank lost $4 billion while ignoring the risk BFR factor. From the real time learning the bank illustrated the guidelines of using technologies, especially blochkchain and ML for assessing the BFR. The bank stressed to apply unsupervised ML types to measure all types of BFR (Luo, 2022).

Xpresso.ai's developed a platform allowing the systematizing feature collection where ML tools could reduce the business complexities, increase profit and identify BFR at its early stage. The charter leveraged the latest ML and DL tools applied for identification, automation and assessment of BFR.

Market research companies such as Emerj are dedicated to applying AI and ML as use cases in businesses and BFR (Prasad, et al. 2019). Figure 21 shows the results of EMERJ for the application of AI and ML in banking functions. The results show for many functions but for the related research we focus only on Risk management. The firm applied the supervised and unsupervised ML types to measure liquidity risk and operational risk (Prasad, et, al. 2019).

Figure 22. Result of EMERJ for the application of AI and ML in banking functions
(Prasad, et, al. 2019)

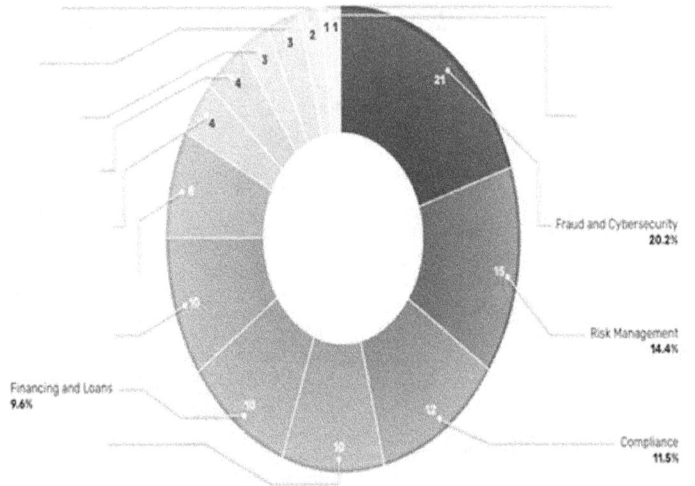

ML can be implemented to automate foremost functions in the banks and Silicon Valley in its use case report mentioning the ML driven model to measure and control BFR (Luo, 2022). Another case study from commercial bank Bay Area sold huge assets to apply ML techniques such as supervised type measure market risk. The bank made major changes in the regulations for risk management and in the application of Information technologies for the compliance of ML tools.

Los Angeles-based Zest Finance announced the Zest Automated ML (ZAML) framework which provided the applications of Unsupervised ML techniques in measuring BFR in the general scenario and also for liquidity risks (Naim et al., 2023). With the application of unsupervised ML type the company was able to reduce 25% annual loss.

Scienaptic Systems, which is a project from the US, introduced application of Classification and Prediction techniques of ML for BFR and was able to save 151 million losses that used to occur due to operational and credit risks (Naim et al., 2023).

Pretesting method is applied to prove RQ2: that refers to the Case studies based on ML applications for reducing BFR and validates the results of RQ1 for the correlation of ML types for BFR.

The referred case studies have not directly identified ML types for BFR but gave a clear vision and relevance on automation practice in measuring the BFR. General suggestions are made for the application of ML types for BFR types and the success rate is also explained. However, the survey results were very specific and gave a clear distinction on the application of ML types for BFR.

CONCLUSION

We studied the definition of ML types and FR types and developed a taxonomy to develop a correlation between ML sub categories and FR types for financial sustainability. The general scenario is developed that gives a simple but definite framework for FR types to be assessed by specific ML types. This research has facilitated the CS experts to apply the ML types to measure FR without doing extensive research and also they do not need to develop a financial understanding to implement the ML type and achieve financial wellbeing.

REFERENCE

Al-Qudah, A. A., Hamdan, A., Al-Okaily, M., & Alhaddad, L. (2022). The impact of green lending on credit risk: Evidence from UAE's banks. *Environmental Science and Pollution Research International*, *30*(22), 1–13. doi:10.1007/s11356-021-18224-5 PMID:35067871

Naim, A. (2022). Benefits of Artificial Intelligence (AI) in Financial Practices. *International Journal of Studies in Business Management, Economics and Strategies, 1*, 1–12. https://scholarsdigest.org/index.php/bmes/article/view/7

Naim, A. (2022). Competitive Trends and Technologies in Business Management (2022). Nova Science. doi:10.52305/VIXO9830

Bhatore, S., Mohan, L., & Reddy, Y. R. (2020). Machine learning techniques for credit risk evaluation: A systematic literature review. *Journal of Banking and Financial Technology*, *4*(1), 111–138. doi:10.1007/s42786-020-00020-3

Gao, B. (2021). The use of machine learning combined with data mining technology in financial risk prevention. *Computational Economics*, 1–21.

Gotoh, J. Y., Takeda, A., & Yamamoto, R. (2014). Interaction between financial risk measures and machine learning methods. *Computational Management Science*, *11*(4), 365–402. doi:10.1007/s10287-013-0175-5

Hoda, N., & Naim, A. (Eds.). (2023). *Social Capital in the Age of Online Networking: Genesis, Manifestations, and Implications: Genesis, Manifestations, and Implications*. IGI Global.

Hoepner, A. G., McMillan, D., Vivian, A., & Wese Simen, C. (2021). Significance, relevance and explainability in the machine learning age: An econometrics and financial data science perspective. *European Journal of Finance*, *27*(1-2), 1–7. doi:10.1080/1351847X.2020.1847725

Kou, G., Chao, X., Peng, Y., Alsaadi, F. E., & Herrera Viedma, E. (2019). *Machine learning methods for systemic risk analysis in financial sectors.*

Lahmiri, S., & Bekiros, S. (2019). Can machine learning approaches predict corporate bankruptcy? Evidence from a qualitative experimental design. *Quantitative Finance*, *19*(9), 1569–1577. doi:10.1080/14697688.2019.1588468

Leo, M., Sharma, S., & Maddulety, K. (2019). Machine learning in banking risk management: A literature review. *Risks*, *7*(1), 29. doi:10.3390/risks7010029

Li, N., Liang, X., Li, X., Wang, C., & Wu, D. D. (2009). Network environment and financial risk using machine learning and sentiment analysis. *Human and Ecological Risk Assessment*, *15*(2), 227–252. doi:10.1080/10807030902761056

Lin, W. Y., Hu, Y. H., & Tsai, C. F. (2011). Machine learning in financial crisis prediction: A survey. *IEEE Transactions on Systems, Man, and Cybernetics. Part C, Applications and Reviews*, *42*(4), 421–436.

Luo, Y. (2022). A general framework of digitization risks in international business. *Journal of International Business Studies*, *53*(2), 344–361. doi:10.1057/s41267-021-00448-9 PMID:34075261

Makhija, P., & Chacko, E. (2021). Efficiency and advancement of artificial intelligence in service sector with special reference to banking industry. Maleh, Y. (eds) Computational Intelligence Techniques for Green Smart Cities. Green Energy and Technology. Springer, Cham. doi:10.1007/978-981-16-3250-1_2

Mashrur, A., Luo, W., Zaidi, N. A., & Robles-Kelly, A. (2020). Machine learning for financial risk management: A survey. *IEEE Access : Practical Innovations, Open Solutions*, *8*, 203203–203223. doi:10.1109/ACCESS.2020.3036322

Naim, A. (2023). Application of machine learning techniques to identify the business financial risks. *Academy of Marketing Studies Journal*, *27*(S5), 1–11.

Naim, A., Alahmari, F., & Rahim, A. (2021). Role of Artificial Intelligence in Market Development and Vehicular Communication. Smart Antennas. *Recent Trends in Design and Applications*, *2*, 28–39. doi: 10.2174/9781681088594121020006

Naim, A., & Alqahtani, H. (2023). Role of IoT in Industrial Transformation and Marketing Management. In *Global Applications of the Internet of Things in Digital Marketing* (pp. 67–93). IGI Global. doi:10.4018/978-1-6684-8166-0.ch005

Naim, A., Alqahtani, H., Fatima, S., & Khan, M. F. (2023). Applications of Artificial Intelligence of Things in Green Supply Chain Management: Challenges and Future Directions. In Blockchain, IoT, and AI Technologies for Supply Chain Management (pp. 99-126). CRC Press.

Naim, A., Alqahtani, H., Muniasamy, A., Bilfaqih, S. M., Mahveen, R., & Mahjabeen, R. (2023). Applications of Information Systems and Data Security in Marketing Management. A. Naim, P. Malik, & F. Zaidi (Eds.), Fraud Prevention, Confidentiality, and Data Security for Modern Businesses (pp. 57-83). IGI Global. doi:10.4018/978-1-6684-6581-3.ch003

Naim, A., & Hassan, S. Z. (2022). Economies of Scale for Security Applications in Business Business Issues. *Competition and Entrepreneurship*, *73*. doi:10.52305/VIXO9830

Naim, A., & Kautish, S. K. (Eds.). (2022). *Building a Brand Image Through Electronic Customer Relationship Management*. IGI Global. doi:10.4018/978-1-6684-5386-5

Naim, A., & Khan, S. A. (2023). Impact and Assessment of Electronic Commerce on Consumer Buying Behaviour. In Global Applications of the Internet of Things in Digital Marketing (pp. 264-289). IGI Global. doi:10.4018/978-1-6684-8166-0.ch014

Naim, A., Malik, P. K., & Zaidi, F. A. (Eds.). (2023). *Fraud Prevention, Confidentiality, and Data Security for Modern Businesses*. IGI Global. doi:10.4018/978-1-6684-6581-3

Naim, A., Muniasamy, A., & Alqahtani, H. (2022). Application of Internet of Things in Marketing Management. *Ambient Intelligence and Internet of Things: Convergent Technologies*, 273-299.

Naim, A., Syed, A., Bilfaqih, S. M., & Muniasamy, A. (2023). Impact of Individual and Organizational Social Capital of Marketing in a Global Network. In Social Capital in the Age of Online Networking: Genesis, Manifestations, and Implications (pp. 1-17). IGI Global. doi:10.4018/978-1-6684-8953-6.ch001

Prasad, R., Rohokale, V., Prasad, R., & Rohokale, V. (2020). Artificial intelligence and machine learning in cyber security. *Cyber security: the lifeline of information and communication technology*, 231-247.

Priya, E. D., Sharmila, A., Rajarajeshwari, K. C., Anand, K. G., & Naim, A. (2023). Wearable Proximity Coupled Antenna for IoT Applications. In *Internet of Things Enabled Antennas for Biomedical Devices and Systems: Impact, Challenges and Applications* (pp. 249–258). Springer Nature Singapore. doi:10.1007/978-981-99-0212-5_20

Sanni-Anibire, M. O., Mohamad Zin, R., & Olatunji, S. O. (2021). Developing a preliminary cost estimation model for tall buildings based on machine learning. *International Journal of Management Science and Engineering Management*, *16*(2), 134–142. doi:10.1080/17509653.2021.1905568

Song, X. P., Hu, Z. H., Du, J. G., & Sheng, Z. H. (2014). Application of machine learning methods to risk assessment of financial statement fraud: Evidence from China. *Journal of Forecasting*, *33*(8), 611–626. doi:10.1002/for.2294

Velte, P. (2022). Meta-analyses on corporate social responsibility (CSR): A literature review. *Management Review Quarterly*, *72*(3), 627–675. doi:10.1007/s11301-021-00211-2

Wang, D., Wu, Z., & Zhu, B. (2022). Controlling Shareholder Characteristics and Corporate Debt Default Risk: Evidence Based on Machine Learning. *Emerging Markets Finance & Trade*, *58*(12), 3324–3339. doi:10.1080/1540496X.2022.2037416

Wong, L. W., Tan, G. W. H., Ooi, K. B., Lin, B., & Dwivedi, Y. K. (2022). Artificial intelligence-driven risk management for enhancing supply chain agility: A deep-learning-based dual-stage PLS-SEM-ANN analysis. *International Journal of Production Research*, 1–21. doi:10.1080/00207543.2022.2063089

Xuan, F. (2021). Regression analysis of supply chain financial risk based on machine learning and fuzzy decision model. *Journal of Intelligent & Fuzzy Systems*, *40*(4), 6925–6935. doi:10.3233/JIFS-189523

Chapter 10
Sustainable Practices for Green Computing and Digital E-Waste Management

K. R. Kalphana
Department of Agricultural Engineering, Mahendra Engineering College (Autonomous), Namakkal, India

K. Madu Mathi
https://orcid.org/0000-0002-4442-8821
Department of Electrical and Electronics Engineering, Mahendra Institute of Technology (Autonomous), Namakkal, India

A. Ranjithkumar
Department of Mechanical Engineering, AVS College of Technology, Salem, India

Sanchita Banerji
Department of HRM, DSIMS, Mumbai, India

S. Anandakumar
Department of Civil Engineering, KPR Institute of Engineering and Technology, Coimbatore, India

ABSTRACT

The chapter explores the integration of circular economy principles into digital operations, focusing on green computing and email waste management. Green computing uses eco-friendly technologies to reduce digital infrastructure's environmental footprint. It involves energy-efficient hardware, virtualization techniques, and cloud computing to reduce energy consumption, carbon emissions, and electronic waste. Email waste management is crucial to address digital waste, which contributes to carbon emissions, energy consumption, and resource depletion. Implementing email optimization, efficient storage practices, and behavioural interventions can help organizations mitigate environmental impacts while improving productivity and information management efficiency.

DOI: 10.4018/979-8-3693-1794-5.ch010

INTRODUCTION

Sustainability is becoming increasingly important in various industries due to environmental concerns and the need for responsible business practices. The rapid pace of technological advancement in the digital realm has increased the urgency of sustainability. The digital footprint of modern society is undeniable, from energy consumption to electronic waste disposal. The principles of the circular economy, which focus on regenerative and restorative practices, are essential for integrating into digital practices. This shift away from the traditional linear "take-make-dispose" model emphasizes waste minimization and resource reuse, aligning with sustainability goals(Raza et al., 2012).

The circular economy can be applied to green computing, which involves implementing eco-friendly technologies to reduce the environmental impact of digital infrastructure. This includes optimizing hardware design, adopting virtualization techniques, and leveraging cloud computing services. By embracing green computing, organizations can reduce their carbon footprint and energy consumption, while also achieving significant long-term cost savings. Sustainable digital practices go beyond hardware and infrastructure, focusing on managing digital resources and reducing digital waste. Email waste is a significant environmental issue, contributing to carbon emissions, energy consumption, and resource depletion. Implementing email optimization techniques and promoting efficient email management practices can minimize environmental footprints, improve productivity, and enhance information management efficiency, ultimately reducing the environmental impact of email usage(S. G. Paul et al., 2023).

The integration of circular economy principles into digital practices can drive innovation, improve competitiveness, and foster resilience. By designing products and services with durability, recyclability, and resource efficiency, businesses can reduce dependence on finite resources and mitigate supply chain disruption risks. Sustainable digital practices can also enhance brand reputation and stakeholder trust, as consumers increasingly scrutinize the environmental and social impact of brands. Sustainability is now a strategic imperative for businesses in the digital age. Integrating circular economy principles into digital practices can reduce environmental footprints, drive innovation, enhance competitiveness, and build a more resilient future. Recognizing the importance of sustainable digital practices and embracing them in all business operations is the first step towards sustainability(Saha, 2018).

The increasing use of digital technologies has led to a significant environmental footprint, necessitating the adoption of sustainable practices. Green computing, which combines sustainability and digital innovation, aims to mitigate the environmental impact of digital infrastructure. This approach includes energy-efficient hardware design, optimization of data centers, and the use of renewable energy sources. Green computing offers a holistic approach to reducing carbon emissions, conserving resources, and promoting environmental stewardship in the digital age, ensuring a circular economy mindset and minimizing ecological impact(Dhaini et al., 2021).

Sustainable digital practices involve managing digital resources and mitigating digital waste, particularly email waste. The daily volume of emails contributes to energy consumption, carbon emissions, and resource depletion. Implementing email waste management strategies and promoting efficient email management practices can minimize environmental footprints, optimize productivity, and improve information management efficiency(Maguluri et al., 2023). This approach goes beyond hardware and infrastructure, ensuring a more sustainable digital future. The circular economy framework promotes waste minimization and resource reuse, aligning with sustainability goals(Mohanty et al., 2023). By incorporating circular economy principles in green computing and email waste management, organizations can reduce environmental impact, drive innovation, enhance competitiveness, and foster long-term resilience, while also enhancing competitiveness(Saha, 2014). This paradigm shift is a significant shift towards a more regenerative approach to technology.

This chapter delves into sustainable circular economy digital practices, specifically green computing and email waste management. It explores environmental challenges and potential solutions within the circular economy framework. The chapter uses case studies, best practices, and forward-looking insights to equip organizations with knowledge and tools to adopt sustainable digital practices and contribute to a more resilient future(Blikra Vea et al., 2018). This chapter delves into green computing principles, strategies, and challenges in sustainable digital infrastructure. It also explores the environmental impact of email waste and discusses practical approaches to managing it. By the end, readers will understand the importance of sustainable circular economy digital practices and how to integrate these principles into their organizations(Butt et al., 2020).

GREEN COMPUTING

Green computing is a important focus in sustainable digital practices, aiming to minimize the environmental impact of digital infrastructure through the adoption of eco-friendly technologies and practices. Significant strategies and innovations in green computing include reducing energy consumption, carbon emissions, and electronic waste generation(S. G. Paul et al., 2023; Raza et al., 2012).

- ***Energy-Efficient Hardware:*** Energy-efficient hardware design is fundamental to reducing the power consumption of computing devices. Manufacturers are increasingly prioritizing energy efficiency in the development of processors, memory modules, and other components. This includes advancements such as low-power CPUs, energy-efficient GPUs, and solid-state drives (SSDs) that consume less energy compared to traditional hard disk drives (HDDs)(Boopathi, 2024b; Revathi et al., 2024a).
- ***Virtualization:*** Virtualization technology allows organizations to optimize server utilization and consolidate physical servers into virtual machines (VMs). By running multiple VMs on a single physical server, virtualization reduces the number of servers required, leading to lower energy consumption, reduced cooling needs, and a smaller physical footprint. Moreover, server virtualization enables dynamic allocation of resources, allowing for more efficient use of computing resources.
- ***Cloud Computing:*** Cloud computing offers significant environmental benefits by enabling resource pooling and dynamic scalability of computing resources. Cloud service providers operate large-scale data centers that are often more energy-efficient and environmentally friendly than on-premises infrastructure. Additionally, cloud computing allows for workload consolidation and optimization, leading to higher resource utilization and reduced energy consumption per unit of computing work(Dhanalakshmi et al., 2024; Nanda et al., 2024; Rahamathunnisa et al., 2023; Sonia et al., 2024).
- ***Renewable Energy Sources:*** Embracing renewable energy sources such as solar, wind, and hydroelectric power is a crucial component of green computing. Many data centers and computing facilities are transitioning to renewable energy to power their operations, thereby reducing reliance on fossil fuels and minimizing carbon emissions. Some organizations are also investing in on-site

renewable energy generation, such as solar panels installed on data center rooftops(Nishanth et al., 2023).

- ***Energy-Efficient Cooling Systems:*** Cooling systems account for a significant portion of energy consumption in data centers. Implementing energy-efficient cooling technologies, such as economizers, evaporative cooling, and hot aisle/cold aisle containment, can help minimize energy usage and improve overall efficiency. Advanced cooling techniques, such as liquid cooling and immersion cooling, offer further opportunities for energy savings by removing heat more efficiently(BOOPATHI et al., 2024).
- ***Energy Management and Monitoring:*** Comprehensive energy management and monitoring systems enable organizations to track and optimize energy usage across their digital infrastructure. By identifying inefficiencies, optimizing workload placement, and implementing power management policies, organizations can achieve significant energy savings without sacrificing performance or reliability(Satav et al., 2023; Syamala et al., 2023; Venkateswaran, Vidhya, et al., 2023).

Green computing minimizes digital infrastructure's environmental impact by utilizing energy-efficient hardware, virtualization, cloud computing, renewable energy sources, cooling systems, and energy management solutions, thereby reducing carbon footprint and promoting sustainability.

ENVIRONMENTAL BENEFITS AND CHALLENGES OF GREEN COMPUTING

Green computing initiatives can significantly reduce the environmental impact of digital infrastructure by reducing energy consumption, reducing carbon emissions, and minimizing electronic waste. However, these initiatives also present challenges and considerations that organizations must address to maximize their effectiveness. This section discusses the environmental benefits and challenges in implementing green computing(Butt et al., 2020; Dhaini et al., 2021; Saha, 2014).

Environmental Benefits

- ***Reduced Energy Consumption:*** Green computing focuses on optimizing energy efficiency across all aspects of digital infrastructure, from hardware design to software optimization. By adopting energy-efficient components, implementing power management policies, and optimizing workload allocation, organizations can significantly reduce energy consumption in data centers and computing facilities. This not only lowers operating costs but also decreases reliance on fossil fuels and reduces greenhouse gas emissions.
- ***Lower Carbon Emissions:*** The reduction in energy consumption achieved through green computing translates directly into lower carbon emissions. By embracing renewable energy sources and implementing energy-efficient technologies and practices, organizations can minimize their carbon footprint and contribute to global efforts to combat climate change. Green computing allows organizations to align their digital operations with sustainability goals and reduce their overall environmental impact(Boopathi, 2022a; Sampath, 2021).
- ***Minimized Electronic Waste Generation:*** Green computing emphasizes the principles of resource efficiency, reuse, and recycling, thereby reducing the generation of electronic waste. By extending the lifespan of computing devices through refurbishment, upgrading, and responsible

end-of-life management, organizations can minimize the environmental impact associated with the disposal of obsolete hardware. Additionally, initiatives such as e-waste recycling programs enable organizations to recover valuable materials from retired IT equipment, further reducing the demand for virgin resources(Janardhana et al., 2023). Challenges and Considerations in Implementation(Czeka\la et al., 2023):

- ***Initial Investment Costs:*** While green computing initiatives offer long-term cost savings through reduced energy consumption and operational efficiencies, the initial investment costs can be significant. Upgrading to energy-efficient hardware, implementing new technologies, and retrofitting existing infrastructure require upfront capital investment, which may pose financial challenges for some organizations. However, it's essential to consider the long-term return on investment and the potential cost savings associated with green computing initiatives(Malathi et al., 2024).
- ***Compatibility and Interoperability:*** Introducing new technologies and implementing energy-saving measures may introduce compatibility and interoperability challenges within existing IT environments. Ensuring seamless integration and compatibility with legacy systems and applications requires careful planning and coordination. Organizations must also consider potential disruptions to business operations during the transition to green computing solutions.
- ***Performance and Reliability:*** Balancing energy efficiency with performance and reliability is a main consideration in green computing implementation. Optimizing energy usage should not come at the expense of system performance or reliability. Organizations must carefully evaluate the trade-offs and implement measures to ensure that green computing initiatives do not compromise the quality of service or user experience(Yupapin et al., 2023).
- ***Regulatory Compliance and Standards:*** Compliance with environmental regulations and adherence to industry standards are essential aspects of green computing implementation. Organizations must stay abreast of evolving regulatory requirements and industry standards related to energy efficiency, emissions reduction, and e-waste management. Non-compliance can result in financial penalties, reputational damage, and legal liabilities.

Green computing initiatives offer environmental benefits like reduced energy consumption, lower carbon emissions, and electronic waste reduction. However, organizations face challenges like initial investment costs, compatibility issues, performance concerns, and regulatory compliance requirements(Dhanalakshmi et al., 2024). By proactively addressing these challenges and adopting a strategic approach, organizations can realize the full potential of sustainable digital practices and contribute to a greener future.

STRATEGIES FOR IMPLEMENTING GREEN COMPUTING

Green computing practices are crucial for organizations to reduce their environmental footprint and promote sustainability in digital operations. By adopting energy-efficient technologies, optimizing data center operations, and promoting sustainable IT procurement, organizations can reduce energy consumption, lower carbon emissions, and contribute to a more sustainable future. Figure 1 illustrates various strategies for implementing green computing(Blikra Vea et al., 2018; Yang et al., 2018).

Green computing practices involve a comprehensive approach that includes assessing energy efficiency, optimizing data center operations, promoting server virtualization, embracing cloud computing, adopting energy-efficient hardware, practicing sustainable IT procurement, and promoting employee engagement(Kumar et al., 2023; Rebecca et al., 2024). These strategies can reduce environmental impact, lower operating costs, and contribute to a more sustainable future.

Figure 1. Strategies for implementing green computing

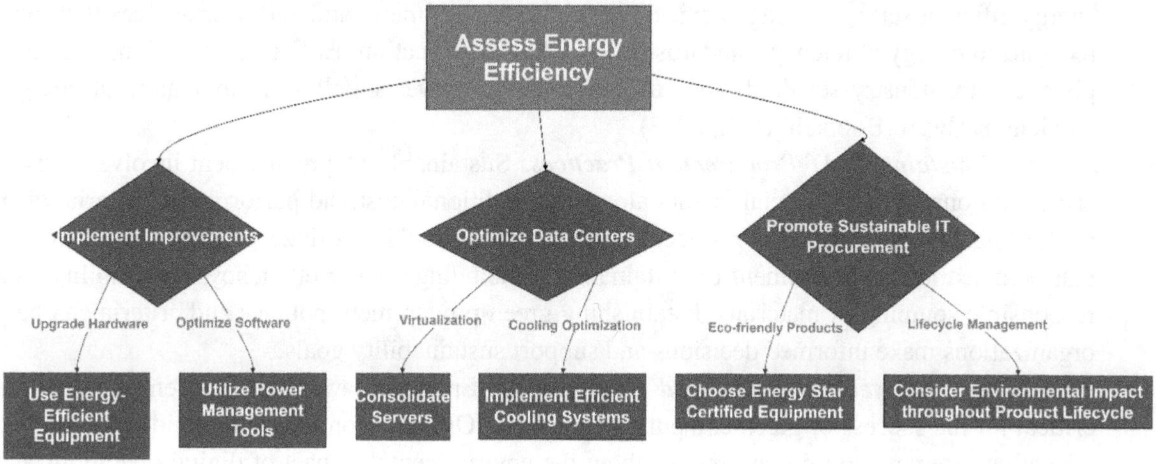

- ***Assessing Energy Efficiency:*** Conducting a comprehensive energy efficiency assessment is the first step towards implementing green computing practices. This involves evaluating the energy consumption of existing hardware, software, and infrastructure components to identify inefficiencies and areas for improvement. Energy audits and monitoring tools can help organizations track energy usage patterns, identify energy-intensive processes, and prioritize energy-saving opportunities.
- ***Optimizing Data Centers:*** Data centers are among the most energy-intensive components of digital infrastructure, making them prime targets for green computing initiatives. Organizations can optimize data center operations by consolidating servers, virtualizing workloads, and implementing energy-efficient cooling systems. Techniques such as hot aisle/cold aisle containment, airflow optimization, and free cooling can significantly reduce energy consumption and improve overall efficiency(Domakonda et al., 2022; Samikannu et al., 2022; Vanitha et al., 2023; Vennila et al., 2022).
- ***Promoting Server Virtualization:*** Server virtualization allows organizations to maximize the utilization of physical server resources by running multiple virtual machines (VMs) on a single physical server. By consolidating workloads and reducing the number of physical servers required, server virtualization reduces energy consumption, hardware costs, and space requirements in data centers. Moreover, virtualization enables dynamic resource allocation, allowing organizations to scale resources up or down based on demand, further optimizing energy usage(Boopathi, 2024a; Sonia et al., 2024).
- ***Embracing Cloud Computing:*** Cloud computing offers significant environmental benefits by providing on-demand access to computing resources without the need for on-premises infrastructure. By leveraging cloud services, organizations can reduce the need for physical servers, minimize energy consumption, and lower carbon emissions associated with data center operations. Cloud

providers often operate energy-efficient data centers powered by renewable energy sources, further enhancing the sustainability of cloud computing(Agrawal et al., 2023; Satav et al., 2023; Venkateswaran, Vidhya, et al., 2023).

- ***Implementing Energy-Efficient Hardware:*** Upgrading to energy-efficient hardware is a crucial component of green computing initiatives. Organizations should prioritize the procurement of energy-efficient servers, storage devices, networking equipment, and end-user devices that meet recognized energy efficiency standards. Energy Star certification, EPEAT registration, and compliance with industry standards such as 80 PLUS for power supplies are indicators of energy-efficient hardware(Boopathi et al., 2023).
- ***Adopting Sustainable IT Procurement Practices:*** Sustainable IT procurement involves considering environmental and social factors alongside traditional cost and performance criteria when purchasing IT equipment and services. Organizations should prioritize products and suppliers that demonstrate a commitment to sustainability, including energy efficiency, recyclability, and responsible sourcing of materials. Establishing green procurement policies and criteria can help organizations make informed decisions and support sustainability goals.
- ***Promoting Employee Awareness and Engagement:*** Employee awareness and engagement are critical for the success of green computing initiatives. Organizations should provide training and education programs to raise awareness about the environmental impact of digital operations and promote sustainable IT practices among employees. Encouraging energy-saving behaviors, such as turning off unused devices, minimizing paper usage, and practicing responsible e-waste disposal, can help foster a culture of sustainability within the organization.

EMAIL WASTE MANAGEMENT

In the digital age, email has become a ubiquitous communication tool, facilitating rapid information exchange and collaboration across organizations and individuals worldwide. However, the convenience and efficiency of email also come with environmental consequences, as the sheer volume of emails exchanged daily contributes to carbon emissions, energy consumption, and resource depletion. In this section, we examine the concept of email waste and its environmental impact, shedding light on the hidden costs of digital communication(Awomeso et al., 2010). Figure 2 illustrates various strategies for managing email waste.

Figure 2. Email waste management strategies

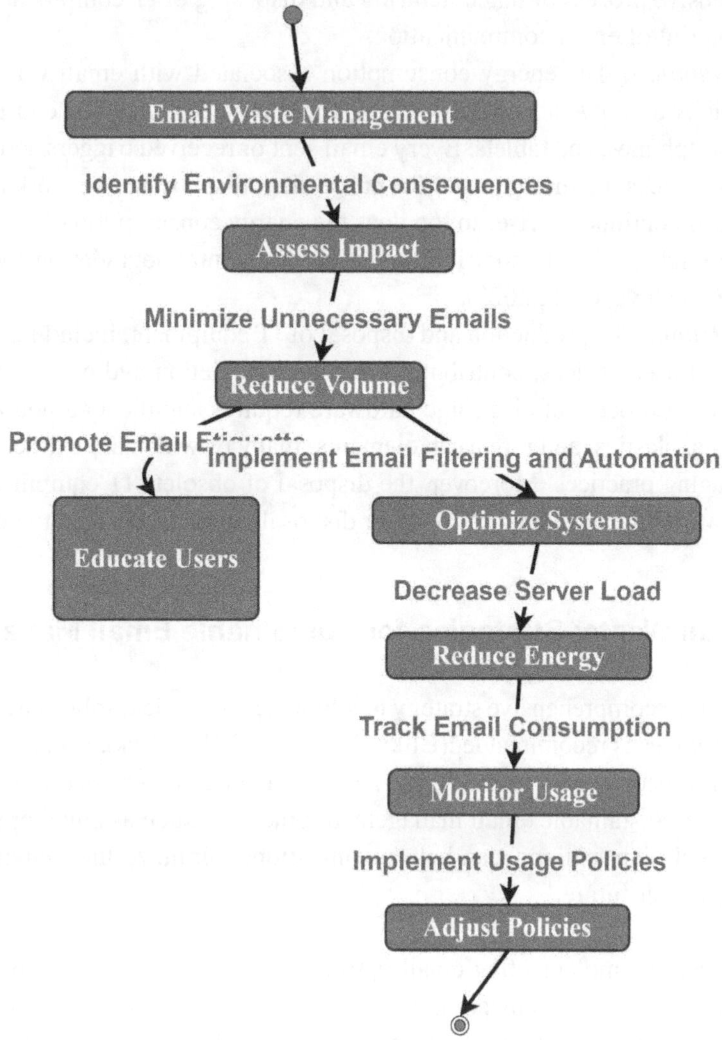

EMAIL WASTE: UNDERSTANDING THE CONCEPT

Email waste is the inefficient use of email, causing environmental harm through carbon emissions, energy consumption, and resource depletion. Despite being a paperless and environmentally friendly communication medium, email requires energy for transmission, storage, and processing, leading to carbon emissions from data center operations, server activity, and resource depletion from hardware production(Yang et al., 2018). Environmental Impact of Email Waste(Blikra Vea et al., 2018):

- ***Carbon Emissions:*** The environmental impact of email waste is primarily driven by the carbon emissions associated with data center operations and energy consumption. Data centers, which host the servers that store and transmit emails, require vast amounts of energy to operate and cool the equipment. The electricity used to power data centers often comes from fossil fuel sources, leading to carbon emissions that contribute to climate change and global warming. Additionally, the energy-intensive process of manufacturing and disposing of IT equipment further exacerbates the carbon footprint of email communication.

- ***Energy Consumption:*** The energy consumption associated with email waste encompasses not only the electricity used by data centers but also the energy consumed by end-user devices such as computers, smartphones, and tablets. Every email sent or received triggers activity on the sender's and recipient's devices, requiring power for processing, display, and network connectivity. As the volume of emails continues to rise, so too does the energy consumption of digital devices, further straining energy resources and contributing to environmental degradation(Syamala et al., 2023; Venkateswaran, Vidhya, et al., 2023).
- ***Resource Depletion:*** The production and disposal of IT equipment, including servers, networking devices, and end-user devices, contribute to resource depletion and environmental degradation. The manufacturing process of electronic hardware requires significant amounts of raw materials, including metals, plastics, and rare earth elements, many of which are extracted through environmentally damaging practices. Moreover, the disposal of obsolete IT equipment poses challenges for electronic waste management, as improper disposal methods can lead to pollution and ecosystem harm.

Addressing Email Waste: Strategies for Sustainable Email Management

To combat email waste, a comprehensive strategy involving technological solutions, behavioral changes, and organizational policies is recommended(Blikra Vea et al., 2018; Czeka\la et al., 2023). Email waste is a significant environmental issue in the digital age, causing carbon emissions, energy consumption, and resource depletion. Sustainable email management practices, such as email optimization, efficient storage, and behavioral interventions, can help organizations minimize their environmental footprint and contribute to a greener future.

- ***Email Optimization:*** Implementing email optimization techniques can help reduce the environmental impact of email communication by minimizing unnecessary emails and optimizing email storage and transmission. Strategies such as reducing email file sizes, consolidating email threads, and implementing email archiving and retention policies can help streamline email communication and reduce energy consumption.
- ***Efficient Storage Practices:*** Adopting efficient email storage practices, such as implementing email archiving solutions and utilizing cloud-based email storage, can help reduce the energy consumption and carbon emissions associated with email servers. By offloading email storage to energy-efficient data centers and leveraging scalable cloud infrastructure, organizations can optimize resource utilization and minimize environmental impact.
- ***Behavioral Interventions:*** Promoting environmentally conscious email behaviors among employees, such as practicing email mindfulness, reducing email volume, and avoiding unnecessary attachments, can help minimize email waste and reduce carbon emissions. Employee education and awareness programs can raise awareness about the environmental impact of email communication and encourage sustainable email practices.

EMAIL OPTIMIZATION TECHNIQUES

Email has become a crucial communication tool for businesses, enabling rapid information exchange and collaboration. However, it also presents challenges like information overload, productivity drains, and environmental impact. To address these, organizations can implement email optimization techniques to minimize waste, enhance productivity, and improve information management efficiency(Dada et al., 2019). Figure 3 depicts the impact of various email optimization techniques.

Figure 3. Effects of email optimization techniques

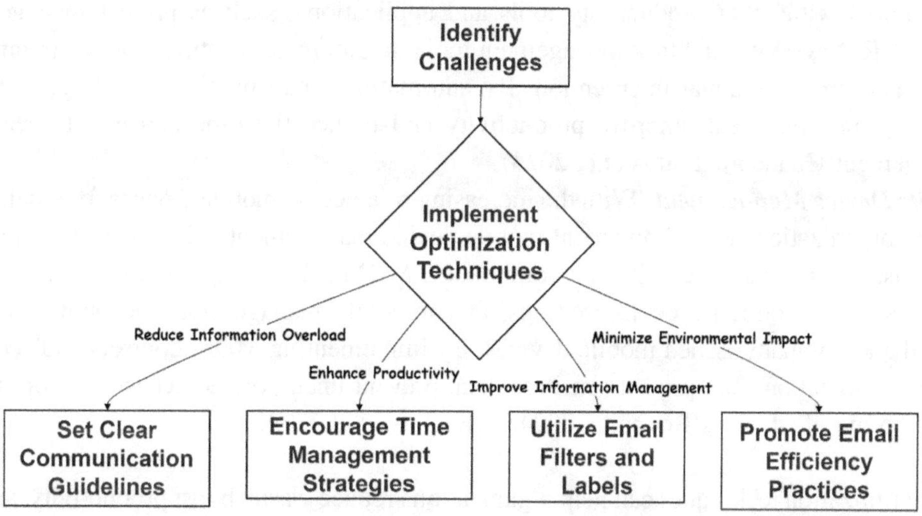

- ***Email Consolidation:*** Email consolidation involves streamlining email communication channels to reduce fragmentation and improve efficiency. Instead of using multiple email accounts across different platforms and devices, organizations can centralize email communication through a single, unified platform or email client. This simplifies email management, reduces duplicate messages, and enhances productivity by providing a centralized hub for all communications.
- ***Storage Optimization:*** Optimizing email storage is essential for managing the growing volume of emails while minimizing resource consumption and environmental impact. Organizations can implement storage optimization techniques such as email archiving, email deduplication, and attachment management to reduce the storage footprint of email servers and improve performance. By archiving older emails to secondary storage systems and removing duplicate messages and attachments, organizations can free up valuable storage space and reduce the energy consumption associated with email storage(A. Paul et al., 2024).
- ***Behavioral Interventions:*** Behavioral interventions focus on promoting environmentally conscious email behaviors among employees to minimize waste and improve productivity. Organizations can encourage employees to adopt email mindfulness practices, such as reducing email volume, avoiding unnecessary attachments, and utilizing email etiquette guidelines. Additionally, providing training and education programs on effective email management techniques can help raise

awareness about the environmental impact of email communication and empower employees to make informed decisions.

- ***Email Filtering and Prioritization:*** Implementing email filtering and prioritization techniques can help employees manage their email inboxes more effectively and focus on high-priority messages. Organizations can use email filtering rules and algorithms to automatically categorize and prioritize incoming emails based on sender, subject, and content. By directing important messages to the inbox and filtering out spam, newsletters, and low-priority emails to separate folders or archives, employees can reduce email clutter and improve productivity.
- ***Automation and Workflow Integration:*** Leveraging automation and workflow integration tools can streamline email management processes and enhance efficiency. Organizations can integrate email with other productivity tools and applications, such as project management platforms, CRM systems, and task management tools, to automate repetitive tasks, streamline workflows, and reduce manual intervention. By automating email processing, filing, and follow-up tasks, organizations can improve productivity and reduce the time and effort spent on email management(Gnanaprakasam et al., 2023).
- ***Mobile Device Management:*** With the increasing reliance on mobile devices for email communication, organizations must implement mobile device management (MDM) strategies to optimize email usage and ensure security and compliance. MDM solutions allow organizations to enforce email policies, configure device settings, and remotely manage email accounts on employee-owned and company-issued mobile devices. By implementing MDM controls and security measures, organizations can protect sensitive data, prevent unauthorized access, and optimize email usage on mobile devices(Boopathi, 2024c; Mohanty et al., 2023).

Email optimization techniques can help organizations reduce waste, boost productivity, and improve information management efficiency in email communication. These techniques include email consolidation, storage optimization, behavioral interventions, email filtering, automation, workflow integration, and mobile device management strategies. These strategies can streamline processes, reduce environmental impact, and enhance digital communication.

INTEGRATING CIRCULAR ECONOMY PRINCIPLES

The adoption of circular economy principles is crucial for organizations to minimize waste, conserve resources, and promote sustainability. While traditional manufacturing and supply chain operations are often applied, there is potential to extend these principles to digital practices. This section explores how organizations can integrate circular economy principles into product lifecycle management and end-of-life strategies for electronic devices, beyond green computing and email waste management(Dhanya et al., 2023; Hussain et al., 2023).

Product Lifecycle Management: Product lifecycle management (PLM) involves managing the entire lifecycle of a product, from design and manufacturing to distribution, use, and disposal. Integrating circular economy principles into PLM processes requires organizations to adopt a holistic approach that considers environmental impact, resource efficiency, and product longevity at every stage of the product lifecycle. Important strategies for integrating circular economy principles into PLM include:

- ***Design for Durability and Repairability:*** Designing products with longevity and repairability in mind can extend their lifespan and reduce the need for premature replacement. This involves using durable materials, modular design principles, and standardized components that are easy to repair, upgrade, and recycle.
- ***Sustainable Materials and Manufacturing:*** Choosing sustainable materials and manufacturing processes can minimize the environmental impact of products throughout their lifecycle. Organizations can prioritize the use of renewable materials, recycled content, and non-toxic substances in product design and production, reducing resource consumption and pollution(Boopathi, 2022b, 2024d, 2024c).
- ***Extended Producer Responsibility (EPR):*** Implementing EPR programs shifts responsibility for managing products at the end of their life from consumers to manufacturers. Organizations can take responsibility for the collection, recycling, and disposal of their products, incentivizing sustainable design and resource recovery.

End-of-Life Strategies for Electronic Devices: Effective end-of-life strategies for electronic devices are essential for minimizing electronic waste and maximizing resource recovery. Circular economy principles can guide organizations in developing sustainable end-of-life strategies that prioritize reuse, refurbishment, and recycling over disposal. Organizations should adopt a lifecycle approach to incorporate circular economy principles into their digital practices, focusing on environmental impact, resource efficiency, and product longevity from design to end-of-life, promoting sustainability(Janardhana et al., 2023; Sangeetha et al., 2023).

- ***Take-Back and Recycling Programs:*** Implementing take-back and recycling programs allows organizations to collect and recycle electronic devices at the end of their useful life. By partnering with certified recyclers and refurbishers, organizations can ensure that electronic devices are responsibly disposed of or refurbished for reuse, reducing landfill waste and conserving resources.
- ***Refurbishment and Resale:*** Refurbishing electronic devices for resale extends their lifespan and reduces the demand for new products. Organizations can refurbish and resell used devices to consumers or business customers, providing cost-effective alternatives to purchasing new products while diverting electronic waste from landfills.
- ***Material Recovery and Recycling:*** Recovering valuable materials from end-of-life electronic devices through recycling processes helps close the loop on resource utilization and reduces the need for virgin materials. Organizations can partner with recycling facilities to extract metals, plastics, and other materials from electronic devices for reuse in new products, contributing to resource conservation and environmental sustainability.
- ***Design for Recycling:*** Designing electronic devices with recyclability in mind facilitates the recovery of materials at the end of their life. Organizations can use standardized components, easily disassembled designs, and labeling for material identification to simplify the recycling process and maximize resource recovery.

BEST PRACTICES

Case studies from various industries showcase successful implementation of sustainable circular economy digital practices, highlighting exemplary practices, innovative strategies, and tangible outcomes achieved through the integration of circular economy principles into digital operations as shown in Figure 4. Real-world case studies showcase successful organizations that have adopted sustainable circular economy digital practices. These organizations have successfully implemented innovative strategies, overcome challenges, and achieved tangible outcomes, demonstrating the transformative potential of circular economy principles in driving environmental sustainability, resource efficiency, and social responsibility. Collaboration, innovation, and commitment to sustainability can pave the way for a more sustainable future(Harikaran et al., 2023; Rahamathunnisa et al., 2024; Venkateswaran, Kumar, et al., 2023).

Figure 4. Case studies from various industries demonstrate the successful implementation of sustainable circular economy digital practices

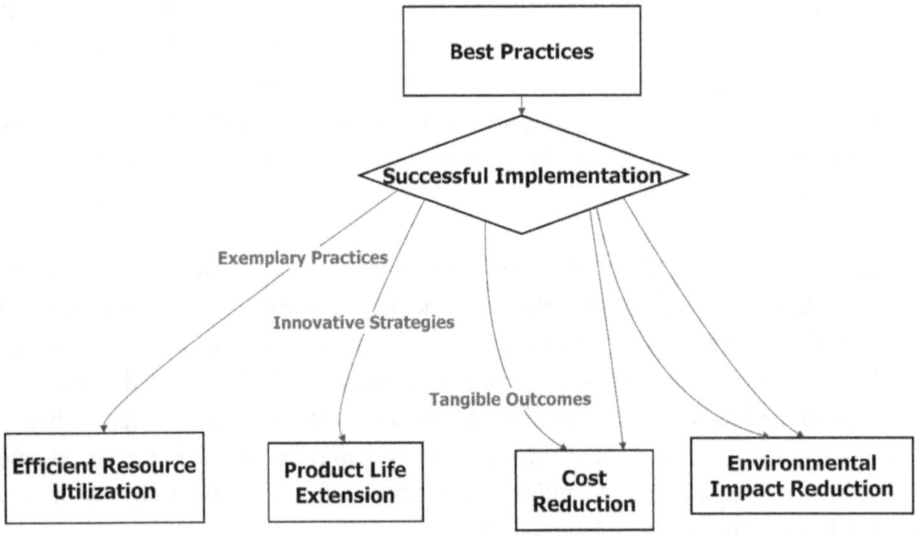

- ***Google's Data Center Efficiency:*** Google is renowned for its commitment to sustainability and environmental stewardship, particularly in its data center operations. The company has implemented numerous initiatives to improve energy efficiency, reduce carbon emissions, and minimize environmental impact across its global network of data centers. Google's data centers utilize advanced cooling technologies, renewable energy sources, and energy-efficient hardware to optimize energy usage and reduce environmental footprint. By leveraging machine learning algorithms and data analytics, Google continuously monitors and optimizes its data center operations to achieve maximum efficiency and sustainability.
- ***Dell's Closed-Loop Recycling:*** Dell, a leading technology company, has demonstrated its commitment to circular economy principles through its closed-loop recycling program. Dell collects end-of-life electronic devices from customers and partners and processes them through its recycling facilities to recover valuable materials such as metals, plastics, and glass. These materials

are then reused in the manufacturing of new products, closing the loop on resource utilization and minimizing waste. Dell's closed-loop recycling program not only reduces environmental impact but also creates economic value by reducing the need for virgin materials and lowering production costs.

- ***Patagonia's Worn Wear Initiative:*** Patagonia, an outdoor apparel company, has implemented the Worn Wear initiative to promote product longevity, repairability, and reuse. The initiative encourages customers to repair, reuse, and recycle their worn-out Patagonia products rather than discarding them. Patagonia offers repair services, DIY repair guides, and trade-in programs to extend the lifespan of its products and minimize waste. Through the Worn Wear initiative, Patagonia fosters a culture of sustainability, encourages responsible consumption, and reduces the environmental footprint of its products.
- ***HP's Sustainable Design Principles:*** HP, a leading technology company, has integrated sustainable design principles into its product development process to minimize environmental impact and promote resource efficiency. HP's Sustainable Impact program focuses on designing products with sustainability in mind, using recycled materials, reducing energy consumption, and optimizing product lifecycle management. HP's sustainable design principles have led to the development of innovative products such as the HP Elite Dragonfly laptop, which features ocean-bound plastics and is designed for easy repair and recycling. By embracing sustainable design principles, HP demonstrates its commitment to environmental sustainability and responsible business practices.
- ***Interface's Net-Works Program:*** Interface, a global leader in commercial flooring, has implemented the Net-Works program to address the issue of marine plastic pollution while promoting social and environmental sustainability. The program collects discarded fishing nets from coastal communities in developing countries and recycles them into carpet tiles. Interface works with local partners to establish community-based collection hubs and provide livelihood opportunities for fishing communities. The Net-Works program not only reduces marine plastic pollution but also creates a sustainable supply chain for recycled materials and fosters social and economic development in coastal communities.

FUTURE TRENDS AND CONSIDERATIONS

As organizations continue to embrace sustainable digital practices within the circular economy framework, several future trends and developments are anticipated to shape the landscape of environmental sustainability, resource efficiency, and social responsibility. Emerging technologies, regulatory changes, and evolving consumer preferences are important future trends for sustainability-conscious organizations, offering both opportunities and challenges(Boopathi, 2023; Ramudu et al., 2023; Revathi et al., 2024b). The figure 5 illustrates the future trends in sustainable digital practices within the circular economy framework.

Figure 5. Sustainable digital practices within the circular economy framework, several future trends

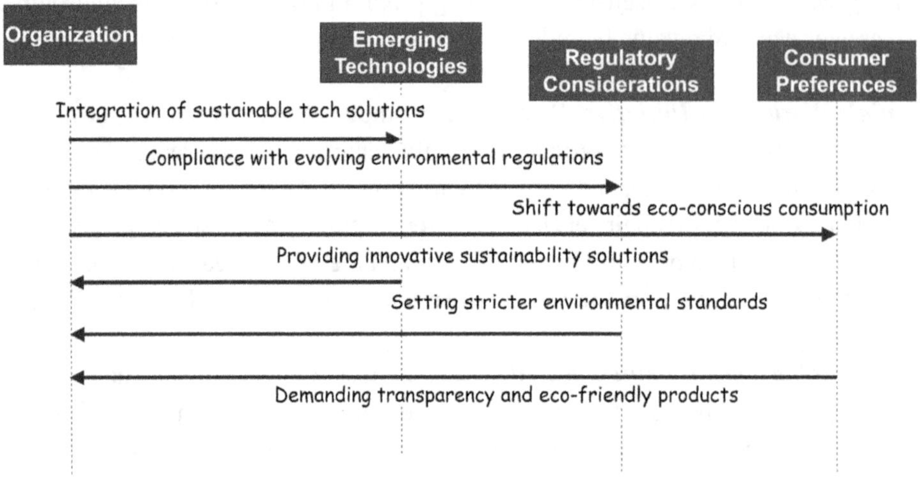

- ***Adoption of Emerging Technologies:*** The adoption of emerging technologies such as artificial intelligence (AI), blockchain, and Internet of Things (IoT) is expected to play a significant role in advancing sustainable digital practices within the circular economy framework. AI-powered analytics and predictive modeling can optimize resource utilization, improve energy efficiency, and enhance product lifecycle management. Blockchain technology can enable transparent and traceable supply chains, facilitating responsible sourcing, recycling, and end-of-life management. IoT devices and sensors can provide real-time monitoring and data insights, enabling proactive environmental management and sustainable decision-making(Sundaramoorthy et al., 2024).
- ***Regulatory Considerations:*** Regulatory frameworks and policies related to environmental sustainability, resource management, and circular economy principles are expected to evolve in response to growing concerns about climate change and resource depletion. Governments and regulatory bodies may introduce new legislation, standards, and incentives to promote sustainable digital practices, encourage renewable energy adoption, and incentivize circular economy initiatives. Organizations will need to stay abreast of regulatory developments and ensure compliance with environmental regulations and reporting requirements to mitigate risks and seize opportunities for sustainability leadership.
- ***Evolving Consumer Preferences:*** Consumer preferences are shifting towards environmentally friendly products, sustainable brands, and responsible consumption practices. As awareness of environmental issues grows, consumers are increasingly demanding transparency, accountability, and sustainability from the organizations they support. Companies that prioritize sustainability, ethical sourcing, and circular economy principles are likely to gain a competitive edge and build trust with environmentally conscious consumers. Organizations will need to align their products, services, and marketing strategies with evolving consumer preferences to remain relevant and resilient in a changing marketplace(Naveeenkumar et al., 2024; Ravisankar et al., 2024).
- ***Circular Business Models:*** Circular business models, such as product-as-a-service (PaaS), sharing economy platforms, and closed-loop supply chains, are gaining traction as organizations seek to minimize waste, maximize resource efficiency, and create value from end-of-life products. PaaS

models offer consumers access to products and services on a subscription or pay-per-use basis, encouraging product longevity, repairability, and reuse. Sharing economy platforms enable peer-to-peer sharing and rental of goods, reducing the need for ownership and promoting collaborative consumption. Closed-loop supply chains prioritize the recovery and recycling of materials, creating a sustainable loop of resource utilization and minimizing waste(Ravisankar et al., 2024).

- ***Collaboration and Partnerships:*** Collaboration and partnerships between industry stakeholders, government agencies, non-profit organizations, and academia are essential for driving collective action and scaling sustainable digital practices within the circular economy framework. Collaborative initiatives such as industry consortia, research partnerships, and multi-stakeholder platforms can facilitate knowledge sharing, innovation, and best practice dissemination. By working together, organizations can overcome common challenges, leverage collective expertise, and accelerate progress towards a more sustainable and resilient future.

To anticipate future trends in sustainable digital practices within the circular economy framework, organizations must embrace innovation, adapt to regulatory changes, respond to evolving consumer preferences, and foster collaboration across sectors. This proactive, agile, and visionary approach can position organizations as leaders in sustainability, driving positive environmental impact and creating value for society.

CONCLUSION

The adoption of sustainable circular economy digital practices offers organizations a transformative opportunity to mitigate environmental impact, conserve resources, and drive long-term economic benefits, with significant findings and recommendations emerging. Sustainability is crucial in the digital age, as digital technologies continue to evolve. Organizations must prioritize sustainability as a core value and integrate circular economy principles into their digital practices. Digital operations, such as green computing, email management, and product lifecycle management, have a significant environmental impact, including carbon emissions, energy consumption, and resource depletion. Embracing circular economy principles like product longevity, repairability, reuse, and recycling is essential for minimizing waste and optimizing resource utilization. Real-world case studies and best practices from leading organizations demonstrate the benefits of implementing sustainable circular economy practices, such as energy-efficient data center operations and closed-loop recycling programs. Future trends, such as emerging technologies, regulatory frameworks, and circular business models, offer opportunities for innovation and sustainability leadership.

Organizations should prioritize sustainability by integrating circular economy principles into digital practices, investing in innovation to optimize resource utilization and reduce environmental impact, fostering collaboration with stakeholders, and educating employees, customers, and stakeholders about the importance of sustainability. Establishing significant performance indicators (KPIs) and metrics can track progress, measure impact, and continuously improve sustainability performance. This approach will accelerate progress towards sustainability goals and foster a culture of environmental consciousness. Organizations can reduce their environmental footprint, drive long-term economic benefits, enhance competitiveness, and build resilience in a rapidly changing world by adopting sustainable circular economy digital practices. Through visionary leadership, strategic investments, and collective action, they can pave the way for a prosperous future.

Abbreviations

- AI: Artificial Intelligence
  - CPU: Central Processing Unit
  - CRM: Customer Relationship Management
  - DIY: Do It Yourself
  - EPEAT: Electronic Product Environmental Assessment Tool
  - EPR: Extended Producer Responsibility
  - GPU: Graphics Processing Unit
  - HDD: Hard Disk Drive
  - HP: Hewlett-Packard (a technology company)
  - IT: Information Technology
  - KPI: Key Performance Indicator
  - MDM: Mobile Device Management
  - PLM: Product Lifecycle Management
  - PLUS: Power Supply Unit
  - SSD: Solid State Drive
  - VM: Virtual Machine

REFERENCES

Agrawal, A. V., Shashibhushan, G., Pradeep, S., Padhi, S., Sugumar, D., & Boopathi, S. (2023). Synergizing Artificial Intelligence, 5G, and Cloud Computing for Efficient Energy Conversion Using Agricultural Waste. In Sustainable Science and Intelligent Technologies for Societal Development (pp. 475–497). IGI Global.

Awomeso, J., Taiwo, A., Gbadebo, A., & Arimoro, A. (2010). Waste disposal and pollution management in urban areas: A workable remedy for the environment in developing countries. *American Journal of Environmental Sciences*, *6*(1), 26–32. doi:10.3844/ajessp.2010.26.32

Blikra Vea, E., Martinez-Sanchez, V., & Thomsen, M. (2018). A review of waste management decision support tools and their ability to assess circular biowaste management systems. *Sustainability (Basel)*, *10*(10), 3720. doi:10.3390/su10103720

Boopathi, S., Karthikeyan, K. R., Jaiswal, C., Dabi, R., Sunagar, P., & Malik, S. (2024). *IoT based Automatic Cooling Tower*. Research Gate.

Boopathi, S. (2022a). An investigation on gas emission concentration and relative emission rate of the near-dry wire-cut electrical discharge machining process. *Environmental Science and Pollution Research International*, *29*(57), 86237–86246. doi:10.1007/s11356-021-17658-1 PMID:34837614

Boopathi, S. (2022b). Performance Improvement of Eco-Friendly Near-Dry wire-Cut Electrical Discharge Machining Process Using Coconut Oil-Mist Dielectric Fluid. *World Scientific:Journal of Advanced Manufacturing Systems.*

Boopathi, S. (2023). Securing Healthcare Systems Integrated With IoT: Fundamentals, Applications, and Future Trends. In Dynamics of Swarm Intelligence Health Analysis for the Next Generation (pp. 186–209). IGI Global.

Boopathi, S. (2024a). Balancing Innovation and Security in the Cloud: Navigating the Risks and Rewards of the Digital Age. In Improving Security, Privacy, and Trust in Cloud Computing (pp. 164–193). IGI Global.

Boopathi, S. (2024b). Energy Cascade Conversion System and Energy-Efficient Infrastructure. In Sustainable Development in AI, Blockchain, and E-Governance Applications (pp. 47–71). IGI Global. doi:10.4018/979-8-3693-1722-8.ch004

Boopathi, S. (2024c). Implementation of Green Manufacturing Practices in Automobile Fields: A Review. *Sustainable Machining and Green Manufacturing*, 221–248.

Boopathi, S. (2024d). Minimization of Manufacturing Industry Wastes Through the Green Lean Sigma Principle. *Sustainable Machining and Green Manufacturing*, 249–270.

Boopathi, S., Kumar, P. K. S., Meena, R. S., Sudhakar, M., & Associates. (2023). Sustainable Developments of Modern Soil-Less Agro-Cultivation Systems: Aquaponic Culture. In Human Agro-Energy Optimization for Business and Industry (pp. 69–87). IGI Global.

Butt, S., Ahmadi, M., & Razavi, M. (2020). Green Computing: Sustainable Design and Technologies. *2020 5th International Conference on Innovative Technologies in Intelligent Systems and Industrial Applications (CITISIA)*, 1–7.

Czekala, W., Drozdowski, J., & \Labiak, P. (2023). Modern technologies for waste management: A review. *Applied Sciences, 13*(15), 8847.

Dada, E. G., Bassi, J. S., Chiroma, H., Adetunmbi, A. O., Ajibuwa, O. E., & ... (2019). Machine learning for email spam filtering: Review, approaches and open research problems. *Heliyon*, *5*(6), e01802. doi:10.1016/j.heliyon.2019.e01802 PMID:31211254

Dhaini, M., Jaber, M., Fakhereldine, A., Hamdan, S., & Haraty, R. A. (2021). Green computing approaches-A survey. *Informatica (Vilnius)*, *45*(1).

Dhanalakshmi, M., Tamilarasi, K., Saravanan, S., Sujatha, G., Boopathi, S., & Associates. (2024). Fog Computing-Based Framework and Solutions for Intelligent Systems: Enabling Autonomy in Vehicles. In Computational Intelligence for Green Cloud Computing and Digital Waste Management (pp. 330–356). IGI Global.

Dhanya, D., Kumar, S. S., Thilagavathy, A., Prasad, D., & Boopathi, S. (2023). Data Analytics and Artificial Intelligence in the Circular Economy: Case Studies. In Intelligent Engineering Applications and Applied Sciences for Sustainability (pp. 40–58). IGI Global.

Domakonda, V. K., Farooq, S., Chinthamreddy, S., Puviarasi, R., Sudhakar, M., & Boopathi, S. (2022). Sustainable Developments of Hybrid Floating Solar Power Plants: Photovoltaic System. In Human Agro-Energy Optimization for Business and Industry (pp. 148–167). IGI Global.

Gnanaprakasam, C., Vankara, J., Sastry, A. S., Prajval, V., Gireesh, N., & Boopathi, S. (2023). Long-Range and Low-Power Automated Soil Irrigation System Using Internet of Things: An Experimental Study. In Contemporary Developments in Agricultural Cyber-Physical Systems (pp. 87–104). IGI Global.

Harikaran, M., Boopathi, S., Gokulakannan, S., & Poonguzhali, M. (2023). Study on the Source of E-Waste Management and Disposal Methods. In *Sustainable Approaches and Strategies for E-Waste Management and Utilization* (pp. 39–60). IGI Global. doi:10.4018/978-1-6684-7573-7.ch003

Hussain, Z., Babe, M., Saravanan, S., Srimathy, G., Roopa, H., & Boopathi, S. (2023). Optimizing Biomass-to-Biofuel Conversion: IoT and AI Integration for Enhanced Efficiency and Sustainability. In Circular Economy Implementation for Sustainability in the Built Environment (pp. 191–214). IGI Global.

Janardhana, K., Singh, V., Singh, S. N., Babu, T. R., Bano, S., & Boopathi, S. (2023). Utilization Process for Electronic Waste in Eco-Friendly Concrete: Experimental Study. In Sustainable Approaches and Strategies for E-Waste Management and Utilization (pp. 204–223). IGI Global.

Kumar, M., Kumar, K., Sasikala, P., Sampath, B., Gopi, B., & Sundaram, S. (2023). Sustainable Green Energy Generation From Waste Water: IoT and ML Integration. In Sustainable Science and Intelligent Technologies for Societal Development (pp. 440–463). IGI Global.

Maguluri, L. P., Ananth, J., Hariram, S., Geetha, C., Bhaskar, A., & Boopathi, S. (2023). Smart Vehicle-Emissions Monitoring System Using Internet of Things (IoT). In Handbook of Research on Safe Disposal Methods of Municipal Solid Wastes for a Sustainable Environment (pp. 191–211). IGI Global.

Malathi, J., Kusha, K., Isaac, S., Ramesh, A., Rajendiran, M., & Boopathi, S. (2024). IoT-Enabled Remote Patient Monitoring for Chronic Disease Management and Cost Savings: Transforming Healthcare. In Advances in Explainable AI Applications for Smart Cities (pp. 371–388). IGI Global.

Mohanty, A., Venkateswaran, N., Ranjit, P., Tripathi, M. A., & Boopathi, S. (2023). Innovative Strategy for Profitable Automobile Industries: Working Capital Management. In Handbook of Research on Designing Sustainable Supply Chains to Achieve a Circular Economy (pp. 412–428). IGI Global.

Nanda, A. K., Sharma, A., Augustine, P. J., Cyril, B. R., Kiran, V., & Sampath, B. (2024). Securing Cloud Infrastructure in IaaS and PaaS Environments. In Improving Security, Privacy, and Trust in Cloud Computing (pp. 1–33). IGI Global. doi:10.4018/979-8-3693-1431-9.ch001

Naveeenkumar, N., Rallapalli, S., Sasikala, K., Priya, P. V., Husain, J., & Boopathi, S. (2024). Enhancing Consumer Behavior and Experience Through AI-Driven Insights Optimization. In *AI Impacts in Digital Consumer Behavior* (pp. 1–35). IGI Global. doi:10.4018/979-8-3693-1918-5.ch001

Nishanth, J., Deshmukh, M. A., Kushwah, R., Kushwaha, K. K., Balaji, S., & Sampath, B. (2023). Particle Swarm Optimization of Hybrid Renewable Energy Systems. In *Intelligent Engineering Applications and Applied Sciences for Sustainability* (pp. 291–308). IGI Global. doi:10.4018/979-8-3693-0044-2.ch016

Paul, A., & Thilagham, K. KG, J.-, Reddy, P. R., Sathyamurthy, R., & Boopathi, S. (2024). Multi-criteria Optimization on Friction Stir Welding of Aluminum Composite (AA5052-H32/B4C) using Titanium Nitride Coated Tool. Engineering Research Express.

Paul, S. G., Saha, A., Arefin, M. S., Bhuiyan, T., Biswas, A. A., Reza, A. W., Alotaibi, N. M., Alyami, S. A., & Moni, M. A. (2023). A comprehensive review of green computing: Past, present, and future research. *IEEE Access : Practical Innovations, Open Solutions*, *11*, 87445–87494. doi:10.1109/ACCESS.2023.3304332

Rahamathunnisa, U., Sudhakar, K., Murugan, T. K., Thivaharan, S., Rajkumar, M., & Boopathi, S. (2023). Cloud Computing Principles for Optimizing Robot Task Offloading Processes. In *AI-Enabled Social Robotics in Human Care Services* (pp. 188–211). IGI Global. doi:10.4018/978-1-6684-8171-4.ch007

Rahamathunnisa, U., Sudhakar, K., Padhi, S., Bhattacharya, S., Shashibhushan, G., & Boopathi, S. (2024). Sustainable Energy Generation From Waste Water: IoT Integrated Technologies. In Adoption and Use of Technology Tools and Services by Economically Disadvantaged Communities: Implications for Growth and Sustainability (pp. 225–256). IGI Global.

Ramudu, K., Mohan, V. M., Jyothirmai, D., Prasad, D., Agrawal, R., & Boopathi, S. (2023). Machine Learning and Artificial Intelligence in Disease Prediction: Applications, Challenges, Limitations, Case Studies, and Future Directions. In Contemporary Applications of Data Fusion for Advanced Healthcare Informatics (pp. 297–318). IGI Global.

Ravisankar, A., Shanthi, A., Lavanya, S., Ramaratnam, M., Krishnamoorthy, V., & Boopathi, S. (2024). Harnessing 6G for Consumer-Centric Business Strategies Across Electronic Industries. In AI Impacts in Digital Consumer Behavior (pp. 241–270). IGI Global.

Raza, K., Patle, V., & Arya, S. (2012). A review on green computing for eco-friendly and sustainable it. *Journal of Computational Intelligence and Electronic Systems*, *1*(1), 3–16. doi:10.1166/jcies.2012.1023

Rebecca, B., Kumar, K. P. M., Padmini, S., Srivastava, B. K., Halder, S., & Boopathi, S. (2024). Convergence of Data Science-AI-Green Chemistry-Affordable Medicine: Transforming Drug Discovery. In *Handbook of Research on AI and ML for Intelligent Machines and Systems* (pp. 348–373). IGI Global.

Revathi, S., Babu, M., Rajkumar, N., Meti, V. K. V., Kandavalli, S. R., & Boopathi, S. (2024a). Unleashing the Future Potential of 4D Printing: Exploring Applications in Wearable Technology, Robotics, Energy, Transportation, and Fashion. In Human-Centered Approaches in Industry 5.0: Human-Machine Interaction, Virtual Reality Training, and Customer Sentiment Analysis (pp. 131–153). IGI Global.

Revathi, S., Babu, M., Rajkumar, N., Meti, V. K. V., Kandavalli, S. R., & Boopathi, S. (2024b). Unleashing the Future Potential of 4D Printing: Exploring Applications in Wearable Technology, Robotics, Energy, Transportation, and Fashion. In Human-Centered Approaches in Industry 5.0: Human-Machine Interaction, Virtual Reality Training, and Customer Sentiment Analysis (pp. 131–153). IGI Global.

Saha, B. (2014). Green computing. [IJCTT]. *International Journal of Computer Trends and Technology*, *14*(2), 46–50. doi:10.14445/22312803/IJCTT-V14P112

Saha, B. (2018). Green computing: Current research trends. *International Journal on Computer Science and Engineering*, *6*(3), 467–469.

Samikannu, R., Koshariya, A. K., Poornima, E., Ramesh, S., Kumar, A., & Boopathi, S. (2022). Sustainable Development in Modern Aquaponics Cultivation Systems Using IoT Technologies. In *Human Agro-Energy Optimization for Business and Industry* (pp. 105–127). IGI Global.

Sampath, B. (2021). *Sustainable Eco-Friendly Wire-Cut Electrical Discharge Machining: Gas Emission Analysis.*

Sangeetha, M., Kannan, S. R., Boopathi, S., Ramya, J., Ishrat, M., & Sabarinathan, G. (2023). Prediction of Fruit Texture Features Using Deep Learning Techniques. *2023 4th International Conference on Smart Electronics and Communication (ICOSEC)*, 762–768.

Satav, S. D., Lamani, D., Harsha, K., Kumar, N., Manikandan, S., & Sampath, B. (2023). Energy and Battery Management in the Era of Cloud Computing: Sustainable Wireless Systems and Networks. In Sustainable Science and Intelligent Technologies for Societal Development (pp. 141–166). IGI Global.

Sonia, R., Gupta, N., Manikandan, K., Hemalatha, R., Kumar, M. J., & Boopathi, S. (2024). Strengthening Security, Privacy, and Trust in Artificial Intelligence Drones for Smart Cities. In *Analyzing and Mitigating Security Risks in Cloud Computing* (pp. 214–242). IGI Global. doi:10.4018/979-8-3693-3249-8.ch011

Sundaramoorthy, K., Singh, A., Sumathy, G., Maheshwari, A., Arunarani, A., & Boopathi, S. (2024). A Study on AI and Blockchain-Powered Smart Parking Models for Urban Mobility. In *Handbook of Research on AI and ML for Intelligent Machines and Systems* (pp. 223–250). IGI Global.

Syamala, M., Komala, C., Pramila, P., Dash, S., Meenakshi, S., & Boopathi, S. (2023). Machine Learning-Integrated IoT-Based Smart Home Energy Management System. In *Handbook of Research on Deep Learning Techniques for Cloud-Based Industrial IoT* (pp. 219–235). IGI Global. doi:10.4018/978-1-6684-8098-4.ch013

Vanitha, S., Radhika, K., & Boopathi, S. (2023). Artificial Intelligence Techniques in Water Purification and Utilization. In *Human Agro-Energy Optimization for Business and Industry* (pp. 202–218). IGI Global. doi:10.4018/978-1-6684-4118-3.ch010

Venkateswaran, N., Kumar, S. S., Diwakar, G., Gnanasangeetha, D., & Boopathi, S. (2023). Synthetic Biology for Waste Water to Energy Conversion: IoT and AI Approaches. *Applications of Synthetic Biology in Health. Energy & Environment*, 360–384.

Venkateswaran, N., Vidhya, K., Ayyannan, M., Chavan, S. M., Sekar, K., & Boopathi, S. (2023). A Study on Smart Energy Management Framework Using Cloud Computing. In 5G, Artificial Intelligence, and Next Generation Internet of Things: Digital Innovation for Green and Sustainable Economies (pp. 189–212). IGI Global. doi:10.4018/978-1-6684-8634-4.ch009

Vennila, T., Karuna, M., Srivastava, B. K., Venugopal, J., Surakasi, R., & Sampath, B. (2022). New Strategies in Treatment and Enzymatic Processes: Ethanol Production From Sugarcane Bagasse. In Human Agro-Energy Optimization for Business and Industry (pp. 219–240). IGI Global.

Yang, H., Ma, M., Thompson, J. R., & Flower, R. J. (2018). Waste management, informal recycling, environmental pollution and public health. *Journal of Epidemiology and Community Health*, *72*(3), 237–243. doi:10.1136/jech-2016-208597 PMID:29222091

Yupapin, P., Trabelsi, Y., Nattappan, A., & Boopathi, S. (2023). Performance improvement of wire-cut electrical discharge machining process using cryogenically treated super-conductive state of Monel-K500 alloy. *Iranian Journal of Science and Technology. Transaction of Mechanical Engineering*, *47*(1), 267–283. doi:10.1007/s40997-022-00513-0

Chapter 11
Next Generation Materials for Sustainable Development

Vijay Bhutani
Amity University, India

Preeti Singh Bahadur
Amity University, India

Sunil Kumar Sansaniwal
https://orcid.org/0000-0002-0997-5096
The Energy and Resources Institute, New Delhi, India

ABSTRACT

In a rapidly evolving world, sustainability has emerged as a central focus, and the exploration of next-generation materials for sustainable development and engineering has become a pivotal endeavour. The exploration of next-generation materials for sustainable development and engineering is a vital pursuit in our quest for a more sustainable future. This book chapter delves into the critical era of "Next Generation Materials for Sustainable Development or Sustainable Engineering." providing an in-depth examination of the evolution of materials, their characteristics, benefits, challenges, and innovative applications across various industries.

INTRODUCTION

The Evolution of Materials: Pioneering Sustainability

Materials have always played a fundamental role in human progress. From the discovery of fire and the use of primitive tools to the Industrial Revolution, advancements in materials have driven innovation. However, the 21st century presents unique challenges, primarily the need to balance progress with environmental responsibility. Traditional materials like steel, concrete, and plastics have been essential

DOI: 10.4018/979-8-3693-1794-5.ch011

to modern society but have also posed significant environmental concerns. Therefore, the evolution of materials has taken a new direction, one that prioritizes sustainability. (Sauvé et al., 2016)

The Historical Significance of Materials

Materials have been integral to human civilization since time immemorial. Early humans used stones, bones, and wood to craft tools and shelters. The discovery of fire allowed for cooking and protection from predators, marking the first breakthrough in the manipulation of materials. The mastery of metalwork marked a significant leap in human history. Bronze and iron transformed tools and weaponry, contributing to the rise of complex societies. The Eiffel Tower, a testament to the power of steel, still stands as an iconic monument from this era.

The Industrial Revolution: A Turning Point

The 18th and 19th centuries ushered in the Industrial Revolution, a period marked by rapid advancements in materials. Steel, a durable and versatile material, fueled the construction of railways, bridges, and skyscrapers. It symbolized human engineering prowess. While these materials brought tremendous progress, they also unleashed environmental issues. The carbon footprint of steel production and the longevity of plastic waste pose daunting challenges. This paradox between development and environmental consequences necessitated a reevaluation of our approach. In recent decades, scientists, engineers, and innovators have been driven by the imperative to find sustainable alternatives to traditional materials. These endeavors have birthed materials like carbon fiber, bamboo, and bio-plastics. These materials are not only strong but also environmentally responsible. (Ashby, n.d.)

Recycling and Circular Economy

Sustainability is not only about inventing new materials but also about reusing existing ones. The concept of the circular economy promotes the recycling and repurposing of materials. It is a promising solution to reduce waste and lower the carbon footprint. The evolution of materials has been a pivotal force in human history, propelling us from primitive tools to the heights of modern engineering. However, with progress came environmental concerns, necessitating a shift towards sustainability. Innovations in sustainable materials, along with a commitment to recycling and a circular economy, are paving the way for a future where progress aligns with environmental responsibility. Sustainable materials are those that have a minimal environmental impact, both in their production and disposal. They are designed to meet human needs without depleting natural resources or causing harm to the environment. The circular economy is an economic system that focuses on minimizing waste and making the most of resources. In the context of materials, it involves recycling, repurposing, and reusing materials to reduce environmental impact. Sustainability is crucial because it ensures that the materials we use today do not compromise the well-being of future generations. It helps us strike a balance between technological advancement and environmental responsibility. Individuals can contribute by reducing their consumption, recycling, and choosing products made from sustainable materials. Additionally, supporting policies and businesses that prioritize sustainability is vital in promoting change. (Sauvé et al., 2016)

Figure 1. Sustainable material management

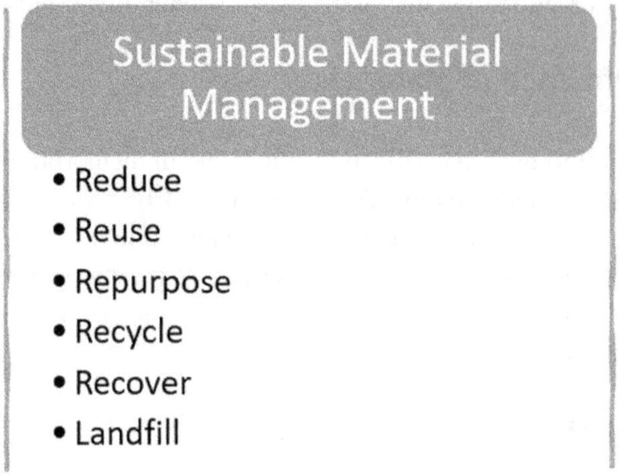

SUSTAINABLE MATERIALS

Mycelium

Mycelium, in the realm of fungal biology, represents the intricate, thread-like network formed by the vegetative part of fungi. It acts as the foundation for cultivating mycelium-based materials with diverse applications. This complex, web-like structure spreads through selected substrates, connecting and surrounding materials. Mycelium possesses the remarkable ability to break down natural polymers, making it an essential element in the creation of eco-friendly biomaterials. Often associated with species like Ganoderma lucidum and Pleurotus ostreatus, mycelium thrives on a variety of substrates, including agricultural materials like straw, grains, and sawdust. This versatile, sustainable material forms the basis for numerous innovative solutions. Mycelium is used in advanced biodegradable packaging for electronics and advanced scientific equipment, offering protection and sustainability. It's also employed in advanced myco-materials used in aerospace as lightweight and fire-resistant components. Mycelium-based materials exhibit impressive attributes, combining exceptional strength surpassing that of EPS, insulation properties akin to fiberglass or sheep's wool (with a thermal conductivity of $\lambda = 0.04$ W/mk), customizable acoustic features, and elevated fire resistance when compared to alternative organic materials. (Haneef et al., 2017; Vandelook et al., 2021)

Figure 2. Mycelium block

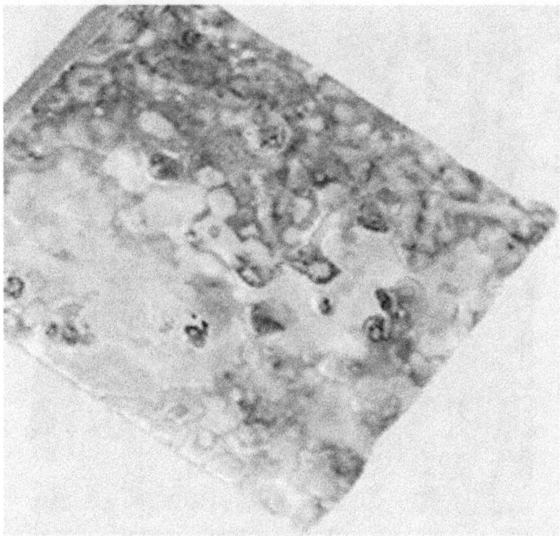

Bamboo

Bamboo boasts a range of remarkable mechanical properties that underscore its status as a sustainable building material. Its compressive strength, twice that of concrete, offers robust support. Tensile strength on par with steel ensures structural integrity, while higher shear stress than wood bolsters stability. With a bending strength of 502.3–1240.3 kgf/cm2 and a modulus of elasticity (MOE) ranging from 57.515–121.334 kgf/cm2, bamboo is incredibly resilient. Additional mechanical attributes, including shear strength (9 MPa), compressive strength (52 MPa), and tensile strength (159 MPa), further enhance its suitability for construction. Bamboo's sustainability is underscored by its renewability, eco-friendliness, widespread availability, durability, reusability, and recyclability, making it a locally accessible and environmentally responsible choice for building projects. Bamboo stands out as a sustainable material due to its rapid growth, renewability, and minimal environmental impact. It's a versatile resource, used in construction, furniture, and textiles, while its efficient regrowth and CO_2 absorption make it an eco-friendly choice for a variety of applications. Bamboo's remarkable strength-to-weight ratio makes it a top choice for advanced applications in the aerospace industry. It's used in lightweight components, such as airplane interiors and turbine blades. Bamboo's natural resilience also contributes to earthquake-resistant building structures. (Bhonde et al., 2014)

Figure 3. Bamboo used as fencing

Cork

Cork is a sustainable material prized for its eco-friendly properties. Harvested from the bark of cork oak trees, it regenerates without harming the trees. Its lightweight, buoyant, and insulating qualities make it ideal for various applications, from flooring to wine stoppers. Cork represents an environmentally responsible choice. Cork's compressibility and resilience make it an ideal choice for advanced gaskets and seals in the automotive and aerospace industries. Its lightweight properties have also led to its use in the construction of advanced composite materials for space exploration.(Anjos'; et al., n.d.; Díaz-Parralejo et al., 2014)

Figure 4. Cork sheets

Reclaimed Wood

Reclaimed wood is a sustainable material, sourced from repurposed timber, old structures, or salvaged wood. It reduces the demand for fresh logging, conserving forests and minimizing environmental impact. Reclaimed wood embodies eco-friendliness, offering character, history, and durability in a variety of construction and design projects. Reclaimed wood's unique appearance and history make it a sought-after material in high-end architectural designs and luxury furniture. In advanced applications, it's used for acoustic panels and sound-absorbing surfaces in concert halls and studios.

Figure 5. Reclaimed wood

Hemp

Hemp is a sustainable material, renowned for its eco-friendly attributes. It grows quickly, requires minimal water and pesticides, and has a wide range of applications, from textiles and building materials to biofuels. Its versatility and low environmental impact make hemp an ideal choice for sustainable and environmentally responsible products. Hemp's durable and lightweight fibers are used in advanced automotive components, like biodegradable car panels. It's also gaining traction in the development of advanced biodegradable composites used in prosthetic limbs and aerospace components. Hemp fibers possess notable mechanical properties, with a density ranging from 1.4 to 1.6 g/cm3. They exhibit a remarkable tensile strength of 200-1000 MPa, a stiffness of 18-66 GPa, and a failure strain of 2-4%. Their strength surpasses many other natural fibers like cotton and flax. These properties can be influenced by factors such as water absorption, hemicellulose, lignin content, and fiber content, impacting their durability and flexibility in various applications. (Suardana et al., 2011)

Figure 6. Hemp material

Recycled Metal

Recycled metal is a sustainable material derived from repurposed metal products, reducing the need for extensive mining and energy-intensive refining processes. This eco-friendly practice conserves natural resources and minimizes environmental pollution. Recycled metal is versatile and durable, making it a responsible choice for various applications while contributing to a greener, more sustainable future. Advanced applications for recycled metals include aerospace components and advanced medical devices due to their strength, malleability, and resistance to corrosion. Recycled aluminum alloys are utilized in lightweight, fuel-efficient aircraft. (Ellis et al., 2014)

Figure 7. Recycled material

Sustainable Fabrics

Sustainable fabrics like Tencel are eco-conscious textiles crafted from renewable resources, typically wood pulp from eucalyptus trees. Their production employs closed-loop processes that minimize waste and chemical usage, promoting environmental responsibility. Tencel and similar fabrics offer comfort, durability, and biodegradability, aligning with a sustainable, planet-friendly approach to fashion and textiles. Sustainable fabrics like Tencel are incorporated into advanced sportswear that wicks moisture efficiently. In the medical field, they're used to create advanced wound dressings with antibacterial properties, promoting faster healing. (Basit et al., 2019)

Figure 8. Tancel fabric

Recycled Plastics

Recycled plastics are a sustainable material created by reprocessing post-consumer or post-industrial plastic waste. This practice reduces the need for new plastic production, conserving resources and minimizing plastic pollution. Recycled plastics can be used in various applications, offering an environmentally responsible solution to plastic-related environmental challenges. In advanced manufacturing, recycled plastics are utilized to create lightweight yet durable components for electric vehicles and advanced solar panel technology. They are also employed in advanced 3D printing for precision medical implants.

Figure 9. Recycled plastic material

Sustainable Concrete

Sustainable concrete is an environmentally responsible building material designed to reduce its carbon footprint. It typically incorporates recycled materials like fly ash or slag, reduces water usage, and lowers greenhouse gas emissions during production. This innovation aims to enhance the durability and environmental impact of concrete construction. Sustainable concrete formulations are used in advanced architectural designs, such as self-healing concrete that repairs cracks autonomously. In infrastructure, it's employed in advanced high-speed rail systems for its exceptional durability. Sustainable concrete boasts various mechanical properties, such as compressive strength, which may decrease with higher plastic waste content but increase with greater sand replacement (0% to 50%). It exhibits flexural strength typically ranging from 3–5 MPa and tensile strength of 2–5 MPa, along with a modulus of elasticity of 14–41 GPa. Additional attributes include splitting tensile strength, thermal conductivity, water absorption, density, slump, and compaction index. Sustainability in concrete extends to its potential for reuse and recycling in the aftermath of demolition, providing a closed-loop, eco-friendly construction solution. (Ahmad et al., 2021)

Figure 10. Sustainable concrete

Biodegradable Plastics

Biodegradable plastics are sustainable materials designed to break down naturally, reducing environmental pollution. They are typically made from renewable resources and can decompose into harmless byproducts. Biodegradable plastics offer an eco-friendly alternative to traditional plastics, minimizing the ecological impact and supporting a more sustainable future. Advanced applications of biodegradable plastics include 3D printed biodegradable medical implants and advanced agricultural films that break down harmlessly in the soil, reducing waste and promoting sustainable farming practices. Biodegradable plastics exhibit diverse mechanical characteristics, with a tensile strength of 0.5766 N/mm2, elongation at 24.2917%, and Young's modulus at 2.3552 N/mm2. They offer transparency, homogeneity, and flexibility. While they are up to 36% as strong in tension and 54% in bending as annealed stainless steel, they are only about 3% as stiff in either mode. Biodegradable plastics compare favorably to petrochemical polyethylene in terms of mechanical strength and offer additional benefits like printability, transparency, barrier properties, heat resistance, and gloss. These plastics are composed of organic compounds that serve as food for natural microorganisms. After their use, recycling, and disposal, these microbes decompose the plastic, leaving behind compost and biogas, contributing to a sustainable and environmentally friendly waste management solution. (*Biodegradable Plastics*, n.d.)

Figure 11. Biodegradable plastic material

Quantitative data of different materials as per its carbon footprint, recyclability, tensile strength, cost water usage and its life span are given in the table 1 and table 2 below.

Table 1. Quantitative table showing different materials carbon footprints, its recyclability and tensile strength

Material Type	Carbon Footprint (kg CO2/kg)	Recyclability (%)	Tensile Strength (MPa)
Biodegradable Plastics	1.5	80	50
Recycled Aluminum	5.0	90	300
Bamboo Fiber Composite	0.8	95	200
Bio-Based Polymers	2.0	75	80
Hempcrete	0.5	100	30
Recycled PET (rPET)	6.0	85	70
Mycelium Leather	1.2	98	40

Table 2. Quantitative table showing different materials cost, water usage and its lifespan

Material Type	Cost ($/kg)	Water Usage (L/kg)	Lifespan (years)
Biodegradable Plastics	2.5	150	5
Recycled Aluminum	1.8	300	50
Bamboo Fiber Composite	3.0	200	20
Bio-Based Polymers	2.2	180	10
Hempcrete	1.2	50	30
Recycled PET (rPET)	2.8	250	20
Mycelium Leather	4.0	120	15

Sustainable Insulation

Sustainable insulation refers to eco-friendly materials used to regulate temperature in buildings. These materials, like recycled glass or natural fibers, reduce energy consumption and environmental impact. They offer efficient thermal performance, contribute to reduced energy bills, and promote a more sustainable, energy-efficient construction industry. In extreme environments like Arctic research stations, they provide advanced insulation to maintain stable indoor temperatures. (Asdrubali et al., 2015; Doyle & Weidlich, 2021)

Figure 12. Sustainable insulation material

Natural Stone

Natural stone, an advanced sustainable material, is a durable and aesthetically appealing choice in construction. It requires minimal processing, reducing carbon emissions, and is highly resilient, promoting long-term use. Its timeless beauty and natural origin make it an eco-conscious option for architectural and design applications. Advanced architectural designs incorporate natural stone in façades and interior spaces. In advanced sculptures and monuments, the enduring beauty of natural stone stands the test of time.

Figure 13. Natural stone material

Straw Bales

Straw bales, an advanced sustainable material, are used in eco-friendly construction. They offer excellent insulation properties and are readily available as an agricultural byproduct. Straw bale construction reduces energy consumption, minimizes waste, and provides a natural, renewable, and cost-effective solution, contributing to environmentally responsible building practices. Advanced eco-housing projects use straw bales for exceptional insulation. In advanced sustainable agriculture, they're employed in vertical farming systems that optimize space and resource use. Straw bales exhibit specific mechanical attributes. Young's modulus, at low stress, remains consistent for densities within the 90–110 kg/m3 range, regardless of bale orientation. Poisson's ratio varies based on bulk density and orientation. Thermal conductivity increases linearly with bulk density ($60 \leq \rho \leq 120$ kg/m3). Straw bales encompass a range of properties such as density (80–150 kg/m3), specific heat (1075–2000 J/(kg·K)), elastic modulus (150–350 kPa), and bulk density (150–200 kg/m3). They are versatile and can serve as load-bearing structural components, offer resistance to wind and seismic loads, or act as insulation substrates, often combined with a separate load-bearing timber frame. (Konečný et al., 2013)

Figure 14. Straw bales material

Sustainable Composites

Sustainable composites are engineered by combining natural or recycled fibers with eco-friendly resins. They offer high strength-to-weight ratios and reduced environmental impact. These composites are versatile, used in automotive, aerospace, and construction, contributing to resource efficiency and sustainability in various industries. Bio-based composites are used in advanced automotive designs, where lightweight, strong materials are crucial for fuel efficiency. Glass-reinforced composites find applications in advanced wind turbine blades, harnessing wind energy efficiently. (Marques et al., 2015)

Figure 15. Sustainable composites

Low VOC Paints

Low VOC (Volatile Organic Compounds) paints, an advanced sustainable material, are environmentally responsible choices for interior and exterior coatings. These paints emit fewer harmful chemicals, promoting better indoor air quality and reducing pollution. They contribute to a healthier living environment while supporting eco-conscious, sustainable building practices. In advanced healthcare facilities and clean rooms, low VOC paints are used to maintain air quality. Advanced urban planning incorporates them into city infrastructure to mitigate air pollution. (Schieweck & Bock, 2015)

Figure 16. Low VOC paints

Solar Tiles

Solar tiles serve as roofing materials while harnessing solar energy. They seamlessly integrate into building designs, providing renewable electricity and reducing energy bills. This innovative technology promotes sustainability, offering an eco-friendly power source with a dual purpose of shelter and energy production in construction projects. Solar tiles are integrated into advanced smart homes, where they contribute to energy generation and storage. In advanced green architecture, they form the basis of energy-efficient, self-sustaining building designs. (Alim et al., 2020)

Figure 17. Solar roof tiles

THE DAWN OF NEXT-GENERATION MATERIALS: ECO-FRIENDLY, DURABLE, AND VERSATILE

The 21st century is witnessing a remarkable shift in the world of materials. The traditional stalwarts like steel, concrete, and plastics are making way for a new era of next-generation materials. These materials are characterized by their eco-friendliness, durability, and versatility. In this article, we'll explore the characteristics and benefits of these innovative materials, highlighting their potential to revolutionize industries and promote sustainability.

Figure 18. Sustainable materials

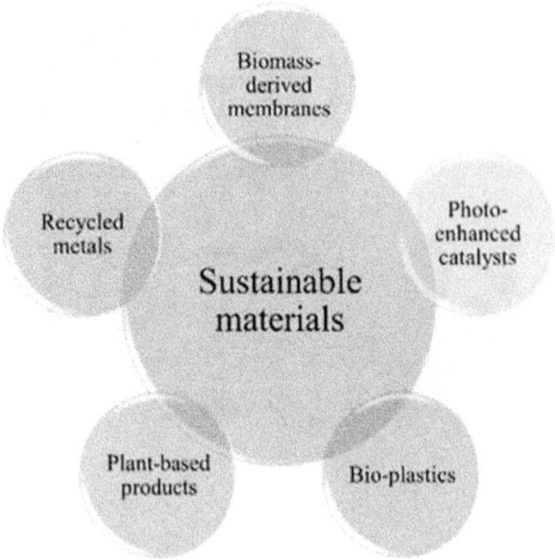

Eco-Friendliness as the Core

One of the defining features of next-generation materials is their commitment to eco-friendliness. These materials are designed to minimize their environmental impact at every stage, from production to disposal. They are the answer to the urgent need for a more sustainable approach to materials. (Baranwal et al., 2022)

Sustainable Sourcing

Next-generation materials often rely on sustainable sourcing, using renewable resources like flax, hemp, or bamboo. This minimizes the depletion of natural resources and reduces the environmental footprint. For instance, bamboo, a rapidly renewable resource, is being used in construction for its strength and sustainability. (Soltani et al., 2015)

Biodegradability

Some of these materials, like advanced composites made from natural fibers, are not only strong but also biodegradable. When their lifecycle ends, they return to nature without leaving a lasting imprint. An example is biodegradable plastics, which break down naturally and have applications in packaging and single-use items. (Irimia-Vladu, 2014)

Durability That Withstands Time

Durability is a non-negotiable factor for any material, and next-generation materials meet this criterion with flying colors. They offer longevity that ensures they can compete with traditional materials head-to-head. Many next-generation materials exhibit exceptional corrosion resistance. They can withstand harsh environmental conditions, making them ideal for outdoor applications, such as infrastructure and transportation.

Low Maintenance

The durability of these materials often translates into lower maintenance costs. Reduced maintenance requirements contribute to cost savings and environmental benefits. For example, composite decking, which requires minimal maintenance, is a sustainable alternative to traditional wood decking. (Khalid et al., 2021)

Advanced Composites

Advanced composites made from natural fibers, such as flax or hemp, are emerging as sustainable alternatives to traditional materials. They are strong, lightweight, and biodegradable, making them an excellent choice for construction and manufacturing. These composites are used in the automotive industry to create lightweight yet sturdy components. (Zeng et al., 2014)

Energy Efficiency

Materials designed with a focus on energy efficiency are crucial for the sustainable future of energy. Lightweight and insulating materials are essential for improving the energy performance of buildings and vehicles. Aerogel, an ultra-light and highly insulating material, is used in building insulation and space exploration. (Doyle & Weidlich, 2021)

REVOLUTIONIZING INDUSTRIES

Sustainable construction is no longer a buzzword but a necessity. Next-generation materials play a pivotal role in eco-friendly building practices. From structural elements to insulation, they are reshaping the construction landscape. Green roofs, made with sustainable materials, reduce energy consumption and improve urban air quality.

Manufacturing Innovation

Manufacturers are incorporating these materials into their processes to reduce waste, improve product quality, and meet the growing demand for sustainable products. From automotive components to consumer goods, next-gen materials are making their mark. Tesla's use of lithium-ion batteries in electric vehicles is a prime example of innovation in the automotive industry. The characteristics and benefits of next-generation materials make them more than just a trend; they are a necessity for a sustainable future. These materials prioritize eco-friendliness, durability, and versatility, and they are already transforming industries. As we embrace the potential of next-gen materials, we take a significant step towards a greener, more sustainable world. Next-generation materials are innovative materials known for their eco-friendliness, durability, and versatility. They are designed to minimize their environmental impact and offer superior performance compared to traditional materials. Next-generation materials promote sustainability by using sustainable sourcing, minimizing environmental impact, and offering properties like biodegradability and energy efficiency. Examples include advanced composites made from natural fibers like flax and hemp, as well as sustainable building materials like bamboo and recycled plastics. Next-gen materials are transforming manufacturing by reducing waste, improving product quality, and meeting the growing demand for sustainable products. Manufacturers are incorporating these materials into various industries. While they offer many benefits, next-gen materials may not replace traditional materials entirely. They are often used alongside traditional materials, especially in applications where their unique properties are most advantageous. (Zhu et al., 2016)

OVERCOMING CHALLENGES IN THE ADOPTION OF NEXT-GENERATION SUSTAINABLE MATERIALS

The promise of next-generation materials, characterized by their eco-friendliness, durability, and versatility, is undeniable. However, their widespread adoption is not without its share of challenges. In this article, we'll explore the hurdles these innovative materials face, the potential solutions, and how overcoming these challenges can lead to a greener and more sustainable future.

The Challenge of Scalability

One of the primary hurdles in the path of next-generation materials is scalability. While these materials hold immense potential, many of them are still in the research and development stage. Scaling up production can be a costly and time-consuming process.

Limited Production Capacities

In many cases, production facilities for these materials are limited, which can result in a shortage of supply. For example, advanced composites made from natural fibers like flax or hemp are not yet produced on a large scale.

Research and Development Costs

Bringing a new material to the market often involves significant research and development costs. This can deter potential manufacturers from investing in the necessary infrastructure.

Resistance from Established Industries

The industries that rely on traditional materials have a well-established infrastructure and processes. Transitioning to new materials can be met with resistance due to the high initial costs and potential disruptions in production.

Automotive Industry Example

The automotive industry, which heavily depends on steel and plastics, faces substantial challenges in integrating next-gen materials. The costs of retooling manufacturing processes and ensuring the safety and quality of the new materials are significant concerns.

Construction Sector Challenges

The construction sector, traditionally reliant on concrete and steel, is cautious about adopting alternative materials. The perceived risks and uncertainties in construction using these new materials can be a barrier.

Regulatory and Certification Issues

To ensure the safety and quality of next-generation materials, regulatory and certification frameworks need to be established. This is a critical step in assuring consumers and industries of the reliability of these materials.

Testing and Certification

Certifying new materials involves rigorous testing to ensure they meet safety standards. Testing can be time-consuming and expensive, and regulatory authorities need to adapt to accommodate the influx of innovative materials.

Environmental Standards

Regulations must also address the environmental impact of these materials throughout their lifecycle. This includes considerations for disposal, recycling, and environmental sustainability.

PAVING THE WAY FOR A GREENER FUTURE

While the challenges of adopting next-generation materials are substantial, the potential benefits far outweigh them. Overcoming these hurdles is vital for a sustainable future. Innovations in scalability, gradual industry transitions, and the establishment of robust regulatory and certification processes are essential steps. By working together, industries, governments, and innovators can lead the way to a greener, more sustainable future where next-generation materials play a central role.

Breaking Boundaries: Innovative Applications of Next-Generation Materials

The world of materials is undergoing a revolution with the advent of next-generation materials. These materials, known for their eco-friendliness, durability, and versatility, are not only transforming industries but also opening up innovative applications. In this article, we will delve into how next-gen materials are being creatively used in various sectors, pushing the boundaries of technology and design. The construction industry is experiencing a significant transformation, thanks to innovative applications of next-generation materials. Mycelium-based composites, derived from the root-like structures of fungi, are revolutionizing sustainable building structures. These composites offer both strength and thermal insulation, making them an ideal choice for eco-friendly construction. Mycelium-based composites are biodegradable, low-energy, and non-toxic, reducing the environmental footprint of construction. They can be used for everything from insulation to structural components, creating buildings that are not just green but genuinely sustainable. The use of mycelium-based materials has the potential to reduce our reliance on traditional concrete, which is energy-intensive and contributes significantly to carbon emissions. By replacing or supplementing concrete with mycelium composites, we can make construction more sustainable. (Akeiber et al., 2016)

Redefining Automotive Industry Efficiency

The automotive industry is embracing next-generation materials to develop energy-efficient vehicles without compromising on safety. Lightweight and incredibly strong, carbon-fibre-reinforced polymers (CFRPs) are driving innovation in automotive design. CFRPs have a high strength-to-weight ratio, making them ideal for constructing lightweight yet sturdy vehicle components. By reducing the weight of vehicles through the use of CFRPs, the automotive industry is making strides in fuel efficiency. Lighter vehicles consume less fuel and produce fewer emissions, contributing to a more sustainable transportation sector. CFRPs are a key component in the production of electric vehicles (EVs), helping extend their battery life and overall efficiency. As the world shifts toward EVs, CFRPs are playing a pivotal role in shaping the future of transportation. (Titirici et al., 2015a, 2015b; Yang et al., 2019)

Pioneering Wearable Technology

The world of electronics is not left behind when it comes to innovative applications of next-generation materials. Flexible substrates that are both biodegradable and electronic are revolutionizing the design of wearable devices. These substrates offer unprecedented comfort and sustainability in the world of wearables. Flexible substrates allow wearables to conform to the body, making them more comfortable to wear for extended periods. Their adaptability ensures a snug fit and enhances the user experience. (Fan et al., 2016; Fic et al., 2018; Oltean et al., 2016; Zeng et al., 2014)

Reduced Electronic Waste

Biodegradable substrates reduce electronic waste, a growing concern in our tech-driven world. When wearables reach the end of their lifecycle, they naturally decompose, leaving a minimal environmental footprint. (Irimia-Vladu, 2014)

Environmental Advantages of Next-Generation Materials

The adoption of next-generation materials is not just a matter of innovation; it's a crucial step towards a more sustainable future. Traditional materials like concrete and steel, while essential for various industries, have a substantial environmental footprint. In this article, we will explore the environmental advantages of next-gen materials, focusing on their sustainability, reduced carbon footprint, and their contribution to a cleaner, healthier environment.

The Burden of Traditional Materials

Traditional materials like concrete and steel are notorious for their resource-intensive production processes. These materials require significant amounts of raw materials, water, and energy, contributing to resource depletion and environmental stress. The production of concrete and steel is a major contributor to carbon emissions. The cement industry alone is responsible for a significant portion of global carbon dioxide emissions, making it a substantial concern for climate change. (Provis, 2018)

Sustainable Materials: A Renewable Approach

One of the most significant environmental advantages of next-generation materials is their reliance on renewable resources. Many sustainable materials are derived from sources like bamboo, flax, or mycelium, which can be rapidly replenished. Sustainable materials typically have a much lower carbon footprint compared to their traditional counterparts. Their production emits fewer greenhouse gases, making them an eco-friendly choice. The use of recycled plastic in manufacturing is a prime example of how next-generation materials contribute to reducing waste and pollution. Instead of adding to the mountains of plastic waste in landfills and oceans, recycling allows for a second life for these materials. Additionally, recycling plastic is an energy-efficient process. It conserves energy compared to the production of new plastic, which requires the extraction and refinement of fossil fuels. (Hu et al., 2010)

A Healthier Environment

The environmental benefits of next-generation materials go beyond reduced carbon emissions. By reducing air and water pollution, these materials contribute to a cleaner, healthier environment. Many traditional materials involve the release of harmful toxins during production. Sustainable materials, on the other hand, are often non-toxic and do not contribute to the release of harmful chemicals into the environment. The environmental advantages of next-generation materials are undeniable. From reducing resource consumption and carbon emissions to minimizing waste and pollution, these materials play a significant role in creating a greener, more sustainable tomorrow. (C. Liu et al., 2010)

PAVING THE WAY FOR SUSTAINABLE INNOVATION: OVERCOMING HURDLES IN NEXT-GENERATION MATERIALS

The world of next-generation materials is teeming with promise, offering eco-friendliness, durability, and versatility. However, the journey towards making these materials a reality is not without its challenges. In this article, we will explore the hurdles faced by researchers and industries in the development and utilization of next-gen materials. From interdisciplinary collaboration to securing funding and navigating regulatory frameworks, the road to sustainable innovation is paved with obstacles that demand commitment and innovation. Developing next-generation materials often requires bringing together experts from various fields, including scientists, engineers, and designers. Interdisciplinary collaboration is essential to harness the full potential of these materials.

Bridging the Gap Between Science and Design

Scientists and designers must work in tandem to ensure that the materials are not only sustainable but also aesthetically appealing and functional. This collaboration ensures that sustainable materials are not just eco-friendly but also user-friendly.

Securing Funding: A Competitive Landscape

Obtaining funding for research and development in the field of next-gen materials can be highly competitive. Innovators must convince investors of the long-term potential and benefits of their projects. Sustainable materials often require long-term investment in research and development before they can be brought to market. This financial commitment can be a significant hurdle.

Navigating Regulatory Frameworks: Safety and Quality Assurance

Industries using next-gen materials must navigate regulatory frameworks to ensure that their products meet safety and quality standards. These standards are critical to ensuring that sustainable materials are safe for consumers and the environment. Navigating complex and evolving regulatory frameworks can be a daunting task. Industries must invest time and resources in understanding and adhering to these standards.

Overcoming Hurdles: The Way Forward

Overcoming the hurdles faced by researchers and industries requires a strong commitment to sustainable innovation. The benefits of next-gen materials in terms of sustainability, eco-friendliness, and durability make the journey worthwhile. Innovation is at the heart of overcoming these challenges. Researchers and industries must constantly push the boundaries of what's possible and seek creative solutions to the problems they encounter. The hurdles faced by researchers and industries in the realm of next-generation materials are not insurmountable. Interdisciplinary collaboration, securing funding, and navigating regulatory frameworks are challenges that can be addressed with determination and innovation. As we pave the way for sustainable innovation, we take a significant step towards a greener and more sustainable future.

UNLEASHING INNOVATION: THE TRANSFORMATIVE POWER OF NANOTECHNOLOGY, BIO-BASED MATERIALS, AND RECYCLED RESOURCES

In the quest for sustainable materials, three innovative approaches stand out: nanotechnology, bio-based materials, and recycled resources. These cutting-edge technologies hold the key to transforming the way we produce and consume materials. In this article, we'll explore their transformative potential and how they are reshaping the landscape of sustainable materials.

Nanotechnology: Precision on a Tiny Scale

Nanotechnology allows us to manipulate materials at the nanoscale, a dimension so small that it's measured in nanometers. This precision offers new possibilities for enhancing material properties and functionalities. Nanocomposites are a prime example of nanotechnology's transformative potential. By incorporating nanoparticles into materials, we can achieve remarkable strength and conductivity. What's more, this can be done using fewer resources, contributing to sustainability. (Ashby et al., 2009; N. Liu et al., 2013; Wagle et al., 2014)

Bio-Based Materials: Harnessing the Power of Nature

Bio-based materials leverage the power of nature to provide a sustainable alternative to traditional materials. These materials are often renewable and biodegradable, making them an eco-friendly choice. Plant-based polymers, derived from sources like corn or sugarcane, offer a renewable alternative to petroleum-based plastics. They can be used in a wide range of applications, from packaging to textiles, reducing our reliance on fossil fuels.(Elias Kaloush et al., 2010; Li et al., 2021; N. Liu et al., 2013; Silva et al., 2014)

Recycled Resources: Extending Lifecycles

Recycling resources, whether in the form of materials or waste, is a powerful way to reduce the environmental footprint of industries. By repurposing materials, we can extend their lifecycle and decrease the demand for virgin resources. Recycling creates a closed-loop system, where materials are used, collected, and reintroduced into the production cycle. This process not only conserves resources but also minimizes waste and pollution.

POTENTIAL DRAWBACKS AND HURDLES IN THE WAY OF NEXT-GENERATION MATERIALS

In the ever-evolving landscape of sustainable materials, the emergence of next-generation options brings both excitement and challenges. While these materials promise to address environmental concerns and promote eco-friendly practices, it's crucial to understand the potential drawbacks that may shape their future. Let's delve into the complexities of next-generation sustainable materials and explore the hurdles they may encounter over the coming years.

High Production Costs

The allure of sustainability often comes at a price – a higher production cost. Industries, particularly those with thin profit margins, may face challenges in adopting these materials on a large scale. The economic feasibility of integrating next-generation sustainable materials becomes a critical consideration.

Availability Woes

The success of sustainable materials relies heavily on a steady supply of raw materials. Some next-generation options may face hurdles in terms of limited availability due to constraints in raw material sourcing or production capacity. This can result in supply chain disruptions and price fluctuations.

Technological Hurdles

The development and implementation of new sustainable materials necessitate technological advancements. Challenges in research, development, and integration into existing production systems may pose hurdles. Overcoming these technological challenges is crucial for the successful adoption of these materials.

Durability and Longevity

A notable concern with next-generation sustainable materials lies in their durability and longevity. If these materials degrade more rapidly than traditional counterparts, it could lead to increased maintenance costs and a shorter lifespan for products. Striking a balance between sustainability and durability becomes imperative.

Performance Discrepancies

In certain industries, materials must meet stringent standards for strength, flexibility, and other mechanical properties. Next-generation sustainable materials may not always align with these performance requirements, posing a significant challenge. Striving for a harmonious blend of sustainability and functionality is key.

Biodegradability Quandaries

While the intention is for sustainable materials to be biodegradable, the conditions required for effective biodegradation may not be universally present. This limitation could impact the effectiveness of these materials, especially in diverse environmental settings. Biodegradability challenges must be addressed to fulfill sustainability goals.

Energy-Intensive Production

Ironically, some sustainable materials may have energy-intensive production processes. If the energy used is derived from non-renewable sources, it could offset the environmental benefits. A holistic approach, considering the entire life cycle of a material, is essential for a genuine eco-friendly impact.

Navigating Regulatory Complexities

As new materials enter the market, navigating the regulatory landscape becomes a challenge. Establishing and adhering to standardized regulations for safety and environmental standards require meticulous scrutiny. Overcoming regulatory hurdles is vital for the seamless integration of next-generation sustainable materials.

Overcoming Resistance to Change

The inertia of industries and consumers to adopt new materials due to familiarity with existing products and processes poses a significant challenge. Convincing stakeholders of the benefits of next-generation sustainable materials requires effective communication and education.

Recycling Infrastructure Gaps

The environmental benefits of sustainable materials heavily rely on efficient recycling infrastructure. Unfortunately, some regions may lack well-developed systems for recycling and disposing of these materials. Addressing these gaps is essential to maximize the potential positive impact on the environment.

In navigating the landscape of next-generation sustainable materials, addressing these challenges is imperative. Ongoing research, innovation, and collaboration among industries, governments, and research institutions are vital for the successful integration of these materials into mainstream use. As we tread the path toward a more sustainable future, acknowledging the potential challenges of next-generation sustainable materials is a step in the right direction. By fostering collaboration, investing in technology, and embracing a mindset of continuous improvement, we can overcome these obstacles and usher in a new era of eco-friendly innovation.

THE SYMBIOSIS OF TRANSFORMATION

The transformative potential of these three approaches is amplified when they work together. For instance, bio-based materials can be enhanced using nanotechnology to create highly efficient and sustainable products. The combination of nanotechnology, bio-based materials, and recycled resources offers a multi-faceted approach to sustainability, addressing various aspects of material production and consumption. Nanotechnology, bio-based materials, and recycled resources are driving innovation in the world of sustainable materials. Their transformative potential extends from enhancing material properties to reducing our reliance on non-renewable resources and minimizing waste. As we embrace these cutting-edge technologies, we take a significant step towards a greener and more sustainable tomorrow.

Promising Future Prospects of Next-Generation Materials

In a world marked by environmental challenges, the significance of next-generation materials for sustainable development and engineering has never been clearer. These materials are not just an option; they are an imperative for a planet grappling with climate change, resource scarcity, and pollution. They bring the promise of transforming various industries, from eco-friendly infrastructure and renewable energy solutions to transportation and consumer goods. Embracing these materials is not only a matter of environmental responsibility but also a path to innovation, resource efficiency, and economic sustainability.

Unveiling the Need for Change

The world faces a host of environmental challenges, from rising temperatures and extreme weather events to the depletion of natural resources and the alarming levels of pollution. These issues underscore the urgency for sustainable solutions. Next-generation materials offer a beacon of hope in this scenario. They present a fundamental shift towards materials that are not just eco-friendly but also highly efficient in their use of resources.

Transportation Reimagined

Next-generation materials are driving innovation in the automotive industry. Lightweight yet robust materials are enabling the development of energy-efficient vehicles, reducing fuel consumption and emissions. Even in aviation, where fuel efficiency is paramount, next-gen materials are making an impact. Aircraft are becoming lighter, leading to lower fuel consumption and reduced carbon emissions.

A Greener Consumer Experience

From electronics to packaging, consumer goods are undergoing a transformation. Next-generation materials, particularly bio-based and recyclable options, are replacing traditional plastics and materials. This shift not only benefits the environment but also reduces waste, as many of these materials can be repurposed or are biodegradable. The promise of next-generation materials is not confined to a single industry or application; it's a global transformation. Embracing these materials is not just an environmental responsibility but also a pathway to innovation, resource efficiency, and economic sustainability.

REFERENCES

Ahmad, J., Zaid, O., Shahzaib, M., Abdullah, M. U., Ullah, A., & Ullah, R. (2021). Mechanical properties of sustainable concrete modified by adding marble slurry as cement substitution. *AIMS Materials Science*, *8*(3), 343–358. doi:10.3934/matersci.2021022

Akeiber, H., Nejat, P., Majid, M. Z. A., Wahid, M. A., Jomehzadeh, F., Zeynali Famileh, I., Calautit, J. K., Hughes, B. R., & Zaki, S. A. (2016). A review on phase change material (PCM) for sustainable passive cooling in building envelopes. In Renewable and Sustainable Energy Reviews (Vol. 60, pp. 1470–1497). Elsevier Ltd. doi:10.1016/j.rser.2016.03.036

Alim, M. A., Tao, Z., Abden, M. J., Rahman, A., & Samali, B. (2020). Improving performance of solar roof tiles by incorporating phase change material. *Solar Energy*, *207*, 1308–1320. doi:10.1016/j.solener.2020.07.053

Asdrubali, F., D'Alessandro, F., & Schiavoni, S. (2015). A review of unconventional sustainable building insulation materials. *Sustainable Materials and Technologies*, *4*, 1–17. doi:10.1016/j.susmat.2015.05.002

Ashby, M. F. (n.d.). *Materials and the Environment: Eco-Informed Material Choice*.

Ashby, M. F., Ferreira, P. J. S. G., & Schodek, D. L. (2009). *Nanomaterials, nanotechnologies and design : an introduction for engineers and architects*. Butterworth-Heinemann.

Baranwal, J., Barse, B., Fais, A., Delogu, G. L., & Kumar, A. (2022). Biopolymer: A Sustainable Material for Food and Medical Applications. In Polymers (Vol. 14, Issue 5). MDPI. doi:10.3390/polym14050983

Basit, A., Latif, W., Ashraf, M., Rehman, A., Iqbal, K., Maqsood, H. S., Jabbar, A., & Baig, S. A. (2019). Comparison of mechanical and thermal comfort properties of tencel blended with regenerated fibers and cotton woven fabrics. *AUTEX Research Journal, 19*(1), 80–85. doi:10.1515/aut-2018-0035

Bhonde, D., Nagarnaik, P. B., Parbat, D. K., & Waghe, U. P. (2014). Physical and Mechanical Properties of Bamboo (Dendrocalmus Strictus). *International Journal of Scientific and Engineering Research, 5*(1). http://www.ijser.org

Biodegradable plastics. (n.d.).

Díaz-Parralejo, A., Cuerda-Correa, E. M., Macías-García, A., Sánchez-González, J., & Díaz-Díez, M. Á. (2014). The Excellent Mechanical Properties of Cork: A Novel Approach through the Analysis of Contact Stress. *ISRN Materials Science, 2014*, 1–6. doi:10.1155/2014/898439

Doyle, L., & Weidlich, I. (2021). Sustainable insulation for sustainable DHC. *Energy Reports, 7*, 150–157. doi:10.1016/j.egyr.2021.08.161

Elias Kaloush, K., Prapoorna Biligiri, K., Abdelaziz Zeiada, W., Carolina Rodezno, M., & Xavier Reed, J. (2010). *Evaluation of Fiber-Reinforced Asphalt Mixtures Using Advanced Material Characterization Tests*. www.astm.org

Ellis, B. L., Knauth, P., & Djenizian, T. (2014). Three-dimensional self-supported metal oxides for advanced energy storage. In Advanced Materials (Vol. 26, Issue 21, pp. 3368–3397). Wiley-VCH Verlag. doi:10.1002/adma.201306126

Fan, F. R., Tang, W., & Wang, Z. L. (2016). Flexible Nanogenerators for Energy Harvesting and Self-Powered Electronics. In Advanced Materials (Vol. 28, Issue 22, pp. 4283–4305). Wiley-VCH Verlag. doi:10.1002/adma.201504299

Fic, K., Platek, A., Piwek, J., & Frackowiak, E. (2018). Sustainable materials for electrochemical capacitors. In Materials Today (Vol. 21, Issue 4, pp. 437–454). Elsevier B.V. doi:10.1016/j.mattod.2018.03.005

Haneef, M., Ceseracciu, L., Canale, C., Bayer, I. S., Heredia-Guerrero, J. A., & Athanassiou, A. (2017). Advanced Materials from Fungal Mycelium: Fabrication and Tuning of Physical Properties. *Scientific Reports, 7*(1), 41292. doi:10.1038/srep41292 PMID:28117421

Hu, B., Wang, K., Wu, L., Yu, S. H., Antonietti, M., & Titirici, M. M. (2010). Engineering carbon materials from the hydrothermal carbonization process of biomass. In Advanced Materials, 22(7). doi:10.1002/adma.200902812

Irimia-Vladu, M. (2014). "Green" electronics: Biodegradable and biocompatible materials and devices for sustainable future. In Chemical Society Reviews, 43(2). Royal Society of Chemistry. doi:10.1039/C3CS60235D

Khalid, M. Y., Al Rashid, A., Arif, Z. U., Ahmed, W., Arshad, H., & Zaidi, A. A. (2021). Natural fiber reinforced composites: Sustainable materials for emerging applications. In *Results in Engineering* (Vol. 11). Elsevier B.V. doi:10.1016/j.rineng.2021.100263

Konečný, P., Teslík, J., & Hamala, M. (2013). Mechanical and physical properties of straw bales. *Advanced Materials Research*, *649*, 250–253. . doi:10.4028/www.scientific.net/AMR.649.250

Li, T., Chen, C., Brozena, A. H., Zhu, J. Y., Xu, L., Driemeier, C., Dai, J., Rojas, O. J., Isogai, A., Wågberg, L., & Hu, L. (2021). Developing fibrillated cellulose as a sustainable technological material. In Nature, 590(7844). doi:10.1038/s41586-020-03167-7

Liu, C., Li, F., Lai-Peng, M., & Cheng, H. M. (2010). Advanced materials for energy storage. In Advanced Materials, 22(8). doi:10.1002/adma.200903328

Liu, N., Huo, K., McDowell, M. T., Zhao, J., & Cui, Y. (2013). Rice husks as a sustainable source of nanostructured silicon for high performance Li-ion battery anodes. *Scientific Reports*, *3*(1), 1919. doi:10.1038/srep01919 PMID:23715238

Marques, M. F. V., Lunz, J. N., Aguiar, V. O., Grafova, I., Kemell, M., Visentin, F., Sartori, A., & Grafov, A. (2015). Thermal and Mechanical Properties of Sustainable Composites Reinforced with Natural Fibers. *Journal of Polymers and the Environment*, *23*(2), 251–260. doi:10.1007/s10924-014-0687-2

Oltean, V. A., Renault, S., Valvo, M., & Brandell, D. (2016). Sustainable materials for sustainable energy storage: Organic Na electrodes. In Materials, 9(3). MDPI AG. doi:10.3390/ma9030142

Provis, J. L. (2018). Alkali-activated materials. In *Cement and Concrete Research* (Vol. 114, pp. 40–48). Elsevier Ltd., doi:10.1016/j.cemconres.2017.02.009

Sauvé, S., Bernard, S., & Sloan, P. (2016). Environmental sciences, sustainable development and circular economy: Alternative concepts for trans-disciplinary research. *Environmental Development*, *17*, 48–56. doi:10.1016/j.envdev.2015.09.002

Schieweck, A., & Bock, M. C. (2015). Emissions from low-VOC and zero-VOC paints - Valuable alternatives to conventional formulations also for use in sensitive environments? *Building and Environment*, *85*, 243–252. doi:10.1016/j.buildenv.2014.12.001

Silva, N. H. C. S., Vilela, C., Marrucho, I. M., Freire, C. S. R., Pascoal Neto, C., & Silvestre, A. J. D. (2014). Protein-based materials: From sources to innovative sustainable materials for biomedical applications. *Journal of Materials Chemistry. B*, *2*(24), 3715–3740. doi:10.1039/c4tb00168k PMID:32261720

Soltani, N., Bahrami, A., Pech-Canul, M. I., & González, L. A. (2015). Review on the physicochemical treatments of rice husk for production of advanced materials. In *Chemical Engineering Journal* (Vol. 264, pp. 899–935). Elsevier. doi:10.1016/j.cej.2014.11.056

Suardana, N. P. G., Piao, Y., & Lim, J. K. (2011). *Mechanical Properties Of Hemp Fibers And HEMP/PP Composites: Effects Of Chemical Surface Treatment.*

Titirici, M. M., White, R. J., Brun, N., Budarin, V. L., Su, D. S., Del Monte, F., Clark, J. H., & MacLachlan, M. J. (2015a). Sustainable carbon materials. In Chemical Society Reviews, 44(1). Royal Society of Chemistry. doi:10.1039/C4CS00232F

Vandelook, S., Elsacker, E., Van Wylick, A., De Laet, L., & Peeters, E. (2021). Current state and future prospects of pure mycelium materials. In Fungal Biology and Biotechnology, 8(1). BioMed Central Ltd. doi:10.1186/s40694-021-00128-1

Wagle, D. V., Zhao, H., & Baker, G. A. (2014). Deep eutectic solvents: Sustainable media for nanoscale and functional materials. *Accounts of Chemical Research*, *47*(8), 2299–2308. doi:10.1021/ar5000488 PMID:24892971

Yang, W., Li, X., Li, Y., Zhu, R., & Pang, H. (2019). Applications of Metal–Organic-Framework-Derived Carbon Materials. In Advanced Materials, 31(6). Wiley-VCH Verlag. doi:10.1002/adma.201804740

Zeng, W., Shu, L., Li, Q., Chen, S., Wang, F., & Tao, X. M. (2014). Fiber-based wearable electronics: A review of materials, fabrication, devices, and applications. In Advanced Materials, 26(31). Wiley-VCH Verlag. doi:10.1002/adma.201400633

Zhu, Y., Romain, C., & Williams, C. K. (2016). Sustainable polymers from renewable resources. In Nature, 540(7633). doi:10.1038/nature21001

APPENDIX

1. High Production Costs

The allure of sustainability often comes at a price – a higher production cost. Industries, particularly those with thin profit margins, may face challenges in adopting these materials on a large scale. The economic feasibility of integrating next-generation sustainable materials becomes a critical consideration.

2. Availability Woes

The success of sustainable materials relies heavily on a steady supply of raw materials. Some next-generation options may face hurdles in terms of limited availability due to constraints in raw material sourcing or production capacity. This can result in supply chain disruptions and price fluctuations.

3. Technological Hurdles

The development and implementation of new sustainable materials necessitate technological advancements. Challenges in research, development, and integration into existing production systems may pose hurdles. Overcoming these technological challenges is crucial for the successful adoption of these materials.

4. Durability and Longevity

A notable concern with next-generation sustainable materials lies in their durability and longevity. If these materials degrade more rapidly than traditional counterparts, it could lead to increased maintenance costs and a shorter lifespan for products. Striking a balance between sustainability and durability becomes imperative.

5. Performance Discrepancies

In certain industries, materials must meet stringent standards for strength, flexibility, and other mechanical properties. Next-generation sustainable materials may not always align with these performance requirements, posing a significant challenge. Striving for a harmonious blend of sustainability and functionality is key.

6. Biodegradability Quandaries

While the intention is for sustainable materials to be biodegradable, the conditions required for effective biodegradation may not be universally present. This limitation could impact the effectiveness of these materials, especially in diverse environmental settings. Biodegradability challenges must be addressed to fulfill sustainability goals.

7. Energy-Intensive Production

Ironically, some sustainable materials may have energy-intensive production processes. If the energy used is derived from non-renewable sources, it could offset the environmental benefits. A holistic approach, considering the entire life cycle of a material, is essential for a genuine eco-friendly impact.

8. Navigating Regulatory Complexities

As new materials enter the market, navigating the regulatory landscape becomes a challenge. Establishing and adhering to standardized regulations for safety and environmental standards require meticulous scrutiny. Overcoming regulatory hurdles is vital for the seamless integration of next-generation sustainable materials.

9. Overcoming Resistance to Change

The inertia of industries and consumers to adopt new materials due to familiarity with existing products and processes poses a significant challenge. Convincing stakeholders of the benefits of next-generation sustainable materials requires effective communication and education.

10. Recycling Infrastructure Gaps

The environmental benefits of sustainable materials heavily rely on efficient recycling infrastructure. Unfortunately, some regions may lack well-developed systems for recycling and disposing of these materials. Addressing these gaps is essential to maximize the potential positive impact on the environment. In navigating the landscape of next-generation sustainable materials, addressing these challenges is imperative. Ongoing research, innovation, and collaboration among industries, governments, and research institutions are vital for the successful integration of these materials into mainstream use. As we tread the path toward a more sustainable future, acknowledging the potential challenges of next-generation sustainable materials is a step in the right direction. By fostering collaboration, investing in technology, and embracing a mindset of continuous improvement, we can overcome these obstacles and usher in a new era of eco-friendly innovation.

Chapter 12
Leveraging High-Performance Computing and Artificial Intelligence in Climate Modeling and Prediction

Asfia Sabahath
https://orcid.org/0000-0002-7820-5976
King Khalid University, Saudi Arabia

Arshiya Begum
King Khalid University, Saudi Arabia

Arshi Naim
https://orcid.org/0000-0003-1325-6964
King Khalid University, Saudi Arabia

ABSTRACT

Climate change is a pressing global challenge that demands improved understanding, accurate predictions, and effective mitigation strategies. High-performance computing (HPC) combined with artificial intelligence (AI) presents a transformative opportunity for climate modeling and prediction. HPC and AI for environmental sustainability have the potential to revolutionize our efforts in addressing pressing environmental challenges. By combining the computational power of HPC with the analytical capabilities of AI, we can accelerate research and develop innovative solutions to promote sustainability.

DOI: 10.4018/979-8-3693-1794-5.ch012

INTRODUCTION

The fields of advanced computing and artificial intelligence have emerged as capable tools in climate modeling and prediction, innovating how researchers examine and comprehend Earth's climate system (Chen et al., 2023). Climate modeling involves simulating intricate connections between various aspects of the climate system, like the atmosphere, oceans, land surface, and ice. These models necessitate tremendous computational power to accurately depict the physical processes and simulate how the climate system acts over space and time.

Advanced computing provides the required computational assets to solve the complex equations that control climate dynamics. With the greater availability of supercomputers and advanced computing clusters, scientists can run more intricate models with higher resolutions, capturing finer details and bettering the exactness of climate projections. HPC allows for the simulation of large ensembles of climate models, enabling researchers to explore uncertainties and quantify the range of possible future climate scenarios (Gadzhev et al., 2021).

Recent technological advances, particularly those relying on data-based insights rather than human programming, have gained significant consideration in climate analysis and projection. These techniques can discern patterns and connections from large amounts of climate information, allowing the development of evidence-driven models that can enhance the portrayal of smaller-scale interactions, better represent generalizations, and heighten the aptitude for weather and climate predictions (Chen et al., 2023). They can also assist in combining observed facts with model results to refine the starting conditions of a projection.

The integration of HPC and AI in climate modeling and prediction holds great promise for advancing our understanding of the Earth's climate system and improving the accuracy and reliability of weather and climate forecasts (Singh, 2022). By leveraging the computational power of HPC and the data-driven capabilities of AI, scientists can tackle complex climate problems, explore new avenues of research, and address critical challenges such as extreme weather events, climate change impacts, and long-term climate projections (Minevich, 2022).

In this context, researchers are developing novel approaches to harness the potential of HPC and AI. This includes using convolutional neural networks (CNNs) and recurrent neural networks (RNNs) for weather nowcasting, downscaling, and regional climate modeling, improving parameterizations of subgrid-scale processes, enhancing ensemble forecasting, and optimizing climate model simulations. These advancements aim to bridge the gap between traditional physics-based modeling and data-driven approaches, leading to more accurate and reliable climate projections (Chattopadhyay et al., 2020).

However, there are challenges and considerations in leveraging HPC and AI in climate modeling. These include the need for massive computational resources, the development of robust and interpretable ML algorithms, the integration of AI techniques into existing modeling frameworks, and the ethical implications of using AI in decision-making processes related to climate policy and adaptation strategies (Fan et al., 2023).

Overall, the combination of HPC and AI offers exciting opportunities to advance climate modeling and prediction capabilities, enhance our understanding of the Earth's climate system, and provide valuable insights for decision-makers and stakeholders to address the challenges posed by climate change. Continued research and collaboration among scientists, computational experts, and policymakers are crucial to harnessing the full potential of HPC and AI in climate science.

Background

Climate change and its associated effects have raised serious concerns worldwide. Understanding and predicting climate patterns is crucial for developing effective mitigation and adaptation strategies. Traditional climate models rely on complex equations and large datasets, which require substantial computational resources and time. High-performance computing (HPC) has been instrumental in addressing these challenges by offering the capability to process vast amounts of data and perform computationally intensive simulations. Additionally, the emergence of artificial intelligence (AI) techniques has revolutionized the field of climate modeling and prediction by providing powerful tools for data analysis and pattern recognition.

Figure 1. Artificial intelligence-based solutions for climate change: a review

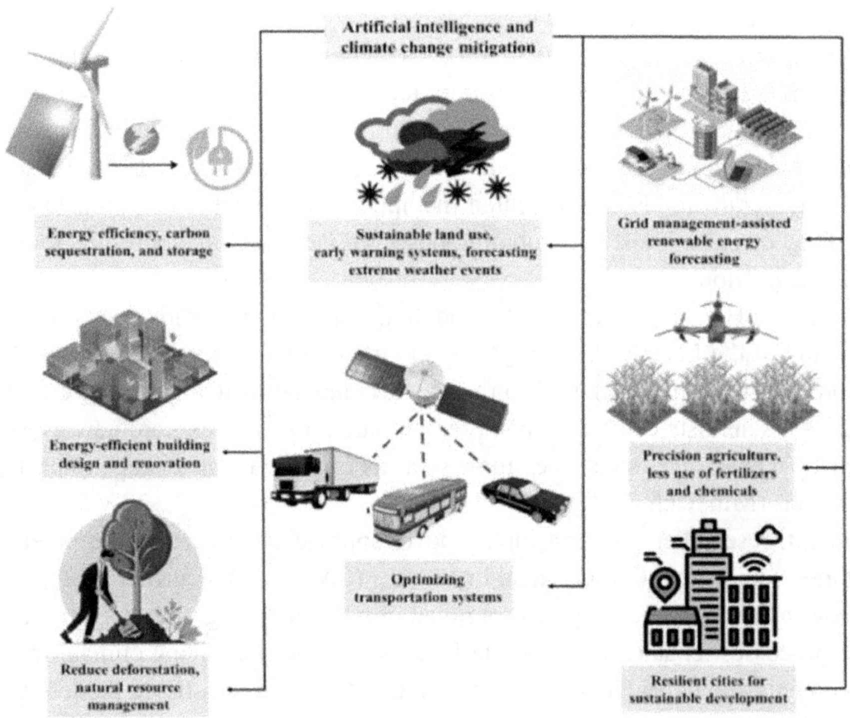

Problem Statement

Despite significant progress in climate modeling, there remain considerable uncertainties and limitations in accurately predicting future climate scenarios. The integration of HPC and AI in climate science aims to overcome these challenges by enhancing the efficiency, accuracy, and reliability of climate models (Shivanna, 2022). However, substantial research is needed to explore the potential benefits and challenges associated with leveraging HPC and AI technologies in climate modeling and prediction.

Objectives

The main objectives of this research paper are as follows:

1. To understand the role of high-performance computing in climate modeling and prediction.
2. To explore the applications of artificial intelligence techniques in climate science.
3. To analyze the potential benefits of integrating HPC and AI in climate modeling.
4. To identify the challenges and limitations of using HPC and AI in climate science.
5. To discuss the future prospects and potential limitations of this integration.

The research paper aims to provide a comprehensive overview of the advancements and potential of leveraging HPC and AI technologies in climate modeling and prediction. By addressing these objectives, this study contributes to the development of improved strategies for climate analysis and projection.

HIGH-PERFORMANCE COMPUTING IN CLIMATE SCIENCE

Overview of High-Performance Computing

High-performance computing (HPC) refers to the use of powerful computing systems capable of processing and analyzing extensive and complex datasets within a significantly reduced timeframe compared to traditional computing approaches (Lee et al., 2011). HPC systems typically consist of parallel processing architectures, large memory capacities, and high-speed input/output capabilities. These systems enable researchers to perform computationally intensive tasks, such as climate modeling, simulations, and data analysis, with exceptional speed and efficiency (Soori, 2023).

Applications of High-Performance Computing in Climate Modeling

HPC has revolutionized the field of climate modeling by enabling more detailed and comprehensive simulations of Earth's climate system. It facilitates the development and execution of complex weather and climate models that incorporate various Earth system components, including the atmosphere, oceans, land surface, and cryosphere (Wang & Yuan, 2020; Yi & Loia, 2019). HPC enables scientists to explore interactions between these components and simulate climate processes with higher spatial and temporal resolutions. This level of detail allows for an improved understanding of climate dynamics and their potential future evolution (Vital et al., 2013).

Furthermore, HPC enables ensemble simulations, where multiple simulations with slight variations are run simultaneously, providing insights into uncertainties associated with climate projections. With HPC, it is also possible to perform scenario-based simulations, exploring the impacts of different greenhouse gas emission scenarios on future climate conditions (Soori et al., 2023).

Benefits of High-Performance Computing in Climate Science

The integration of HPC in climate science offers several significant benefits:

1. Enhanced Model Resolution: HPC systems enable higher computational power, allowing climate models to operate at finer spatial and temporal resolutions. This enhances the accuracy and realism of simulations by capturing more detailed climate processes, local-scale phenomena, and regional climate variations (Grama et al., 2020).
2. Improved Predictability: With the increased computational power of HPC, climate models can better capture the complex interactions and feedback mechanisms within the climate system. This improved representation leads to enhanced predictive capabilities for short-term weather forecasts and long-term climate projections (André, 2014).
3. Efficient Data Processing: HPC facilitates the analysis of vast amounts of climate data from observational records, remote sensing, and model output. Advanced data analytics techniques running on HPC systems enable efficient identification of patterns, trends, and anomalies in climate data, aiding in understanding climate change patterns and informing decision-making processes (Yi & Loia, 2019).
4. Scalability and Flexibility: HPC systems can scale up their computing power to handle larger and more complex climate models, accommodating the growing demand for higher-resolution simulations. Additionally, HPC systems can be reconfigured to run different types of climate models and experiments, providing flexibility in addressing various research questions and scenarios.
5. Collaboration and Sharing: HPC resources can be accessed by researchers worldwide, promoting collaboration and knowledge sharing in the climate science community. This fosters the development of standardized methodologies, model intercomparisons, and data sharing, leading to more robust, reliable, and transparent climate projections (André, 2014).

ARTIFICIAL INTELLIGENCE IN CLIMATE SCIENCE

Overview of Artificial Intelligence

Artificial intelligence (AI) refers to the development of intelligent systems that can perform tasks traditionally requiring human intelligence. AI techniques, such as machine learning, deep learning, and data mining, have transformed numerous fields, including climate science. These techniques enable computers to analyze vast amounts of data, identify patterns, learn from experience, and make predictions or decisions (Vital et al., 2013).

Figure 2. Science in the age of artificial intelligence

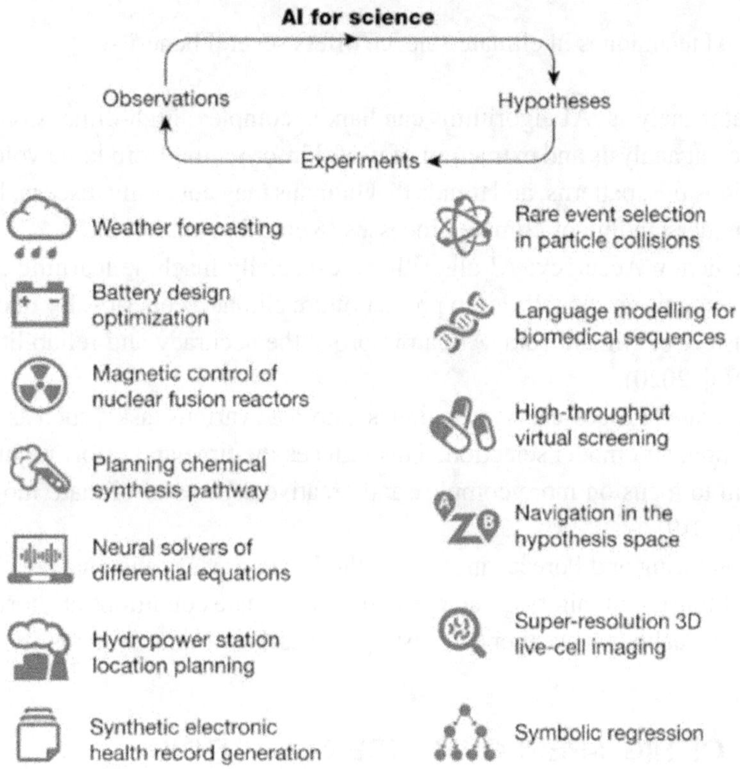

Applications of Artificial Intelligence in Climate Modeling

Artificial intelligence techniques have found various applications in climate science:

a. Climate Data Analysis: AI algorithms can analyze climate data, identify complex patterns, and extract valuable insights. They can analyze large datasets, including satellite imagery, climate observations, and model outputs, to detect anomalies, trends, and spatial-temporal patterns.
b. Statistical Downscaling: AI algorithms can downscale coarse-grained climate model outputs to generate higher-resolution spatial and temporal predictions. This technique is particularly useful for regional climate modeling, where local-scale climate information is required.
c. Extreme Weather Event Prediction: By analyzing historical climate data, AI algorithms can identify precursors and patterns associated with extreme weather events such as hurricanes, heat waves, or heavy rainfall. This can aid in the early prediction and mitigation of such events.
d. Climate Pattern Recognition: AI techniques can identify and classify climate patterns, such as El Niño events, monsoonal cycles, or teleconnections. This helps in understanding the drivers and influences behind these patterns and their impact on global and regional climate variability (Yi & Loia, 2019).

Benefits of Artificial Intelligence in Climate Science

The integration of AI techniques in climate science offers several benefits:

1. Enhanced Data Analysis: AI algorithms can handle complex, high-dimensional climate datasets, enabling efficient analysis and extraction of useful information from large volumes of data. It can identify relationships, patterns, and trends that humans may not easily discern, leading to improved insights and understanding of climate processes (Kelechi et al., 2020).
2. Increased Prediction Accuracy: AI algorithms, especially machine learning models, can benefit from pattern recognition capabilities to predict future climate scenarios. By uncovering underlying patterns in historical climate data, AI can improve the accuracy and reliability of climate model projections (Raj, 2020).
3. Efficient Automated Processes: AI techniques automate various tasks, such as data preprocessing, feature extraction, and model selection. This reduces the time and effort required by researchers, allowing them to focus on more complex and creative aspects of climate modeling and analysis (Huntingford, 2019).
4. Real-time Monitoring and Forecasting: AI methods can process and analyze climate data in near-real-time, enabling the monitoring and prediction of climate conditions at shorter time scales. This is particularly valuable for weather forecasting and early warning systems (Filho et al., 2022).

INTEGRATION OF HIGH-PERFORMANCE COMPUTING AND ARTIFICIAL INTELLIGENCE

Challenges and Limitations

The integration of HPC and AI in climate science also presents challenges and limitations:

a. Computational Resource Requirements: AI algorithms, especially deep learning techniques, demand significant computational resources to train and run models. The integration of AI with HPC systems requires adequate hardware capabilities and efficient parallelization strategies to effectively handle computational demands (Allen, 2003).
b. Data Quality and Quantity: AI techniques heavily rely on high-quality, diverse, and large-scale climate data for accurate predictions. Ensuring data quality, availability, and access can be a challenge, particularly in regions and variables with limited observations (Sarker, 2022).
c. Interpretability and Explain ability: AI models often lack interpretability, making it challenging to understand the underlying factors influencing their predictions. Interpreting AI outputs in climate science is crucial to gaining the trust of policymakers and stakeholders, who require transparent decision-making processes (Sarker, 2022).

Potential Benefits

The integration of HPC and AI in climate science unlocks several opportunities:

a. Faster Model Development and Optimization: HPC enables rapid training and optimization of AI models, reducing the time needed to develop and test different AI-based climate models. This accelerates the improvement and refinement of climate predictions (Dewitte et al., 2021; Wang et al., 2023).
b. Improved Model Performance: Combining HPC and AI techniques can enhance the performance of climate models, leveraging AI's ability to capture complex patterns and uncertainties, and HPC's capability to perform computationally intensive simulations at high resolutions (Milojicic et al., 2021).
c. Data-Driven Model Calibration: AI algorithms can utilize historical climate data to calibrate and optimize model parameters, leading to more accurate simulations and projections. HPC facilitates the efficient processing of large datasets required for model calibration (Gupta, 2019).

Algorithms and Techniques for Integrating HPC and AI in Climate Modeling and Prediction

Several algorithms and techniques can be leveraged for integrating HPC and AI in climate modeling and prediction (Jiang et al., 2018):

a. Deep Learning: Utilizing deep neural networks can capture intricate patterns and nonlinear relationships within climate data. HPC systems accelerate deep learning model training by parallelizing computations, facilitating the exploration of complex climate processes (Konya & Nematzadeh, 2024).
b. Ensemble Methods: Running ensemble simulations with variations in input parameters or model configurations can be computationally demanding. HPC enables efficient ensemble simulations, providing a more comprehensive exploration of uncertainties in climate projections (Sukeshini et al., 2021).
c. Hybrid Models: Integrating physics-based climate models with AI techniques, such as data assimilation or emulators, can improve the representation of complex climate processes while incorporating real-time observational data (Tripathy & Mishra, 2023).

In summary, integrating HPC and AI in climate modeling and prediction holds great potential to enhance our understanding of climate dynamics, improve predictive accuracy, and inform decision-making for climate mitigation and adaptation strategies (Manubens-Gil et al., 2016). However, careful consideration of computational resources, data quality, interpretability, and model performance is necessary to fully leverage the benefits of this integration. Continued research and development in this area are crucial for advancing climate science and addressing the challenges posed by climate change.

FUTURE CHALLENGES AND LIMITATIONS

While the integration of HPC and AI in climate science brings many opportunities, several challenges and limitations should be addressed. The computational demands of AI algorithms require substantial resources, including energy consumption, which needs to be managed efficiently. Additionally, the interpretation and explainability of AI-driven climate models remain crucial, as stakeholders need transparent decision-making processes (Ali et al., 2023). Furthermore, the availability and quality of climate data, especially in data-sparse regions, can limit the effectiveness of AI techniques. Researchers must also address issues of bias and uncertainty when using AI-driven approaches for climate modeling and prediction (Karanth et al., 2023).

CONCLUSION

Summary of Findings

The integration of high-performance computing (HPC) and artificial intelligence (AI) techniques in climate modeling and prediction presents substantial benefits. HPC enables higher computational power, leading to enhanced model resolution, improved predictability, efficient data analysis, and increased collaboration and sharing among researchers (Yi & Loia, 2019). Meanwhile, AI techniques offer advanced data analysis, pattern recognition, and optimization capabilities, enhancing the accuracy and efficiency of climate models.

Contribution to Climate Science

The integration of HPC and AI contributes significantly to climate science by advancing our understanding of climate dynamics, improving predictive accuracy, and informing decision-making processes. It allows for more detailed and realistic climate simulations, aiding in the development of effective climate mitigation and adaptation strategies (Changlani & Thakore, 2023).

Recommendations for Future Research

To further harness the potential of HPC and AI in climate modeling and prediction, future research should focus on addressing challenges such as computational resource requirements and data availability. Efforts should be made to enhance the interpretability and explainability of AI-driven models, ensuring transparency in decision-making processes. Collaboration among researchers and policymakers is crucial in promoting the development and adoption of standardized methodologies, data sharing, and model intercomparisons to maximize the benefits of this integration.

REFERENCES

Ali, S., Abuhmed, T., El-Sappagh, S., Muhammad, K., Alonso-Moral, J. M., Confalonieri, R., Guidotti, R., Del Ser, J., Díaz-Rodríguez, N., & Herrera, F. (2023). Explainable Artificial Intelligence (XAI): What we know and what is left to attain Trustworthy Artificial Intelligence. *Information Fusion, 99*. doi:10.1016/j.inffus.2023.101805

Alizadeh, O. (2022). Advances and challenges in climate modeling. *Climatic Change*, *170*(1-2), 18. doi:10.1007/s10584-021-03298-4

Allen, M. (2003). Liability for climate change. *Nature*, *421*(6926), 891–892. doi:10.1038/421891a PMID:12606972

André, J.C. (2014). High-performance Computing for limate Modeling. AMS Journal. doi:10.1175/BAMS-D-13-00098.1

APEC. (2023). *APEC Climate Symposium: Enhancing APEC Resilience through AI Applications and Climate Change Adaption*. APEC. https://www.apec.org/docs/default-source/publications/2023/1/apec-climate-symposium-enhancing-apec-resilience-through-ai-applications-in-climate-change-adaptation/223_ppsti_apec-climate-symposium.pdf?sfvrsn=7565b6cb_2

ChanglaniK.ThakoreR. (2023). Artificial Intelligence for Climate Action. doi:10.13140/RG.2.2.21102.18245

Chattopadhyay, A., Hassanzadeh, P., & Pasha, S. (2020). Predicting clustered weather patterns: A test case for applications of convolutional neural networks to spatio-temporal climate data. *Scientific Reports*, *10*(1), 1317. doi:10.1038/s41598-020-57897-9 PMID:31992743

Chen, L., Chen, Z., Zhang, Y., Liu, Y., Osman, A. I., Farghali, M., Hua, J., Al-Fatesh, A., Ihara, I., Rooney, D. W., & Yap, P.-S. (2023). Artificial intelligence-based solutions for climate change: A review. *Environmental Chemistry Letters*, *21*(5), 2525–2557. doi:10.1007/s10311-023-01617-y

Dell Technologies. (2020). *HPC Enhances Weather and Climate Modeling, Research, and Proactive Responses to Dangerous Conditions*. Dell Technologies.

Dewitte, S., Cornelis, J. P., Müller, R., & Munteanu, A. (2021). Artificial Intelligence Revolutionises Weather Forecast, Climate Monitoring and Decadal Prediction. *Remote Sensing (Basel)*, *13*(16), 3209. doi:10.3390/rs13163209

Ejarque, J., Badia, R. M., Albertin, L., Aloisio, G., Baglione, E., Becerra, Y., Boschert, S., Berlin, J. R., D'Anca, A., Elia, D., Exertier, F., Fiore, S., Flich, J., Folch, A., Gibbons, S. J., Koldunov, N., Lordan, F., Lorito, S., Løvholt, F., & Volpe, M. (2022). Enabling dynamic and intelligent workflows for HPC, data analytics, and AI convergence. *Future Generation Computer Systems*, *134*. doi:10.1016/j.future.2022.04.014

Fan, Z., Yan, Z., & Wen, S. (2023). Deep Learning and Artificial Intelligence in Sustainability: A Review of SDGs, Renewable Energy, and Environmental Health. *Sustainability (Basel)*, *15*(18), 13493. doi:10.3390/su151813493

Filho, W. L., Wall, T., Serafino, A. R. M., Nagy, G. J., Balogun, A.-L., Luetz, J. M., Ng, A. W., Kovaleva, M., Fardous, M. S. A., Alves, F., & Guevara, Z. (2022). Deploying artificial intelligence for climate change adaptation. *Technological Forecasting and Social Change*, *180*. doi:10.1016/j.techfore.2022.121662

Gadzhev, G., Ivanov, V., & Valcheva, R. (2021). HPC Simulations of the Present and Projected Future Climate of the Balkan Region. Springer. doi:10.1007/978-3-030-55347-0_20

Gupta, G. (2019). *The Benefits of Bringing Artificial Intelligence to High Performance Computing*. High Performance. https://high-performance-computing.cioreview.com/cxoinsight/the-benefits-of-bringing-artificial-intelligence-to-high-performance-computing-nid-26530-cid-84.html

Huntingford, C. (2019). *Machine learning and artificial intelligence to aid climate change research and preparedness*. IOP Publishing. doi:10.1088/1748-9326/ab4e55

Jiang, Z., Gao, W., Wang, L., Xiong, X., Zhang, Y., Wen, X., Luo, C., Ye, H., Lu, X., & Zhang, Y. (2018). *HPC AI500: A Benchmark Suite for HPC AI Systems*. Springer. https://link.springer.com/chapter/10.1007/978-3-030-32813-9_2

Karanth, S., Benefo, E. O., Patra, D., & Pradhan, A. K. (2023). Importance of artificial intelligence in evaluating climate change and food safety risk. *Journal of Agriculture and Food Research, 11*. doi:10.1016/j.jafr.2022.100485

Kelechi, A. H., Alsharif, M. H., Bameyi, O. J., Ezra, P. J., Joseph, I. K., Atayero, A.-A., Geem, Z. W., & Hong, J. (2020). Artificial Intelligence: An Energy Efficiency Tool for Enhanced High performance computing. *Symmetry*, *12*(6), 1029. doi:10.3390/sym12061029

Konya, A., & Nematzadeh, P. (2024). Recent applications of AI to environmental disciplines: A review. *Science of The Total Environment, 906*. doi:10.1016/j.scitotenv.2023.167705

Lee, C. A., Gasster, S. D., Plaza, A., Chang, C.-I., & Huang, B. (2011, September). Recent Developments in High Performance Computing for Remote Sensing: A Review. *IEEE Journal of Selected Topics in Applied Earth Observations and Remote Sensing*, *4*(3), 508–527. doi:10.1109/JSTARS.2011.2162643

Manubens-Gil, D., Vegas-Regidor, J., Prodhomme, C., Mula-Valls, O., & Doblas-Reyes, F. J. (2016). Seamless management of ensemble climate prediction experiments on HPC platforms. *2016 International Conference on High Performance Computing & Simulation (HPCS)*, Innsbruck, Austria. 10.1109/HPCSim.2016.7568429

Michalakes, J. (2020). HPC for Weather Forecasting. In A. Grama & A. Sameh (Eds.), *Parallel Algorithms in Computational Science and Engineering*. Modeling and Simulation in Science, Engineering and Technology. Birkhäuser. doi:10.1007/978-3-030-43736-7_10

Milojicic, D., Faraboschi, P., Dube, N., & Roweth, D. (2021). Future of HPC: Diversifying Heterogeneity. 2021 Design, Automation & Test in Europe Conference & Exhibition. IEEE. doi:10.23919/DATE51398.2021.9474063

Minevich, M. (2022). How To Fight Climate Change Using A. I. *Forbes*. https://www.forbes.com/sites/markminevich/2022/07/08/how-to-fight-climate-change-using-ai/?sh=5682c32a838e

Palmer, T. (2014). Climate forecasting: Build high-resolution global climate models. *Nature*, *515*(7527), 338–339. doi:10.1038/515338a PMID:25409812

Raj, R. (2020). *High-Performance Computing Education: Current Challenges and Future Directions ITiCSE-WGR '20: Proceedings of the Working Group Reports on Innovation and Technology in Computer Science Education*. ACM. doi:10.1145/3437800.3439203

Sarker, I. H. (2022). AI-Based Modeling: Techniques, Applications and Research Issues Towards Automation, Intelligent and Smart Systems. *SN Computer Science*, *3*(2), 158. doi:10.1007/s42979-022-01043-x PMID:35194580

See, S., & Adie, J. (2021). Challenges and opportunities for a hybrid modelling approach to earth system science. *CCF Transactions on High Performance Computing*, *3*(3), 320–329. doi:10.1007/s42514-021-00071-y

Shivanna, K. R. (2022). Climate change and its impact on biodiversity and human welfare. *Proceedings of the Indian National Science Academy. Part A, Physical Sciences*, *88*(2), 160–171. doi:10.1007/s43538-022-00073-6

Singh, M. (2022). *Quantum Artificial Intelligence for the Science of Climate Change*. Springer. . doi:10.1201/9781003250357-13

Soori, M. (2023). Artificial intelligence, machine learning, and deep learning in advanced robotics, a review. *Cognitive Robotics, 3*. doi:10.1016/j.cogr.2023.04.001

Soori, M., Arezoo, B., & Dastres, R. (2023). *Artificial intelligence, machine learning and deep learning in advanced robotics, a review* (Vol. 3). Cognitive Robotics. doi:10.1016/j.cogr.2023.04.001

Sukeshini, S. P., Ved, M., Chintalapti, J., & Pal, S.N. (2021). Big Data Analytics and Machine Learning Technologies for HPC Applications. In Evolving Technologies for Computing, Communication and Smart World. Lecture Notes in Electrical Engineering. Springer. doi:10.1007/978-981-15-7804-5_31

Tripathy, K. P., & Mishra, A. K. (2023). Deep Learning in Hydrology and Water Resources Disciplines: Concepts, Methods, Applications, and Research Directions. *Journal of Hydrology*. doi:10.1016/j.jhydrol.2023.130458

Vital, J.-A., Gaurut, M., Lardy, R., Viovy, N., Soussana, J.-F., Bellocchi, G., & Martin, R. (2013). High-Performance computing for climate change impact studies with the Pasture Simulation model. *Computers and Electronics in Agriculture*, *98*, 131–135. doi:10.1016/j.compag.2013.08.004

Wang, D. & Yuan, F. (2020). *High-Performance Computing for Earth System Modeling*. Springer. . doi:10.1007/978-3-030-47998-5_10

Wang, H., Fu, T., Du, Y., Gao, W., Huang, K., Liu, Z., Chandak, P., Liu, S., Van Katwyk, P., Deac, A., Anandkumar, A., Bergen, K., Gomes, C. P., Ho, S., Kohli, P., Lasenby, J., Leskovec, J., Liu, T.-Y., Manrai, A, & Zitnik, M. (2023). Scientific discovery in the age of artificial intelligence. *Nature*, *620*(7972), 47–60. doi:10.1038/s41586-023-06221-2 PMID:37532811

Yi, G., & Loia, V. (2019). High-performance computing systems and applications for AI. *The Journal of Supercomputing*, *75*(8), 4248–4251. doi:10.1007/s11227-019-02937-z

Chapter 13
Energy Optimization and Smart Grids:
IoT-Based Smart Grid Solution and Smart Grids Applications

Sana Dahmani
Independent Researcher, USA

ABSTRACT

Grid computing, utilizing distributed resources, achieves goals in handling workloads in a distributed system. Unlike traditional high-performance computing, grid computing assigns tasks to nodes, showcasing heterogeneity without physical coupling. Grids, with variable sizes based on applications, use general-purpose middleware. In modern energy systems, integrating smart grids and grid computation is vital. Emphasizing resource efficiency, grid computation enhances smart grid performance; by enabling real-time monitoring, rapid adjustments for optimization, and outage prevention. Grid computation supports demand response strategies for electricity management and integrates distributed generation resources, such as solar panels and wind turbines, including autonomous microgrid supervision.

INTRODUCTION

Smart Grids, often referred to as "intelligent" or "advanced" grids, describe a modern and technologically sophisticated evolution of electrical networks, with the purpose of incorporating cutting-edge digital technologies aimed at automation combined with efficiency, sustainability, accountability and adaptability for electricity generation, distribution and consumption optimization. This dynamization and smartening incentive of the electric power grid consists of a transformative evolution by leveraging advanced digital technologies and communication frameworks to create a responsive and efficient ecosystem through the seamless integration of renewable resources, the optimized telecommunication and the meaningful data management by harnessing the Internet of Things (IoT) technologies. Modern energy systems require the implementation of Smart Grids and fundamental concepts serving as a ground for those Smart Grids such as the Distributed Energy Resources (DERs) which are small-scale power

DOI: 10.4018/979-8-3693-1794-5.ch013

generation units shaped at a local level and interlinked with a border grid, and integrating technologies such as renewable energies and variable energy resources (VERs) generators. Smart Grids are founded on six fundamental pillars which are: (1) the Advanced Metering Infrastructure (AMI): which consists in a bidirectional communication channel between users and utilities, thus, transferring, collecting, assessing and storing energy consumption data for an efficient remote monitoring and potential anomaly identification. (2) Demand Response (DR) is a flexibility catalyst involving collaborative efforts of energy between providers and users, and aimed at balancing the demand-supply driven by their incentive to lower energy usage during peak consumption. It fosters the monitoring of smart electrical demand loads in Smart Grids through instantaneous communication. (Siddiquee et al., 2020) researches showed that the DR concept-even though sustained by the integration of renewable energies reducing energy consumption burden, for instance, thanks to EVs reinjection electrical power back to the grid-still faces challenges in tackling a sufficiently enhanced user comfort and high costs, which require alternative strategies to incentivize consumer participate and engage in the cooperation endeavor, in addition to the importance of the need for adequate mathematical modeling to harness the renewable energy resources through their best integration in a DR context. (3) Grid Automation Systems represent a key factor of smartening through data-based monitoring on a real-time basis, founded on automated control and optimization of electricity flow distribution. (4) Supervisory Control and Data Acquisition (SCADA) serve as user-friendly interfaces for supervision and control. Finally, (5) Big Data Analytics and Cybersecurity are associated with the smart aspect of grids. While data analysis and application in forecasting purposes stands behind the system adjustments and relates to decision making, Cybersecurity remains a challenge to be addressed. (6) Communication Networks Infrastructure Robustness is crucial for the information exchange flow between the Smart Grids' elements/components and the Machine to Machine (M2M) communication which excludes human intervention. Moreover, Transmission Enhancement involves distribution streamlining, selective load control, and microgrid dynamics. The integration of innovative technologies in the traditional electrical grid enables data collection and monitoring, grid automation, electricity systems control, renewable energy integration, and the management of the infrastructure for electric vehicle (EV) charging. It also contributes to supply-demand balance, energy consumption reduction, and environmental preservation. The Smart Grid Technology Market Size Worldwide between 2021- 2026 is prone to grow from 36.9 to 55.9 billion U.S Dollar as shown by **Figure 1**.

Smart Grids hold a fundamental role in the real-time monitoring and enable optimized adjustments by utility providers in order to control downtime and fulfill the DR strategies. Moreover, they empower a smooth integration of distributed generation resources, in addition to the supervision of autonomous microgrids. The sound management of Smart Grid components extends to sensors, communication systems, data centers and control systems. They aim to ensure cautious and optimal resource consumption, cost-effectiveness and enhanced reliability. Wireless Sensor Networks (WSNs)-which are nodes with sensing, analyzing and communication abilities-are crucial components of Smart Grids which collect data dedicated to solving problems, in the world of power data bandwidth, i.e. the data amount able to be transferred by means of a communication channel in given time period. Those networks are essential for real-time communication. In modern energy systems, the integration of Smart Grids significantly addresses challenges strengthened by the effect of population growth; triggering escalating energy demands. The evolution of a "Smart Grid" enhances user-friendly interface and resilience against physical and cyber threats, contributing to asset optimization, environmental friendliness, advanced sensors, and distributed computing technology. They aim at fostering efficiency, reliability on both hardware and cyber safety; aligning with the principles of the Industry 4.0 revolution (Tuttokmaği and Kaygusuz, 2018; Routray, 2021; Joy et al., 2013).

Figure 1. Forcasted smart grid technology market size worldwide between 2021- 2026 (Statistica, 2022)

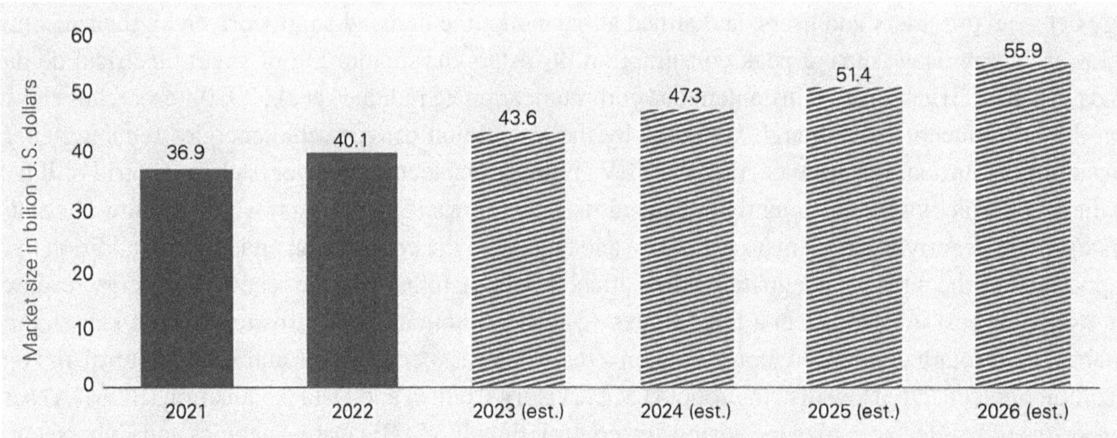

Regardless of the conception of new networks from scratch or the adoption of existing ones, Smart Grids serve as a facilitator for cost-effectiveness and sustainability (UNECE, 2015). The implementation of Smart Metering (SM) represents the primary initiative to achieve the goals associated with Smart Grids development, thus, addressing the challenges of balancing renewable energy, optimizing customer operations, efficient peak management and various further critical issues. Researchers have proposed diverse algorithms and models to address challenges in Smart Grids which contribute to the optimization of energy generation, transmission, and consumption; as highlighted in 1.1.

Energy Optimization and Smart Grids

1. IoT-based Smart Grid Solution

The Internet of Things (IoT)-a term coined by computer scientist Kevin Ashton in 1999-embodies a network of hardware devices designed for autonomous data transmission without human intervention. IoT-enabled real-time data collection and transmission in critical infrastructures proves to be an innovative, safer, more reliable and efficient solution. This approach includes a myriad of items beyond traditional computers or machinery, and harnesses sensors to collect information and undergo data assessment for prediction purposes. The integration of IoT into Smart Grids enables dynamic adaptation to changing conditions; thus allowing a synergy for a resilient and intelligent grid infrastructure (Ghayoor et al., 2022).

The traditional electric delivery system highlights specific features such as centralized power generation, unidirectional energy flow, passive consumer intervention in the grid management, restricted real-time monitoring and control; and a lack of adaptability to the integration of alternative energy sources, or the ability to support new services and emerging technologies. Conversely, the Smart Grid of the future aims to integrate renewable energy sources, promote consumer engagement, create safe communication networks prone to protecting against cyberattacks, maximizing asset utilization, mitigating energy loss and system collapse, integrating electric vehicles, and deploying distributed energy storage. With a focus on security and operational flexibility, this cutting-edge, novel grid model is well-positioned to promote compatibility and open systems (Anderson et al., 2018). **Table 1** provides a comparison between the two paradigms.

Table 1. Traditional versus smart grids comparison

Feature	Traditional Grid	Smart Grid
Energy flow	Unidirectional	Bidirectional and interactive
Power generation	Centralized	Decentralized for a more sustainable infrastructure thanks to a facilitated integration of diverse small-scale and decentralized energy sources such as solar panels, wind turbines, and further local generation resources. The decentralized approach of power generation within Smart Grids enhances the incorporation of renewable energy sources, promotes localized energy production; and engages consumers in energy management strategies and consumption decisions through intervening during peak times when an increase in load demand is detected.
Consumer participation	Passive	Active. The Smart Grid fosters active consumer engagement by providing real-time energy usage data, enabling data-grounded decisions and encouraging the user participation in demand-response programs with incentives for reducing energy consumption, thus, empowering consumers' role in upgrading the grid's efficiency and reliability.
Real-time monitoring	Limited or absent. Grid operators face challenges due to its limited real-time visibility into the current state of the grid. Indeed, it alternatively relies on historical data or periodic inspections. This data handling approach leads to potential delays in problem response and the risk of outages or disruptions.	Extensive
Flexibility	Low: In respect of the grids' ability of learn and calibrate its models to embrace emerging incidents and needs, the conventional grid is designed with limited flexibility, making it challenging to integrate renewable energy sources to it. Moreover, including distributed energy resources, or addressing the increasing electricity demand is deemed to be difficult. This constraint arises from the traditional grid's design, which is dedicated to a unidirectional flow of electricity from large power plants to consumers.	High: The flexibility allows for the integration of compelling renewable energy sources, the optimal utilization of distributed energy resources (DERs), the management of peak demand, the enhancement of grid resilience against disruption; and active consumer empowerment through real-time data management and demand-response programs; which contribute to the overall grid efficiency.
Metering	Electro-mechanical	Sophisticated: a metering system enabling the facility to monitor and control energy consumption.
Integration of renewable energy	Limited	Easy
Cybersecurity	Weak	Strong
Cost-effectiveness	Relatively high: Manual inspections, relying on workforce especially in holidays and non-business days triggers extra costs.	Potentially lower: Reduced during peak hours thanks to the distributed generation and control of power consumption.
Reliability	Moderate: Susceptible to breakdowns and cascading outages stemming from punctual defections.	High: Proactive, instantaneous and islanding-driven; i.e. Distributed Generation (DG) system continues power supply to a particular zone in the absence of electricity availability in the external grid (Goudarzi et al., 2022).
Sustainability	Low	High: thanks to the digitalization, electrification and renewable energy integration trends.

Architecture of IoT Technologies

The challenges encountered in traditional power grids are prone to being addressed by the implementation of Smart Grid technology. As a matter of fact, endowed with advanced bidirectional communication and real-time monitoring capabilities, the Smart Grid fosters active consumer involvement, reinforces grid resilience and argues in favor of a more flexible and adaptive electricity infrastructure. Fortunately, those traditional grids have the potential of transition into a Smart Grid system thanks to IoT integration. This way, traditional grid quality and liability are strengthened by the integration of intelligent data-processing abilities within the AMI, as well as Smart Grids (SM) technologies. IoT, through the integration of sensing and operation systems within the AMI, offers significant perspectives in optimizing energy consumption. Moreover, its interconnected system helps manage ubiquitous data covering energy usage, voltage, current readings, and phase measurements; and applies technology to efficiently transmit and smartly assess this data, thereby enhancing energy grid capabilities. The applications of IoT technologies extend to various domains, including power generation infrastructure, SCADA systems, energy transmission and distribution operations; and environmental impact monitoring. Technology distribution approaches can be harnessed by Smart Grids. Indeed, Cloud and Edge Computing technologies further contribute to the distributed monitoring and management of decentralized energy resources, through tackling Cybersecurity vulnerabilities connected with traditional centralized SCADA systems. While Cloud Computing offers a centralized approach gathering under its umbrella several systems and consolidating many resources; Grid Computation segregates resources in several systems working together within a distributed architecture of nodes operating in an independent way. Like Cloud Computation, Cluster Computation aggregates resources towards a centralized approach with a multitude of nodes associated together, but having the failure risk if one node is faulty, while this threat is low at the Crid's Computation level. Grid Computation can be applied to Smart Grids as a data-oriented field of application. Thanks to a user-friendly user interface and a modular Computing where resources (including but not limited to storage/memory, GPUs and networking) are separated. In Grid Computation, data is managed with a Grid Access to Secondary Storage (GASS) which allows remote data access and storage from DERs including batteries and solar panels. Moreover, Globus Security Infrastructure also known as Grid Security Infrastructure (GSI), leverages secure authentication and communication across an unsecured network thanks to the use of a pair of keys in the framework of asymmetric cryptography, also quoted as public-key cryptography. As for Cluster Computing, it has been harnessed for the efficient handling of real-time data in order to maintain a stable grid control and big data management.

Furthermore, the integration of IoT-enabled Smart Grids within further intelligent entities such as smart appliances, homes, buildings, and cities, enhances the efficiency of operating and managing the electrical grid. While providing substantial advantages on the efficiency level, the implementation of IoT-enabled Smart Grids presents challenges consisting in the need for ongoing research and development (Geneiatakis et al., 2017; Chang et al., 2020; Li et al., 2021; Goudarzi et al., 2022) to face Cybersecurityattack and data-analysis-related issues such as misleading signals and alerts-leading to potential errors in decision-making. Smart grids with IoT capabilities show an increasing trend thanks to their contribution to sensing, actuation, communication and processing. Additionally, the integration of 5G technology into Smart Grids represents prominent progress, in particular in the context of the so-called DR strategy, towards smart and energy provider-consumer, jointly managing energy consumption. The advent of 5G networks and, thus, 5G-based IoTs, offers benefits including rapid data transfer speeds, high reliability, robust security, low electrical energy use and the ability to connect multiple devices, thereby enhancing DR capabilities. Research works have confirmed that the integration of 5G technology into Smart Grids would have a huge revolutionary influence across numerous industries (Huang et al., 2020).

The IoT-enabled Smart Grid has a four-layered architecture composed of (1) a Physical Layer or Sensing layer which is the upper level, including the hardware responsible for data collection. Upper to it, (2) the Communication Layer plays a crucial role in the IoT-driven Smart Grid, by enabling the efficient exchange of information among the extensive network of devices and systems comprising the grid. (3) The Cyber Layer is responsible for decision-making while the highest decision-making level is the so-called (4) Application Layer also known as the Management and Control Layer. **Figure 2** and **Table 2** illustrate this architecture.

Figure 2. Four-layered IoT architecture for smart grids

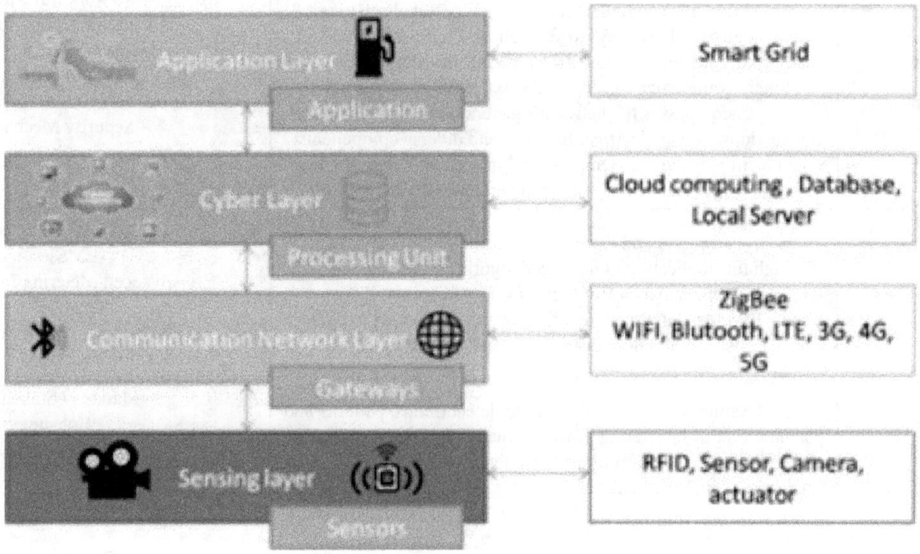

Table 2. IoT-based smart grid architecture layer

Architecture Layer	Description	Examples of the Layer's Components
Physical layer or Sensing layer	- Constituted by physical devices responsible for gathering data related to the infrastructure and operations of the grid such as electrical components. - Data collected by these devices is wirelessly transmitted using diverse communication protocols such as Zigbee, Wi-Fi, and LTE. - The two-way exchange of energy between the power generation systems and the customers occurs within this layer (Goudarzi et al., 2022).	- Sensors - Meters - Actuators that measure parameters such as voltage, current, power, temperature and humidity.
Communication Network Layer	- Furnishes a human-computer interface (HMI) to the upper layer, allowing it to coordinate lower levels by creating and issuing appropriate commands - Links the physical layer to the cyber layer - Integrates varied communication protocols, network technologies, and gateways to securely and efficiently transmit data from field devices to control centers, ensuring reliable and timely communication across the extensive grid network. - Encompasses the network's general activities, involving interactions among electrical facilities and diverse components; and facilitates the transfer of control instructions from higher layers and collected data from lower layers.	- Cloud-based central - Network Infrastructure - Communication Protocols - Gateways and Aggregators - Quality of Service (QoS) - Fault Detection and Recovery - Security Mechanisms
Cyber Layer = Control Layer	- Receives and assesses data from the sensing layer, through the application of advanced algorithms and analytics to enhance the efficiency of grid operations. - Executes intelligent decision-making processes to oversee power generation, distribution and consumption.	- Energy Management Systems (EMS) - Advanced Metering Infrastructure (AMI) - Supervisory Control and Data Acquisition (SCADA) systems
Application layer	- Examines data from the grid to detect trends, patterns and potential issues; facilitating proactive maintenance and enhancing overall grid efficiency.	- Markets / Stakeholders - Consumers - Grid Operators - Service Providers

Figure 3 presents the interference of the Smart Grid four-layered structure with the Cloud, Edge, and Fog Computation deployment. The complex interconnections among the physical, cyber, application, cloud, edge, and fog layers in the network ecosystem are established through a series of important relationships. Firstly, physical network data is transmitted to the cyber network, where it undergoes processing before being relayed to the application layer. Smart Grid users receive outputs from the application layer, with the potential of feedback from the cyber network, where further refinement is integrated. The cloud layer assumes the pivotal role of storing and processing data which stems from both the physical and cyber networks. Meanwhile, the edge layer undertakes localized data processing, effectively restraining the volume of data transmitted to the cloud. Lastly, the fog layer, constituting a distributed computing environment, confers scalability and flexibility to the overall network architecture, ensuring adaptability and efficient operation.

Figure 3. Layered smart grid architecture and enabling technologies of the cyber-physical power system (CPPs)
(Liu et al., 2021)

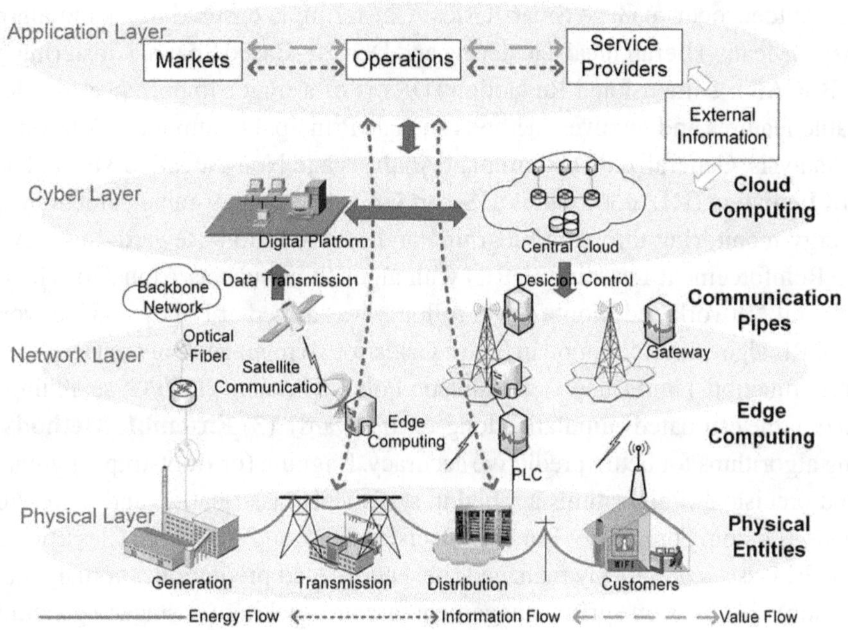

Cutting-Edge-Technology and Smart Grids (Based on (Jiao, 2020; Omitaomu and Niu, 2021))

AI serves as a key element in the integration of renewable energies, secure stability within energy grids and the mitigation of costs associated with instability risks. As a matter of fact, the capability of ML to utilize data to train models helps take control over potential failures and adjust the imbalance of production-consumption burdens distribution. AI intervenes at the optimization of daily operations of electrical grids. In general, AI was proved to be a backbone for Smart Grids, enhancing their efficiency. Indeed, while traditional grids show a limitation for data-based predictions, Artificial Neural Networks (ANNs) algorithms helped image processing and pattern recognition. Using one single hidden layer, the Extreme Learning Machines (ELMs)-part of ANN algorithms- help solve electricity systems' stability issues and anomaly detection. In the Neural Network Framework, the Back-Propagation Neural Network (BPNN) contributes to the recurrent and consistent adjustment of the network's weights to minimize the margin of error between the input data and ground truth (assessed as true data obtained by observations rather than by inference). A further Neural Network is the Probabilistic Neural Network (PNN), efficient in classification and pattern recognition. Deep Learning (DL) was harnessed through further technologies such as Deep Belief Network (DBN), Convolution Neural Networks (CNNs), Recurrent Neural Networks (RNNs), Generative Adversarial Networks, and autoencoder. AI also helps solve regression and classification problems by using SVM which showed robustness in classification, KNN which was harnessed for regression and classification; and Decision Tree Learning model and Logistic Regression algorithms dedicated to anomaly detection. Forecasting and Demand Response (DR) rely on regression models such as SVR, Multivariate Adaptive Regression Spline (MARS), Gaussian Process Regression

(GPR) and Linear Regression (LR). Anomaly detection, demand forecasting, stability examination etc. are the common achievements of AI tools. Indeed, they are ensured through (1) **Unsupervised Learning**. For instance, Neural Networks Algorithms including Boltzmann Machine, Autoencoder, and Variational Autoencoder manage Smart Grids. Clustering is ensured thanks to algorithms such as K-Means, Fuzzy c-Means, Hierarchical Clustering and Density-Based Spatial Clustering of Applications with Noise (DBSCAN). Dimensional Reduction (DR) is of a major importance for the diminution of recurrent variable features and involves methods such as Principal Component Analysis (PCA), Linear Discriminant Analysis, Generalized Discriminant Analysis and Non-Negative Matrix Factorization. (2) **Reinforcement Learning (RL)** contributes to Smart Grids technology enhancement by the detection of attacks and energy monitoring through Q-learning and State–Action–Reward–State–Action (SARSA) methods. Deep Reinforcement Learning (DRL) with algorithms such as AlphaGo-which is a combination of Deep Neural Network and Monte Carlo rollouts-was aimed at Policy and Network; and Rollout Policy. Further DRL algorithms common in Smart Grids sound management are Deep Q network (useful in approximating functions) and Deep Deterministic Policy Gradient (DDPG, searching for an optimal policy maximizing the estimated cumulative long-term reward). (3) **Ensemble Methods** bring together several learning algorithms for better predictive accuracy. Bagging (or Bootstrap Aggregation) improves the stability and precision of algorithms applied in statistical classification and regression. Among the Bagging methods, Random Forest fuses Random Decision Trees and Advanced Classification Algorithms to tackle Smart Grid issues previously managed (i.e. energy load prediction, anomaly identification and stability evaluation). For the same grid management targets, stacking (or stacked generalization) generates predictions based on the different weak learners, put together parallelly and then joint by means of training a meta-model. Finally, Boosting models learn from sequent weak learners and combine them in a preset strategy. They help rectify previous model misclassification, which is prominent in optimizing smart grids.

- **Energy consumption forecasting:** The prediction of power demand within the grid can be broken down into short, medium and long-time prediction; expanding from a few minutes to over a year time. Artificial Neural Networks (ANNs) were harnessed for power load prediction. Among those algorithms we cite: The Back Propagation algorithm (BP algorithm) used in early stages, thanks to one single hidden node, however with limited performance in respect of complex functions. Long and Short Time Memory (LSTM) DL models are based on a Recursive Neural Network (RNN) and serve as an alternative to the BP. The LSTM model ensures that the gradient is not altered over time, which is the issue encountered during the traditional RNN training. The data analyzed consists in power consumption and load monitoring, and aims at providing a theoretical framework for energy consumption detection, implementating of fair pricing, and enhancing the energy system infrastructure and the bidirectional data flow. Machine Learningis applied in the clustering of unlabeled data and classification for labeled data. Further ML algorithms were used such as ANN, SVM, RNN, KNN, and GPR. Apache Sparks showed a better hyperparameter tuning by harnessing LSTM and Gated Recurrent Unit (GRU). Both ANN and LR models were efficient in accurate predictions of correlation between load demand and the environmental factors. Moreover, due to the volatility of renewable energies and their lack of stability, accurate prediction is of crucial importance, for instance, Deep learning such as LSTM model are used for power generation prediction. The traditional estimation method showed limitations especially with wind or light data with both nonlinear and non-stationary aspects.

- **Energy stability monitoring:** Helps harness the real-time data to supervise and predict potential issues. ELMs including single hidden layers showed efficiency in solving grid stability issues. Transient Stability Assessment (TSA) involves the robustness of the grid system in the presence of disturbances and potential disruptions. Indeed, data analysis harvested by Phasor Measurement Units (PMU) and Wide Area Measurement System (WAMS) are analyzed for further decision making. A combination of Trajectory Fitting (TF) and ELM demonstrated performance and consistency (Tang et al., 2017). Combined RNN-LSTM model mitigated the dependency on input data (James et al., 2017). Applying Deep Belief Networks (DBN) showed a great precision optimization (Wang et al., 2020). SVMs were also used for classification, image recognition and regression; ANNs for predictive modeling and complex data patterns recognition; and Decision Trees for classification, regression and anomaly detection as well.
- **Anomaly / Fault Detection:** Electric Power Systems (EPS) can be described as flexible as they manage several tasks including energy transmission and power supply, thus, equipment anomaly detection helps detect issues and limit the harm's scope. Extreme Learning Machines (ELMs) showed efficiency. Moreover High-Impedance Fault Detection (HIFD), based on an ANN method, helped solve the issue in question with an accuracy of 98.67% (Sirojan et al., 2018). LSTM networks, SVM, stacked sparse autoencoder-based networks with SVM and PCA showed reliability in fault detection (Wang et al., 2019).
- **Leveraging Cybersecurity:** The bidirectional real-time data flow can expose inflowing and outflowing data to cyberthreats. The nature of the power grid system attack is hidden. Even if the physical material remains spared from danger, attacks can reach the secondary level which spreads to the hardware equipment. Deep Learning helps detect abnormal behaviors thanks to its training process that copes with the effect of limited sample size. Some of its used algorithms are ANNs, SVMs, Stacked Denoising Autoencoders (SDAE) and Reinforcement Learning (RL). They have shown reliability in their risk attack identification and classification.

Smart Grids Applications

There are several applications of Smart Grids as shown in **Figure 4** and **Figure 5** which visualize the Smart Grids Technologies by subsector.

Figure 4. Smart grids applications

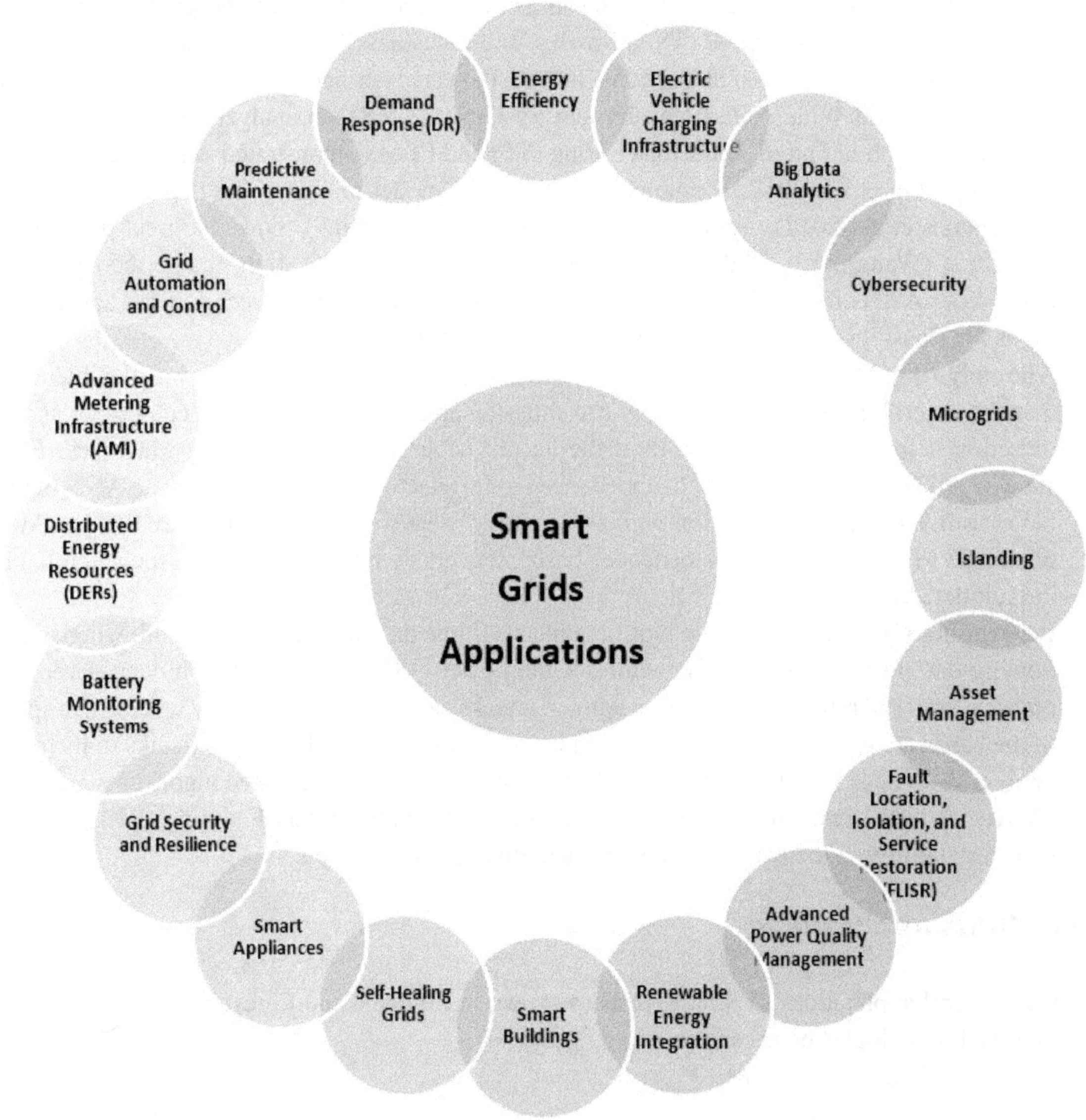

Figure 5. Smart grids technologies by subsector
(Moreno Escobar et al., 2021)

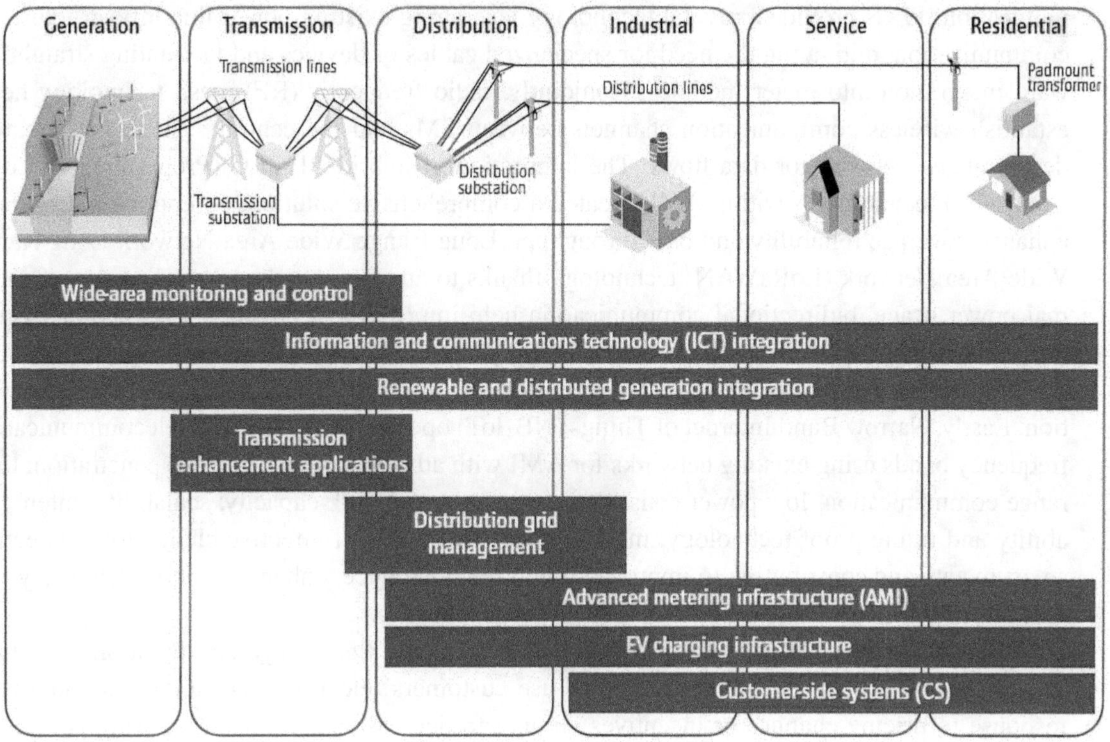

- **Advanced Metering Infrastructure (AMI):** Stemming from Automatic Meter Reading (AMR) which automates tasks such as monthly invoicing, AMR evolution progressed to Advanced Meter Management (AMM), focusing on automated processes and data-informed decision-making. AMI represents an integrated system with advanced communication and data management technologies to enhance the efficiency and reliability of electricity supply. Indeed, through shaping the AMM concept, AMI allows utilities to rapidly respond to changing consumer demand in addition to power resource integration and the growing consistency of electric vehicles. Smart Meters, integral to AMI, surpass traditional meters by their capabilities of measuring electrical parameters and tracking detailed data. AMI software enhances controls and functionalities, thanks to SM which involve diverse hardware components for prompt data processing. Customized for monitoring energy power flow, those meters contain modular hardware components and apply advanced communication technologies, including ZigBee, Wi-Fi, or GPS. Moreover, they ensure adaptability to specific applications (kpolat & Dursun, 2017). Despite the multiple challenges such as Cybersecurity risks, data privacy compliance concerns and doubts about the profitability behind initial investments in this field; continuous research and development position AMI as a transformative technology with valuable benefits such as reduced energy costs, enhanced grid reliability and energy transparency; and improved environmental sustainability. It is also to be highlighted that the integration of emerging communication technologies facilitates real-time streaming and interpretation of data, enabling quick evaluation of metering and grid-related issues for a faster grid modernization. Electronic energy meters, provided with communication modules, are crucial for the data collection from diverse nodes, and emphasize the importance of exchanging these modules due to technological progress and meter lifespan. In the context of AMI power line communication, PLCs provide a pivotal technology leveraging existing power line infrastructure for communication, mitigating the need for specialized cables or devices and facilitating straightforward integration into meter modules. Conjointly, radio frequency (RF) mesh technology helps establish wireless communication channels between SMs and the central system; and create a decentralized network for data flows. The integration of hybrid RF-PLC (Programmable Logic Controller) technologies within AMI, created a comprehensive solution, combining strengths to enhance coverage reliability and data throughput. Long Range Wide Area Network Long Range Wide Area Network (LoRaWAN) technology-thanks to attributes such as broad coverage, minimal power usage, bidirectional communication help improve cost effectiveness, support secure data transmission scalability and enhance flexibility; thus, fostering extensive low power communication between SMs and the central system for real-time energy monitoring and outage detection. Lastly, Narrow Band Internet of Things (NB-IoT) operates on authorized telecommunication frequency bands using existing networks for AMI with advantages such as deep penetration, long range communication, low power consumption, massive network capacity, scalability, interoperability and future proof technology; making it an ideal and cost-effective choice for connecting smart meters and contributing to improved reliability, resilience, enhanced energy efficiency and reduced environmental impact (Clou, 2022).

- **Demand Response (DR):** Is a strategy aimed at reducing the energy consumption especially at peak hours. It involves alterations in end-use customers' electricity consumption patterns in response to pricing changes or incentives, aiming to decrease electricity use during periods of high market prices or system reliability challenges: by fostering customer interaction and responsiveness. DR brings short term impacts on electricity markets, yielding economic benefits for both customers and utilities. It enhances power system reliability, lowers peak demand, shrinks overall plant and capital costs; and defers the need for network upgrades. In the long term, DR programs coordinate electricity use with power system operations, employing various resource types such as distributed generation dispatchable load and storage. Customers can participate in DR by curtailing energy consumption, shifting consumption to different periods or using onsite standby-generated energy with automation and control technologies-crucial for managing energy use processes and minimizing inconvenience for customers (Siano, 2014).
- **Distributed Energy Resources (DER):** Also known as on-site generation (OSG) or district decentralized energy, DER refers to the generation and storage of electricity carried out by various small devices connected to the grid or to the distribution system, collectively known as distributed energy resources. According to the (IEA, 2022), distributed energy resources (DERs) encompass small-scale energy facilities often located in close proximity to electricity consumption sites including rooftop solar panels and battery storage. The rapid expansion of the DERs is not only about to reshape the electricity generation landscape, but is also revolutionizing the dynamics of trading delivery and consumption. Consequently, DERs introduce new possibilities for power systems and can also present challenges when the grid lacks adequate preparation. Many jurisdictions are just commencing on the journey of comprehending the role of DERs in the broader energy framework, understanding their grid impacts and potential contributions to enhancing system reliability and reducing overall energy costs. In contrast, some regions have accumulated experience with de DERs, showcasing their capacity to offer valuable grid services through the efficient mix

and package of technologies, policies and regulations. DERs typically have a capacity of less than 10 mw. Examples encompass diesel or natural gas generators, microturbines, solar arrays, small wind farms and battery energy storage systems. The ownership and operation of these assets may rest with the electric utility, independent power producers or local businesses. The utility manages its operation similarly to overseeing larger central power plants providing directives for starts and stops as needed (Cummins Inc., Global Power Technology Leader, 2021). Transmission System Operators (TSOs) and Distribution System Operators (DSOs) encounter challenges that extend beyond the impact of large-scale renewable generation and the rapid integration of small-scale renewable energy facilities such as residential solar. Indeed, they have disrupted the traditional hierarchical structure of electricity transmission and distribution. Unlike the conventional grid characterized by centralization, orderliness and predictability; the current landscape reflects decentralized and dynamic grids where power flows diversely in multiple directions (Energy Storage World Forum, 2023).

- **Electric Vehicles (EVs):** The anticipated shift from internal combustion engine vehicles (ICEVs) to electric vehicles (EVs) is driven by economic analyses and the necessity to address challenges encompassing climate change, energy security and geopolitical concerns related to fossil fuels. Electric vehicles (EVs) not only offer a solution to these challenges but also hold the potential to function as distributed energy sources for the electric grid, through the concept of vehicle-to-grid (V2G) technology introduced by Kempton and Letendre in 1997. V2G envisions supply stored energy to the grid contributing to backup services such as peak power shaving, spinning reserve and voltage; and frequency regulations. As the dependence on renewable energy sources (RES) grows, EVs emerge as dynamic and responsive Energy Storage Solutions (ESS) under the framework of V2G; presenting opportunities for aggregated EV fleets to act as Virtual Power Plants (VPP). This dual role of EVs dynamic loads and emergency backups of the power grid emphasizes the need for a collaborative load, and sharing arrangement between vehicle owners and utilities. While perceived as operational, expenses remain a concern for EV ownership, their strategic integration into the smart grid promises a more sustainable and efficient energy; accelerating DR, energy storage and grid stability. However, ensuring the harmonious integration of EVs into the Smart Grid requires standardizing the V2G technology, implementing robust Cybersecurity measures, modernizing the grid and providing incentives for consumers (Mwasilu et al., 2014; Sultan et al., 2022).
- **Self-Healing Grids:** The expansion of the power system grid, coupled with a disproportionate demand for load; has prompted challenges related to the integration of renewable energy, energy efficiency, susceptibility to faults and the overall complexity of the power grid in response. The Smart Grid concept has emerged as a solution aiming to establish a secure, efficient, economical and intelligent grid. A pivotal feature of the Smart Grid is its self-healing capability, denoting the network's autonomous recovery from outages. This aspect becomes crucial for resilience and adaptability in the face of power system challenges. The integration of Smart Grids introduces complexities to the power restoration process attributed to factors like distributed generation storage and mobile loads including electric vehicles; however, this heightened complexity is offset by improved observability, facilitated by Smart Meters and Intelligent Electronic Devices (IEDs). The dynamic synergy between challenges and benefits, underscores the evolving nature of power restoration in the contemporary energy landscape. Self-healing, coupled with Smart Grids, enhances power system resilience by combining automated fault detection with intelligent monitoring and control. While self-healing grids autonomously mitigate faults using sensors, Smart Grids

leverage real-time data to enhance the overall efficiency resulting in benefits such as improved resilience, cost reduction, enhanced power quality, faster recovery times after outages, increased situational awareness and optimized resource allocation. Despite challenges, decreasing costs and evolving technologies anticipate a widespread future adoption, thus, paving the way for a more resilient reliable and sustainable energy over the course of the years and with the progress of innovative technologies (Elgenedy et al., 2015; Oualmakran et al., 2012).

- **Islanding:** Occurs when a Distributed Generator (DG) continues supplying power in the absence of external electrical grid power, a state that poses risks to unresponsive utility workers and hinders automatic device reconnection. Indeed, without a rigorous frequency control, imbalances between load and generation may lead to abnormal frequencies and voltages, thus mitigating these risks requires distributed generators to promptly detect Islanding and disconnect. Known as anti-Islanding in Smart Grids, smart islanding enhances resilience and reliability by utilizing advanced technologies for smooth transitions and optimized grid operations. Smart Islanding involves automated fault detection, intelligent control, power flow optimization and seamless reconnection. Its related benefits include improved grid resilience, reduced costs, enhanced power quality, increased reliability and support for renewable energy integration. Despite challenges, both the genuine and potential advantages including a broader adoption integration with other Smart Grid technologies and improved data analytics; drive the evolution and cost reduction of smart islanding technologies.
- **Microgrids:** The prominence of microgrids has risen due to distributed generators' advantages, offering a heightened reliability during grid outages; caused by various factors. Microgrids must exhibit intelligence and operate autonomously to ensure uninterrupted power supply. Energy management is crucial for Smart Microgrid self-governance, addressing challenges in consistent energy supply faced by distributed generators. The Distributed Energy Management System (DEMS) systematically addresses outage problems, manages demand response and makes data-based Judicious decisions for grid reconnection. Microgrids and Smart Grids, although being two distinct concepts, are interconnected for a sustainable and resilient electrical grid. Their synergy supports local resilience, broadens energy resources and facilitates the integration of renewable energy sources on a larger scale, for a distributed dynamic and efficient energy system. Smart Grid denotes the evolving electricity grid landscape as defined by European Regulators Group for Electricity and Gas (ERGEG) and European Technology Platforms (ETPS). A Smart Grid seamlessly integrates all users including generators and consumers creating an economically efficient and sustainable power system. Despite numerous benefits, it still encounters challenges like bi-directional communication issues, the possibility and efficiency of renewable energy integration, suboptimal utilization and inadequate infrastructure coordination. Digital distribution involves localized grouping of electricity generation, energy storage and loads. Microgrids-small local systems with microsources and storage-offer efficient solutions by connecting to the utility grid (grid mode) or operating independently in island mode. AC microgrids use shared AC bus linking mixed loads, while DC microgrids-connected via an AC / DC converter offer higher efficiency and stability. Hybrid AC /DC microgrids accumulate advantages through: minimizing costs and enhancing reliability of direct connections of DG (Yoldaş et al., 2017).
- **Big Data Analytics:** In the dynamic and growing field of Smart Grids, the strategic application of Big Data analytics is crucial for optimizing energy management and unlocking intelligence. The analysis of data sourced from Smart Meters, sensors and infrastructure yields valuable in-

sights; enhancing demand forecasting, enabling proactive fault detection and improving consumer energy efficiency. It also helps facilitate seamless renewable energy integration and identify real and potential Cybersecurity threats. Leveraging cloud-based storage and advanced analytics including AI facilitates such as ML; help empower well-informed decision-making in respect of electricity grid monitoring strategies. It is imperative, however, to address points such as robust data security, compliance with privacy regulations, efficient data management, standardized formats interoperability and Cybersecurity. Despite these challenges, the potential of Big Data mining, uncovering hidden correlations between internal and environmental factors is substantial; promising the development of intelligent and automated grids, the empowerment of consumers and the promotion of a cleaner sustainable energy in the future. The integration of Big Data in Smart Grids encounters various challenges, including the management of large volumes of data requiring extensive storage and computing resources. Issues related to data quality may result in erroneous decisions in addressing security concerns that are paramount to safeguard consumer privacy. Moreover, financial transactions and time synchronization challenges can be mitigated by synchronizing devices based on the same reference clocks. Employing data indexing techniques like Rtrees, B-trees, and Quadtrees can reduce computational complexity and processing time. Additionally, evaluating the technical and economic value of Big Data for key stakeholders, such as consumers, system operators and utilities is crucial for widespread adoption. Regulatory entities should establish guidelines for data sharing and exchange and standards must ensure technical compliance with regulatory aspects, thus facilitating the integration of Big Data technologies in Smart Grids (Bhattarai et al., 2019).

- **Renewable Energy Integration:** The substantial integration of renewable energy sources (RES) such as wind and photovoltaic (PV) solar energy into the power system has notably increased due to the trend towards Net-Zero emissions and Greener energy resources, alternative to gas, coal and fossil fuel. These RES possess intermittent characteristics posing challenges for accurate forecasting. The heightened consideration of RES into the power market aligns with stringent energy policies and addresses energy security concerns (Mwasilu et al., 2014). To accommodate this current and growing significant integration of RES into the power system, extensive Energy Storage Systems (ESS) must be deployed. These systems ensure the support of electric grids meeting electrical power demand and consistently maintaining the operational standards (IEC, 2012).
- **Smart Buildings:** A smart building integrates various building-wise systems such as Heating, Ventilation, and Air Conditioning (HVAC), lighting alarms and security into a unified managed network infrastructure often achieved through foundational technologies like power over Ethernet (PoE) (CISCO, 2023). The collaboration between smart buildings and Smart Grids involves real-time energy usage adjustments for peak demand reduction and stability enhancement. Those buildings equipped with energy storage, contribute to the stabilization of grids by storing and releasing surplus energy in real-time data exchange, thus, optimizing energy usage and grid operations. Moreover, microgrids enable independent operation during the main grid outages. Challenges include interoperability, Cybersecurity costs and consumer awareness; therefore, the collaboration offers widespread adoption potential for a more intelligent and efficient energy ecosystem reinforcing grid resilience and promoting decentralized energy generation. The transformative shift towards Smart Grids and smart buildings holds significant potential for a greener, more efficient power system, improved air quality and reduced greenhouse gas emissions (Al Abri et al., 2021). (Wurtz and Delinchant, 2017) emphasized the integration of smart buildings with Smart Grids for

the purpose of mitigating dependence on fossil fuels and amplifying the use of renewable energy sources within climate-related constraints. Indeed, buildings, as substantial energy consumers, can contribute significantly to renewable energy generation, particularly through the adoption of the energy-positive building concepts. Energy-positive buildings manage energy demand and generate surplus renewable energy which makes them an active factor in the broader energy transition framework. To maximize the benefits of Smart Grid building, owners must align their internal infrastructure with Smart Grid capabilities and further key elements such as grid connection, the feed in the comprehensive control information, the separation of metering, and the access of data from controllable components (Van Bruggen, 2012).

- **Advanced Power Quality Management:** In the evolving energy landscape, ensuring reliable power delivery is a priority; and the collaboration between advanced power quality management (PQM) and Smart Grids proves to be a mutually beneficial alliance. Advanced (PQM) actively monitors and enhances power quality offering advantages such as improved equipment reliability and heightened grid resilience. Smart Grids leverage advanced technology for intelligent monitoring and energy flow optimization, leading to enhanced reliability efficiency and enriched data insights. The synergy between advanced (PQM) and Smart Grids facilitates. Real-time data exchange improved stability and empowered consumers overcoming challenges like Cybersecurity and interoperability. This integrated approach holds the promise of an improved power quality and a more sustainable energy future to address frequent power quality issues. The incorporation of Smart Grid technology becomes imperative thanks to its monitoring communication and digital computing technologies, enhancing the power delivery systems' efficiency. Various power quality problems such as voltage variation, frequency variation voltage unbalance, current unbalance waveform distortion harmonics interharmonics notching noise, and power delivery system transients can be categorized and mitigated through Smart Grid technologies. Systems like Volt/Var management (VVM), optimize voltage levels, improve phase balancing and reduce peak demand. Distribution Automation (DA) systems provide monitoring control; fault location, isolation and restoration (FLIR), real-time power flow analysis and auto transfer functions. Advanced metering infrastructure (AMI) enables the two-way communication, real-time measurement, outage detection and monitoring of power quality parameters. Integrating AMI data into a power quality analyzer (PQA) allows for a comprehensive investigation of disturbances, while Smart Grid technologies effectively address various power quality issues and unavoidable challenges, such as voltage sags which necessitate additional solutions like true on-line UPS. Noise problems can be tackled using filters, isolation transformers and line conditioners (Boloorchi, 2019).
- **Fault Location, Isolation, and Service Restoration (FLISR):** FLISR comprises three functions which are: (1) sectioning and restoration (2) automatic circuit reconfiguration and (3) distributed automation (DA) operations. These applications work together to achieve the DA goals by coordinating the actions of field devices, software and dedicated communication networks. FLISR plays a role in the dynamics of grids as it greatly influences their resilience and reliability, by utilizing sensors and intelligent algorithms to detect faults in time accurately, pinpoint their location promptly, isolate affected areas and efficiently restore service. This innovative technology brings advantages such as improved grid reliability, reduced operational costs, enhanced power quality, faster service restoration and increased awareness of the situation (U.S. Department of Energy 2014). To harvest these benefits, FLISR easily integrates with DMS, utilizes Intelligent Electronic Devices (IEDs) throughout the grid, establishes a communication infrastructure and applies ad-

vanced software algorithms. Although there are challenges involving investments, Cybersecurity risks, complexities, data management and the requirement for smooth integration; the significant potential advantages of FLISR still include widespread adoption thanks to the increased automation, improved self-healing capabilities and enhanced integration, with renewable energy sources (Spalding et al., 2016).

- **Grid Automation and Control:** Smart Grids represent progress in the field of electric power systems as they blend information technologies in a fluid way.AI-enabled systems are efficient in the real time-monitoring of DR thanks to the AI-related accuracy and capabilities. The DR integration and energy consumption control is a crucial aspect of Smart Grid, allowing users to adjust their energy usage; and ultimately leading to reduced operational costs and delayed capacity additions. This comprehensive study by (Meliani et al., 2020) delves into the various concepts, advantages and components of SGs with a focus on DR and its associated control systems from a global perspective. It highlights the significance of DR as a key feature of SG-driven power systems, while also addressing recent challenges such as heuristic optimization and scheduling structures. The evolution of the Smart Grid has opened avenues to enhance efficiency within electric powered energy grids and microgrids. SCADA structures mounted inside the enterprise provide an attractive answer for remotely monitoring Renewable Energy Resources (RES) and improving gadget performance. However, challenges persist in ensuring the reliability, safety and safety of SCADA systems in the electricity era. Transmission and distribution integrating SCADA structures into the Smart Grid's layout support the remote control of electrical microgrids and grids, contributing to expanded software reliability and efficiency. (Sayed and Gabbar, 2017) highlighted the cutting-edge device structure security measures and the importance of simplifying requirements, tracking the unpredictable traits of RES, such as wind generators and solar PV arrays. It is vital for green power utilization SCADA structures to accumulate and utilize statistics from these sources optimizing and enhancing the performance of wind farms, underscoring the potential of the clever grid.
- **Asset Management (AM):** In the framework of Smart Grids, Asset Management helps track, assess and optimize grid components utilizing technologies such as Smart Meters and data analytics. AM contributes to improved reliability, cost reduction, enhanced safety; and facilitates data-driven decision-making. Despite encountering challenges, the adoption of these practices offers benefits such as large-scale acceptance of advanced maintenance strategies, the development of self healing grids and improved decision-making processes, contributing to a resilient and secure energy. Future asset management holds particular significance in Smart Grids, addressing issues like aging assets, uncertain load demand and the integration of renewable energy sources. It encompasses systematic processes for maintaining, operating and upgrading assets to minimize costs and risks; while ensuring optimal electrical grid performance. Within this field, the Smart Grid introduces new technologies and adds a sophisticated dimension. Therefore, continuous improvements in equipment and techniques are required. Indeed, challenges, opportunities, services and tools available in Smart Grid asset management are being and should be further emphasized, considering the importance of continuous improvement to meet evolving standards and expectations (Jung et al., 2019).
- **Smart Grid Security and Resilience:** Smart Grids characterized by its advanced technology and interconnected design; underscore the key significance of security and resilience through establishing uninterrupted electricity flow and mitigating their associated threats. Grid security defends

against both cyber and physical threats and helps safeguard data privacy. While grid resilience focuses on efficient recovery from disturbances and outages, the synergy between improved security and resilience generates enhanced reliability, safety, cost reduction and sustainability; and fosters the establishment of public trust. Despite challenges, the potential benefits encompass innovative solutions and AI-powered security measures, the development of smarter grids and collaborative efforts toward a secure energy future. Prioritizing both security and resilience unlocks the hidden potential of data for a reliable, efficient and sustainable energy future, especially as power systems. Transitions into Smart Grids anticipate increased renewable energy sources (RES) and power electronics and contribute to low-carbon power systems, a heightened penetration of RES, a rise in security margins and resilience during extreme events. Moreover, the integration of information and communication technologies (ICT) introduces vulnerabilities and security risks. This topic emphasizes the urgent need for practical methods to enhance power system security and resilience addressing challenges associated with high RES penetration risk and the deep (ICT) integration. (Liu et al., 2022) focused on the resilient control of grid-tied inverters stability through the analysis of voltage source converters, the resilience of synchrophasor communication, and the network security power systems.

- **Battery Monitoring Systems (BMS):** Ensure the secure and efficient operation of large-scale energy storage systems. These systems function as vigilant overseers, continuously monitoring crucial battery parameters to enhance overall performance, grid stability, and safety and informed decision-making. Key components of a BMS include sensor data acquisition mechanisms-a central processing unit (CPU) communication interfaces and dedicated software. Despite facing challenges related to standardization, Cybersecurity and data integration, embracing BMS presents considerable potential benefits such as the widespread adoption and implementation of AI-driven analytics, the realization of self-healing battery capabilities and the improved integration within the Grid framework. The incorporation of BMS into energy storage infrastructure unlocks the latent potential for a cleaner, more resilient and sustainable energy future. The escalating demand for energy driven by ongoing human development requires alternative solutions such as renewable energy resources paired with battery storage systems, to meet the surging load demand. To address this challenge effectively, precise and stable data acquisition from batteries is indispensable; necessitating a robust device capable of providing accurate information while being compact for portability to remote locations; and equipped for instant data logging in removable storage. This data is subsequently analyzed to assess critical battery parameters including capacity, Energy State of Charge (SOC) and discharge characteristics which are pivotal indicators of battery health. Practical applications of this approach include battery testing in electric vehicles, remote solar power grids and research sites where users can log data using a portable battery-monitoring device onto an SD card for subsequent analysis. The proposed system introduces a unique user input method allowing users to input current and cut off voltage parameters in a .txt file on the micro SD card (Saeed et al., 2017) provided a comprehensive account of the design and development of a portable microcontroller based device adept at recording battery parameters such as voltage current and time while drawing a user inputted constant current from the battery until it reaches a user defined cut off voltage.
- **Energy Efficiency and Storage:** Smart Grids related concepts stand behind the core purpose of achieving energy efficiency, gathering underlying principles which cover waste minimization, grid performance optimization, renewable energies integration and both technical and compliance

industrial frameworks. The importance of this technology resides in coping with the increasing demand for energy, and climate change; through the development and implementation of adequate solutions adapted to the current market challenges and turbulences. Smart Grids, equipped with advanced monitoring and control, synergize with efficiency and storage, thus yielding benefits such as reduced energy consumption, heightened reliability, increased RES integration, and empowered consumers. Addressing and overcoming challenges is necessary for widespread adoption. As technology progresses, costs show decreasing trends. While policies evolve triggering greater accessibility, automation, consumer involvement, and grid-integrated buildings can be expected. Adopting this cooperative synergetic approach offers a future in energy with is cleaner, more dependable and sustainable aspect. In parallel, Energy Storage Technologies (EST), including electrochemical batteries, ultra-capacitors and kinetic energy storage systems, are leading the integration of renewable energy sources into modern power systems and Smart Grids. They enhance system flexibility, thus, enabling the coherent integration of intermittent RES and providing further services. (Ozdemir et al., 2016; Dedes et al., 2021) presented an overview highlighting diverse EST applications, exploring opportunities and challenges in the integration of RES into Smart Grids. They demonstrated mechanical, electrochemical, and thermal energy storage systems, emphasizing the widespread adoption of cost-effective battery energy storage. They also investigated alternative ES technologies like hydro, pump hydroelectric, and thermal energy storage, encompassing latent, sensible, and thermochemical storage systems.

- **Grid Balancing Technologies:** It is vital for Smart Grids to balance energy supply and demand for a reliable electricity system. Technologies like demand-side and supply-side management, grid automation and distributed energy resources help manage variations in electricity needs. Demand-side management encourages consumers to adjust energy use while supply side management adapts energy production. Grid automation monitors and controls the grid in real-time and helps distributed energy resources integrate small-scale energy sources. These technologies bring benefits like better reliability, lower energy costs, increased use of renewable energy, positive environmental impact and a more resilient grid. Despite challenges such as costs and data security, the potential benefits encourage widespread adoption of advanced automation, a flexible grid and empowered consumers ensuring a stable, reliable and sustainable Smart Grid for clean and affordable energy.
- **Predictive Maintenance:** Smart Grids provide a proactive and cost-effective approach to ensure grid reliability and minimize operational costs by leveraging data analytics and advanced technologies. It actively collects real-time data from sensors, Smart Meters and intelligent devices, and processes them with AI methodologies to predict potential failures. This predictive approach yields various benefits including reduced downtime, prolonged asset lifespan, optimized maintenance resource allocation, improved grid reliability and cost reduction. The key technologies involved encompass sensors, Smart Meters, data analytics, Cloud Computing and mobile applications, however challenges exist as related to Cybersecurity investment, and both costs and model accuracy. Despite these obstacles, the potential benefits drive widespread integration with other Smart Grid technologies, enhanced model accuracy and increased standardization of data formats; paving the way for a more efficient, reliable and sustainable energy future. Across diverse sectors such as utilities, power companies, energy service providers and transmission companies; the adoption of data analytics and predictive maintenance is transforming Smart Grid management. By proactively detecting potential issues and optimizing asset utilization through data analysis

entities in the energy; industry is achieving tangible advantages including reduced maintenance costs, improved system reliability and heightened safety through the identification and mitigation of potential hazards. The active utilization of data analytics and predictive maintenance underscores a significant shift in enhancing the overall performance of Smart Grids within the energy industry (Günther, 2023).

- **Cybersecurity:** Addressing Cybersecurity challenges is vital in the era of Industry 4.0, particularly with the increasing interconnection of Smart Grids to the internet. The potential cascading effects of a cyberattack on a Smart Grid can have severe repercussions for the stability of interconnected grid infrastructures. (Ghayoor et al., 2022) emphasized the need to tackle security concerns before the widespread implementation of IoT-based devices in large-scale energy systems. (Goudarzi et al., 2022) classified the damages induced by Cybersecurity into operational economic and system security categories. Despite the deployment of technologies such as AI Multifactor Authentication Systems (MFA) and Blockchain for the purpose of countering; cyber threats challenges persist. AI plays a pivotal role in preventing Cybersecurity attacks in Smart Grids through anomaly detection, behavioral analysis, intrusion detection systems and incident response automation. It ensures secure communication within Smart Grids by implementing robust cryptographic measures; facilitating adaptive security measures through ML; and conducting vulnerability assessments to detect and remediate potential breaches. MFA encompassing knowledge possession and inherence elements adds an additional layer of defense in Smart Grids by minimizing vulnerability and unauthorized access; however, challenges like user adoption inconvenience, interoperability complexities and associated costs need consideration. Blockchain emerges as a crucial Cybersecurity strategy providing a secure and tamper proof system for data exchange and transactions in Smart Grids. Indeed, it encrypts and stores information on a distributed ledger, enhancing authentication, controlling access, securing meter to meter communication and supporting transparent peer to peer energy trading through smart contracts. In summary, AI, MFA and Blockchain are integral to fortifying Smart Grids, protecting critical infrastructure, ensuring data integrity and promoting resilience and efficiency in the Smart Grid ecosystem.

CONCLUSION

Smart Grids are characterized by the coordinated utilization of distributed computer resources for the purpose of reaching common objectives by managing electricity power user-supply, injecting electricity generated from renewable resources in the grid, harnessing innovative technologies for the grid optimization and continuously identifying and efficiently tackling the upcoming challenges. The smart Grid Computing allocates distinct tasks to each node, allowing for increased heterogeneity and geographical dispersion. Smart Grids can fulfill wide-ranging purposes and exhibit diverse sizes contingent on their intended applications such as cities and buildings, smart houses, electric vehicles, factories and data centers; power plants like nuclear, solar, thermal and wind-powered ones.

Smart Grids offer three main solutions driving this cutting-edge paradigm. Efficiency concept leveraging which resides in greenhouse emissions and CO_2 footprint mitigation; harnessing real-time bidirectional power data to foster energy efficiency-i.e. the required energy amount to achieve services-, and consequently reducing environmental impacts, cost-effectiveness and business competitiveness. Smart Grids combine renewable energy and energy efficiency towards sustainable energy supply and management. Reliability is proven to be enhanced by Smart Grids which showed a smaller frequency of failure, hence the absence of necessity to add any reserve power to the grid systems (Cepin, 2019). Grid resilience is achieved thanks to anomaly detection and Cybersecurity robustness is mandatory to reach Smart Grids' targets. Smart Grids, through grid balancing, reduce downtime; allows remote monitoring and Fault Location, Isolation, and Service Restoration (FLISR).The trajectory of smart Grids unfolds a future highlighted by a cleaner, more resilient, and efficient energy system, steered by key technological trends. Artificial intelligence, Machine Learning and advanced sensors. Data analytics will optimize grid performance, predict failures and augment automation. Standardization efforts in data formats and communication protocols are poised to enhance interoperability among grid components and facilitate accurate data analysis for decision-making. The integration of renewable energies, combined with distributed energy resources (DERs) and virtual power plants (VPPs), are prone to assuming larger roles. Advanced Metering Infrastructure (AMI) integrated Smart Meters, communication networks and data management systems allow the bidirectional communication between users' data and grids. Consumer empowerment through smart home technologies, participation in energy markets and incentives such as lower bills and better energy supply quality at the peak hours; fosters Smart Grids efficiency.

With the growing energy demand, the transformation of traditional electrical grids into "Smart Grids'' becomes indispensable. Smart Grids, even though holding substantial promise, still encounter notable challenges that can hinder their adoption at a larger scale. Technological obstacles encompass interoperability issues occurring due to the absence of standardization, Cybersecurity threats and the complexities of data management including the possibility of delivering wrong information such as true positives and false negatives whose respective effects depend on the data analysis purpose and decision-making scope behind. Economic barriers include high initial costs, and the necessity to demonstrate long-term economic sustainability driven by a fair and dynamic pricing model. On the regulatory side, supportive policies, data privacy regulations and the establishment of industry-wide standards are mandatory to involve consumers in responsible electricity handling and in creating a reliable framework that ensures data protection, reliability and accuracy. Furthermore, social behavior is very important as raising awareness throughout communities and fostering cooperation for responsible power usage. Cybersecurity-oriented mindset is a further important pillar where user engagement is crucial. To overcome these obstacles and fully harness Smart Grid potential, cooperation amongst a variety of stakeholders is essential to ensuring a sustainable energy future. Enhanced grid resilience will be achieved through advanced monitoring, microgrids for backup power and predictive maintenance. The lead on policy support should emerge from governments and establish rules such as data privacy safeguards and international standardization initiatives. It is critical to go beyond setbacks from large investments and raise public awareness perspectives. Despite these obstacles, Smart Grid technology seems to have a bright future ahead with a focus on consumer empowerment; renewable integration, technological advancements and resilient grid development for safe and secure energy sources.AI algorithm robustness and accuracy in addition to Cybersecurityissues related to the online aspect of Smart Grids are situated at the core challenges. Furthermore, the integration of renewable energy resources remains challenging due to the volatility and abrupt behavior of their inputs to electrical grids. Scalability consists in the efficiency of AI algorithms to handle the real-time and bidirectional demand-supply data flow. Explainable AI (XAI) is important to understand the exact functioning of the implemented algorithms and risks related to it to avoid potential harms and failures. Furthermore, the interaction between AI and user has to be thoroughly clarified with a comprehensive and rigorous explanation of the responsibilities of every side. The future of AI-empowered Smart Grids predicts a progress, parallel to IoT technology progress. Indeed, the integration of Cloud

Computing is prone to enhancing security, algorithm robustness (reflecting their ability to perform efficiently notwithstanding the input errors) and mitigating system failures. Fog Computing, thanks to its ability to process data locally before transmitting it to the Cloud, showed several advantages such as data scalability, flexibility and energy efficiency. Moreover, transfer learning ML technology-which harnesses pre-trained models to solve problems for an improved learning performance-is applied in case of lack of data or labels, the latter being a major issue of Smart Grid data (Omitaom and Niu, 2021).

Leveraging 5G technology, smart grids are advancing to new levels of intelligence. This evolution involves two-way communication for rapid response and digitalization for efficient operation. 5G technology stands out for addressing the critical need for reliable, high-bandwidth communication, 5G technology; thanks to their speed, flexibility and wide-area coverage which fosters a prompt the real-time monitoring of diverse power sources and field devices, and propels the Smart Grid to be even smarter and more efficient (Ghosh, 2021). Furthermore, Quantum Computing is prone to offering enhanced computation ability and coping with scalability issues, thanks to their leveraged data processing abilities, prediction accuracy and quantum encryption, bringing more data security and reliability. In respect of Edge Computing technology, it is still a young and immature concept with no universal definition, however, this technology intercepts the Smart Grids on pivotal functions (4C: computation, communication, caching, and control). Cloud Computing helps improve resilience, optimization and accuracy of simulation (Fang et al., 2016; Feng et al., 2021). Hybrid Cloud-Edge analytics involves a combination of Cloud and Edge Computing attributes allowing robust data pre-processing and transmission. This infrastructure allows the execution of advanced analytics. This combined technology secures a harmonious balance between response time minimization and optimized results.

REFERENCES

Akpolat, A., & Dursun, E. (2017). *Advanced Metering Infrastructure (AMI)*. Smart Meters and New Technologies.

Al Abri, D., Malik, A., Al-Saadi, S., Albadi, M., Charabi, Y., & Hosseinzadeh, N. (2021). Smart Grids and Smart Buildings. In Handbook of Climate Change Mitigation and Adaptation (pp. 1–20). Springer. doi:10.1007/978-1-4614-6431-0_78-2

Anderson, R., Ghafurian, R., & Gharavi, H. (2018). *Smart Grid: The Future of the Electric Energy System.*

Bhattarai, B. P., Paudyal, S., Luo, Y., Mohanpurkar, M., Cheung, K., Tonkoski, R., Hovsapian, R., Myers, K. S., Zhang, R., Zhao, P., Manic, M., Zhang, S., & Zhang, X. (2019). Big data analytics in smart grids: State-of-the-art, challenges, opportunities, and future directions. *IET Smart Grid*, *2*(2), 141–154. doi:10.1049/iet-stg.2018.0261

Cepin, M. (2019). Reliability of Smart Grids. In *2019 International Conference on Information and Digital Technologies (IDT)* (pp. 1-6). IEEE. DOI: 10.1109/DT.2019.8813387

Dedes, I. C., Tsampasis, E., Elias, C., & Gkonis, P. (2021). Energy Storage in Smart Electrical Grids. In *2021 10th Mediterranean Conference on Embedded Computing (MECO)* (pp. 1–4). IEEE. 10.1109/MECO52532.2021.9460169

Ehsani, M., Falahi, M., & Lotfifard, S. (2012). Vehicle to grid services: Potential and applications. *Energies, 5*(10), 4076–4090. doi:10.3390/en5104076

Elgenedy, M. A., Massoud, A. M., & Ahmed, S. (2015). Smart grid self-healing: Functions, applications, and developments. In *2015 First Workshop on Smart Grid and Renewable Energy (SGRE)* (pp. 1–6). IEEE. 10.1109/SGRE.2015.7208737

Fang, B., Yin, X., Tan, Y., Li, C., Gao, Y., Cao, Y., & Li, J. (2016). The contributions of cloud technologies to smart grid. *Renewable & Sustainable Energy Reviews, 59*, 1326–1331. doi:10.1016/j.rser.2016.01.032

Feng, C., Wang, Y., Chen, Q., Ding, Y., Strbac, G., & Kang, C. (2021). Smart grid encounters edge computing: Opportunities and applications. *Advances in Applied Energy, 1*, 100006. doi:10.1016/j.adapen.2020.100006

Geneiatakis, D., Kounelis, I., Neisse, R., Nai-Fovino, I., Steri, G., & Baldini, G. (2017). Security and privacy issues for an IoT based smarthome. In *Proceedings of the 2017 40th International Convention on Information and Communication Technology, Electronics and Microelectronics (MIPRO)* (pp. 1292–1297). IEEE.

Ghayoor, F., Fahad, S., Traore, I., Waseem, M., & Goudarzi, A. (2022). A Survey on IoT-Enabled Smart Grids: Emerging, Applications, Challenges, and Outlook. *Energies, 15*(19), 6984. doi:10.3390/en15196984

Hamdaoui, Y., & Maach, A. (2016). *Smart islanding in smart grids. In 2016 IEEE Smart Energy Grid Engineering*. SEGE., doi:10.1109/SEGE.2016.7589521

Hui, H., Ding, Y., Shi, Q., Li, F., Song, Y., & Yan, J. (2020). 5G network-based Internet of Things for demand response in smart grid: A survey on application potential. *Applied Energy, 257*, 113972. doi:10.1016/j.apenergy.2019.113972

Jiao, J. (2020). Title of the Paper. *IOP Conference Series. Earth and Environmental Science, 510*(2), 022012. doi:10.1088/1755-1315/510/2/022012

Joy, J., Jasmin, D. E. A., & John, V. R. (2013). Challenges of Smart Grid. International Journal of Advanced Research in Electrical. *Electronics and Instrumentation Engineering, 2*, 976–981.

Jung, C. M., Ray, P., & Salkuti, S. R. (2019). Asset management and maintenance: A smart grid perspective. [IJECE]. *Iranian Journal of Electrical and Computer Engineering, 9*(5), 3391. doi:10.11591/ijece.v9i5.pp3391-3398

Kempton, W., & Letendre, S. (1997). Electric vehicles as a new power source for electric utilities. *Transportation Research Part D, Transport and Environment, 2*(3), 157–175. doi:10.1016/S1361-9209(97)00001-1

Meliani, M., el Barkany, A., el Abbassi, I., Darcherif, A. M., & Mahmoudi, M. (2020). Control system in the smart grid: State of the art and opportunities. In *2020 IEEE 13th International Colloquium of Logistics and Supply Chain Management (LOGISTIQUA)* (pp. 1–6). IEEE. 10.1109/LOGISTIQUA49782.2020.9353878

Moreno Escobar, J. J., Morales Matamoros, O., Tejeida Padilla, R., Lina Reyes, I., & Quintana Espinosa, H. (2021). A Comprehensive Review on Smart Grids: Challenges and Opportunities. *Sensors (Basel), 21*(21), 6978. doi:10.3390/s21216978 PMID:34770285

Mwasilu, F., Justo, J. J., Kim, E.-K., Do, T. D., & Jung, J.-W. (2014). Electric vehicles and smart grid interaction: A review on vehicle to grid and renewable energy sources integration. *Renewable & Sustainable Energy Reviews*, *34*, 501–516. doi:10.1016/j.rser.2014.03.031

Omitaomu, O. A., & Niu, H. (2021). Artificial Intelligence Techniques in Smart Grid: A Survey. *Smart Cities*, *4*(2), 548–568. doi:10.3390/smartcities4020029

Oualmakran, Y., Melendez, J., & Herraiz, S. (2012). *Self-healing for smart grids: Problem formulation and considerations. In 2012 3rd IEEE PES Innovative Smart Grid Technologies Europe (ISGT Europe).* IEEE., doi:10.1109/ISGTEurope.2012.6465856

Ozdemir, E., Ozdemir, S., Erhan, K., & Aktas, A. (2016). Energy storage technologies opportunities and challenges in smart grids. In *2016 International Smart Grid Workshop and Certificate Program (ISGWCP)* (pp. 1–6). IEEE. 10.1109/ISGWCP.2016.7548263

Saeed, S., Shezan, S. K. A., Arbab, M. I., & Rana, S. (2017). Battery monitoring system for the smart grid applications. In *2017 Australasian Universities Power Engineering Conference (AUPEC)* (pp. 1–5). IEEE. 10.1109/AUPEC.2017.8282490

Savazzi, S., Nicoli, M., & Rampa, V. (2020). Federated learning with cooperating devices: A consensus approach for massive IoT networks. *IEEE Internet of Things Journal*, *7*(5), 4641–4654. doi:10.1109/JIOT.2020.2964162

Sayed, K., Gabbar, H. A., & Krishnamoorthy, S. (2017). SCADA and smart energy grid control automation. In *Smart Energy Grid Engineering* (pp. 481–514). Academic Press. doi:10.1016/B978-0-12-805343-0.00018-8

Siano, P. (2014). Demand response and smart grids—A survey. *Renewable & Sustainable Energy Reviews*, *30*, 461–478. doi:10.1016/j.rser.2013.10.022

Siddiquee, S. M. S., Howard, B., O'Sullivan, D. T. J., & Bruton, K. (2020). Demand Response in Smart Grid – A Systematic Mapping Study. In 2020 2nd International Conference on Smart Power & Internet Energy Systems (SPIES) (pp. 1-6). IEEE. 10.1109/SPIES48661.2020.9243107

Spalding, R. A. (2016). Fault Location, Isolation and service restoration (FLISR) functionalities tests in a Smart Grids laboratory for evaluation of the quality of service. In *2016 17th International Conference on Harmonics and Quality of Power (ICHQP)* (pp. 879–884). IEEE. 10.1109/ICHQP.2016.7783370

Sultan, V., Aryal, A., Chang, H., & Kral, J. (2022). Integration of EVs into the smart grid: A systematic literature review. *Energy Informatics*, *5*(1), 65. doi:10.1186/s42162-022-00251-2

Vasirani, M., Kota, R., Cavalcante, R. L. G., Ossowski, S., & Jennings, N. R. (2013). An agent-based approach to virtual power plants of wind power generators and electric vehicles. *IEEE Transactions on Smart Grid*, *4*(3), 1314–1322. doi:10.1109/TSG.2013.2259270

Wurtz, F., & Delinchant, B. (2017). "Smart buildings" integrated in "smart grids": A key challenge for the energy transition by using physical models and optimization with a "human-in-the-loop" approach. *Comptes Rendus Physique*, *18*(7–8), 428–444. doi:10.1016/j.crhy.2017.09.007

Yoldaş, Y., Önen, A., Muyeen, S. M., Vasilakos, A. V., & Alan, İ. (2017). Enhancing smart grid with microgrids: Challenges and opportunities. *Renewable & Sustainable Energy Reviews*, *72*, 205–214. doi:10.1016/j.rser.2017.01.064

Yu, X., & Xue, Y. (2016). Smart Grids: A Cyber–Physical Systems Perspective. *Proceedings of the IEEE*, *104*(5), 1058–1070. doi:10.1109/JPROC.2015.2503119

ADDITIONAL REFERENCES

Boloorchi, A., & Sharma, V. (2019, June 10). *Smart grid and power quality management.* IEEE Smart-Grid Bulletin.

CISCO. (2023, March 28). *What is a smart building?* Cisco.

Clou. (2022, January 04). *Advanced metering infrastructure and smart grid.* Smart-Energy.com.

Cummins Inc. (2021, November 04). What are distributed energy resources and how do they work? Cummins.

Energy Storage World Forum. (2023, July 12). *Can energy storage systems reduce investment expenditure?* EnergyStorageForum.com.

European Commission (EC). (2023, May 10). *Smart grids and meters.* EC.

Ghosh, S. K. (2021, April 22). *Making smart grids more smart with 5g communication.* IEEE SmartGrid Bulletin.

Günther, M. (2023, March 31). *How data analytics and predictive maintenance are boosting efficiency in smart grids.* ClouGlobal.com.

IEA. (2022). *Unlocking the Potential of Distributed Energy Resources.* IEA. https://www.iea.org/reports/unlocking-the-potential-of-distributed-energy-resources

International Electrotechnical Commission (IEC). (2012, July 17). *Grid integration of large-capacity renewable energy sources and use of large-capacity electrical energy storage.* (White paper, 3). IEC.

International Energy Agency (IEA). (2023, May 04). *Unlocking the Potential of Distributed Energy Resources.* IEA.

Liu, C., Chen, C., Xie, Z., & Zhu, Y. (2021). Evaluating the Cybersecurityof smart grid communication systems: A systematic review. *Frontiers in Communication*, *6*, 689590. doi:10.3389/frcmn.2021.689590

Liu, C., Li, B., Li, M., & Li, C. (2022). The impact of distributed energy resources on distribution systems: A review. *Frontiers in Energy*, *9*, 950523. doi:10.3389/fenrg.2022.950523

Routray, P. (2021, August 09). *ICT for measurement, control, and monitoring in smart grids.* IEEE SmartGrid Bulletin.

UNECE. (2015). *Smart grids overview.* UNECE.

U.S. Department of Energy. (2014, December 18). *B5 draft report (PDF)*. SmartGrid.gov.

Van Bruggen, H. (2012, June 28). *Smart buildings for smart grids*. Europea. build-up.ec.europa.eu

Chapter 14
Cloud Computing Adoption for Small and Medium Enterprises in Engineering and Environmental Aspects

S. Abirami
Department of Artificial Intelligence and Data Science, Jai Shriram Engineering College, Tirupur, India

Santosh K. C.
https://orcid.org/0000-0001-5012-8449
Department of Computer Science and Engineering, Bapuji Institute of Engineering and Technology, Davangere, India

R. Somasundaram
https://orcid.org/0000-0002-0377-9094
Department of Management Studies, Kongu Engineering College, Erode, India

Kavitha K. S.
Department of Computer Science and Engineering, Dayananda Sagar College of Engineering, Bangalore, India

Harshita Gangadhar Patil
Department of Computer Science and Engineering, S.B. Jain Institute of Technology, Management, and Research, Nagpur, India

Sureshkumar Myilsamy
Bannari Amman Institute of Technology, India

ABSTRACT

Cloud computing is revolutionizing small and medium enterprises (SMEs) by offering cost savings, scalability, collaboration, and security. However, it also presents challenges like security concerns, regulatory compliance, and migration complexities. The chapter explores the adoption of cloud technologies in engineering domains like CAD/CAM, simulation, project management, and remote monitoring, highlighting their role in enhancing efficiency and innovation. It also discusses the environmental impact of cloud computing, highlighting its potential to reduce carbon emissions through energy-efficient data centers and sustainable practices. The chapter provides case studies and best practices for SMEs to embrace cloud computing while navigating environmental responsibilities and identifies future trends in integrating cloud computing with emerging technologies and sustainable infrastructure.

DOI: 10.4018/979-8-3693-1794-5.ch014

INTRODUCTION

Cloud computing is revolutionizing the engineering sector by providing SMEs with access to powerful computing resources and advanced technologies. This chapter explores the transformative potential of cloud computing for engineering SMEs, focusing on integrating environmental concepts into operations. Cloud-based solutions can streamline processes, enhance collaboration, and achieve efficiency while contributing to environmental sustainability. Cloud services offer scalability, flexibility, and cost-effectiveness, making them attractive for SMEs competing in a rapidly evolving marketplace. This scalability is especially beneficial for engineering SMEs, which often face fluctuating demands and project requirements(Khayer et al., 2021).

Cloud computing helps SMEs overcome barriers like high upfront costs, complex maintenance requirements, and limited scalability in traditional IT infrastructure. By outsourcing IT to cloud service providers, SMEs can focus on core business activities while relying on experts for resource management. This shift from capital-intensive infrastructure to pay-as-you-go cloud services has democratized access to advanced technologies, leveling the playing field for SMEs. In the engineering sector, cloud computing offers benefits like cloud-based software tools for CAD, simulation, and project management, enabling better collaboration with clients and partners. Centralizing data and applications in the cloud improves productivity and remote work(Vajjhala & Ramollari, 2021).

The engineering industry is increasingly adopting cloud computing to promote environmental sustainability. Cloud service providers are investing in energy-efficient data centers and renewable energy sources, reducing the carbon footprint associated with computing infrastructure. This can help engineering SMEs reduce energy consumption, minimize hardware waste, and lower their environmental impact. However, SMEs face challenges such as security concerns, data privacy issues, and regulatory compliance requirements. Migrating existing systems and data to the cloud can be complex and time-consuming, requiring careful planning and execution(Modisane & Jokonya, 2021).

This chapter delves into the adoption of cloud computing for engineering SMEs, discussing its advantages, challenges, and best practices. It provides real-world case studies and examples of successful cloud technology use to enhance operations and achieve environmental sustainability goals. The chapter also discusses emerging trends and future opportunities in cloud computing, highlighting the potential for innovation and growth in the engineering sector. By embracing cloud-driven solutions, engineering SMEs can thrive in a competitive market and contribute to a sustainable future(Liu et al., 2021).

Cloud computing is a shift in computing resource provisioning, access, and management. It offers computing services like servers, storage, databases, networking, software, and analytics over the internet on a pay-as-you-go basis. This model eliminates the need for costly on-premises infrastructure and allows organizations to leverage shared resources from cloud service providers. Cloud computing's scalability allows organizations to adjust their computing resources dynamically and efficiently, particularly beneficial for SMEsin the engineering sector, which often face fluctuating project requirements and resource needs(Raut et al., 2019).

Cloud computing offers various deployment models, including public, private, and hybrid clouds. Public clouds are owned and operated by third-party service providers, while private clouds are dedicated to a single organization and can be hosted on-premises or by a third-party provider. Hybrid clouds combine elements of both public and private clouds, allowing organizations to leverage scalability and cost-effectiveness while maintaining control over sensitive data and applications. Cloud computing also includes service models like Infrastructure as a Service (IaaS), Platform as a Service (PaaS), and Software as a Service (SaaS)(Attaran & Woods, 2019; Raut et al., 2019). IaaS provides virtualized computing resources over the internet, PaaS allows developers to build, deploy, and manage applications without the complexity of underlying infrastructure, and SaaS delivers software applications over the internet on a subscription basis.

Cloud computing provides SMEs in the engineering sector a flexible, scalable, and cost-effective alternative to traditional IT infrastructure, enabling them to innovate, collaborate, and compete more effectively in the digital economy. It streamlines processes, facilitates collaboration, and enhances innovation. Engineers often work on complex projects requiring extensive computational resources, sophisticated software tools, and efficient team collaboration. Cloud computing enables on-demand access to computing resources, deployment of specialized software applications, and location-independent collaboration(Attaran & Woods, 2019).

Cloud computing is playing a crucial role in engineering, as it reduces the need for on-premises hardware infrastructure, reducing energy consumption, electronic waste, and promoting resource efficiency. Cloud providers are investing in energy-efficient data centers and renewable energy sources, aligning with sustainability goals. Cloud-based tools and platforms enable engineers to incorporate environmental considerations into their designs more effectively, such as simulation and modeling software. Collaborative platforms allow engineers to share data, collaborate on projects, and integrate environmental factors into decision-making processes more seamlessly(Mujinga, 2020).

Cloud computing is revolutionizing engineering firms by enhancing efficiency, innovation, and sustainability. Engineers can access necessary resources, collaborate more effectively, and incorporate environmental considerations into their work. As the industry adapts to technological and environmental challenges, cloud computing will play a crucial role in driving progress and promoting sustainability(Abdullah et al., 2020).

CLOUD COMPUTING

Definition and Characteristics

Cloud computing is the internet-based delivery of computing services, offering users access to a shared pool of configurable resources like servers, storage, networks, applications, and services. Users can rent computing resources from cloud service providers on a pay-as-you-go basis, typically billed on usage or subscription basis. This approach is particularly attractive for small and medium enterprises (SMEs) (Khayer et al., 2021; Vajjhala & Ramollari, 2021).

i. *Scalability:* Cloud computing allows SMEs to scale their computing resources up or down based on demand, without the need for significant upfront investment in additional hardware. This scalability is essential for SMEs with fluctuating resource requirements, enabling them to adapt quickly to changing business needs.
ii. *Flexibility:* Cloud computing provides SMEs with the flexibility to choose from a range of computing services and deployment models to suit their specific needs. Whether they require infrastructure resources (Infrastructure as a Service - IaaS), development platforms (Platform as a Service - PaaS), or software applications (Software as a Service - SaaS), SMEs can select the services that best align with their business objectives.

iii. *Cost-effectiveness:* Cloud computing offers SMEs a cost-effective alternative to traditional IT infrastructure, as it eliminates the need for upfront capital expenditure on hardware and software. Instead, SMEs pay only for the resources they use, avoiding the costs associated with purchasing, maintaining, and upgrading on-premises infrastructure.
iv. *Accessibility: Cloud* computing provides SMEs with ubiquitous access to computing resources and applications from any location with an internet connection. This accessibility enables remote work, collaboration among distributed teams, and greater flexibility in how SMEs manage their operations.
v. *Security and Reliability:* Cloud service providers invest heavily in security measures to protect their infrastructure and data centers, often employing advanced encryption, access controls, and monitoring tools. Additionally, cloud services typically offer high levels of reliability and uptime, with built-in redundancy and failover mechanisms to ensure continuous availability of services.

Cloud computing provides SMEs with a cost-effective, scalable solution for accessing computing resources, enabling them to compete effectively in the digital economy and adapt to evolving business needs. Its scalability, flexibility, and accessibility drive innovation, enhance productivity, and achieve business objectives efficiently. The figure 1 depicts various deployment models and cloud services for SMEs.

Figure 1. Deployment models and cloud services for SMEs

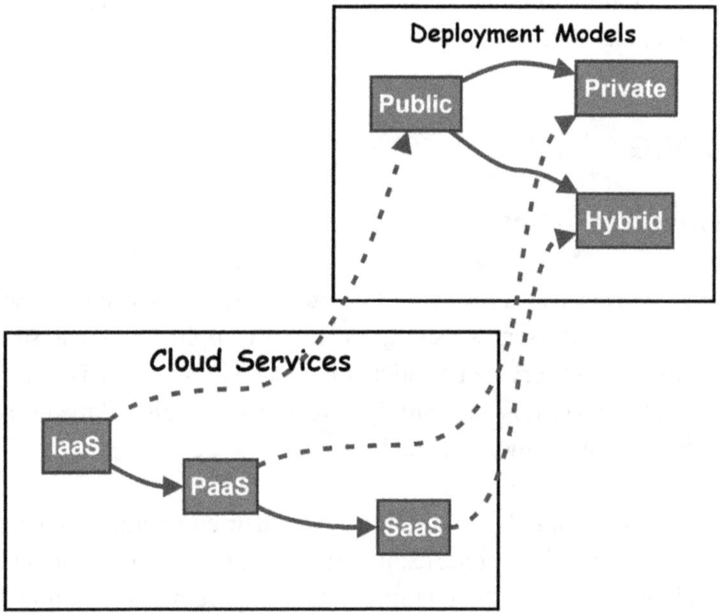

Types of Cloud Services (IaaS, PaaS, SaaS) for SMEs

SMEs can greatly benefit from various cloud services that cater to their specific business needs, including(Nanda et al., 2024a): The integration of IaaS, PaaS, and SaaS offers SMEs a comprehensive suite of cloud services, enabling them to manage infrastructure, development, and software applications, reduce costs, enhance agility, and drive innovation and business growth.

- **Infrastructure as a Service (IaaS):** Infrastructure as a Service (IaaS) provides SMEs with virtualized computing resources over the internet. Instead of investing in physical hardware such as servers, storage, and networking equipment, SMEs can rent these resources from cloud service providers on a pay-as-you-go basis. With IaaS, SMEs have the flexibility to scale their infrastructure up or down based on demand, eliminating the need for costly upfront investments in hardware. This model is particularly beneficial for SMEs with dynamic resource requirements or those looking to avoid the complexities of managing on-premises infrastructure(Malathi et al., 2024; Nanda et al., 2024b; Pachiappan et al., 2024; Paul et al., 2024).
- **Platform as a Service (PaaS):** Platform as a Service (PaaS) offers SMEs a platform for developing, deploying, and managing applications without the complexity of underlying infrastructure. PaaS providers offer a comprehensive set of tools and services, including development frameworks, databases, middleware, and application management services, accessible through a web-based interface(M. Sharma et al., 2024a; Sundaramoorthy et al., 2024; Verma et al., 2024). SMEs can leverage PaaS to accelerate the software development process, reduce time-to-market, and focus on building innovative applications without worrying about the underlying infrastructure. PaaS is ideal for SMEs looking to streamline their development processes and scale their applications efficiently.
- **Software as a Service (SaaS):** Software as a Service (SaaS) delivers software applications over the internet on a subscription basis, eliminating the need for SMEs to install, maintain, and update software locally. SaaS applications are hosted and managed by third-party providers, who handle all aspects of software maintenance, including updates, patches, and security. SMEs can access SaaS applications through a web browser or mobile app, making them accessible from any location with an internet connection(Boopathi, 2024a, 2024d, 2024e). SaaS offers SMEs a cost-effective and scalable solution for accessing a wide range of productivity tools, business applications, and collaboration platforms without the need for upfront investments in software licenses or hardware infrastructure.

Deployment Models (Public, Private, Hybrid)

SMEs have various cloud computing deployment models, each offering distinct advantages and considerations, including(Boopathi, 2024b; Nanda et al., 2024a; M. Sharma et al., 2024b). The choice of cloud deployment model depends on an organization's IT strategy, budget, security requirements, and regulatory compliance. Understanding public, private, and hybrid cloud deployments helps SMEs make informed decisions for innovation, efficiency, and business growth.

- **Public Cloud:** Public cloud deployment involves hosting computing resources and services on infrastructure owned and operated by third-party cloud service providers, such as Amazon Web

Services (AWS), Microsoft Azure, or Google Cloud Platform (GCP). SMEs access these resources over the internet on a pay-as-you-go basis, with the provider handling all aspects of infrastructure management, including hardware provisioning, maintenance, and security. Public cloud environments are highly scalable, cost-effective, and accessible from anywhere with an internet connection, making them well-suited for SMEs with dynamic resource requirements, limited IT budgets, or a need for rapid scalability(Boopathi, 2024c; Nanda et al., 2024b; M. Sharma et al., 2024a).

- **Private Cloud:** Private cloud deployment involves hosting computing resources and services on infrastructure dedicated to a single organization, either on-premises or in a data center operated by a third-party provider. Unlike public cloud environments, which are shared among multiple tenants, private clouds offer greater control, security, and customization options, making them suitable for SMEs with strict compliance requirements, sensitive data, or specific performance needs. Private clouds can be managed internally by the organization's IT team or outsourced to a managed service provider, offering flexibility in deployment and management(Agrawal et al., 2023a; Hema et al., 2023; Rahamathunnisa, Sudhakar, et al., 2023).
- **Hybrid Cloud:** Hybrid cloud deployment combines elements of both public and private cloud environments, allowing SMEs to leverage the benefits of each while addressing specific business needs and requirements. In a hybrid cloud setup, SMEs can run certain workloads or applications in a public cloud environment while keeping others in a private cloud or on-premises infrastructure. This flexibility enables SMEs to optimize performance, security, and cost-effectiveness based on the unique characteristics of each workload(Satav et al., 2023a; Srinivas et al., 2023; Syamala et al., 2023a). For example, SMEs may use the public cloud for scalable web applications or development environments while keeping sensitive data or critical workloads in a private cloud for enhanced control and security.

CLOUD ADOPTION FOR SMEs

Cloud adoption offers SMEs numerous benefits including cost savings, scalability, accessibility, flexibility, enhanced collaboration, efficiency, improved security, and disaster recovery, all of which contribute to their overall operations(Hema et al., 2023; Rahamathunnisa, Sudhakar, et al., 2023; Syamala et al., 2023a). The figure 2 illustrates the various advantages of implementing cloud technology in small and medium-sized enterprises (SMEs).

Cost Savings and Scalability

- Pay-as-you-go Model: Cloud services operate on a pay-as-you-go basis, allowing SMEs to pay only for the resources they consume. This eliminates the need for upfront capital investment in hardware and software, resulting in significant cost savings.
- Economies of Scale: Cloud service providers benefit from economies of scale, spreading infrastructure costs across multiple users. As a result, SMEs can access computing resources at lower rates than if they were to build and maintain their own on-premises infrastructure.
- Scalability: Cloud computing offers SMEs the ability to scale their computing resources up or down based on demand. This elasticity allows SMEs to align their IT resources with business needs, avoiding over-provisioning and underutilization of resources.

- Resource Optimization: Cloud adoption enables SMEs to optimize resource allocation by dynamically adjusting computing resources in response to changing workloads. This ensures that SMEs can efficiently utilize their IT resources, further enhancing cost savings.
- Reduced Maintenance Costs: Cloud service providers handle infrastructure maintenance, updates, and upgrades, reducing the burden on SMEs' IT staff and lowering operational expenses associated with maintaining on-premises hardware and software.

Accessibility and Flexibility

- Anytime, Anywhere Access: Cloud computing enables SMEs to access computing resources and applications from anywhere with an internet connection, providing employees with flexibility to work remotely or on the go.
- Device Agnostic: Cloud-based applications and services are accessible from a wide range of devices, including laptops, tablets, and smartphones. This flexibility allows SMEs to support diverse work styles and accommodate employees' preferred devices.
- Rapid Deployment: Cloud services can be provisioned and deployed rapidly, allowing SMEs to quickly launch new applications, services, or infrastructure resources without the need for lengthy procurement or setup processes.
- Global Reach: Cloud computing enables SMEs to expand their operations globally by leveraging cloud service providers' global infrastructure and data center locations. This allows SMEs to reach customers and employees worldwide with minimal latency and downtime.
- Customization and Integration: Cloud services offer SMEs the flexibility to customize and integrate applications and services to meet their specific business requirements. This allows SMEs to adapt and evolve their IT environment as their business needs change over time.

Figure 2. Various benefits of cloud adoption in SMEs

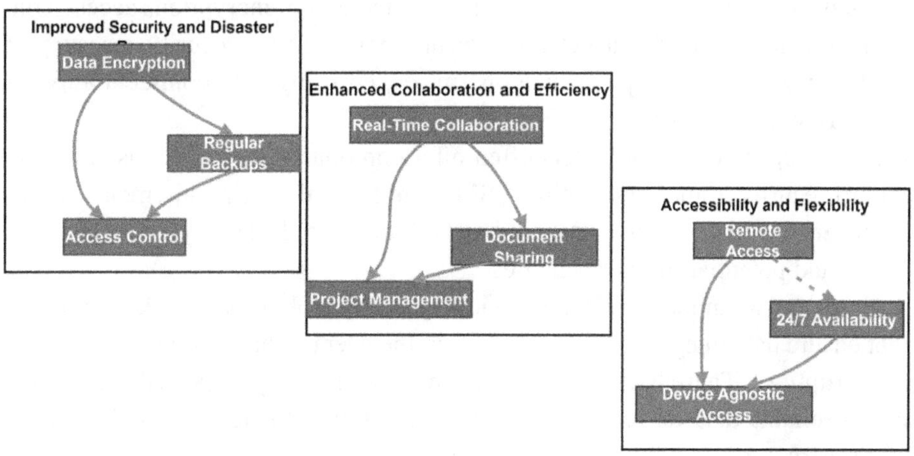

Enhanced Collaboration and Efficiency

- **Real-time Collaboration Tools:** Cloud-based collaboration tools such as document sharing platforms, project management software, and communication applications facilitate real-time collaboration among SME teams. These tools allow employees to work together on projects, share documents, and communicate seamlessly, regardless of their location(Rahamathunnisa, Subhashini, et al., 2023; Vennila et al., 2022).
- **Centralized Data Access:** Cloud storage solutions provide SMEs with centralized data repositories accessible to authorized users from anywhere with an internet connection. This enables employees to access the latest documents, files, and project resources, promoting collaboration and reducing the risk of version control issues.
- **Remote Work Enablement:** Cloud adoption enables SMEs to support remote work arrangements by providing employees with access to essential tools and resources from any location. This flexibility allows employees to work from home, on the road, or in distributed team environments, improving work-life balance and productivity.
- **Streamlined Workflows:** Cloud-based workflow automation tools automate repetitive tasks, streamline business processes, and eliminate manual intervention. This reduces the time and effort required to complete tasks, enhances productivity, and ensures consistency and accuracy across operations.
- **Integration Capabilities:** Cloud services often offer integration capabilities that allow SMEs to connect disparate systems, applications, and data sources. This integration streamlines data exchange, improves information visibility, and enhances collaboration between different departments or teams within the organization.

Improved Security and Disaster Recovery

- Robust Security Measures: Cloud service providers implement robust security measures to protect data and infrastructure against cyber threats, including encryption, access controls, multifactor authentication, and threat detection technologies. These security measures exceed what many SMEs can afford to implement on-premises, ensuring data confidentiality, integrity, and availability(Boopathi & Khang, 2023).
- Compliance Support: Cloud providers often offer compliance certifications and adhere to industry-standard security frameworks, helping SMEs meet regulatory requirements and industry standards. This ensures that sensitive data is handled and stored securely, reducing the risk of noncompliance and potential fines or penalties.
- Data Backups and Redundancy: Cloud services typically include data backup and redundancy features that ensure data integrity and availability in the event of hardware failures, natural disasters, or other disruptions. These backups are stored in geographically dispersed data centers, providing SMEs with reliable disaster recovery capabilities without the need for additional investment in backup infrastructure.
- Continuous Monitoring and Updates: Cloud providers continuously monitor their infrastructure for security threats and vulnerabilities, applying patches and updates as needed to mitigate risks. This proactive approach to security reduces SMEs' exposure to emerging threats and ensures that

their data and applications remain secure and up-to-date(Boopathi et al., 2023; Samikannu et al., 2022; Sampath, 2021).

- Disaster Recovery Planning: Cloud adoption enables SMEs to implement comprehensive disaster recovery plans that leverage cloud-based infrastructure and services. This includes replicating critical data and applications to multiple geographic regions, implementing failover mechanisms, and conducting regular disaster recovery testing to ensure readiness in the event of a disruption.

CHALLENGES AND CONSIDERATIONS

SMEs are grappling with security and privacy concerns when considering cloud adoption. Despite the benefits of cost savings, scalability, and accessibility, SMEs must also address potential security risks and privacy implications, including(Otuka et al., 2014).

- Data Breaches: Cloud environments are attractive targets for cybercriminals due to the large volumes of sensitive data stored and processed in the cloud. SMEs face the risk of data breaches, unauthorized access, and theft of confidential information, which can result in financial losses, reputational damage, and legal liabilities.
- Compliance with Data Protection Regulations: SMEs must comply with data protection regulations and privacy laws governing the collection, storage, and processing of personal and sensitive data. This includes regulations such as the General Data Protection Regulation (GDPR) in the European Union and the California Consumer Privacy Act (CCPA) in the United States. Ensuring compliance with these regulations requires SMEs to implement robust data protection measures, including encryption, access controls, and data residency requirements.
- Shared Responsibility Model: Cloud service providers operate under a shared responsibility model, wherein they are responsible for securing the underlying infrastructure, while SMEs are responsible for securing their data and applications. SMEs must understand their responsibilities under this model and implement appropriate security controls to protect their assets in the cloud.
- Identity and Access Management: Managing user identities and access rights is critical for ensuring the security of cloud-based resources. SMEs must implement strong authentication mechanisms, such as multi-factor authentication (MFA) and role-based access control (RBAC), to prevent unauthorized access to sensitive data and applications.
- Vendor Lock-in: SMEs may face vendor lock-in when relying on a single cloud service provider for their computing needs. This can limit their ability to switch providers or migrate to alternative solutions in the future. SMEs should carefully evaluate vendor lock-in risks and consider strategies for mitigating this risk, such as adopting open standards and using multi-cloud or hybrid cloud architectures.

SMEs face significant challenges in complying with regulatory requirements for cloud computing, as these environments often span multiple jurisdictions with their own set of regulations, posing important considerations and challenges(Vasiljeva et al., 2017).

- Data Sovereignty: Cloud computing introduces challenges related to data sovereignty, as data stored in the cloud may be subject to the laws and regulations of the jurisdiction in which it is located. SMEs must ensure that they understand the data residency requirements of their regulatory authorities and choose cloud providers that can meet these requirements.

- Industry-specific Regulations: SMEs operating in regulated industries, such as healthcare, finance, or government, must comply with industry-specific regulations and standards governing the handling of sensitive data. Cloud providers may offer compliance certifications and attestations to demonstrate adherence to these regulations, but SMEs remain ultimately responsible for ensuring compliance with industry standards(Boopathi & Kumar, 2024; P. Das et al., 2024; Revathi et al., 2024; Vijayakumar et al., 2024).
- Data Transfer and Export Controls: SMEs must consider data transfer and export controls when storing and processing data in the cloud. Transferring data across international borders may be subject to export control regulations, import restrictions, or data localization requirements, which can vary depending on the nature of the data and the countries involved.
- Audit and Reporting Requirements: Cloud adoption may impact SMEs' ability to meet audit and reporting requirements imposed by regulatory authorities or industry standards bodies. SMEs must ensure that they have access to audit trails, logs, and compliance reports provided by their cloud service providers to demonstrate compliance with regulatory requirements.
- Contractual Agreements: SMEs should carefully review and negotiate contractual agreements with cloud service providers to ensure that they adequately address compliance and regulatory issues. Important considerations include data protection clauses, liability provisions, audit rights, and indemnification clauses to protect SMEs' interests and mitigate legal risks associated with cloud adoption.

The integration of cloud services with existing systems for SMEs presents significant challenges, despite the benefits of scalability, flexibility, and cost savings. Migrating existing applications, data, and workflows to the cloud can be complex and challenging(Widyastuti & Irwansyah, 2018a).

- Legacy Systems Compatibility: SMEs often rely on legacy systems and applications that may not be compatible with cloud environments. Migrating these legacy systems to the cloud may require significant modifications or redevelopment to ensure compatibility with cloud platforms and services.
- Data Migration Complexity: Migrating large volumes of data to the cloud can be time-consuming, resource-intensive, and prone to errors. SMEs must carefully plan and execute data migration strategies, considering factors such as data volume, latency, bandwidth constraints, and data residency requirements.
- Application Refactoring: Some applications may need to be refactored or redesigned to take full advantage of cloud-native capabilities, such as scalability, elasticity, and fault tolerance. SMEs must assess their existing applications and determine whether they need to be modernized or rearchitected for the cloud.
- Integration with On-premises Systems: Integrating cloud services with existing on-premises systems and applications can be challenging due to differences in technology stacks, protocols, and data formats. SMEs must develop integration strategies and implement middleware solutions to facilitate seamless communication and data exchange between cloud and on-premises environments(D. M. Sharma et al., 2024a).

- Interoperability and Vendor Lock-in: SMEs may face interoperability challenges when integrating cloud services from multiple vendors or cloud providers. Vendor-specific APIs, proprietary formats, and compatibility issues can hinder interoperability and lead to vendor lock-in. SMEs should carefully evaluate interoperability requirements and choose cloud providers that support open standards and interoperability frameworks.

Skills and Training Requirements

SMEs need new skills and expertise for cloud adoption, but many lack the necessary training to navigate the complexities of cloud computing, posing important challenges and considerations in this area(Kreslins et al., 2018; Kumalo & Van Der Poll, 2015). SMEs must address migration and integration challenges, invest in skills development and training, and successfully adopt cloud computing to drive innovation, efficiency, and business growth.

- Cloud Architecture and Design: SMEs need expertise in cloud architecture and design principles to architect scalable, resilient, and secure cloud solutions. This includes knowledge of cloud services, deployment models, networking, security, and compliance best practices.
- Cloud Management and Operations: Managing cloud environments requires skills in cloud administration, monitoring, optimization, and troubleshooting. SMEs need to train their IT staff or hire cloud specialists with expertise in cloud management platforms, automation tools, and DevOps practices.
- Security and Compliance: Ensuring the security and compliance of cloud-based solutions requires expertise in cloud security, data protection, and regulatory compliance. SMEs need to train their IT staff on security best practices, encryption technologies, identity and access management (IAM), and compliance frameworks such as GDPR, HIPAA, or PCI DSS.
- Vendor Management and Governance: SMEs must develop skills in vendor management and governance to effectively manage relationships with cloud service providers. This includes negotiating contracts, monitoring service-level agreements (SLAs), and ensuring compliance with contractual obligations and regulatory requirements.
- Continuous Learning and Development: Cloud computing is a rapidly evolving field, with new technologies, services, and best practices emerging regularly. SMEs need to invest in continuous learning and development programs to keep their IT staff updated on the latest trends and advancements in cloud computing.

CLOUD APPLICATIONS IN ENGINEERING

Cloud computing has significantly impacted engineering by providing CAD/CAM software and simulation/modeling tools, enhancing collaboration, streamlining workflows, and improving productivity, making them essential tools in the field. The figure 3 depicts a diverse range of cloud applications used by SMEs(Kratzke, 2018).

CAD/CAM Software in the Cloud: Computer-Aided Design (CAD) and Computer-Aided Manufacturing (CAM) software are essential tools for engineers involved in product design, prototyping, and manufacturing. Traditionally, CAD/CAM software has been installed and operated on local machines, requiring significant upfront investment in hardware and software licenses(Truong & Dustdar, 2015).

- Accessibility: Cloud-based CAD/CAM software allows engineers to access design and manufacturing tools from anywhere with an internet connection. This enables remote collaboration between team members located in different geographic locations, facilitating real-time design reviews, feedback, and iteration.
- Scalability: Cloud-based CAD/CAM solutions offer scalability, allowing engineering SMEs to scale their computing resources up or down based on project requirements. This flexibility eliminates the need for costly hardware upgrades and allows SMEs to adapt to changing workload demands.
- Collaboration: Cloud-based CAD/CAM software facilitates collaboration among multiple stakeholders, including engineers, designers, manufacturers, and clients. These tools provide features such as version control, commenting, and annotation, enabling seamless collaboration throughout the design and manufacturing process.
- Cost Savings: Cloud-based CAD/CAM solutions operate on a subscription-based pricing model, eliminating the need for upfront capital investment in hardware and software licenses. This pay-as-you-go pricing model offers cost savings for engineering SMEs, particularly those with limited budgets or fluctuating project requirements(Malathi et al., 2024).

Simulation and Modeling Tools: Simulation and modeling tools play a crucial role in engineering, enabling engineers to analyze and optimize designs, predict performance, and validate assumptions before physical prototypes are built. Cloud-based simulation and modeling tools offer several advantages for engineering SMEs(Chhabra et al., 2010). Cloud computing has significantly improved engineering projects by enhancing collaboration, streamlining workflows, and improving efficiency through various cloud-based applications, particularly in project management platforms and remote monitoring/control systems.

- Computational Power: Cloud-based simulation tools leverage the computational power of cloud infrastructure, enabling engineers to perform complex simulations and analyses that would be impractical or prohibitively expensive using on-premises resources alone.
- Scalability: Cloud-based simulation tools offer scalability, allowing engineering SMEs to scale their computing resources to accommodate large-scale simulations or multiple concurrent simulations. This flexibility enables SMEs to optimize resource utilization and reduce simulation turnaround times.
- Collaboration: Cloud-based simulation tools facilitate collaboration among multidisciplinary teams, enabling engineers, analysts, and domain experts to collaborate on simulation projects in real-time. These tools provide features such as shared workspaces, version control, and access controls, ensuring seamless collaboration throughout the simulation lifecycle.
- Accessibility: Cloud-based simulation tools allow engineers to access simulation capabilities from anywhere with an internet connection, enabling remote work and distributed collaboration. This accessibility ensures that engineers can perform simulations and analyses regardless of their location or the devices they use.

Cloud-based CAD/CAM software and simulation tools provide engineering SMEs with improved collaboration, scalability, accessibility, and cost savings, enabling efficient design innovation and optimization. These applications accelerate product development, reduce time-to-market, and drive business growth in a competitive market(Bahrami, 2015).

Figure 3. Wide range of cloud applications in SMEs

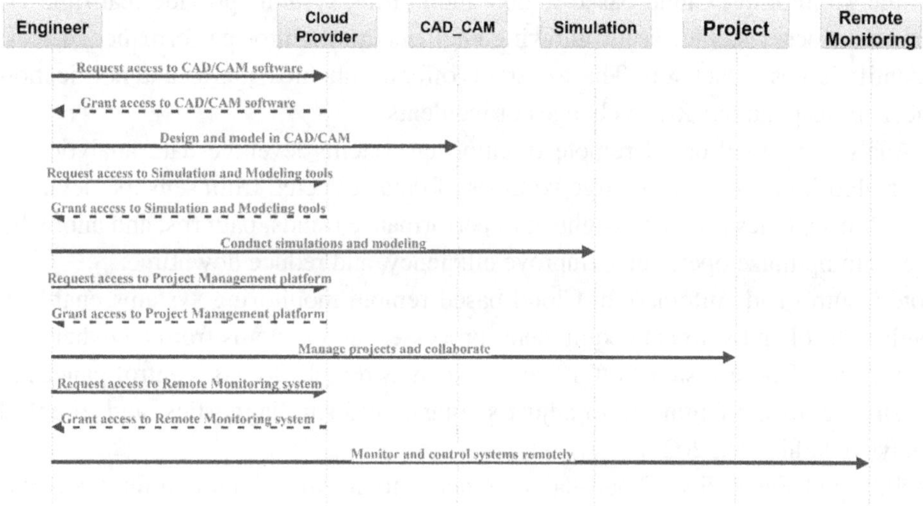

Project Management and Collaboration Platforms: Cloud-based project management platforms provide numerous benefits for engineering SMEs, enhancing coordination, resource allocation, and timely communication, thereby ensuring project success(Kreslins et al., 2018; Kumalo & Van Der Poll, 2015; Tamburri et al., 2020; Widyastuti & Irwansyah, 2018a).

- Centralized Communication: Cloud-based project management platforms provide a centralized hub for communication, allowing team members to collaborate, share updates, and coordinate tasks in real-time. These platforms offer features such as messaging, file sharing, task assignment, and progress tracking, enabling seamless communication and collaboration across distributed teams.
- Document Management: Cloud-based project management platforms offer robust document management capabilities, allowing engineering SMEs to store, organize, and share project documents securely in the cloud. These platforms provide version control, access controls, and audit trails, ensuring that team members have access to the latest project documents and revisions.
- Task Tracking and Scheduling: Cloud-based project management platforms enable engineering SMEs to track tasks, milestones, and deadlines effectively. These platforms offer features such as Gantt charts, Kanban boards, and task dependencies, allowing project managers to plan, schedule, and allocate resources efficiently and monitor progress in real-time.
- Collaboration Tools: Cloud-based project management platforms offer collaboration tools such as discussion forums, wikis, and virtual whiteboards, enabling team members to brainstorm ideas,

share insights, and solve problems collaboratively. These tools foster innovation, creativity, and knowledge sharing among project teams, driving better outcomes for engineering projects.

Remote Monitoring and Control Systems: Remote monitoring and control systems are crucial in engineering, allowing engineers to remotely manage equipment, processes, and infrastructure, offering numerous advantages for engineering SMEs(Boopathi & Myilsamy, 2021; Subha et al., 2023).

- Real-time Monitoring: Cloud-based remote monitoring systems provide real-time visibility into equipment, processes, and assets, allowing engineers to monitor performance, detect anomalies, and identify issues proactively. These systems offer dashboards, alerts, and notifications, enabling engineers to respond quickly to changes or incidents.
- Data Analytics: Cloud-based remote monitoring systems leverage data analytics and machine learning algorithms to analyze large volumes of data collected from sensors, devices, and equipment. These analytics provide insights into performance trends, patterns, and anomalies, enabling engineers to optimize operations, improve efficiency, and reduce downtime.
- Remote Control and Automation: Cloud-based remote monitoring systems enable engineers to remotely control and automate equipment, processes, and systems from anywhere with an internet connection. These systems offer features such as remote access, control panels, and automation scripts, allowing engineers to adjust settings, perform diagnostics, and troubleshoot issues remotely(Malathi et al., 2024).
- Scalability and Flexibility: Cloud-based remote monitoring systems offer scalability and flexibility, allowing engineering SMEs to scale their monitoring capabilities up or down based on demand. These systems can accommodate changes in equipment, processes, and infrastructure, ensuring that engineering SMEs can adapt to evolving business needs and requirements.

Cloud-based project management platforms and remote monitoring systems enhance collaboration, efficiency, and flexibility for engineering SMEs. These applications improve productivity, reduce costs, and drive innovation in a competitive marketplace, allowing them to effectively manage projects and monitor operations remotely.

ENVIRONMENTAL CONSIDERATIONS IN CLOUD COMPUTING

The growing demand for computing resources and rising climate change and sustainability concerns have highlighted the need for environmental considerations in cloud computing. While cloud computing offers benefits like scalability, cost-effectiveness, and flexibility, it also has environmental implications, particularly in terms of energy efficiency and reducing carbon footprint(S. Kumar & Buyya, 2012). The figure 4 highlights the important factors for implementing cloud computing in environmental aspects.

Energy Efficiency: Data centers, housing cloud computing infrastructure, consume significant energy for powering and cooling servers, networking equipment, and other hardware. Improving energy efficiency measures is crucial for minimizing consumption and reducing environmental impact, and important strategies include(Schniederjans & Hales, 2016):

- Hardware Optimization: Cloud service providers invest in energy-efficient hardware, including servers, storage devices, and networking equipment, to minimize power consumption and maximize performance per watt. This hardware is designed to operate efficiently under varying workloads and utilization levels, reducing energy waste.
- Virtualization and Consolidation: Virtualization technology enables data centers to consolidate multiple virtual machines (VMs) onto a single physical server, reducing the number of servers required and optimizing resource utilization. This consolidation improves energy efficiency by reducing idle capacity and eliminating underutilized hardware.
- Dynamic Resource Management: Data center operators use dynamic resource management techniques, such as workload balancing, server consolidation, and power management, to optimize resource allocation and energy usage in real-time. These techniques ensure that resources are allocated efficiently based on workload demand, minimizing energy waste and reducing operating costs.
- Renewable Energy Sources: Many cloud service providers are investing in renewable energy sources, such as solar, wind, and hydroelectric power, to power their data centers and reduce reliance on fossil fuels. By sourcing renewable energy, cloud providers can reduce their carbon footprint and contribute to global efforts to combat climate change(Boopathi, 2024d; Rahamathunnisa et al., 2024).
- Energy-Efficient Cooling Systems: Data centers use advanced cooling systems, such as air-side economizers, liquid cooling, and hot/cold aisle containment, to maintain optimal temperature and humidity levels while minimizing energy consumption. These cooling systems are designed to remove heat efficiently from server racks and equipment, reducing the energy required for cooling(Gunasekaran et al., 2022; Myilsamy & Sampath, 2021).

Figure 4. Important factors for implementing cloud computing in environmental aspects

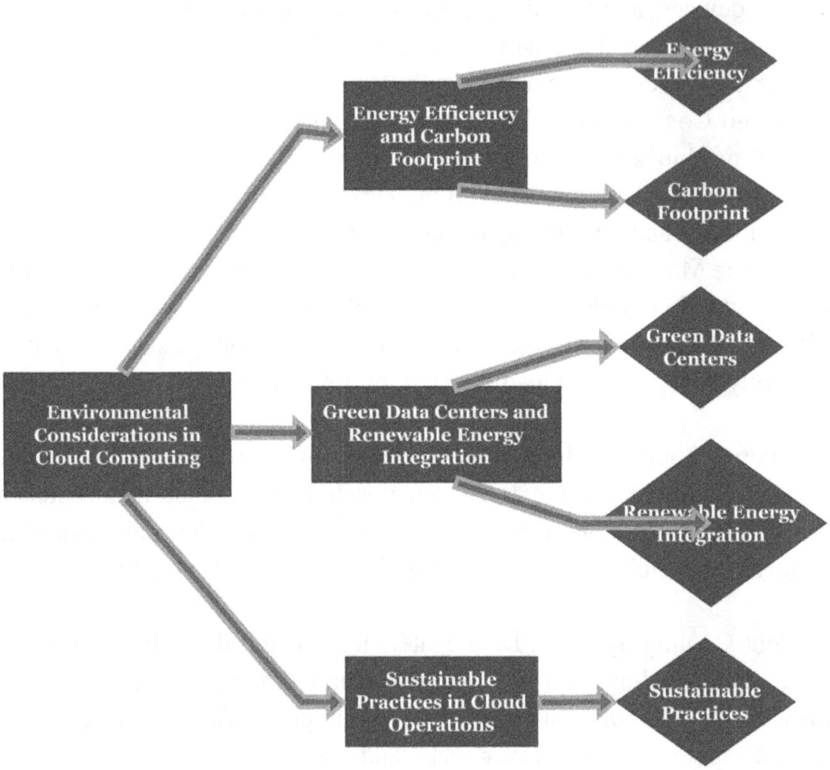

Energy efficiency measures are crucial for minimizing the environmental impact of cloud computing and its contribution to climate change. Cloud service providers can adopt energy-efficient technologies and practices, promoting sustainability and supporting global efforts towards a low-carbon economy. Businesses and individuals can choose providers that prioritize energy efficiency and renewable energy sourcing to minimize their environmental impact.

Green Data Centers and Renewable Energy Integration

The growing demand for computing resources and climate change concerns have led to a growing need for environmental considerations in cloud computing. This includes adopting green data centers and integrating renewable energy sources to minimize the environmental impact of data centers and reduce the carbon footprint associated with cloud computing(Gutierrez et al., 2015; Schniederjans & Hales, 2016).

Green Data Centers: Green data centers are designed and operated with a focus on energy efficiency, sustainability, and environmental responsibility. These data centers implement various technologies and practices to minimize energy consumption, reduce waste, and optimize resource utilization. Some significant features of green data centers include:

- Energy-Efficient Infrastructure: Green data centers utilize energy-efficient hardware, including servers, storage devices, and networking equipment, to minimize power consumption and maximize performance per watt. These hardware components are designed to operate efficiently under varying workloads and utilization levels, reducing energy waste.

- Cooling Efficiency: Cooling systems account for a significant portion of energy consumption in data centers. Green data centers use advanced cooling technologies, such as air-side economizers, liquid cooling, and hot/cold aisle containment, to maintain optimal temperature and humidity levels while minimizing energy usage. These cooling systems are designed to remove heat efficiently from server racks and equipment, reducing the energy required for cooling.
- Renewable Energy Integration: Green data centers prioritize the use of renewable energy sources, such as solar, wind, and hydroelectric power, to power their operations. By sourcing renewable energy, data centers can reduce their reliance on fossil fuels and minimize their carbon footprint. Some green data centers generate their renewable energy on-site through solar panels or wind turbines, while others purchase renewable energy credits or contracts from external sources.

Renewable Energy Integration: The integration of renewable energy sources into data center operations is a important strategy for reducing the environmental impact of cloud computing. By harnessing renewable energy, data centers can lower their greenhouse gas emissions and contribute to global efforts to combat climate change. Some common approaches to integrating renewable energy into data center operations include(Nishanth et al., 2023; Saravanan et al., 2022):

- On-Site Renewable Energy Generation: Data centers can generate their renewable energy on-site through solar panels, wind turbines, or other renewable energy technologies. On-site generation allows data centers to directly offset their energy consumption with renewable energy production, reducing their reliance on grid power and fossil fuels.
- Power Purchase Agreements (PPAs): Data centers can enter into power purchase agreements (PPAs) with renewable energy developers to purchase renewable energy directly from renewable energy projects. PPAs provide a stable and predictable source of renewable energy at competitive prices, enabling data centers to reduce their carbon footprint while supporting the growth of renewable energy infrastructure.
- Renewable Energy Credits (RECs): Data centers can purchase renewable energy credits (RECs) to offset their carbon emissions from grid power consumption. RECs represent the environmental attributes of renewable energy generation and can be purchased from renewable energy projects that generate renewable energy. By purchasing RECs, data centers can support renewable energy development and demonstrate their commitment to sustainability.

Green data centers and renewable energy integration are crucial for minimizing the environmental impact of cloud computing. By investing in energy-efficient infrastructure and renewable technologies, data centers can reduce their carbon footprint, promote sustainability, and support global transition to a low-carbon economy. Businesses and individuals can choose cloud providers that prioritize green data center practices(M. Kumar et al., 2023; Rebecca et al., 2024; Venkateswaran et al., 2023a).

Sustainable Practices in Cloud Operations

The increasing demand for computing resources has raised concerns about the environmental impact of cloud computing. To mitigate this, cloud service providers are adopting sustainable practices to minimize energy consumption, reduce carbon emissions, and promote environmental responsibility throughout the service's lifecycle(Agrawal et al., 2023a; Satav et al., 2023a; M. Sharma et al., 2024b; Venkateswaran et al., 2023b).

Sustainable practices in cloud operations are crucial for minimizing the environmental impact of cloud computing and promoting environmental responsibility in the technology industry. By adopting energy-efficient technologies, sourcing renewable energy, optimizing software and algorithms, and investing in carbon offsetting and emissions reduction initiatives, cloud service providers can contribute to a more sustainable future. Businesses and individuals can choose providers that prioritize sustainability and environmental responsibility.

- Cloud service providers are investing in energy-efficient data centers to minimize the energy consumption of their infrastructure. This includes the use of energy-efficient hardware, such as servers, storage devices, and networking equipment, which are designed to optimize performance per watt. Additionally, data centers are implementing advanced cooling technologies and infrastructure design principles to improve energy efficiency and reduce the environmental impact of cooling systems.
- Many cloud service providers are sourcing renewable energy to power their data centers and operations. This includes investing in solar, wind, and hydroelectric power projects, purchasing renewable energy credits (RECs), and entering into power purchase agreements (PPAs) with renewable energy developers. By sourcing renewable energy, cloud service providers can reduce their reliance on fossil fuels, lower their carbon footprint, and support the growth of renewable energy infrastructure.
- Cloud service providers are enhancing energy efficiency and resource utilization by developing energy-efficient software architectures, optimizing code for performance and energy consumption, and implementing dynamic resource allocation and workload scheduling algorithms. This approach aims to reduce energy consumption and enhance the sustainability of cloud computing.
- Cloud service providers are focusing on minimizing environmental impact by selecting data centers in regions with renewable energy sources, favorable climate conditions, and efficient cooling options. They are also utilizing modular construction, free-air cooling, and efficient power distribution to reduce energy consumption, increase operational efficiency, and minimize environmental footprint.
- Cloud service providers are focusing on carbon offsetting and emissions reduction initiatives to reduce their environmental impact. This includes funding reforestation, renewable energy development, and energy efficiency improvements in their communities. This move demonstrates their commitment to environmental sustainability and contributes to global efforts to combat climate change.

BEST PRACTICES AND RECOMMENDATIONS

Planning and Strategy for Cloud Adoption The task involves a thorough evaluation of current IT infrastructure, applications, and business processes to identify areas suitable for cloud migration, setting clear business objectives and success criteria, and creating a detailed roadmap with phased migration plans, timelines, and milestones(Carcary et al., 2014).

Choosing the Right Cloud Service Providers: Selecting a cloud service provider involves assessing factors like reliability, performance, scalability, security, compliance, and cost-effectiveness, while considering your organization's specific needs like workload, location, industry, and budget constraints(Hentschel et al., 2019). It emphasizes the significance of reviewing cloud provider certifications, compliance standards, audit reports, customer support, service level agreements, and uptime guarantees to ensure industry best practices and regulatory requirements.

Figure 5. Best practices in security and compliance measures

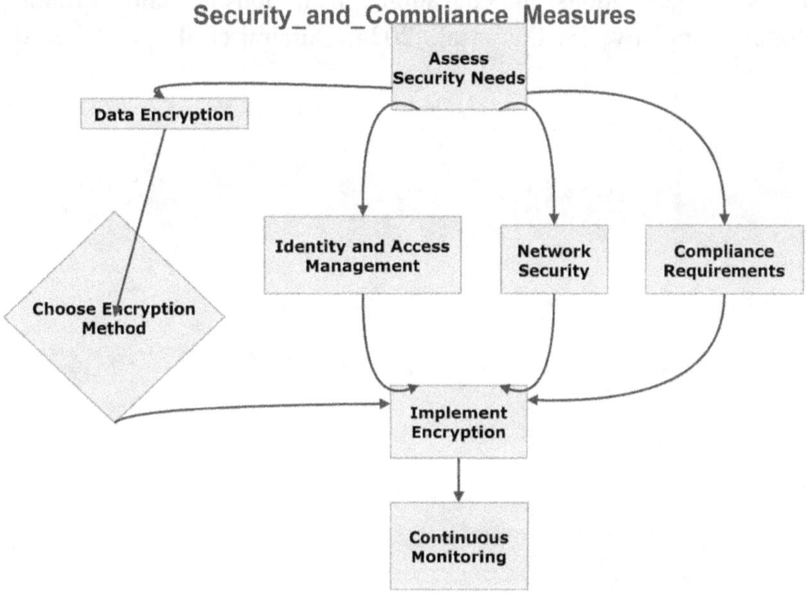

Security and Compliance Measures

The figure 5 outlines the best practices in security and compliance measures.

- To protect cloud data and applications, implement robust security controls like encryption, access controls, IAM, and network security. Regularly assess security posture, conduct vulnerability assessments, and perform penetration testing to identify risks. Adhere to industry-specific compliance requirements and regulatory standards like GDPR, HIPAA, PCI DSS, and SOC 2 for sensitive data protection(Vidhyalakshmi & Kumar, 2016).
- Implement clear policies for cloud data governance, privacy, and compliance management, and offer continuous security awareness training to employees to foster a culture of security and compliance.

Training and Upskilling Initiatives

- Invest in training and upskilling initiatives to equip IT staff with the necessary skills for planning, deploying, and managing cloud-based solutions. Offer training programs, certifications, and hands-on workshops on cloud technologies, platforms, and best practices. Encourage continuous learning through online courses, webinars, conferences, and community forums(Widyastuti & Irwansyah, 2018b).
- Create a culture of innovation and knowledge sharing within the organization to encourage employees to explore new ideas and technologies for cloud adoption. Collaborate with cloud service providers, industry organizations, and educational institutions for tailored training resources and talent development initiatives(S. Das et al., 2024; Durairaj et al., 2023; D. M. Sharma et al., 2024b).

Figure 6. Planning and adaptation for cloud computing in SMEs

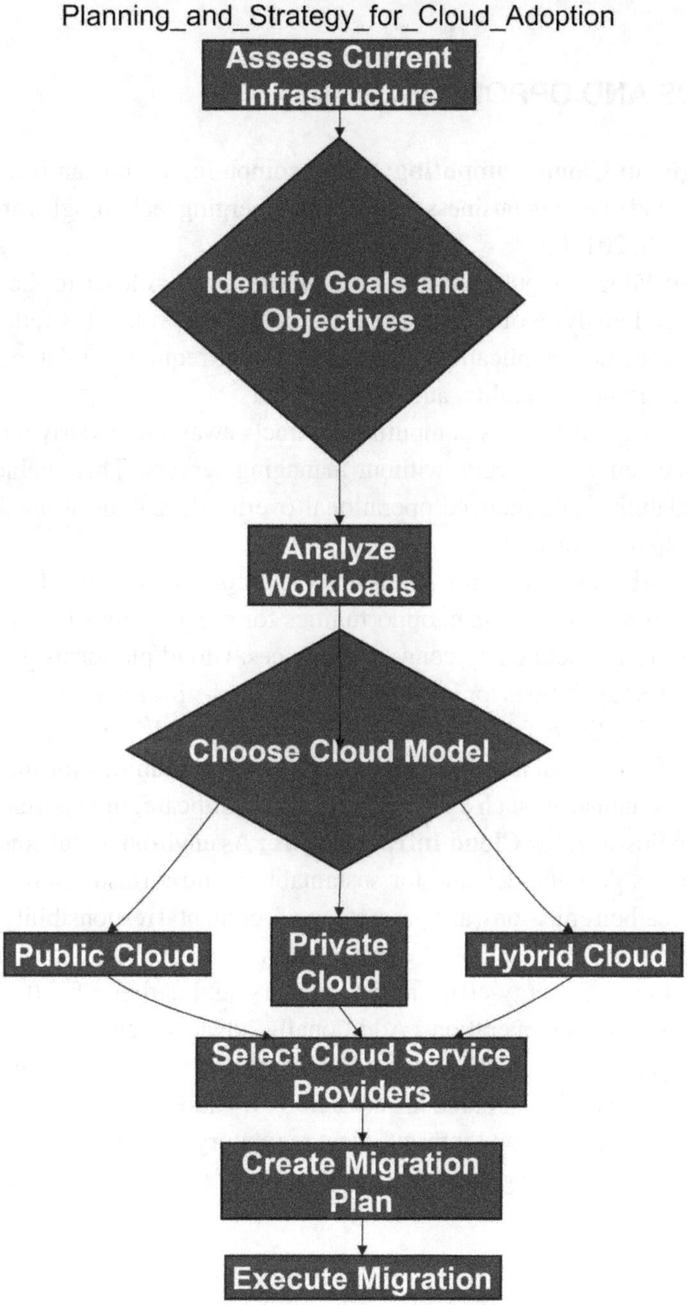

The figure illustrates the planning and adaptation process for SMEs to adopt cloud computing. Successful implementation involves careful planning, selecting the right service providers, implementing robust security measures, and providing IT staff training and upskilling opportunities. This approach ensures successful cloud adoption and helps businesses achieve their objectives in the digital landscape (Shetty & Panda, 2021).

FUTURE TRENDS AND OPPORTUNITIES

Evolving Technologies in Cloud Computing: Cloud computing is constantly evolving due to technological advancements and evolving business needs, with emerging technologies are(Carcary et al., 2014; Widyastuti & Irwansyah, 2018b):

Edge Computing: Edge computing brings compute resources closer to the data source, enabling real-time processing and analysis of data at the edge of the network. This reduces latency, improves performance, and enables new applications and services that require low-latency processing, such as autonomous vehicles, augmented reality, and industrial IoT.

Serverless Computing: Serverless computing abstracts away the underlying infrastructure, allowing developers to focus on writing code without managing servers. This enables faster development cycles, improved scalability, and reduced operational overhead, making it ideal for event-driven and microservices-based applications.

Integration with IoT and Big Data Analytics: The convergence of cloud computing with IoT and big data analytics presents significant opportunities for organizations to derive actionable insights from vast amounts of data generated by connected devices. Cloud platforms provide scalable storage, processing, and analytics capabilities for IoT data, enabling real-time monitoring, analysis, and decision-making(Hema et al., 2023; Syamala et al., 2023a). By integrating IoT devices with cloud services and big data analytics platforms, organizations can unlock new opportunities for innovation, optimization, and cost savings across industries such as manufacturing, healthcare, transportation, and smart cities.

Advancements in Sustainable Cloud Infrastructure: As environmental concerns become increasingly prominent, there is growing demand for sustainable cloud infrastructure that minimizes energy consumption, reduces carbon emissions, and promotes environmental responsibility(Agrawal et al., 2023b; Satav et al., 2023b; Syamala et al., 2023b; Venkateswaran et al., 2023a). Cloud providers are investing in renewable energy sources, energy-efficient data centers, and carbon offsetting initiatives to reduce the environmental impact of their operations. Additionally, advancements in green technologies, such as renewable energy integration, energy-efficient hardware, and data center cooling technologies, are driving improvements in the sustainability of cloud infrastructure. By adopting sustainable cloud infrastructure, organizations can reduce their carbon footprint, meet regulatory requirements, and contribute to global efforts to combat climate change(Agrawal et al., 2023a; Satav et al., 2023a; M. Sharma et al., 2024b).

CONCLUSION

Cloud computing has revolutionized SMEsin engineering, offering cost savings, scalability, collaboration, and security. However, it also presents challenges like security concerns, regulatory compliance, and migration complexities. This chapter explores the adoption of cloud technologies in CAD/CAM, simulation, project management, and remote monitoring. Cloud-based solutions have improved workflows, collaboration, and business growth, while also addressing security concerns and regulatory compliance.

The chapter explores the environmental impact of cloud computing and its potential to reduce carbon emissions through energy-efficient data centers and sustainable practices. It highlights the role of SMEs in combating climate change by embracing renewable energy sources, optimizing infrastructure, and implementing green technologies. It provides insights, case studies, and best practices for SMEs to adopt cloud computing while fulfilling their environmental responsibilities. Future trends include integrating cloud computing with emerging technologies and sustainable infrastructure. The adoption of cloud computing provides SMEs with unique opportunities to innovate, grow, and compete in the digital landscape, while also contributing to a more environmentally sustainable future by utilizing best practices and embracing sustainability.

ABBREVIATIONS

AI: Artificial Intelligence
AWS: Amazon Web Services
CAD: Computer-Aided Design
CAD/CAM: Computer-Aided Design/Computer-Aided Manufacturing
CCPA: California Consumer Privacy Act
GCP: Google Cloud Platform
GDPR: General Data Protection Regulation
HIPAA: Health Insurance Portability and Accountability Act
IaaS: Infrastructure as a Service
IAM: Identity and Access Management
IT: Information Technology
MFA: Multi-Factor Authentication
PaaS: Platform as a Service
PCI DSS: Payment Card Industry Data Security Standard
PPAs: Power Purchase Agreements
RBAC: Role-Based Access Control
RECs: Renewable Energy Certificates
SaaS: Software as a Service
SMEs: Small and Medium-sized Enterprises
VMs: Virtual Machines

REFERENCES

Abdullah, P. Y., Zeebaree, S., Jacksi, K., & Zeabri, R. R. (2020). An hrm system for small and medium enterprises (sme) s based on cloud computing technology. *International Journal of Research-GRANTHAALAYAH*, *8*(8), 56–64.

Agrawal, A. V., Shashibhushan, G., Pradeep, S., Padhi, S., Sugumar, D., & Boopathi, S. (2023a). Synergizing Artificial Intelligence, 5G, and Cloud Computing for Efficient Energy Conversion Using Agricultural Waste. In Sustainable Science and Intelligent Technologies for Societal Development (pp. 475–497). IGI Global.

Agrawal, A. V., Shashibhushan, G., Pradeep, S., Padhi, S., Sugumar, D., & Boopathi, S. (2023b). Synergizing Artificial Intelligence, 5G, and Cloud Computing for Efficient Energy Conversion Using Agricultural Waste. In Sustainable Science and Intelligent Technologies for Societal Development (pp. 475–497). IGI Global.

Attaran, M., & Woods, J. (2019). Cloud computing technology: Improving small business performance using the Internet. *Journal of Small Business and Entrepreneurship*, *31*(6), 495–519.

Bahrami, M. (2015). Cloud computing for emerging mobile cloud apps. *2015 3rd IEEE International Conference on Mobile Cloud Computing, Services, and Engineering*, (pp. 4–5). IEEE.

Boopathi, S. (2024a). Advancements in Machine Learning and AI for Intelligent Systems in Drone Applications for Smart City Developments. In *Futuristic e-Governance Security With Deep Learning Applications* (pp. 15–45). IGI Global.

Boopathi, S. (2024b). Balancing Innovation and Security in the Cloud: Navigating the Risks and Rewards of the Digital Age. In Improving Security, Privacy, and Trust in Cloud Computing (pp. 164–193). IGI Global.

Boopathi, S. (2024c). Balancing Innovation and Security in the Cloud: Navigating the Risks and Rewards of the Digital Age. In Improving Security, Privacy, and Trust in Cloud Computing (pp. 164–193). IGI Global.

Boopathi, S. (2024d). Energy Cascade Conversion System and Energy-Efficient Infrastructure. In Sustainable Development in AI, Blockchain, and E-Governance Applications (pp. 47–71). IGI Global.

Boopathi, S. (2024e). Sustainable Development Using IoT and AI Techniques for Water Utilization in Agriculture. In Sustainable Development in AI, Blockchain, and E-Governance Applications (pp. 204–228). IGI Global.

Boopathi, S., Alqahtani, A. S., Mubarakali, A., & Panchatcharam, P. (2023). Sustainable developments in near-dry electrical discharge machining process using sunflower oil-mist dielectric fluid. *Environmental Science and Pollution Research International*, 1–20.

Boopathi, S., & Khang, A. (2023). AI-Integrated Technology for a Secure and Ethical Healthcare Ecosystem. In *AI and IoT-Based Technologies for Precision Medicine* (pp. 36–59). IGI Global.

Boopathi, S., & Kumar, P. (2024). Advanced bioprinting processes using additive manufacturing technologies: Revolutionizing tissue engineering. *3D Printing Technologies: Digital Manufacturing, Artificial Intelligence, Industry 4.0*, 95.

Boopathi, S., & Myilsamy, S. (2021). Material removal rate and surface roughness study on Near-dry wire electrical discharge Machining process. *Elsevier: Materials Today: Proceedings Elsevier*, *45*(9), 8149–8156.

Carcary, M., Doherty, E., & Conway, G. (2014). The adoption of cloud computing by irish smes an exploratory study. *Electronic Journal of Information Systems Evaluation*, *17*(1), 3–14.

Chhabra, B., Verma, D., & Taneja, B. (2010). Software engineering issues from the cloud application perspective. *International Journal of Information Technology and Knowledge Management*, *2*(2), 669–673.

Das, P., Ramapraba, P., Seethalakshmi, K., Mary, M. A., Karthick, S., & Sampath, B. (2024). Sustainable Advanced Techniques for Enhancing the Image Process. In *Fostering Cross-Industry Sustainability With Intelligent Technologies* (pp. 350–374). IGI Global.

Das, S., Lekhya, G., Shreya, K., Shekinah, K. L., Babu, K. K., & Boopathi, S. (2024). Fostering Sustainability Education Through Cross-Disciplinary Collaborations and Research Partnerships: Interdisciplinary Synergy. In Facilitating Global Collaboration and Knowledge Sharing in Higher Education With Generative AI (pp. 60–88). IGI Global.

Durairaj, M., Jayakumar, S., Karpagavalli, V., Maheswari, B. U., Boopathi, S., & ... (2023). Utilization of Digital Tools in the Indian Higher Education System During Health Crises. In *Multidisciplinary Approaches to Organizational Governance During Health Crises* (pp. 1–21). IGI Global.

Gunasekaran, K., Boopathi, S., & Sureshkumar, M. (2022). Analysis of a Cryogenically Cooled Near-Dry WEDM Process using Different Dielectrics. *Mater-Tehnol.Si. Materials Technology*, *56*(2), 179–186.

Gutierrez, A., Boukrami, E., & Lumsden, R. (2015). Technological, organisational and environmental factors influencing managers' decision to adopt cloud computing in the UK. *Journal of Enterprise Information Management*, *28*(6), 788–807.

Hema, N., Krishnamoorthy, N., Chavan, S. M., Kumar, N., Sabarimuthu, M., & Boopathi, S. (2023). A Study on an Internet of Things (IoT)-Enabled Smart Solar Grid System. In *Handbook of Research on Deep Learning Techniques for Cloud-Based Industrial IoT* (pp. 290–308). IGI Global.

Hentschel, R., Leyh, C., & Baumhauer, T. (2019). *Critical success factors for the implementation and adoption of cloud services in SMEs.*

Khayer, A., Jahan, N., Hossain, M. N., & Hossain, M. Y. (2021). The adoption of cloud computing in small and medium enterprises: A developing country perspective. *VINE Journal of Information and Knowledge Management Systems*, *51*(1), 64–91.

Kratzke, N. (2018). A brief history of cloud application architectures. *Applied Sciences (Basel, Switzerland)*, *8*(8), 1368.

Kreslins, K., Novik, D., & Vasiljeva, T. (2018). Challenge of cloud computing for SMEs: A case of Baltic countries. *Journal of Innovation Management in Small & Medium Enterprises, 2018.*

Kumalo, N., & Van Der Poll, J. A. (2015). The role of cloud computing in addressing SME challenges in South Africa. *2015 International Conference on Cloud Computing Research and Innovation (Icccri)*, (pp. 139–147). IEEE.

Kumar, M., Kumar, K., Sasikala, P., Sampath, B., Gopi, B., & Sundaram, S. (2023). Sustainable Green Energy Generation From Waste Water: IoT and ML Integration. In Sustainable Science and Intelligent Technologies for Societal Development (pp. 440–463). IGI Global.

Kumar, S., & Buyya, R. (2012). Green cloud computing and environmental sustainability. *Harnessing Green IT: Principles and Practices*, 315–339.

Liu, Y., Ni, Z., Karlsson, M., & Gong, S. (2021). Methodology for digital transformation with internet of things and cloud computing: A practical guideline for innovation in small-and medium-sized enterprises. *Sensors (Basel)*, *21*(16), 5355.

Malathi, J., Kusha, K., Isaac, S., Ramesh, A., Rajendiran, M., & Boopathi, S. (2024). IoT-Enabled Remote Patient Monitoring for Chronic Disease Management and Cost Savings: Transforming Healthcare. In Advances in Explainable AI Applications for Smart Cities (pp. 371–388). IGI Global.

Modisane, P., & Jokonya, O. (2021). Evaluating the benefits of cloud computing in small, medium and micro-sized enterprises (SMMEs). *Procedia Computer Science*, *181*, 784–792.

Mujinga, M. (2020). Cloud Computing Inhibitors Among Small and Medium Enterprises. *2020 3rd International Conference on Intelligent Sustainable Systems (ICISS)*, 1385–1391.

Myilsamy, S., & Sampath, B. (2021). Experimental comparison of near-dry and cryogenically cooled near-dry machining in wire-cut electrical discharge machining processes. *IOP:Surface Topography: Metrology and Properties*, *9*(3), 035015.

Nanda, A. K., Sharma, A., Augustine, P. J., Cyril, B. R., Kiran, V., & Sampath, B. (2024a). Securing Cloud Infrastructure in IaaS and PaaS Environments. In Improving Security, Privacy, and Trust in Cloud Computing (pp. 1–33). IGI Global.

Nanda, A. K., Sharma, A., Augustine, P. J., Cyril, B. R., Kiran, V., & Sampath, B. (2024b). Securing Cloud Infrastructure in IaaS and PaaS Environments. In Improving Security, Privacy, and Trust in Cloud Computing (pp. 1–33). IGI Global.

Nishanth, J., Deshmukh, M. A., Kushwah, R., Kushwaha, K. K., Balaji, S., & Sampath, B. (2023). Particle Swarm Optimization of Hybrid Renewable Energy Systems. In *Intelligent Engineering Applications and Applied Sciences for Sustainability* (pp. 291–308). IGI Global.

Otuka, R., Preston, D., & Pimenidis, E. (2014). The use and challenges of cloud computing services in SMEs in Nigeria. *Proceedings of the European Conference on Information Management*, 325. IEEE.

Pachiappan, K., Anitha, K., Pitchai, R., Sangeetha, S., Satyanarayana, T., & Boopathi, S. (2024). Intelligent Machines, IoT, and AI in Revolutionizing Agriculture for Water Processing. In *Handbook of Research on AI and ML for Intelligent Machines and Systems* (pp. 374–399). IGI Global.

Paul, A., & Thilagham, K. KG, J.-, Reddy, P. R., Sathyamurthy, R., & Boopathi, S. (2024). Multi-criteria Optimization on Friction Stir Welding of Aluminum Composite (AA5052-H32/B4C) using Titanium Nitride Coated Tool. Engineering Research Express.

Rahamathunnisa, U., Subhashini, P., Aancy, H. M., Meenakshi, S., Boopathi, S., & ... (2023). Solutions for Software Requirement Risks Using Artificial Intelligence Techniques. In *Handbook of Research on Data Science and Cybersecurity Innovations in Industry 4.0 Technologies* (pp. 45–64). IGI Global.

Rahamathunnisa, U., Sudhakar, K., Murugan, T. K., Thivaharan, S., Rajkumar, M., & Boopathi, S. (2023). Cloud Computing Principles for Optimizing Robot Task Offloading Processes. In *AI-Enabled Social Robotics in Human Care Services* (pp. 188–211). IGI Global.

Rahamathunnisa, U., Sudhakar, K., Padhi, S., Bhattacharya, S., Shashibhushan, G., & Boopathi, S. (2024). Sustainable Energy Generation From Waste Water: IoT Integrated Technologies. In Adoption and Use of Technology Tools and Services by Economically Disadvantaged Communities: Implications for Growth and Sustainability (pp. 225–256). IGI Global.

Raut, R. D., Gardas, B. B., Narkhede, B. E., & Narwane, V. S. (2019). To investigate the determinants of cloud computing adoption in the manufacturing micro, small and medium enterprises: A DEMATEL-based approach. *Benchmarking*, *26*(3), 990–1019.

Rebecca, B., Kumar, K. P. M., Padmini, S., Srivastava, B. K., Halder, S., & Boopathi, S. (2024). Convergence of Data Science-AI-Green Chemistry-Affordable Medicine: Transforming Drug Discovery. In *Handbook of Research on AI and ML for Intelligent Machines and Systems* (pp. 348–373). IGI Global.

Revathi, S., Babu, M., Rajkumar, N., Meti, V. K. V., Kandavalli, S. R., & Boopathi, S. (2024). Unleashing the Future Potential of 4D Printing: Exploring Applications in Wearable Technology, Robotics, Energy, Transportation, and Fashion. In Human-Centered Approaches in Industry 5.0: Human-Machine Interaction, Virtual Reality Training, and Customer Sentiment Analysis (pp. 131–153). IGI Global.

Samikannu, R., Koshariya, A. K., Poornima, E., Ramesh, S., Kumar, A., & Boopathi, S. (2022). Sustainable Development in Modern Aquaponics Cultivation Systems Using IoT Technologies. In *Human Agro-Energy Optimization for Business and Industry* (pp. 105–127). IGI Global.

Sampath, B. (2021). *Sustainable Eco-Friendly Wire-Cut Electrical Discharge Machining: Gas Emission Analysis.*

Saravanan, A., Venkatasubramanian, R., Khare, R., Surakasi, R., Boopathi, S., Ray, S., & Sudhakar, M. (2022). POLICY TRENDS OF RENEWABLE ENERGY AND NON. *Renewable Energy.*

Satav, S. D., Lamani, D., Harsha, K., Kumar, N., Manikandan, S., & Sampath, B. (2023a). Energy and Battery Management in the Era of Cloud Computing: Sustainable Wireless Systems and Networks. In Sustainable Science and Intelligent Technologies for Societal Development (pp. 141–166). IGI Global.

Satav, S. D., Lamani, D., Harsha, K., Kumar, N., Manikandan, S., & Sampath, B. (2023b). Energy and Battery Management in the Era of Cloud Computing: Sustainable Wireless Systems and Networks. In Sustainable Science and Intelligent Technologies for Societal Development (pp. 141–166). IGI Global.

Schniederjans, D. G., & Hales, D. N. (2016). Cloud computing and its impact on economic and environmental performance: A transaction cost economics perspective. *Decision Support Systems*, *86*, 73–82.

Sharma, D. M., Ramana, K. V., Jothilakshmi, R., Verma, R., Maheswari, B. U., & Boopathi, S. (2024a). Integrating Generative AI Into K-12 Curriculums and Pedagogies in India: Opportunities and Challenges. *Facilitating Global Collaboration and Knowledge Sharing in Higher Education With Generative AI*, 133–161.

Sharma, D. M., Ramana, K. V., Jothilakshmi, R., Verma, R., Maheswari, B. U., & Boopathi, S. (2024b). Integrating Generative AI Into K-12 Curriculums and Pedagogies in India: Opportunities and Challenges. *Facilitating Global Collaboration and Knowledge Sharing in Higher Education With Generative AI*, 133–161.

Sharma, M., Sharma, M., Sharma, N., & Boopathi, S. (2024a). Building Sustainable Smart Cities Through Cloud and Intelligent Parking System. In *Handbook of Research on AI and ML for Intelligent Machines and Systems* (pp. 195–222). IGI Global.

Sharma, M., Sharma, M., Sharma, N., & Boopathi, S. (2024b). Building Sustainable Smart Cities Through Cloud and Intelligent Parking System. In *Handbook of Research on AI and ML for Intelligent Machines and Systems* (pp. 195–222). IGI Global.

Shetty, J. P., & Panda, R. (2021). An overview of cloud computing in SMEs. *Journal of Global Entrepreneurship Research, 11*(1), 175–188.

Srinivas, B., Maguluri, L. P., Naidu, K. V., Reddy, L. C. S., Deivakani, M., & Boopathi, S. (2023). Architecture and Framework for Interfacing Cloud-Enabled Robots. In *Handbook of Research on Data Science and Cybersecurity Innovations in Industry 4.0 Technologies* (pp. 542–560). IGI Global.

Subha, S., Inbamalar, T., Komala, C., Suresh, L. R., Boopathi, S., & Alaskar, K. (2023). A Remote Health Care Monitoring system using internet of medical things (IoMT). *IEEE Explore*, (pp. 1–6). IEEE.

Sundaramoorthy, K., Singh, A., Sumathy, G., Maheshwari, A., Arunarani, A., & Boopathi, S. (2024). A Study on AI and Blockchain-Powered Smart Parking Models for Urban Mobility. In *Handbook of Research on AI and ML for Intelligent Machines and Systems* (pp. 223–250). IGI Global.

Syamala, M., Komala, C., Pramila, P., Dash, S., Meenakshi, S., & Boopathi, S. (2023a). Machine Learning-Integrated IoT-Based Smart Home Energy Management System. In *Handbook of Research on Deep Learning Techniques for Cloud-Based Industrial IoT* (pp. 219–235). IGI Global.

Syamala, M., Komala, C., Pramila, P., Dash, S., Meenakshi, S., & Boopathi, S. (2023b). Machine Learning-Integrated IoT-Based Smart Home Energy Management System. In *Handbook of Research on Deep Learning Techniques for Cloud-Based Industrial IoT* (pp. 219–235). IGI Global.

Tamburri, D. A., Miglierina, M., & Di Nitto, E. (2020). Cloud applications monitoring: An industrial study. *Information and Software Technology, 127*, 106376.

Truong, H.-L., & Dustdar, S. (2015). Principles for engineering IoT cloud systems. *IEEE Cloud Computing, 2*(2), 68–76.

Vajjhala, N. R., & Ramollari, E. (2021). *Big data using cloud computing-opportunities for small and medium-sized enterprises.*

Vasiljeva, T., Shaikhulina, S., & Kreslins, K. (2017). Cloud computing: Business perspectives, benefits and challenges for small and medium enterprises (case of Latvia). *Procedia Engineering, 178*, 443–451.

Venkateswaran, N., Vidhya, K., Ayyannan, M., Chavan, S. M., Sekar, K., & Boopathi, S. (2023a). A Study on Smart Energy Management Framework Using Cloud Computing. In 5G, Artificial Intelligence, and Next Generation Internet of Things: Digital Innovation for Green and Sustainable Economies (pp. 189–212). IGI Global.

Venkateswaran, N., Vidhya, K., Ayyannan, M., Chavan, S. M., Sekar, K., & Boopathi, S. (2023b). A Study on Smart Energy Management Framework Using Cloud Computing. In 5G, Artificial Intelligence, and Next Generation Internet of Things: Digital Innovation for Green and Sustainable Economies (pp. 189–212). IGI Global.

Vennila, T., Karuna, M., Srivastava, B. K., Venugopal, J., Surakasi, R., & Sampath, B. (2022). New Strategies in Treatment and Enzymatic Processes: Ethanol Production From Sugarcane Bagasse. In Human Agro-Energy Optimization for Business and Industry (pp. 219–240). IGI Global.

Verma, R., Christiana, M. B. V., Maheswari, M., Srinivasan, V., Patro, P., Dari, S. S., & Boopathi, S. (2024). Intelligent Physarum Solver for Profit Maximization in Oligopolistic Supply Chain Networks. In *AI and Machine Learning Impacts in Intelligent Supply Chain* (pp. 156–179). IGI Global.

Vidhyalakshmi, R., & Kumar, V. (2016). Determinants of cloud computing adoption by SMEs. *International Journal of Business Information Systems*, *22*(3), 375–395.

Vijayakumar, G. N. S., Domakonda, V. K., Farooq, S., Kumar, B. S., Pradeep, N., & Boopathi, S. (2024). Sustainable Developments in Nano-Fluid Synthesis for Various Industrial Applications. In Adoption and Use of Technology Tools and Services by Economically Disadvantaged Communities: Implications for Growth and Sustainability (pp. 48–81). IGI Global.

Widyastuti, D., & Irwansyah, I. (2018a). Benefits and challenges of cloud computing technology adoption in small and medium enterprises (SMEs). *Bandung Creative Movement (BCM), 4*(1).

Widyastuti, D., & Irwansyah, I. (2018b). Benefits and challenges of cloud computing technology adoption in small and medium enterprises (SMEs). *Bandung Creative Movement (BCM), 4*(1).

Chapter 15

Assessment of Water Quality Using Multi-Layered Mamdani Fuzzy Inference Expert System

Gulzar Ahmad
https://orcid.org/0009-0000-5869-1164
Minhaj University, Lahore, Pakistan

Sehrish Aqeel
University of Malaysia, Sarawak, Malaysia

Zobia Zafar
University of South Asia, Pakistan

Kanza Fatima
University of South Asia, Pakistan

ABSTRACT

In this chapter, a new multi-layered mamdani fuzzy inference system (ML-MFIS) is proposed for the assessment of water quality (AWQ). The proposed AWQ-ML-MFIS expert system can categorize the level of water quality into excellent, normal, or dangerous for health (polluted). AWQ-ML-MFIS expert system for drinking water is developed by the guidelines from the World Health Organization (WHO) and Pakistan's Punjab environmental quality standard. AWQ-ML-MFIS expert system uses input water parameters such as bacterial, physical, chemical, and radioactive for different layers. The chemical parameters are further divided into essential inorganics, toxic inorganics, and organics. This chapter also analyses the intensities of the parameters and the results achieved by using the proposed AWQ-ML-MFIS expert system. All these parameters and results are discussed with the experts of the Pakistan Council of Scientific and Industrial Research (PCSIR), Lahore. The accuracy of the proposed AWQ-ML-MFIS Expert System is more accurate as compared to others approaches used for it.

DOI: 10.4018/979-8-3693-1794-5.ch015

INTRODUCTION

Water is significant for the endurance of any form of life. Human health and happiness are closely related to the quality of water used for use where water consumption per capita is an indication of their living standard, their economic and social situation. Any kind of animal, plant, and human being can't survive without water (Boulay et al. 2015). All biochemical reactions take place in an aqueous medium. It is necessary for the maintenance of body temperature and fluids in living organisms. Water is responsible for the transport of different materials in organisms. Water is a medium of life for aquatic organisms. Water is vital for a human being in many ways like drinking, industrial manufacturing, washing, irrigation, fisheries, power generation, shipping, etc. due to this, the water demand is increasing enormously.

0.002% of the water out of about 70% of Earth's surface is available for human consumption (Alrumman et al. 2016). Most of the water is polluted, which is not suitable for human health. Human Health can affect by using polluted water (Beer et al. 2015). Treatment and assessment of water quality are essential for drinking water (Melita et al. 2014). Different diseases like Typhoid, Hepatitis, Cholera, Paratyphoid Fever, Diarrhea, Malaria, Amoebiasis, etc. are the primary source of contaminated water (Alrumman et al. 2016, Rahmanian et al. 2015, Verma and Gupta 2015). The excess of different elements like Total Hardness as CaCo3, TDS, pH, Aluminum(Al), Antimony (Sb), bacterial, etc. are the main cause to make water contaminated (Abtahi et al. 2015, Tiwari et al. 2015, El-Chakhtoura et al. 2015, Haydar et al. 2016). So, the quality of drinking water must be good to save human life from such diseases. Harmless water is not available for humans from one-sixth of the world's population, presented in WHO report (Alrumman et al. 2016, WHO 2017). To provide the clean water for a human is the big issue for all over the world specially developing countries like Pakistan. There are different methods to assess the water quality like Water Quality Index (WQI) introduced by Ott (1976), Weighted Index Number (WIN), etc.

Various countries have a framework to appraisal, screen, looking at independent extends of freshwater regarding their synthetic, organic, and supplement constituents and general aesthetic condition for a wide range of water like river and groundwater (Jacobs et al. 1965, Chang et al. 2001, Sii et al. 1993, Dojlido et al. 1994).

Horton (1965) first time used Weighted Index Number (WIN) technique to measure the water quality. Water Quality Index (WQI) is another assessment of water quality technique is developed by National Sanitation Foundation (NSF) (Ott 1978, Abtahi et al. 2015). WQI using the Delphi Methodology(DM) as a tool in a formal valuation process. By utilizing these assessment methods, differences typically arise from the lack of apparent discrepancies between each mode. The ambiguity, inaccuracy, or fuzziness appears in the decision-making output values. Sometimes, it is tough to find the right results of water quality on a seasonal basis. The conflicting values can look in the analysis by using these techniques. To remove the ambiguity in the results, an advanced method is needed which may capable of accounting for vague information has long been recognized.

The vulnerabilities associated with utilizing fuzzy membership with values going from 0 to 1 to shape an appropriate fuzzy set rather than the 0 to 100 scales used as a part of standard rating curves in WQI first time utilized by Sii et al. in 1993 (Sii et al. 1993, Yalcuk and Postalcioglu 2015). Semiromi et al. discussed the river water quality by using fuzzy logic based on some water parameters like turbidity, fecal coliform, total dissolved solids (TDS), dissolved oxygen (DO), and pH (Semiromi et al. 2011).

In this research, a new Assessment Water Quality Multi-Layered MFIS based (AWQ-ML-MFIS) Expert System is proposed after the comparative study of WQI and fuzzy synthetic evaluation approaches. There are a total of thirty-three bacterial, chemical, radioactive, and physical water parameters, and their numerical values are used (shown in table-1) to assess the more accurate water quality. The Proposed Expert System is developed based on above mentioned desired parameters. These Parameter values are taken from The Punjab Standards for Drinking Water Quality approved by Pakistan Environmental Protection Council (PEPC) & WHO guidelines (GOP 2016). This proposed Expert System is useful for assessing the water quality of all kinds of drinking water like wall water, Groundwater, river water, and mineral water companies.

PROPOSED ASSESSMENT WATER QUALITY MULTI-LAYERED MFIS BASED EXPERT SYSTEM (AWQ-ML-MFIS ES)

In this article, a novel Multi-Layered MFIS based Expert System is proposed for the assessment of water quality as shown in Fig. 1. In the proposed Assessment Water Quality Multi-Layered MFIS based Expert System (AWQ-ML-MFIS-ES) use all possible suitable parameters about Pakistan. Some other water elements are not used which are not applicable in Pakistan Water Quality Standard. By water elements characteristics, the expert system is divided into multiple layers which present the complete structure of the proposed method shown in Fig. 2. Level-1 layers exams the existence of chemical elements (Essential-Inorganic, Toxic-Inorganic, Organic), physical and radioactive parameters. The final layer calculates the water Quality based on level-1 layers output & detection of bacteria as shown in Fig. 2.

The values of Bacterial, Physical, Chemical (Essential Inorganic, Toxic Inorganic, and Organic), and Radioactive parameters are also used to build up a lookup table of proposed AWQ-ML-MFIS-ES as shown in Table-1. In the proposed system thirty-three, input parameters are used to Assessment Water Quality. Proposed AWQ-ML-MFIS Expert System mathematically can be written as

$$\mu_{B \cap P \cap E \cap T \cap O \cap R}(b, p, e, t, o, r) = \min[\mu_B(b), \mu_P(p), \mu_E(e), \mu_T(t), \mu_O(o), \mu_R(r)] \quad (1)$$

Figure 1. Proposed AWQ-ML-MFIS expert system methodology

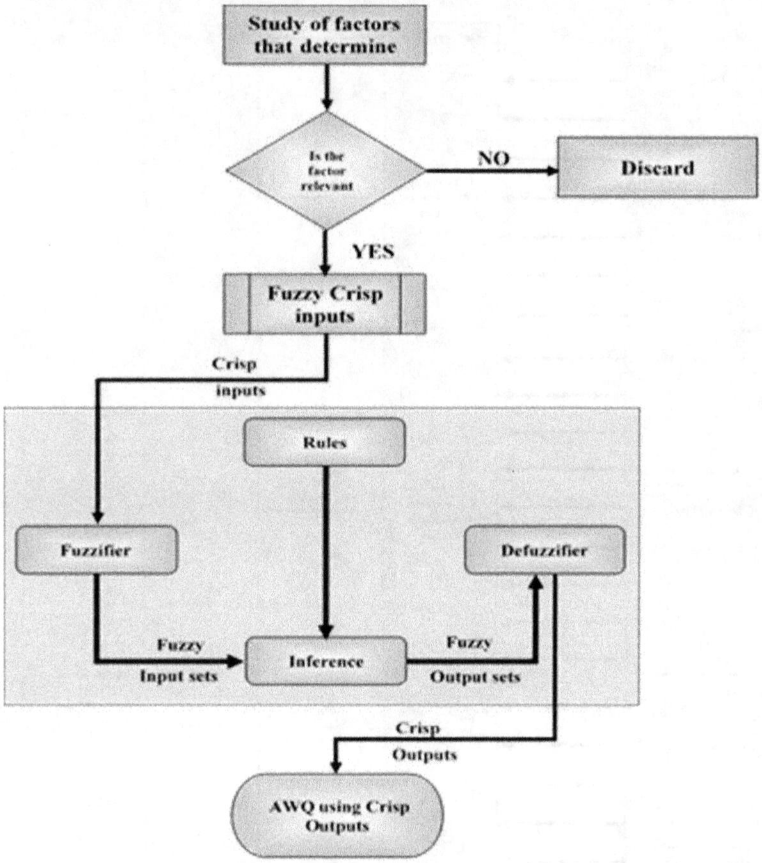

Input Fuzzy Sets

To Assessment the Water quality the statistical values of fuzzy input variables are used. The numerical values are divided into three categories which are Excellent, Normal, and Bad (polluted and dangerous for health) after the discussion with a biochemistry expert. The description of all input variables with their numerical values is shown in Table 1.

Figure 2. Proposed AWQ-ML-MFIS expert system

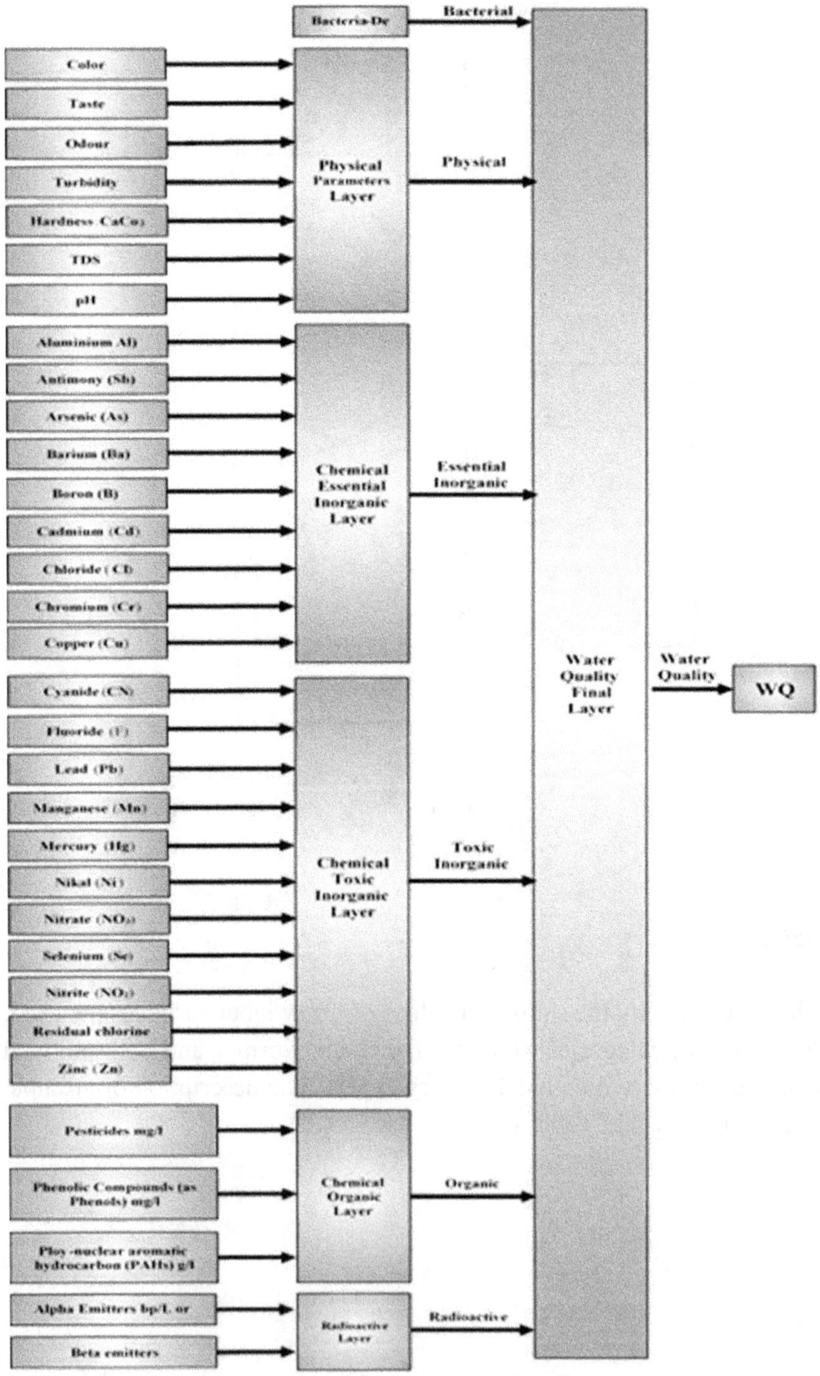

Table 1. Input variable of proposed AWQ-ML-MFIS expert system (Abtahi et al. 2015, Wasana et al. 2016, Shahid et al. 2017)

Bacterial			
Properties/Parameter	**Excellent**		**Bad**
Bacterial	Negative detection in 100 ml Sample		yes
Physical			
Properties/Parameter	**Excellent**	**Normal**	**Bad**
Color	≤13*TCU*	12*TCU*≤N≤16*TCU*	15*TCU*≥
Taste	No	No	Yes
Odour	No	No	Yes
Turbidity	≤3*TCU*	2.5*NTU*≤*T*≤5.5*NTU*	5*NTU*≥
Total Hardness as $CaCo_3$	≤200mg/l	170mg/l≤H≤530mg/l	500mg/l≥
TDS	≤600	550≤DS≤1050	1000mg/l≥
pH	6.5–8.5	6.5 – 8.5	≥6.5 – 8.5≤
Chemical: Essential Inorganic			
Properties/Parameter	**Excellent**	**Normal**	**Bad**
Aluminium(Al)	0.14 – 0.22	≤0.16	≥0.2
Antimony (Sb)	0.0038 – 0.0052	≤0.004(*P*)	≥0.005
Arsenic (As)	0.038 – 0.052	≤0.04(*P*)	≥0.05(*P*)
Barium (Ba)	0.47 – 0.73	≤0.5	≥0.7
Boron (B)	0.18 – 0.32	≤0.2	≥0.3
Cadmium (Cd)	0.0008 – 0.0011	≤0.0009	≥0.001
Chloride (Cl)	180 – 270	≤200	≥250
Chromium (Cr)	0.038 – 0.052	≤0.04(*P*)	≥0.05(*P*)
Copper (Cu)	0.8 – 2.2	≤1	≥2
Chemical: Toxic Inorganic			
Properties/Parameter	**Excellent**	**Normal**	**Bad**
Cyanide (CN)	≤0.01	0.008 – 0.052	≥0.05
Fluoride (F)	≤0.6	0.5 – 1.6	≥1.5
Lead (Pb)	≤0.012	0.01 – 0.052	≥0.05
Manganese (Mn)	≤0.3	0.28 – 1.52	≥0.5
Mercury (Hg)	≤0.0007	0.0006 – 0.0011	≥0.004
Nikal (Ni)	≤0.01	0.009 – 0.021	≥0.02
Nitrate (NO_3)	≤20	18 – 52	≥50
Nitrite (NO_2)	≤1(*P*)	0.9 – 3.1	≥3
Selenium (Se)	≤0.0015(*P*)	0.001 – 0.0105(*P*)	≥0.01(*P*)
Residual chlorine	≤0.2	0.18 – 0.52	≥0.5
Zinc (Zn)	≤2.0	1.8 – 5.2	≥5
Chemical: Organic			
Properties/Parameter	**Excellent**	**Normal**	**Bad**
Pesticides *μg/l*	≤0.2	0.18-0.52	≥0.5

Continued on following page

Table 1. Continued

Bacterial				
Phenolic Compounds (as Phenols) *mg/l*		≤0.001	0.0008 – 0.0022	≥0.002
Ploy-nuclear aromatic hydrocarbon (PAHs) g/l		≤0.003	0.002 – 0.011	≥0.01
Radioactive				
Properties/Parameter	**Excellent**	**Normal**		**Bad**
Alpha Emitters bp/L or pCi	≤0.003	0.002 – 0.11		≥0.1
Beta Emitters	≤0.3	0.2 – 1.1		≥1

Fuzzy Output Variable

In this research, multi-layered architecture is proposed to Assessment the water quality. The output variables of all level-1 layers & the final layer are shown in Table 2.

Table 2. Output variable of proposed AWQ-ML-MFIS expert system

Sr #	Layers	Output Variables	Semantic Sign
1	Level-1 Layers	Physical, Essential-Inorganic, Toxic-Inorganic, Organic, and Radioactive	Excellent
			Normal
			Bad (Polluted)
			Normal
			Bad (Polluted)
2	Final Layer	Water Quality	Excellent
			Normal
			Bad (Polluted)

Membership Functions

The membership function of this system gives curve value between 0, 1 and also provides a mathematical function which offers statistical values of input and output variable. Input and Output Variables Membership Functions used in Proposed AWQ-ML-MFIS Expert System are given in Table 3.

Table 3. Input and output variables membership functions used in proposed AWQ-ML-MFIS expert system

Variables	Membership Function(MF)	Graphical Representation of MF
Bacterial=B $\mu_B(b)$	$\mu_{B,No}(b) = \begin{Bmatrix} 1, & if\ b = No \\ 0, & else \end{Bmatrix}$	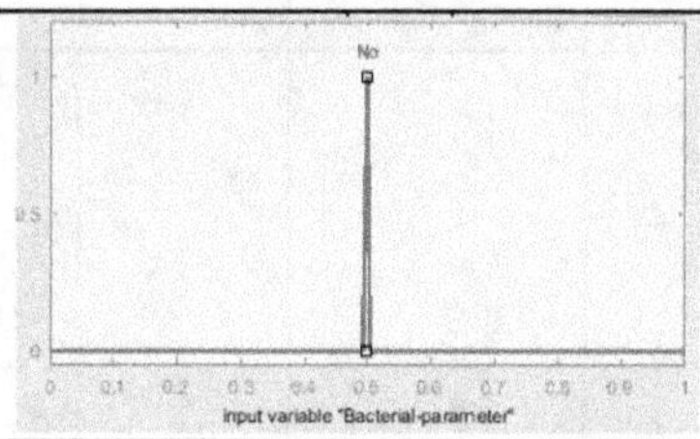
Physical=P $\mu_P(p)$	$\mu_{P,B}(p) = \begin{cases} 1, & p \le 0.3 \\ 1 - 2\left(\frac{p-0.3}{0.1}\right)^2, & 0.3 \le p \le 0.35 \\ 2\left(\frac{p-0.4}{0.1}\right)^2, & 0.35 \le p \le 0.4 \\ 0, & p \ge 0.4 \end{cases}$ $\mu_{P,N}(p) = \begin{cases} \frac{p-0.3}{0.1}, & 0.3 \le p \le 0.4 \\ 1, & 0.4 \le p \le 0.6 \\ \frac{0.7-p}{0.1}, & 0.6 \le p \le 0.7 \end{cases}$ $\mu_{P,E}(p) = \begin{cases} 0, & p \le 0.6 \\ 2\left(\frac{p-0.6}{0.1}\right)^2, & 0.6 \le p \le 0.65 \\ 1 - 2\left(\frac{p-0.7}{0.1}\right)^2, & 0.65 \le p \le 0.7 \\ 1, & p \ge 0.7 \end{cases}$	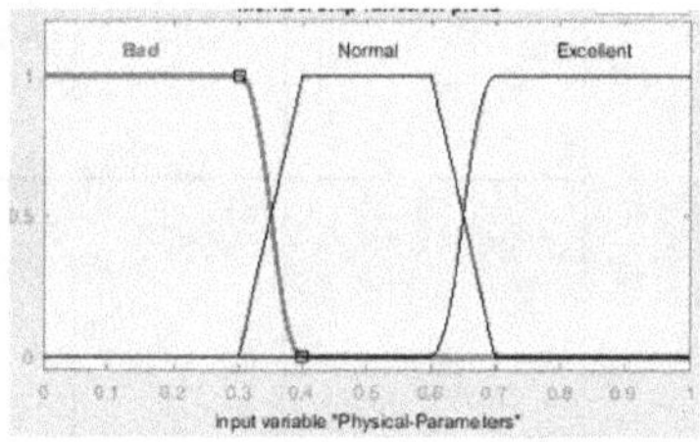
Chemical: Essential Inorganic=E $\mu_E(e)$	$\mu_{phy.B}(e) = \begin{cases} 1, & e \le 0.3 \\ 1 - 2\left(\frac{e-0.3}{0.1}\right)^2, & 0.3 \le e \le 0.35 \\ 2\left(\frac{e-0.4}{0.1}\right)^2, & 0.35 \le e \le 0.4 \\ 0, & P \ge 0.4 \end{cases}$ $\mu_{phy.N}(e) = \begin{cases} \frac{e-0.3}{0.1}, & 0.3 \le e \le 0.4 \\ 1, & 0.4 \le e \le 0.6 \\ \frac{0.7-e}{0.1}, & 0.6 \le e \le 0.7 \end{cases}$ $\mu_{phy.E}(e) = \begin{cases} 0, & e \le 0.6 \\ 2\left(\frac{e-0.6}{0.1}\right)^2, & 0.6 \le e \le 0.65 \\ 1 - 2\left(\frac{e-0.7}{0.1}\right)^2, & 0.65 \le e \le 0.7 \\ 1, & e \ge 0.7 \end{cases}$	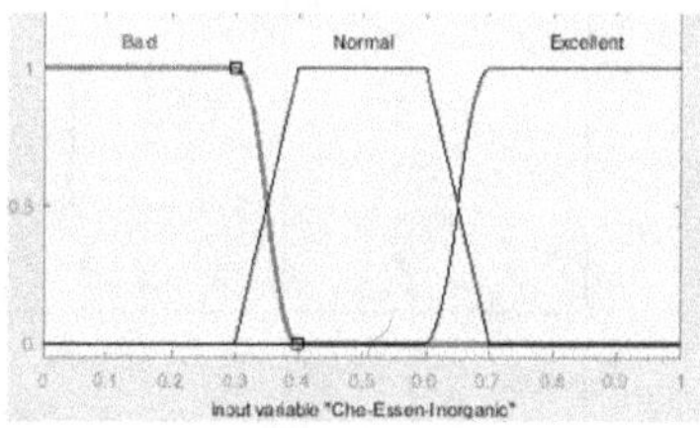
Chemical: Toxic Inorganic=T $\mu_T(t)$	$\mu_{phy.B}(t) = \begin{cases} 1, & t \le 0.3 \\ 1 - 2\left(\frac{t-0.3}{0.1}\right)^2, & 0.3 \le t \le 0.35 \\ 2\left(\frac{t-0.4}{0.1}\right)^2, & 0.35 \le t \le 0.4 \\ 0, & t \ge 0.4 \end{cases}$ $\mu_{phy.N}(t) = \begin{cases} \frac{t-0.3}{0.1}, & 0.3 \le t \le 0.4 \\ 1, & 0.4 \le t \le 0.6 \\ \frac{0.7-t}{0.1}, & 0.6 \le t \le 0.7 \end{cases}$ $\mu_{phy.E}(t) = \begin{cases} 0, & t \le 0.6 \\ 2\left(\frac{t-0.6}{0.1}\right)^2, & 0.6 \le t \le 0.65 \\ 1 - 2\left(\frac{t-0.7}{0.1}\right)^2, & 0.65 \le t \le 0.7 \\ 1, & t \ge 0.7 \end{cases}$	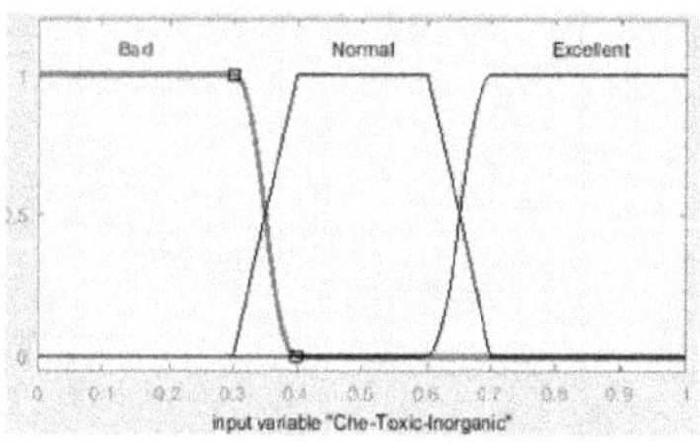

Continued on following page

Table 3. Continued

Variables	Membership Function(MF)	Graphical Representation of MF
Chemical: Organic=O $\mu_O(o)$	$\mu_{phy.B}(o)=\begin{cases}1, & o\le 0.3\\ 1-2\left(\frac{o-0.3}{0.1}\right)^2, & 0.3\le o\le 0.35\\ 2\left(\frac{o-0.4}{0.1}\right)^2, & 0.35\le o\le 0.4\\ 0, & o\ge 0.4\end{cases}$ $\mu_{phy.N}(o)=\begin{cases}\frac{o-0.3}{0.1}, & 0.3\le o\le 0.4\\ 1, & 0.4\le o\le 0.6\\ \frac{0.7-o}{0.1}, & 0.6\le o\le 0.7\end{cases}$ $\mu_{phy.E}(o)=\begin{cases}0, & o\le 0.6\\ 2\left(\frac{o-0.6}{0.1}\right)^2, & 0.6\le o\le 0.65\\ 1-2\left(\frac{o-0.7}{0.1}\right)^2, & 0.65\le o\le 0.7\\ 1, & o\ge 0.7\end{cases}$	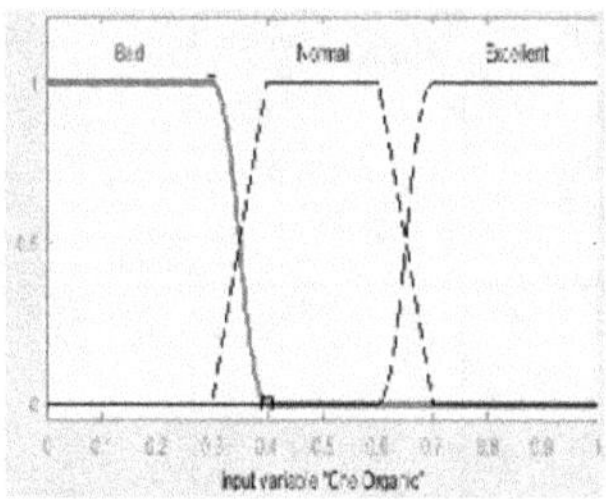
Radioactive=R $\mu_R(r)$	$\mu_{phy.B}(r)=\begin{cases}1, & r\le 0.3\\ 1-2\left(\frac{r-0.3}{0.1}\right)^2, & 0.3\le r\le 0.35\\ 2\left(\frac{r-0.4}{0.1}\right)^2, & 0.35\le r\le 0.4\\ 0, & r\ge 0.4\end{cases}$ $\mu_{phy.N}(r)=\begin{cases}\frac{r-0.3}{0.1}, & 0.3\le r\le 0.4\\ 1, & 0.4\le r\le 0.6\\ \frac{0.7-r}{0.1}, & 0.6\le r\le 0.7\end{cases}$ $\mu_{phy.E}(r)=\begin{cases}0, & r\le 0.6\\ 2\left(\frac{r-0.6}{0.1}\right)^2, & 0.6\le r\le 0.65\\ 1-2\left(\frac{r-0.7}{0.1}\right)^2, & 0.65\le r\le 0.7\\ 1, & r\ge 0.7\end{cases}$	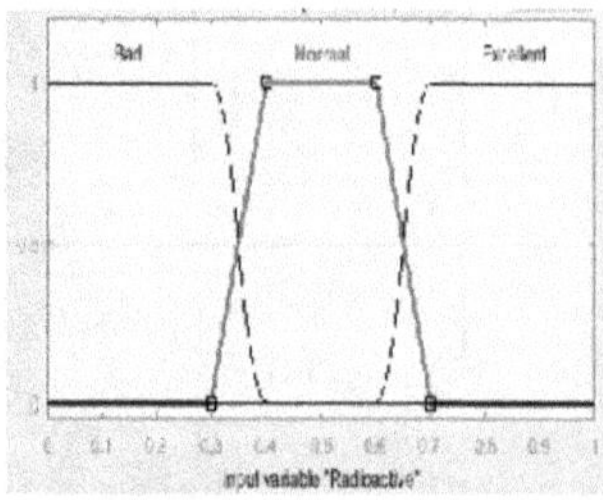
Output $\left(\mu_{AWQ}(W_q)\right)$	$\mu_{AWQ.B}(W_q)=\begin{cases}1, & W_q\le 0.3\\ 1-2\left(\frac{W_q-0.3}{0.1}\right)^2, & 0.3\le W_q\le 0.35\\ 2\left(\frac{W_q-0.4}{0.1}\right)^2, & 0.35\le W_q\le 0.4\\ 0, & P\ge 0.4\end{cases}$ $\mu_{AWQ.N}(W_q)=\begin{cases}\frac{W_q-0.3}{0.1}, & 0.3\le W_q\le 0.4\\ 1, & 0.4\le W_q\le 0.6\\ \frac{0.7-W_q}{0.1}, & 0.6\le W_q\le 0.7\end{cases}$ $\mu_{AWQ.E}(W_q)=\begin{cases}0, & W_q\le 0.6\\ 2\left(\frac{W_q-0.6}{0.1}\right)^2, & 0.6\le W_q\le 0.65\\ 1-2\left(\frac{W_q-0.7}{0.1}\right)^2, & 0.65\le W_q\le 0.7\\ 1, & W_q\ge 0.7\end{cases}$	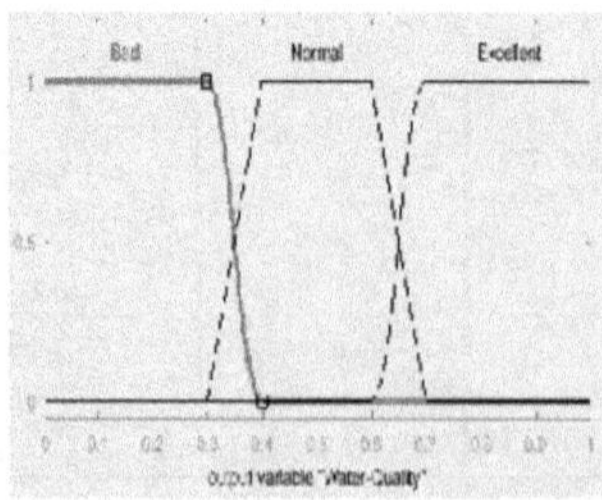

Fuzzy Prepositions

A fuzzy compound proposition is a composition of atomic fuzzy propositions using the connectives "or," "and," & "not" which represent the fuzzy union, intersections, and complement, respectively. Here, b, p, e, t, o & r variables represent Bacterially, Chemical Essential Inorganic, Chemical Toxic Inorganic, Chemical Organic, and Radioactive. Then the following fuzzy propositions hold:

$$t : b \times p \times e \times t \times o \times r \rightarrow W_q \tag{2}$$

All input and output variable values are mapped from real range to probability ranges because the fuzzy expert system works on probability (range 0-1).

Here the function t-norm for the final layer in Eq. (2) is defined as:

$$t : [0,1] \times [0,1] \times [0,1] \times [0,1] \times [0,1] \times [0,1] \rightarrow [0,1] \tag{3}$$

Eq. (3) transforms the membership functions of fuzzy sets of Bacterial, Chemical Essential Inorganic, Chemical Toxic Inorganic, Chemical Organic and Radioactive for a Final layer of proposed fuzzy inference system among membership function of the intersection of Bacterial, Chemical Essential Inorganic, Chemical Toxic Inorganic, Chemical Organic and Radioactive that is:

$$\begin{aligned} & t\left[\mu_B(b), \mu_P(p), \mu_E(e), \mu_T(t), \mu_O(o), \mu_R(r)\right] \\ & = \min\left[\mu_B(b), \mu_P(p), \mu_E(e), \mu_T(t), \mu_O(o), \mu_R(r)\right] \end{aligned} \tag{4}$$

In Eq. (3), for the function t; getting qualified as an intersection, the following axioms must be satisfied and will be called as t-norm:

Axiom t1: Bounded Condition

$$t(0,0) = 0; t(\gamma,1) = t(1,\gamma) = \gamma$$

Axiom t2: Commutativity

$$t(\alpha,\beta) = t(\beta,\alpha)$$

Axiom t3: Nondecreasing

If $\alpha \leq \alpha' and \beta \leq \beta'$, then $t(\alpha,\beta) \leq t(\alpha',\beta')$

Axiom t4: Associativity

$$t\left[t(\alpha,\beta),\gamma\right]=t\left[\alpha,t({}^{2},\gamma)\right]$$

Eq. (4) can be written regarding t-norm as:

$$\mu_{B\cap P\cap E\cap T\cap O\cap R}(b,p,e,t,o,r)=t[\mu_B(b),\mu_P(p),\mu_E(e),\mu_T(t),\mu_O(o),\mu_R(r)] \tag{5}$$

From Eq. (4) & (5)

$$\mu_{B\cap P\cap C\cap R}(b,p,e,t,o,r)=\min[{}^{1}\!/\!_{4_B}(b),{}^{1}\!/\!_{4_P}(p),{}^{1}\!/\!_{4_E}(e),{}^{1}\!/\!_{4_T}(t),{}^{1}\!/\!_{4_O}(o),{}^{1}\!/\!_{4_R}(r)] \tag{6}$$

Lookup Table

The lookup table for the proposed AWQ-ML-MFIS Expert System contains 20 input-output rules from 729 as shown in Table 5. This lookup table is developed with the help of a bio-chemist and The Punjab Standards for Drinking Water Quality Department, Pakistan.

Table 4. Lookup table for proposed AWQ-ML-MFIS expert systems

Rules	Bacterial	Physical	Che-Essen-Inorganic	Che-Toxic-Inorganic	Che-Organic	Radioactive	AWQ
1	NO	E	E	E	E	N	E
2	NO	E	E	E	E	B	B
3	NO	E	E	E	N	E	E
4	NO	E	E	N	N	E	E
5	NO	E	E	N	N	N	N
6	NO	E	N	N	E	N	N
7	NO	E	N	N	N	E	N
8	NO	N	E	E	E	E	E
9	NO	N	E	E	E	N	E
10	NO	N	E	E	E	B	B
11	NO	N	E	E	N	E	E
12	NO	N	E	E	N	N	N
13	NO	N	E	N	N	B	B
14	NO	N	E	N	B	E/N/B	B
15	NO	N	E	N	E	E	E
16	NO	N	E	N	E	N	N
17	NO	N	N	B	E/N/B	E/N/B	B
18	NO	N	N	E	E	E	E
19	NO	N	N	E	E	N	N
20	YES	E/N/B	E/N/B	E/N/B	E/N/B	E/N/B	B

Fuzzy IF-THEN rules are the conditional statement applied to the membership functions. These rules are elements of the fuzzy rule base. Others components like rules surface, rules viewer, etc. are depend upon fuzzy rule base so fuzzy rules base is a major element of FIS. The fuzzy rule base of our expert system has 729 rules at the final layer. Rules are denoted by Rw^n, where $1 \leq n \leq 729$.

Rw^1= **IF** Bacteria is no AND Physical elements are excellent AND Chemical essential Inorganic are excellent, AND Chemical Toxic Inorganics are excellent AND Chemical Organics are excellent AND Radioactive elements are Excellent **THEN** Water Quality is Excellent

Rw^2= **IF** Bacteria is no AND Physical elements are excellent AND Chemical essential Inorganic are excellent, AND Chemical Toxic Inorganics are excellent AND Chemical Organics are excellent AND Radioactive elements are Normal **THEN** Water Quality is Excellent

Rw^3= **IF** Bacteria is no AND Physical elements are excellent AND Chemical essential Inorganic are excellent, AND Chemical Toxic Inorganics are excellent AND Chemical Organics are excellent AND Radioactive elements are Bad **THEN** Water Quality is Not good

⋮

Rw^{729}= **IF** Bacteria is Yes AND Physical elements are bad AND Chemical essential Inorganic is bad, AND Chemical Toxic Inorganics are bad AND Chemical Organics are bad AND Radioactive elements are bad **THEN** Water Quality is not Good

Inference Engine

Fuzzy inference is the way toward mapping from an offered contribution to a yield utilizing fuzzy logic. Its main component of Fuzzy inference is membership functions, fuzzy logic operators, and if-then rules. A single fuzzy relation is created by all rules in the fuzzy rule base. It lies under the inner product on the input which can be seen as only fuzzy IF-THEN rule.

All rules in the fuzzy rule base are combined into a single fuzzy relation that lies under the inner product on input universes of discourse, which is then viewed as only fuzzy IF-THEN rule. A reasonable operator for combining the rules is union.

Let Rw^n be a fuzzy relation that represents fuzzy IF-THEN rule of the final layer of the proposed AWQ-ML-MFIS expert system; which is,

$$Rw^n = B^n \times P^n \times E^n \times T^n \times O^n \times R^n \rightarrow W_q^n \tag{7}$$

Eq. (7) can be written as

$$\mu_{B\cap P\cap E\cap T\cap O\cap R}(b,p,e,t,o,r) = \mu_B(b) \cap ¼_P(p) \cap ¼_E(e) \cap ¼_T(t) \cap ¼_O(o) \cap ¼_R(r)] \tag{8}$$

The rules of the final layer are interpreted as a single fuzzy relation defined by

$$R_{729} = \bigcap_{n=1}^{729} Rw^n \tag{9}$$

This combination of rules is called the Mamdani combination. Assume i and Ψ be any two fuzzy sets and also input and output of fuzzy inference engine respectively. To view R_{729} as a single fuzzy IF-THEN rule by using the generalized modus ponens (Hellendoorn, February 1992), we obtain the output of the fuzzy inference engine as

$$\mu_{Execellent \cap Normal \cap Bad}\left(\cdot\cdot \right) = sup_{i \in (B,P,E,T,O,R)} t\left[\mu_i \left(b,p,e,t,o,r\right), \mu_{R_{729}}\left(b,p,e,t,o,r,W_q\right)\right] \tag{10}$$

The Product Inference Engine (PIE) of the proposed AWQ-ML-MFIS expert system can be written as

$$\mu_{\cdot\cdot}\left(WaterQuality\right) = \max_{1 \le n \le 729}\left[sup_{i \in (B,P,E,T,O,R)}\left(\prod_{j=1}^{81}\left(\mu_{B_j,P_j,E_j,T_j,O_j,R_j}\left(b,p,e,t,o,r\right), \mu_{A_{1i}A_{2i},A_{3i}}\left(a_1,a_2,a_3\right)\right)\right)\right] \tag{11}$$

De-Fuzzifier

One of the most essential components of an Expert system is Defuzzifier. It the process of mapping the fuzzy sent to the crisp output. There are three types of the fuzzified center of gravity Defuzzifier, the center of average Defuzzifier, and Maximum Defuzzifier. From these, the best demulsifier is the "center of gravity defuzzifier". In our proposed AWQ-ML-MFIS expert system the center of gravity Defuzzifier is used. The center of gravity defuzzifier specifies the λ* as the center of the area covered by the membership function of Ψ, that is,

$$\lambda^* = \frac{\int \Psi \mu_\Psi \left(\Psi\right) d\Psi}{\int \mu_\Psi \left(\Psi\right) d\Psi} \tag{12}$$

The graphical representation of fuzzified of all layers of proposed AWQ-ML-MFIS Expert System is shown in Figs 3(a)-3(f).

Fig.3(a) shows water quality concerning Physical components (Turbidity and color). It was observed that in Figure 3(a) Water Quality is Excellent (Yellow shade) when color is $\le$ 13 TCU and Turbidity is $\le$ 3 NTU. And Water Quality is Normal (Greenish Shade) when color lies between 12TCU to 16 TCU & Turbidity is occupied between 2.5 NTU to 5.5 NTU. Another wise water quality is bad (Polluted, Bluish Shade).

Figure 3. Rules surface of water quality based upon different input parameters

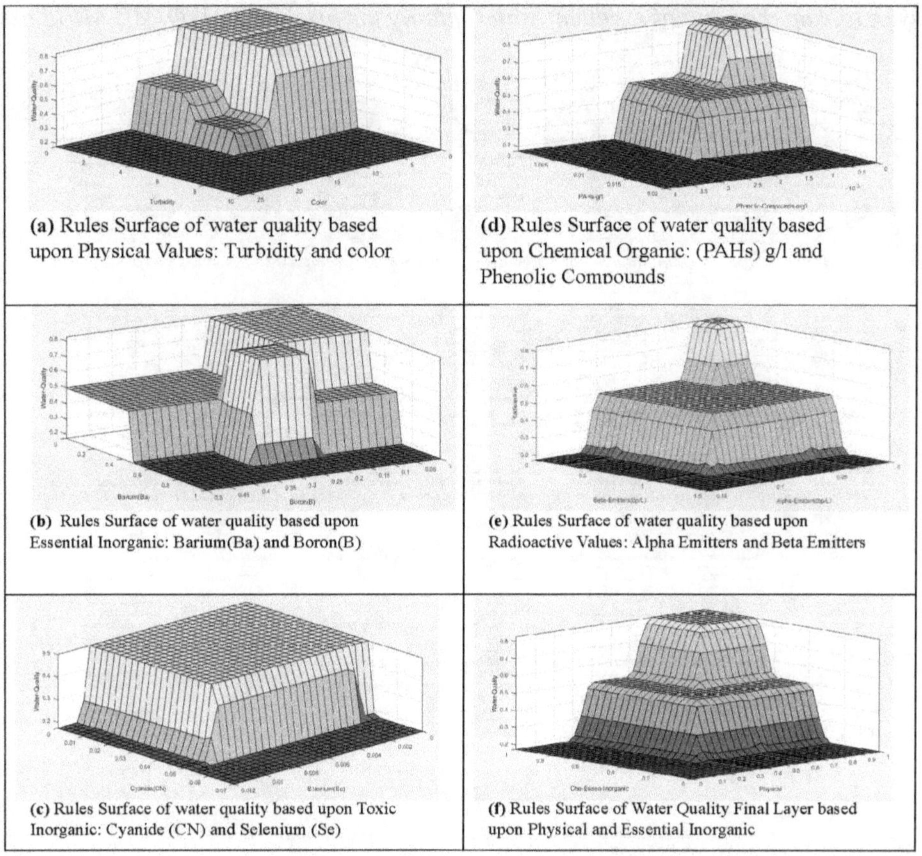

(a) Rules Surface of water quality based upon Physical Values: Turbidity and color

(d) Rules Surface of water quality based upon Chemical Organic: (PAHs) g/l and Phenolic Compounds

(b) Rules Surface of water quality based upon Essential Inorganic: Barium(Ba) and Boron(B)

(e) Rules Surface of water quality based upon Radioactive Values: Alpha Emitters and Beta Emitters

(c) Rules Surface of water quality based upon Toxic Inorganic: Cyanide (CN) and Selenium (Se)

(f) Rules Surface of Water Quality Final Layer based upon Physical and Essential Inorganic

Figure 3(f) shows water quality concerning Physical & Essential Inorganic components. It was observed that in Fig. 3(f) Water Quality is Excellent (Yellow shade) when the Probability of both component's intensity is $\geq$ 0.65. And Water Quality is Normal (Greenish Shade) when the Probability of both components intensity is lies between 0.35 to 0.65. Otherwise water quality is bad (Polluted, Bluish Shade). Similarly, the remaining Figs 3(b)-4(e), are also representing the water quality by prevailing different input parameter values. Yellow, Greenish & Bluish shade represents water quality as Excellent, Normal & Bad(polluted) respectively.

Simulation Results

Figure 4(a) shows that if Bacterial detection value is No, physical value is excellent, Essential Inorganic is Excellent, Toxic Inorganic is Excellent, Chemical Organic is Excellent and Radioactive value is Bad then the Water Quality is bad (Not Good). Fig. 4(b) shows that if Bacterial detection value is No, physical value is excellent, Essential Inorganic is Normal, Toxic Inorganic is Excellent, Chemical Organic is Normal and Radioactive value is Normal then the Water Quality is Normal. Fig. 4(c) shows that if Bacterial detection value is No, physical value is excellent, Essential Inorganic is Excellent, Toxic Inorganic is Normal, Chemical Organic is Excellent and Radioactive value is Normal then Water Quality is Excellent.

Figure 4. (a) Final layer, lookup diagram of bad water quality for proposed AWQ-ML-MFIS expert system; (b) Final layer, lookup diagram of normal water quality for proposed AWQ-ML-MFIS expert system; (c) Final layer, lookup diagram of excellent water quality for proposed AWQ-ML-MFIS expert system

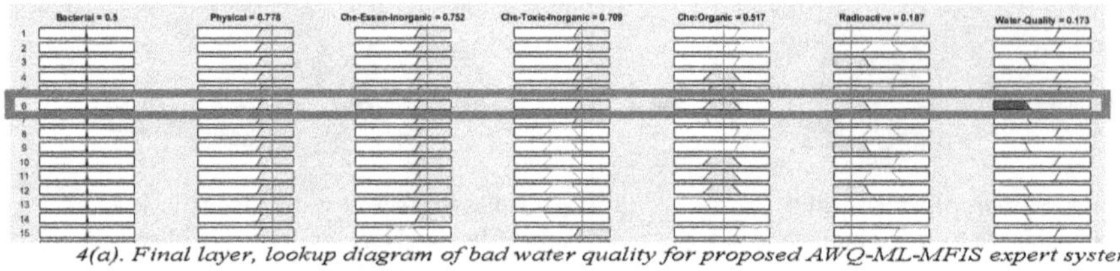

4(a). Final layer, lookup diagram of bad water quality for proposed AWQ-ML-MFIS expert system

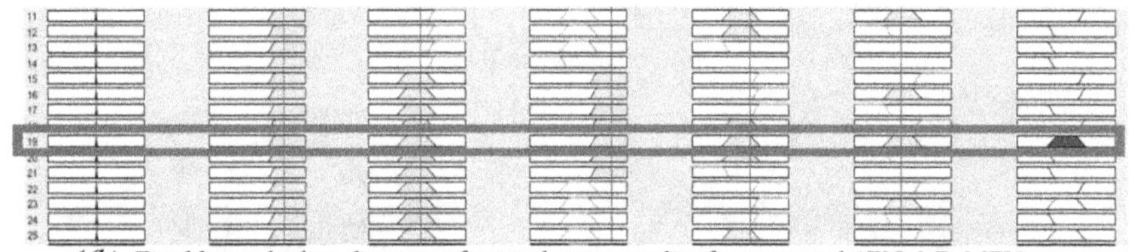

4(b). Final layer, lookup diagram of normal water quality for proposed AWQ-ML-MFIS expert system

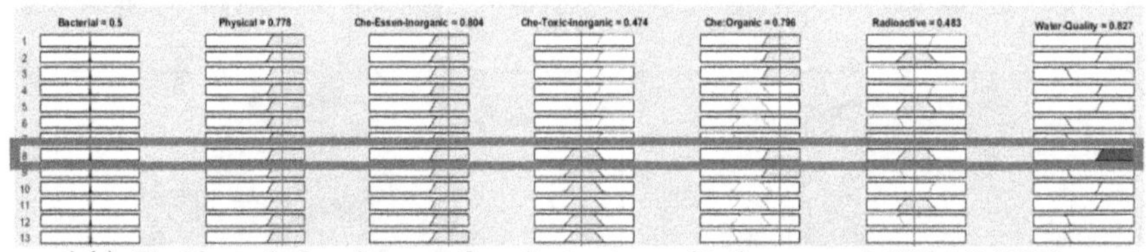

4 (c). Final layer, lookup diagram of excellent water quality for proposed AWQ-ML-MFIS expert system

Table 5. Accuracy of proposed AWQ-ML-MFIS expert system

Rules	Bacterial	Physical	Che-Essen-Inorganic	Che-Toxic-Inor	Che-Organic	Radioactive	Human Expert Decision	Proposed AWQ-ML-MFIS-ES Decision	Probability of Correctness(P_c)	The probability of Error's (P_e=1-P_c)
1	NO(0.5)	E(0.87)	E(0.91)	E(0.79)	E(0.85)	N(0.51)	E (Healthy)	E(Healthy)		
2	NO (0.5)	N(0.61)	N(0.62)	E(0.91)	E(0.79)	E(0.86)	E (Healthy)	E(Healthy)		
3	NO (0.5)	E(0.92)	E(0.77)	E(0.76)	N(0.51)	E(0.83)	E(Healthy)	E(Healthy)		
4	NO (0.5)	E(0.88)	E(0.79)	N(0.56)	N(0.63)	E(0.89)	E(Healthy)	E(Healthy)		
5	NO (0.5)	N(0.61)	E(0.67)	N(0.36)	E(0.68)	E(0.71)	E(Healthy)	N(Healthy)		
6	NO (0.5)	N(0.48)	E(0.75)	E(0.87)	N(0.42)	E(0.87)	E(Healthy)	E(Healthy)		
7	NO (0.5)	N(0.64)	E(0.78)	E(0.86)	E(0.57)	N(0.47)	E(Healthy)	E(Healthy)	1	0
8	NO (0.5)	N(0.56)	E(0.96)	E(0.93)	E(0.85)	E(0.75)	E(Healthy)	E(Healthy)		
9	NO (0.5)	N(0.55)	N(0.49)	E(0.79)	E(0.91)	N(0.54)	N(Healthy)	N(Healthy)		
10	NO (0.5)	E(0.69)	E(0.72)	N(0.48)	N(0.59)	N(0.53)	N(Healthy)	N(Healthy)		
11	NO (0.5)	E(0.95)	N(0.46)	N(0.59)	E(0.86)	N(0.59)	N(Healthy)	N(Healthy)		
12	NO (0.5)	E(0.85)	N(0.64)	N(0.68)	N(0.67)	E(0.93)	N(Healthy)	E(Healthy)		
13	NO (0.5)	N(0.53)	E(0.92)	E(0.80)	N(0.47)	N(0.46)	N(Healthy)	N(Healthy)		
14	NO (0.5)	N(0.55)	E(0.79)	N(0.61)	E(0.78)	N(0.55)	N(Healthy)	N(Healthy)		
15	NO (0.5)	N(0.47)	E(0.97)	N(0.51)	B(0.09)	E(0.75)	B (Unhealthy)	B (Unhealthy)		
16	NO (0.5)	N(0.49)	N(0.51)	B(0.19)	N(0.48)	B(0.21)	B(Unhealthy)	B(Unhealthy)		
17	NO (0.5)	N(0.61)	E(0.82)	N(0.44)	N(0.38)	B(0.11)	B(Unhealthy)	B(Unhealthy)	1	0
18	YES (0.1)	E(0.89)	B(0.02)	N(0.42)	E(0.89)	E(0.79)	B(Unhealthy)	B(Unhealthy)		
19	NO (0.5)	N(0.54)	E(0.81)	E(0.78)	E(0.59)	B(0.01)	B(Unhealthy)	B(Unhealthy)		
20	NO(0.5)	E(0.71)	E(0.91)	E(0.81)	E(0.90)	B(0.21)	B(Unhealthy)	B(Unhealthy)		

CONCLUSION

The primary objective of this research is to design an expert system for the assessment of water quality developed by the guidelines from the World Health Organization (WHO) and Pakistan's Punjab Environmental Quality Standard. The proposed expert system is straightforward to use for professional and non-professionals. The proposed AWQ-ML-MFIS Expert System can categorize the level of water quality into Excellent, Normal, or Dangerous for Health (Polluted). AWQ-ML-MFIS Expert System uses input water parameters such as bacterial, physical, chemical, and radioactive for different layers. It's also observed that Proposed AWQ-ML-MFIS Expert System gives more accurate results shown in Table 4. The efficiency of the proposed system can be improved using other soft computing approaches like Neural networks, Neuro-Fuzzy, etc.

REFERENCES

Abtahi, M., Golchinpour, N., Yaghmaeian, K., Rafiee, M., Jahangiri-rad, M., Keyani, A., & Saeedi, R. (2015). A modified drinking water quality index (DWQI) for assessing drinking source water quality in rural communities of Khuzestan Province, Iran. *Ecological Indicators*, *53*, 283–291. doi:10.1016/j.ecolind.2015.02.009

Alrumman, S. A., El-kott, A. F., & Keshk, S. M. (2016). Water pollution: Source & treatment. *American Journal of Environmental Engineering*, *6*(3), 88–98.

Aqeel, S. (2017).. . *Study on Enhancement on Software Quality by Scheduling Techniques of Real Time Systems European Journal of Advances in Engineering and Technology*, *4*(3), 201–208.

Aqeel, S., Khan, A. S., Ahmad, Z., & Abdullah, J. (2021). *A comprehensive study on DNA based Security scheme Using Deep Learning in Healthcare*. EDPACS. doi:10.1080/07366981.2021.1958742

Beer, K. D., Gargano, J. W., Roberts, V. A., Hill, V. R., Garrison, L. E., Kutty, P. K., Hilborn, E. D., Wade, T. J., Fullerton, K. E., & Yoder, J. S. (2015). Surveillance for Waterborne Disease Outbreaks Associated with Drinking Water-United States, 2011-2012. *MMWR. Morbidity and Mortality Weekly Report*, *64*(31), 842–848. doi:10.15585/mmwr.mm6431a2 PMID:26270059

Boulay, A. M., Motoshita, M., Pfister, S., Bulle, C., Muñoz, I., Franceschini, H., & Margni, M. (2015). Analysis of water use impact assessment methods (part A): Evaluation of modeling choices based on a quantitative comparison of scarcity and human health indicators. *The International Journal of Life Cycle Assessment*, *20*(1), 139–160. doi:10.1007/s11367-014-0814-2

Chang, N. B., Chen, H. W., & Ning, S. K. (2001). Identification of river water quality using the fuzzy synthetic evaluation approach. *Journal of Environmental Management*, *63*(3), 293–305. doi:10.1006/jema.2001.0483 PMID:11775501

Dojlido, J., Raniszewski, J., & Woyciechowska, J. (1994). Water quality index–application for rivers in Vistula river basin in Poland. *Water Science and Technology*, *30*(10), 57–64. doi:10.2166/wst.1994.0511

El-Chakhtoura, J., Prest, E., Saikaly, P., van Loosdrecht, M., Hammes, F., & Vrouwenvelder, H. (2015). *Dynamics of bacterial communities before and after distribution in a full-scale drinking water network. water research, 74, 180-190.GOP (2016).* Government of Punjab. http://epd.punjab.gov.pk/system/files/Punjab%20Environmental%20Quality%20Standards%20for%20%20Drinking%20Water.pdf

Haydar, S., Arshad, M., & Aziz, J. A. (2016). Evaluation of drinking water quality in urban areas of Pakistan: A case study of Southern Lahore. Pakistan. *Journal of Engineering and Applied Sciences (Asian Research Publishing Network).*

Horton, R. K. (1965). An index number system for rating water quality. *Journal - Water Pollution Control Federation*, *37*(3), 300–306.

Jacobs, H. L., Gabrielson, I. N., Horton, R. K., Lyon, W. A., Hubbard, E. C., & McCallum, G. E. (1965). Water Quality Criteria-Stream vs. Effluent Standards. *Journal - Water Pollution Control Federation*, 292–315.

Melita, L., Gumrah, F., & Amareanu, M. (2014). *Issue 2, 2014* (Vol. 5). "Porous polymer membranes used for wastewater treatment", Membrane Water Treatment. Techno-Press., doi:10.12989/mwt.2014.5.2.147

Muniasamy, A., Tabassam, S., Hussain, M. A., Sultana, H., Muniasamy, V., & Bhatnagar, R. Deep Learning for Predictive Analytics in Healthcare. In: A H, A A, T G, R B, M FT, eds. The International Conference on Advanced Machine Learning Technologies and Applications (AMLTA2019). Cairo, Egypt: Springer International Publishing; 2019: 32–42.

Ott, W. (1978). *Water quality indices: a survey of indices used in the United States* (Vol. 1). Environmental Protection Agency, Office of Research and Development, Office of Monitoring and Technical Support.

Rahmanian, N., Ali, S. H. B., Homayoonfard, M., Ali, N. J., Rehan, M., Sadef, Y., & Nizami, A. S. (2015). Analysis of physiochemical parameters to evaluate the drinking water quality in the State of Perak, Malaysia. *Journal of Chemistry, 2015*, 2015. doi:10.1155/2015/716125

Semiromi, F. B., Hassani, A. H., Torabian, A., Karbassi, A. R., & Lotfi, F. H. (2011). Water quality index development using fuzzy logic: A case study of the Karoon River of Iran. *African Journal of Biotechnology, 10*(50), 10125–10133. doi:10.5897/AJB11.1608

Shahid, M. K., Pyo, M., & Choi, Y. G. (2017). Carbonate scale reduction in reverse osmosis membrane by CO2 in wastewater reclamation. *Membrane Water Treatment, 8*(2), 125–136. doi:10.12989/mwt.2017.8.2.125

Sii, H. I., Sherrard, J. H., & Wilson, T. E. (1993). A water quality index based on fuzzy set theory. In *ENVIRONMENTAL ENGINEERING-CONFERENCE* (pp. 1727–1727). American Society Of Civil Engineers.

Tabassam, S., & Al-Qahtane, E. 21. S. Tabassam and E. Al-Qahtane, "Comparative Analysis on Requirement Engineering Modelling Techniques Case Study of Personal Al-Haj E-Guide," 2019 2nd International Conference on Computer Applications & Information Security (ICCAIS), 2019, pp. 1-8, 10.1109/CAIS.2019.8769549

Tabassam, S., Al-Saeed, W., Almughram, O., & Alghamdi, K. (2019). Scalable Data Analysis and Query Processing. *International Journal on Engineering Applications, 7*(3), 81–87. doi:10.15866/irea.v7i3.17012

Tabassam, S., Hassan, O., Al-Qahtnae, E., & Al-Ahmary, N. (2019). Goal Question Metrics and Its Application to Process Management and Improvement. *International Journal on Engineering Applications, 7*(2), 52–58. doi:10.15866/irea.v7i2.17013

Tabassam, S., Shah, H., Alghamdi, K., & Badshah, A. (2019). Social Networks and Digital Security. *2019 International Conference on Electrical, Communication, and Computer Engineering (ICECCE)*, Swat, Pakistan. 10.1109/ICECCE47252.2019.8940808

Tiwari, A. K., De Maio, M., Singh, P. K., & Mahato, M. K. (2015). Evaluation of surface water quality by using GIS and a heavy metal pollution index (HPI) model in a coal mining area, India. *Bulletin of Environmental Contamination and Toxicology, 95*(3), 304–310. doi:10.1007/s00128-015-1558-9 PMID:25962697

Verma, R., & Gupta, B. D. (2015). Detection of heavy metal ions in contaminated water by surface plasmon resonance based optical fibre sensor using conducting polymer and chitosan. *Food Chemistry*, *166*, 568–575. doi:10.1016/j.foodchem.2014.06.045 PMID:25053095

Compilation of References

A Naim. (2022). Applications of E-Learning Tools for Achieving Students Learning Outcomes for Business Programs. *Competitive Trends and Technologies in Business Management*, 125. https://. doi:10.52305/VIXO9830

Abdullah, P. Y., Zeebaree, S., Jacksi, K., & Zeabri, R. R. (2020). An hrm system for small and medium enterprises (sme) s based on cloud computing technology. *International Journal of Research-GRANTHAALAYAH, 8*(8), 56–64.

Abtahi, M., Golchinpour, N., Yaghmaeian, K., Rafiee, M., Jahangiri-rad, M., Keyani, A., & Saeedi, R. (2015). A modified drinking water quality index (DWQI) for assessing drinking source water quality in rural communities of Khuzestan Province, Iran. *Ecological Indicators*, *53*, 283–291. doi:10.1016/j.ecolind.2015.02.009

Adamchuk, V. I., Hummel, J. W., Morgan, M. T., & Upadhyaya, S. K. (2004). On-the-go soil sensors for precision agriculture. *Computers and Electronics in Agriculture*, *44*(1), 71–91. doi:10.1016/j.compag.2004.03.002

Adamides, G., Kalatzis, N., Stylianou, A., Marianos, N., Chatzipapadopoulos, F., Giannakopoulou, M., Papadavid, G., Vassiliou, V., & Neocleous, D. (2020). Smart Farming Techniques for Climate Change Adaptation in Cyprus. *Atmosphere (Basel)*, *11*(6), 557. doi:10.3390/atmos11060557

Adebayo, S., Ogunti, E. O., Akingbade, F. K., & Oladimeji, O. (2018). A review of decision support system using mobile applications in the provision of day-to-day information about farm status for improved crop yield. *Period. Eng. Nat. Sci.*, *6*(2), 89–99. doi:10.21533/pen.v6i2.183

Aggarwal, K., Mijwil, M. M., Al-Mistarehi, A. H., Alomari, S., Gök, M., Alaabdin, A. M. Z., & Abdulrhman, S. H. (2022). Has the future started? The current growth of artificial intelligence, machine learning, and deep learning. *Iraqi J. Comput. Sci. Math.*, *3*, 115–123.

Agostinelli, S., Cumo, F., Guidi, G., & Tomazzoli, C. (2021). Cyber-physical systems improving building energy management: Digital twin and artificial intelligence. *Energies*, *14*(8), 2338. doi:10.3390/en14082338

Agrawal, A. V., Magulur, L. P., Priya, S. G., Kaur, A., Singh, G., & Boopathi, S. (2023). Smart Precision Agriculture Using IoT and WSN: In T. Murugan & N. E. (Eds.), Advances in Information Security, Privacy, and Ethics (pp. 524–541). IGI Global. https://doi.org/ doi:10.4018/978-1-6684-8145-5.ch026

Agrawal, A. V., Shashibhushan, G., Pradeep, S., Padhi, S., Sugumar, D., & Boopathi, S. (2023). Synergizing Artificial Intelligence, 5G, and Cloud Computing for Efficient Energy Conversion Using Agricultural Waste. In Sustainable Science and Intelligent Technologies for Societal Development (pp. 475–497). IGI Global.

Agrawal, A. V., Shashibhushan, G., Pradeep, S., Padhi, S., Sugumar, D., & Boopathi, S. (2023a). Synergizing Artificial Intelligence, 5G, and Cloud Computing for Efficient Energy Conversion Using Agricultural Waste. In Sustainable Science and Intelligent Technologies for Societal Development (pp. 475–497). IGI Global.

Agrawal, A. V., Shashibhushan, G., Pradeep, S., Padhi, S., Sugumar, D., & Boopathi, S. (2023b). Synergizing Artificial Intelligence, 5G, and Cloud Computing for Efficient Energy Conversion Using Agricultural Waste. In Sustainable Science and Intelligent Technologies for Societal Development (pp. 475–497). IGI Global.

Agrawal, A. V., Shashibhushan, G., Pradeep, S., Padhi, S. N., Sugumar, D., & Boopathi, S. (2024). Synergizing Artificial Intelligence, 5G, and Cloud Computing for Efficient Energy Conversion Using Agricultural Waste. In B. K. Mishra (Ed.), (pp. 475–497). Practice, Progress, and Proficiency in Sustainability. IGI Global. doi:10.4018/979-8-3693-1186-8.ch026

Ahmad, J., Zaid, O., Shahzaib, M., Abdullah, M. U., Ullah, A., & Ullah, R. (2021). Mechanical properties of sustainable concrete modified by adding marble slurry as cement substitution. *AIMS Materials Science*, *8*(3), 343–358. doi:10.3934/matersci.2021022

Ahmad, T., Zhang, D., Huang, C., Zhang, H., Dai, N., Song, Y., & Chen, H. (2021). Artificial intelligence in sustainable energy industry: Status Quo, challenges and opportunities. *Journal of Cleaner Production*, *289*, 125834.

Ahmed, A. T., El Gohary, F., Tzanakakis, V. A., & Angelakis, A. N. (2020). Egyptian and Greek Water Cultures and Hydro-Technologies in Ancient Times. *Sustainability (Basel)*, *12*(22), 9760. doi:10.3390/su12229760

Akçay, H. G., Kabasakal, B., Aksu, D., Demir, N., Öz, M., & Erdoğan, A. (2020). Automated bird counting with deep learning for regional bird distribution mapping. *Animals (Basel)*, *10*(7), 1207. doi:10.3390/ani10071207 PMID:32708550

Akeiber, H., Nejat, P., Majid, M. Z. A., Wahid, M. A., Jomehzadeh, F., Zeynali Famileh, I., Calautit, J. K., Hughes, B. R., & Zaki, S. A. (2016). A review on phase change material (PCM) for sustainable passive cooling in building envelopes. In Renewable and Sustainable Energy Reviews (Vol. 60, pp. 1470–1497). Elsevier Ltd. doi:10.1016/j.rser.2016.03.036

Akhavan, M., & Doroodgar, A. (2020). Smart Farming: An Overview of Precision Agriculture Using IoT, Robotics, and AI. *Journal of Ambient Intelligence and Humanized Computing*, *11*(12), 5497–5521.

Akkaş, M. A., & Sokullu, R. (2017). An IoT-based greenhouse monitoring system with Micaz motes. *Procedia Computer Science*, *113*, 603–608. doi:10.1016/j.procs.2017.08.300

Akpolat, A., & Dursun, E. (2017). *Advanced Metering Infrastructure (AMI).* Smart Meters and New Technologies.

Al Abri, D., Malik, A., Al-Saadi, S., Albadi, M., Charabi, Y., & Hosseinzadeh, N. (2021). Smart Grids and Smart Buildings. In Handbook of Climate Change Mitigation and Adaptation (pp. 1–20). Springer. doi:10.1007/978-1-4614-6431-0_78-2

Alafif, T., Tehame, A. M., Bajaba, S., Barnawi, A., & Zia, S. (2021). Machine and deep learning towards COVID-19 diagnosis and treatment: Survey, challenges, and future directions. *International Journal of Environmental Research and Public Health*, *18*(3), 1117. doi:10.3390/ijerph18031117 PMID:33513984

Alahmari, F., Naim, A., & Alqahtani, H. (2023). E-Learning Modeling Technique and Convolution Neural Networks in Online Education. In IoT-enabled Convolutional Neural Networks: Techniques and Applications (pp. 261-295). River Publishers.

Ali, S., Abuhmed, T., El-Sappagh, S., Muhammad, K., Alonso-Moral, J. M., Confalonieri, R., Guidotti, R., Del Ser, J., Díaz-Rodríguez, N., & Herrera, F. (2023). Explainable Artificial Intelligence (XAI): What we know and what is left to attain Trustworthy Artificial Intelligence. *Information Fusion, 99*. doi:10.1016/j.inffus.2023.101805

Ali, H. M. M., El-Gorashi, T. E. H., Lawey, A. Q., & Elmirghani, J. M. H. (2017). Future energy efficient data centers with disaggregated servers. *Journal of Lightwave Technology*, *35*(24), 5361–5380. doi:10.1109/JLT.2017.2767574

Ali, M. N., Senthil, T., Ilakkiya, T., Hasan, D. S., Ganapathy, N. B. S., & Boopathi, S. (2024). IoT's Role in Smart Manufacturing Transformation for Enhanced Household Product Quality. In *Advanced Applications in Osmotic Computing* (pp. 252–289). IGI Global. doi:10.4018/979-8-3693-1694-8.ch014

Alimi, O. A., Ouahada, K., & Abu-Mahfouz, A. M. (2020). A review of machine learning approaches to power system security and stability. *IEEE Access : Practical Innovations, Open Solutions*, *8*, 113512–113531. doi:10.1109/ACCESS.2020.3003568

Alim, M. A., Tao, Z., Abden, M. J., Rahman, A., & Samali, B. (2020). Improving performance of solar roof tiles by incorporating phase change material. *Solar Energy*, *207*, 1308–1320. doi:10.1016/j.solener.2020.07.053

Ali, S. S., & Choi, B. J. (2020). State-of-the-art artificial intelligence techniques for distributed smart grids: A review. *Electronics (Basel)*, *9*(6), 1030.

Alizadeh, O. (2022). Advances and challenges in climate modeling. *Climatic Change*, *170*(1-2), 18. doi:10.1007/s10584-021-03298-4

Al-Kodmany, K. (2018). The Vertical Farm: A Review of Developments and Implications for the Vertical City. *Buildings*, *8*(2), 24. doi:10.3390/buildings8020024

Allen, M. (2003). Liability for climate change. *Nature*, *421*(6926), 891–892. doi:10.1038/421891a PMID:12606972

Alqahtani, H., & Naim, A. (2022). Critical Success Factors for Transforming CRM to SCRM for building E-CRM. In A. Naim & S. Kautish (Eds.), *Building a Brand Image Through Electronic Customer Relationship Management* (pp. 139–168). IGI Global. doi:10.4018/978-1-6684-5386-5.ch007

Al-Qudah, A. A., Hamdan, A., Al-Okaily, M., & Alhaddad, L. (2022). The impact of green lending on credit risk: Evidence from UAE's banks. *Environmental Science and Pollution Research International*, *30*(22), 1–13. doi:10.1007/s11356-021-18224-5 PMID:35067871

Alrumman, S. A., El-kott, A. F., & Keshk, S. M. (2016). Water pollution: Source & treatment. *American Journal of Environmental Engineering*, *6*(3), 88–98.

Alvarez, J., & Peter, N. (2006). Adoption of computer based information systems: The case of dairy farmers in Canterbury, NZ, and Florida, Uruguay. *Computers and Electronics in Agriculture*, *50*(1), 48–60. doi:10.1016/j.compag.2005.08.013

Amato, G., & Moscatelli, A. (2018). Internet of Things in Precision Agriculture for Crop Monitoring. *Sensors (Basel)*, *18*(5), 1515. PMID:29751623

Anayat, S. (2023). Human Resources Management After Industry 4.0: Blending AI and HRM. In Strategic Human Resource Management in the Hospitality Industry: A Digitalized Economic Paradigm (pp. 97-115). IGI Global.

Anderson, R., Ghafurian, R., & Gharavi, H. (2018). *Smart Grid: The Future of the Electric Energy System.*

André, J.C. (2014). High-performance Computing for limate Modeling. AMS Journal. doi:10.1175/BAMS-D-13-00098.1

Anthony, B., Majid, M. A., & Romli, A. (2018). A Descriptive Study towards Green Computing Practice Application for Data Centers in IT Based Industries. *MATEC Web of Conferences, 150*, 05048. 10.1051/matecconf/201815005048

Anthony, L. F. W., Kanding, B., & Selvan, R. (2020). Carbontracker: Tracking and predicting the carbon footprint of training deep learning models. *arXiv*, arXiv:2007.03051.

APEC. (2023). *APEC Climate Symposium: Enhancing APEC Resilience through AI Applications and Climate Change Adaption*. APEC. https://www.apec.org/docs/default-source/publications/2023/1/apec-climate-symposium-enhancing-apec-resilience-through-ai-applications-in-climate-change-adaptation/223_ppsti_apec-climate-symposium.pdf?sfvrsn=7565b6cb_2

Aqeel, S. (2017)*... Study on Enhancement on Software Quality by Scheduling Techniques of Real Time Systems European Journal of Advances in Engineering and Technology*, *4*(3), 201–208.

Aqeel, S., Khan, A. S., Ahmad, Z., & Abdullah, J. (2021). *A comprehensive study on DNA based Security scheme Using Deep Learning in Healthcare*. EDPACS. doi:10.1080/07366981.2021.1958742

Arora, N. K. (2019). Impact of climate change on agriculture production and its sustainable solutions. *Environmental Sustainability*, *2*(2), 95–96. doi:10.1007/s42398-019-00078-w

Asdrubali, F., D'Alessandro, F., & Schiavoni, S. (2015). A review of unconventional sustainable building insulation materials. *Sustainable Materials and Technologies*, *4*, 1–17. doi:10.1016/j.susmat.2015.05.002

Ashby, M. F. (n.d.). *Materials and the Environment: Eco-Informed Material Choice*.

Ashby, M. F., Ferreira, P. J. S. G., & Schodek, D. L. (2009). *Nanomaterials, nanotechnologies and design : an introduction for engineers and architects*. Butterworth-Heinemann.

Aslam, S., Khalid, A., & Javaid, N. (2020). Towards Efficient Energy Management in Smart Grids Considering Microgrids with Day-ahead Energy Forecasting. *Electric Power Systems Research*, *182*, 106232. doi:10.1016/j.epsr.2020.106232

Attaran, M., & Woods, J. (2019). Cloud computing technology: Improving small business performance using the Internet. *Journal of Small Business and Entrepreneurship*, *31*(6), 495–519.

Awomeso, J., Taiwo, A., Gbadebo, A., & Arimoro, A. (2010). Waste disposal and pollution management in urban areas: A workable remedy for the environment in developing countries. *American Journal of Environmental Sciences*, *6*(1), 26–32. doi:10.3844/ajessp.2010.26.32

Ayaz, M., Ammad-uddin, M., Baig, I., & Aggoune, H. M. (2018). Wireless Sensor's Civil Applications, Prototypes, and Future Integration Possibilities: A Review. *IEEE Sensors Journal*, *18*(1), 4–30. doi:10.1109/JSEN.2017.2766364

Babu, B. S., Kamalakannan, J., Meenatchi, N., Karthik, S., & Boopathi, S. (2022). Economic impacts and reliability evaluation of battery by adopting Electric Vehicle. *IEEE Explore*, (pp. 1–6). IEEE.

Bacco, M., Barsocchi, P., Ferro, E., Gotta, A., & Ruggeri, M. (2019). The digitization of agriculture: A survey of research activities on smart farming. *Array (New York, N.Y.)*, *3–4*, 100009. doi:10.1016/j.array.2019.100009

Bahceci, S. (2018). Sensing Technologies for Precision Agriculture and Farming. *Biosensors (Basel)*, *8*(3), 67. PMID:30011807

Bahrami, M. (2015). Cloud computing for emerging mobile cloud apps. *2015 3rd IEEE International Conference on Mobile Cloud Computing, Services, and Engineering*, (pp. 4–5). IEEE.

Bais, P. (2019). Study of workers in small scale industries. *Reports on Economics and Finance*, *5*(1), 49–54. doi:10.12988/ref.2019.911

Bais, P., & Bahadur, P. S. (2023). Agriculture in Indian Economy and Contribution of Science and Technology. [AJAST]. *Asian Journal of Applied Science and Technology*, *7*(1), 167–176. doi:10.38177/ajast.2023.7116

Baranwal, J., Barse, B., Fais, A., Delogu, G. L., & Kumar, A. (2022). Biopolymer: A Sustainable Material for Food and Medical Applications. In Polymers (Vol. 14, Issue 5). MDPI. doi:10.3390/polym14050983

Barrero, O., & Perdomo, S. A. (2018). RGB and multispectral UAV image fusion for Gramineae weed detection in rice fields. *Precision Agriculture*, *19*(5), 809–822. doi:10.1007/s11119-017-9558-x

Basit, A., Latif, W., Ashraf, M., Rehman, A., Iqbal, K., Maqsood, H. S., Jabbar, A., & Baig, S. A. (2019). Comparison of mechanical and thermal comfort properties of tencel blended with regenerated fibers and cotton woven fabrics. *AUTEX Research Journal*, *19*(1), 80–85. doi:10.1515/aut-2018-0035

Batte, M. T., & VanBuren, F. N. (1999). *Precision farming—Factor influencing productivity*. In Proceedings of the Northern Ohio Crops Day Meeting, Wood County, OH, USA.

BeckT.PamukH.RamrattanR.UrasB. (2015). Mobile money, trade credit and economic development: Theory and evidence. CentER Discussion Paper Series No. *2015-023, European Banking Center Discussion Paper No. 2015-005*. https://ssrn.com/abstract=2588392

Beer, K. D., Gargano, J. W., Roberts, V. A., Hill, V. R., Garrison, L. E., Kutty, P. K., Hilborn, E. D., Wade, T. J., Fullerton, K. E., & Yoder, J. S. (2015). Surveillance for Waterborne Disease Outbreaks Associated with Drinking Water-United States, 2011-2012. *MMWR. Morbidity and Mortality Weekly Report*, *64*(31), 842–848. doi:10.15585/mmwr.mm6431a2 PMID:26270059

Ben Ayed, R., & Hanana, M. (2021). Artificial intelligence to improve the food and agriculture sector. *Journal of Food Quality*, *2021*, 1–7.

Benke, K., & Tomkins, B. (2017). Future food-production systems: Vertical farming and controlled-environment agriculture. *Sustainability*, *13*(1), 13–26. doi:10.1080/15487733.2017.1394054

Benos, L., Tagarakis, A. C., Dolias, G., Berruto, R., Kateris, D., & Bochtis, D. (2021). Machine learning in agriculture: A comprehensive updated review. *Sensors (Basel)*, *21*(11), 3758. PMID:34071553

Berntsen, J., Thomsen, A., Schelde, K., Hansen, O. M., Knudsen, L., Broge, N., Hougaard, H., & Hørfarter, R. (2006). Algorithms for sensor-based redistribution of nitrogen fertilizer in winter wheat. *Precision Agriculture*, *7*(2), 65–83. doi:10.1007/s11119-006-9000-2

Bhambri, P. (2021). Electronic Evidence. In Textbook of Cyber Heal (pp. 86-120). AGAR Saliha Publication, Tamil Nadu.

Bhambri, P. (2020). Green Compliance. In S. Agarwal (Ed.), *Introduction to Green Computing* (pp. 95–125). AGAR Saliha Publication.

Bhambri, P., Kaur, H., Gupta, A., & Singh, J. (2020). Human Activity Recognition System. *Oriental Journal of Computer Science and Technology*, *13*(2-3), 91–96.

Bhambri, P., Singh, M., Jain, A., Dhanoa, I. S., Sinha, V. K., & Lal, S. (2021). Classification of the GENE Expression Data with the Aid of Optimized Feature Selection. *Turkish Journal of Physiotherapy and Rehabilitation*, *32*(3), 1158–1167.

Bhatore, S., Mohan, L., & Reddy, Y. R. (2020). Machine learning techniques for credit risk evaluation: A systematic literature review. *Journal of Banking and Financial Technology*, *4*(1), 111–138. doi:10.1007/s42786-020-00020-3

Bhattarai, B. P., Paudyal, S., Luo, Y., Mohanpurkar, M., Cheung, K., Tonkoski, R., Hovsapian, R., Myers, K. S., Zhang, R., Zhao, P., Manic, M., Zhang, S., & Zhang, X. (2019). Big data analytics in smart grids: State-of-the-art, challenges, opportunities, and future directions. *IET Smart Grid*, *2*(2), 141–154. doi:10.1049/iet-stg.2018.0261

Bhattarai, T. N., Ghimire, S., Mainali, B., Gorjian, S., Treiche, H., & Paudel, S. R. (2022). Applications of smart grid technology in Nepal: Status, challenges, and opportunities. *Environmental Science and Pollution Research International*, *30*(10), 1–25. doi:10.1007/s11356-022-19084-3 PMID:35138545

Bhonde, D., Nagarnaik, P. B., Parbat, D. K., & Waghe, U. P. (2014). Physical and Mechanical Properties of Bamboo (Dendrocalmus Strictus). *International Journal of Scientific and Engineering Research*, *5*(1). http://www.ijser.org

Biodegradable plastics. (n.d.).

Blikra Vea, E., Martinez-Sanchez, V., & Thomsen, M. (2018). A review of waste management decision support tools and their ability to assess circular biowaste management systems. *Sustainability (Basel)*, *10*(10), 3720. doi:10.3390/su10103720

Blumenstock, J., Cadamuro, G., & On, R. (2015). Predicting poverty and wealth from mobile phone metadata. *Science*, *350*(6264), 1073–1076. doi:10.1126/science.aac4420 PMID:26612950

Bochtis, D., Sørensen, C. A. G., & Kateris, D. (2019). *Operations Management in Agriculture*. Elsevier. [Google Scholar] [CrossRef]

Boopathi, S. (2022). Effects of Cryogenically-treated Stainless Steel on Eco-friendly Wire Electrical Discharge Machining Process. Preprint : Springer.

Boopathi, S. (2022b). Performance Improvement of Eco-Friendly Near-Dry wire-Cut Electrical Discharge Machining Process Using Coconut Oil-Mist Dielectric Fluid. *World Scientific:Journal of Advanced Manufacturing Systems*.

Boopathi, S. (2023). Securing Healthcare Systems Integrated With IoT: Fundamentals, Applications, and Future Trends. In Dynamics of Swarm Intelligence Health Analysis for the Next Generation (pp. 186–209). IGI Global.

Boopathi, S. (2024a). Balancing Innovation and Security in the Cloud: Navigating the Risks and Rewards of the Digital Age. In Improving Security, Privacy, and Trust in Cloud Computing (pp. 164–193). IGI Global.

Boopathi, S. (2024b). Balancing Innovation and Security in the Cloud: Navigating the Risks and Rewards of the Digital Age. In Improving Security, Privacy, and Trust in Cloud Computing (pp. 164–193). IGI Global.

Boopathi, S. (2024b). Energy Cascade Conversion System and Energy-Efficient Infrastructure. In Sustainable Development in AI, Blockchain, and E-Governance Applications (pp. 47–71). IGI Global.

Boopathi, S. (2024c). Balancing Innovation and Security in the Cloud: Navigating the Risks and Rewards of the Digital Age. In Improving Security, Privacy, and Trust in Cloud Computing (pp. 164–193). IGI Global.

Boopathi, S. (2024c). Implementation of Green Manufacturing Practices in Automobile Fields: A Review. *Sustainable Machining and Green Manufacturing*, 221–248.

Boopathi, S. (2024c). Sustainable Development Using IoT and AI Techniques for Water Utilization in Agriculture. In Sustainable Development in AI, Blockchain, and E-Governance Applications (pp. 204–228). IGI Global.

Boopathi, S. (2024d). Energy Cascade Conversion System and Energy-Efficient Infrastructure. In Sustainable Development in AI, Blockchain, and E-Governance Applications (pp. 47–71). IGI Global.

Boopathi, S. (2024d). Energy Cascade Conversion System and Energy-Efficient Infrastructure. In Sustainable Development in AI, Blockchain, and E-Governance Applications (pp. 47–71). IGI Global. doi:10.4018/979-8-3693-1722-8.ch004

Boopathi, S. (2024d). Minimization of Manufacturing Industry Wastes Through the Green Lean Sigma Principle. *Sustainable Machining and Green Manufacturing*, 249–270.

Boopathi, S. (2024e). Sustainable Development Using IoT and AI Techniques for Water Utilization in Agriculture. In Sustainable Development in AI, Blockchain, and E-Governance Applications (pp. 204–228). IGI Global.

Boopathi, S. (2024g). Sustainable Development Using IoT and AI Techniques for Water Utilization in Agriculture. In Sustainable Development in AI, Blockchain, and E-Governance Applications (pp. 204–228). IGI Global. doi:10.4018/979-8-3693-1722-8.ch012

Boopathi, S., & Khare, R. KG, J. C., Muni, T. V., & Khare, S. (2023). Additive Manufacturing Developments in the Medical Engineering Field. In Development, Properties, and Industrial Applications of 3D Printed Polymer Composites (pp. 86–106). IGI Global.

Boopathi, S., & Kumar, P. (2024). Advanced bioprinting processes using additive manufacturing technologies: Revolutionizing tissue engineering. *3D Printing Technologies: Digital Manufacturing, Artificial Intelligence, Industry 4.0*, 95.

Boopathi, S., & Kumar, P. (2024a). Advanced bioprinting processes using additive manufacturing technologies: Revolutionizing tissue engineering. *3D Printing Technologies: Digital Manufacturing, Artificial Intelligence, Industry 4.0*, 95.

Boopathi, S., Karthikeyan, K. R., Jaiswal, C., Dabi, R., Sunagar, P., & Malik, S. (2024). *IoT based Automatic Cooling Tower*. Research Gate.

Boopathi, S., Kumar, P. K. S., Meena, R. S., Sudhakar, M., & Associates. (2023). Sustainable Developments of Modern Soil-Less Agro-Cultivation Systems: Aquaponic Culture. In Human Agro-Energy Optimization for Business and Industry (pp. 69–87). IGI Global.

Boopathi, S., Kumar, P. K. S., Meena, R. S., Sudhakar, M., & Associates. (2023a). Sustainable Developments of Modern Soil-Less Agro-Cultivation Systems: Aquaponic Culture. In Human Agro-Energy Optimization for Business and Industry (pp. 69–87). IGI Global.

Boopathi, S., Kumar, P. K. S., Meena, R. S., Sudhakar, M., & Associates. (2023b). Sustainable Developments of Modern Soil-Less Agro-Cultivation Systems: Aquaponic Culture. In Human Agro-Energy Optimization for Business and Industry (pp. 69–87). IGI Global.

Boopathi, S. (2022a). An investigation on gas emission concentration and relative emission rate of the near-dry wire-cut electrical discharge machining process. *Environmental Science and Pollution Research International*, *29*(57), 86237–86246. doi:10.1007/s11356-021-17658-1 PMID:34837614

Boopathi, S. (2023a). An Investigation on Friction Stir Processing of Aluminum Alloy-Boron Carbide Surface Composite. In R. V. Vignesh, R. Padmanaban, & M. Govindaraju (Eds.), *Advances in Processing of Lightweight Metal Alloys and Composites* (pp. 249–257). Springer Nature Singapore. doi:10.1007/978-981-19-7146-4_14

Boopathi, S. (2023a). Internet of Things-Integrated Remote Patient Monitoring System: Healthcare Application. In A. Suresh Kumar, U. Kose, S. Sharma, & S. Jerald Nirmal Kumar (Eds.), (pp. 137–161). Advances in Healthcare Information Systems and Administration. IGI Global. doi:10.4018/978-1-6684-6894-4.ch008

Boopathi, S. (2023b). Securing Healthcare Systems Integrated With IoT: Fundamentals, Applications, and Future Trends. In A. Suresh Kumar, U. Kose, S. Sharma, & S. Jerald Nirmal Kumar (Eds.), (pp. 186–209). Advances in Healthcare Information Systems and Administration. IGI Global. doi:10.4018/978-1-6684-6894-4.ch010

Boopathi, S. (2024a). Advancements in Machine Learning and AI for Intelligent Systems in Drone Applications for Smart City Developments. In *Futuristic e-Governance Security With Deep Learning Applications* (pp. 15–45). IGI Global.

Boopathi, S., Hari Balaji, V., Mageswari, M., & Asif, M. M. (2022). Influences of Boron Carbide Particles on the Wear Rate and Tensile Strength of AA2014 Surface Composite Fabricated by Friction-Stir Processing. *Mater-Tehnol.Si. Materials Technology*, *56*(3), 263–270.

Boopathi, S., Alqahtani, A. S., Mubarakali, A., & Panchatcharam, P. (2023). Sustainable developments in near-dry electrical discharge machining process using sunflower oil-mist dielectric fluid. *Environmental Science and Pollution Research International*, 1–20.

Boopathi, S., Balasubramani, V., Kumar, R. S., & Singh, G. R. (2021). The influence of human hair on kenaf and Grewia fiber-based hybrid natural composite material: An experimental study. *Functional Composites and Structures*, *3*(4), 045011. doi:10.1088/2631-6331/ac3afc

Boopathi, S., & Davim, J. P. (2023). Applications of Nanoparticles in Various Manufacturing Processes. In S. Boopathi & J. P. Davim (Eds.), (pp. 1–31). Advances in Chemical and Materials Engineering. IGI Global. doi:10.4018/978-1-6684-9135-5.ch001

Boopathi, S., & Davim, J. P. (2023b). *Sustainable Utilization of Nanoparticles and Nanofluids in Engineering Applications*. IGI Global. doi:10.4018/978-1-6684-9135-5

Boopathi, S., & Kanike, U. K. (2023). Applications of Artificial Intelligent and Machine Learning Techniques in Image Processing. In B. K. Pandey, D. Pandey, R. Anand, D. S. Mane, & V. K. Nassa (Eds.), (pp. 151–173). Advances in Computational Intelligence and Robotics. IGI Global. doi:10.4018/978-1-6684-8618-4.ch010

Boopathi, S., & Khang, A. (2023). AI-Integrated Technology for a Secure and Ethical Healthcare Ecosystem. In *AI and IoT-Based Technologies for Precision Medicine* (pp. 36–59). IGI Global.

Boopathi, S., & Myilsamy, S. (2021). Material removal rate and surface roughness study on Near-dry wire electrical discharge Machining process. *Elsevier: Materials Today: Proceedings Elsevier*, *45*(9), 8149–8156.

Boopathi, S., Umareddy, M., & Elangovan, M. (2023). Applications of Nano-Cutting Fluids in Advanced Machining Processes. In S. Boopathi & J. P. Davim (Eds.), (pp. 211–234). Advances in Chemical and Materials Engineering. IGI Global. doi:10.4018/978-1-6684-9135-5.ch009

Bougdah, H., & Sharples, S. (2009). Environment, Technology and Sustainability. Taylor & Francis eBooks. doi:10.4324/9780203878408

Boulay, A. M., Motoshita, M., Pfister, S., Bulle, C., Muñoz, I., Franceschini, H., & Margni, M. (2015). Analysis of water use impact assessment methods (part A): Evaluation of modeling choices based on a quantitative comparison of scarcity and human health indicators. *The International Journal of Life Cycle Assessment*, *20*(1), 139–160. doi:10.1007/s11367-014-0814-2

Boykin, L. M., Ghalab, A., Rossitto De Marchi, B., Savill, A., Wainaina, J. M., Kinene, T., Lamb, S., Rodrigues, M., Kehoe, M. A., & Ndunguru, J. (2018). Real time portable genome sequencing for global food security. *F1000 Research*, *7*, 1101. doi:10.12688/f1000research.15507.1

Butt, S., Ahmadi, M., & Razavi, M. (2020). Green Computing: Sustainable Design and Technologies. *2020 5th International Conference on Innovative Technologies in Intelligent Systems and Industrial Applications (CITISIA)*, 1–7.

Campero-Jurado, I., Márquez-Sánchez, S., Quintanar-Gómez, J., Rodríguez, S., & Corchado, J. M. (2020). Smart helmet 5.0 for industrial internet of things using artificial intelligence. . *Sensors (Basel)*, *20*(21), 6241. doi:10.3390/s20216241 PMID:33139608

Cao, D., Hu, W., Zhao, J., Zhang, G., Zhang, B., Liu, Z., Chen, Z., & Blaabjerg, F. (2020). Reinforcement learning and its applications in modern power and energy systems: A review. *Journal of Modern Power Systems and Clean Energy*, *8*(6), 1029–1042. doi:10.35833/MPCE.2020.000552

Cao, L. (2012). Social security and social welfare data mining: An overview. *IEEE Trans. Syst. Man. Cybern. Part C*, *42*, 837–853.

Carcary, M., Doherty, E., & Conway, G. (2014). The adoption of cloud computing by irish smes an exploratory study. *Electronic Journal of Information Systems Evaluation*, *17*(1), 3–14.

Carletti, V., Greco, A., Percannella, G., & Vento, M. (2020). Age from Faces in the Deep Learning Revolution. *IEEE Transactions on Pattern Analysis and Machine Intelligence*, *42*(9), 2113–2132. doi:10.1109/TPAMI.2019.2910522 PMID:30990174

Cepin, M. (2019). Reliability of Smart Grids. In *2019 International Conference on Information and Digital Technologies (IDT)* (pp. 1-6). IEEE. DOI: 10.1109/DT.2019.8813387

Chandhini, K. (2016). A Literature Study on Agricultural Production System Using IoT as Inclusive Technology. [Google Scholar]. *Int. J. Innov. Technol. Res.*, *4*, 2727–2731.

Chang, A., Jung, J., Maeda, M., & Landivar, J. (2017). Crop height monitoring with digital imagery from unmanned aerial system (UAV). *Computers and Electronics in Agriculture*, *141*, 232–237. doi:10.1016/j.compag.2017.07.008

ChanglaniK.ThakoreR. (2023). Artificial Intelligence for Climate Action. doi:10.13140/RG.2.2.21102.18245

Chang, N. B., Chen, H. W., & Ning, S. K. (2001). Identification of river water quality using the fuzzy synthetic evaluation approach. *Journal of Environmental Management*, *63*(3), 293–305. doi:10.1006/jema.2001.0483 PMID:11775501

Chattopadhyay, A., Hassanzadeh, P., & Pasha, S. (2020). Predicting clustered weather patterns: A test case for applications of convolutional neural networks to spatio-temporal climate data. *Scientific Reports*, *10*(1), 1317. doi:10.1038/s41598-020-57897-9 PMID:31992743

Chen, F., Kissel, D. E., West, L. T., Adkins, W., Clark, R., Rickman, D., & Luvall, J. C. (2004). Field Scale Mapping of Surface Soil Clay Concentration. *Precision Agriculture*, *5*(1), 7–26. doi:10.1023/B:PRAG.0000013617.96272.9e

Chen, L., Chen, Z., Zhang, Y., Liu, Y., Osman, A. I., Farghali, M., Hua, J., Al-Fatesh, A., Ihara, I., Rooney, D. W., & Yap, P.-S. (2023). Artificial intelligence-based solutions for climate change: A review. *Environmental Chemistry Letters*, *21*(5), 2525–2557. doi:10.1007/s10311-023-01617-y

Chen, S., & Liu, C. (2020). An Intelligent Agriculture System with Wireless Sensor Networks. *Computers and Electronics in Agriculture*, *171*, 105322.

Chen, Z., & He, Y. (2023). Artificial Intelligence and Environmental Protection of Buildings. *Probl. Ekorozwoju/Probl. Sustainable Development (Bradford)*, *18*, 254–262.

Chhabra, B., Verma, D., & Taneja, B. (2010). Software engineering issues from the cloud application perspective. *International Journal of Information Technology and Knowledge Management*, 2(2), 669–673.

Chlingaryan, A., & Sukkarieh, S. (2018). Field Robotics in Precision Agriculture: A Review. *Precision Agriculture*, *19*(2), 152–169.

Chou, S. F., Horng, J. S., Liu, C. H., Yu, T. Y., & Kuo, Y. T. (2022). Identifying the critical factors for sustainable marketing in the catering: The influence of big data applications, marketing innovation, and technology acceptance model factors. *Journal of Hospitality and Tourism Management*, *51*, 11–21. doi:10.1016/j.jhtm.2022.02.010

Chowdhury, M. E. H., Khandakar, A., Ahmed, S., Al-Khuzaei, F., Hamdalla, J., Haque, F., Reaz, M. B. I., Al Shafei, A., & Al-Emadi, N. (2020). Design, Construction and Testing of IoT Based Automated Indoor Vertical Hydroponics Farming Test-Bed in Qatar. *Sensors (Basel)*, *20*(19), 5637. doi:10.3390/s20195637 PMID:33023097

Cirillo, D., Catuara-Solarz, S., Morey, C., Guney, E., Subirats, L., Mellino, S., Gigante, A., Valencia, A., Rementeria, M. J., Chadha, A. S., & Mavridis, N. (2020). Sex and gender differences and biases in artificial intelligence for biomedicine and healthcare. *NPJ Digital Medicine*, *3*(1), 81. doi:10.1038/s41746-020-0288-5 PMID:32529043

Collins, C., Dennehy, D., Conboy, K., & Mikalef, P. (2021). Artificial intelligence in information systems research: A systematic literature review and research agenda. *International Journal of Information Management*, *60*, 102383. doi:10.1016/j.ijinfomgt.2021.102383

Coulter, J. K. (2004). World Agriculture: Towards 2015/2030. An FAO Perspective. Experimental Agriculture, 40(2), 269–269. doi:10.1017/S0014479704211796

Cravero, A., Pardo, S., Sepúlveda, S., & Muñoz, L. (2022). Challenges to Use Machine Learning in Agricultural Big Data: A Systematic Literature Review. *Agronomy (Basel), 12*(3), 748. doi:10.3390/agronomy12030748

Czekala, W., Drozdowski, J., & \Labiak, P. (2023). Modern technologies for waste management: A review. *Applied Sciences, 13*(15), 8847.

D'Adamo, I., Mazzanti, M., Morone, P., & Rosa, P. (2022). Assessing the relation between waste management policies and circular economy goals. *Waste Management (New York, N.Y.), 154*, 27–35. doi:10.1016/j.wasman.2022.09.031 PMID:36206656

Dada, E. G., Bassi, J. S., Chiroma, H., Adetunmbi, A. O., Ajibuwa, O. E., & ... (2019). Machine learning for email spam filtering: Review, approaches and open research problems. *Heliyon, 5*(6), e01802. doi:10.1016/j.heliyon.2019.e01802 PMID:31211254

Daniel, K. W., Tripathi, N. K., Honda, K., & Apisit, E. (2004). Analysis of VNIR (400–1100 nm) spectral signatures for estimation of soil organic matter in tropical soils of Thailand. *International Journal of Remote Sensing, 25*(3), 643–652. doi:10.1080/0143116031000139944

Daniel, K., Tripathi, N. K., Honda, K., & Apisit, E. (2011). Analysis of spectral reflectance and absorption patterns of soil organic matter. In *Proceedings of the 22nd Asian Conference on Remote Sensing*.

Das, S., Lekhya, G., Shreya, K., Shekinah, K. L., Babu, K. K., & Boopathi, S. (2024). Fostering Sustainability Education Through Cross-Disciplinary Collaborations and Research Partnerships: Interdisciplinary Synergy. In Facilitating Global Collaboration and Knowledge Sharing in Higher Education With Generative AI (pp. 60–88). IGI Global.

Dash, G., Sharma, C., & Sharma, S. (2023). Sustainable marketing and the role of social media: An experimental study using natural language processing (NLP). *Sustainability (Basel), 15*(6), 5443. doi:10.3390/su15065443

Das, P., Ramapraba, P., Seethalakshmi, K., Mary, M. A., Karthick, S., & Sampath, B. (2024). Sustainable Advanced Techniques for Enhancing the Image Process. In *Fostering Cross-Industry Sustainability With Intelligent Technologies* (pp. 350–374). IGI Global. doi:10.4018/979-8-3693-1638-2.ch022

Das, S., Lekhya, G., Shreya, K., Lydia Shekinah, K., Babu, K. K., & Boopathi, S. (2024). Fostering Sustainability Education Through Cross-Disciplinary Collaborations and Research Partnerships: Interdisciplinary Synergy. In P. Yu, J. Mulli, Z. A. S. Syed, & L. Umme (Eds.), (pp. 60–88). Advances in Higher Education and Professional Development. IGI Global. doi:10.4018/979-8-3693-0487-7.ch003

Dedes, I. C., Tsampasis, E., Elias, C., & Gkonis, P. (2021). Energy Storage in Smart Electrical Grids. In *2021 10th Mediterranean Conference on Embedded Computing (MECO)* (pp. 1–4). IEEE. 10.1109/MECO52532.2021.9460169

Dell Technologies. (2020). *HPC Enhances Weather and Climate Modeling, Research, and Proactive Responses to Dangerous Conditions*. Dell Technologies.

Deng, L., & Yu, D. (2014). *Deep Learning: methods and applications*. IEEE. doi:10.1561/9781601988157

DenilM.ShakibiB.DinhL.RanzatoM.De FreitasN. (2013). Predicting parameters in deep learning. *Neural Information Processing Systems, 26*, 2148–2156. doi:10.14288/1.0165555

Devadutta, K., Bhambri, P., Gountia, D., Mehta, V., Mangla, M., Patan, R., Kumar, A., Agarwal, P. K., Sharma, A., Singh, M., & Gadicha, A. B. (2020). *Method for Cyber Security in Email Communication among Networked Computing Devices* [Patent application number 202031002649].

Dewitte, S., Cornelis, J. P., Müller, R., & Munteanu, A. (2021). Artificial Intelligence Revolutionises Weather Forecast, Climate Monitoring and Decadal Prediction. *Remote Sensing (Basel), 13*(16), 3209. doi:10.3390/rs13163209

Dhaini, M., Jaber, M., Fakhereldine, A., Hamdan, S., & Haraty, R. A. (2021). Green computing approaches-A survey. *Informatica (Vilnius)*, *45*(1).

Dhanalakshmi, M., Tamilarasi, K., Saravanan, S., Sujatha, G., Boopathi, S., & Associates. (2024). Fog Computing-Based Framework and Solutions for Intelligent Systems: Enabling Autonomy in Vehicles. In Computational Intelligence for Green Cloud Computing and Digital Waste Management (pp. 330–356). IGI Global.

Dhanya, D., Kumar, S. S., Thilagavathy, A., Prasad, D., & Boopathi, S. (2023). Data Analytics and Artificial Intelligence in the Circular Economy: Case Studies. In Intelligent Engineering Applications and Applied Sciences for Sustainability (pp. 40–58). IGI Global.

Dhanya, D., Kumar, S. S., Thilagavathy, A., Prasad, D., & Boopathi, S. (2023a). Data Analytics and Artificial Intelligence in the Circular Economy: Case Studies. In Intelligent Engineering Applications and Applied Sciences for Sustainability (pp. 40–58). IGI Global.

Dhanya, D., Kumar, S. S., Thilagavathy, A., Prasad, D. V. S. S. S. V., & Boopathi, S. (2023). Data Analytics and Artificial Intelligence in the Circular Economy: Case Studies. In B. K. Mishra (Ed.), (pp. 40–58). Advances in Civil and Industrial Engineering. IGI Global. doi:10.4018/979-8-3693-0044-2.ch003

Dhiman, G., & Aulakh, N. S. (2020). Artificial Intelligence in Agriculture: A Review. *Journal of Agricultural Engineering and Biotechnology*, *3*(2), 75–83.

Di Vaio, A., Palladino, R., Hassan, R., & Escobar, O. (2020). Artificial intelligence and business models in the sustainable development goals perspective: A systematic literature review. *Journal of Business Research*, *121*, 283–314. doi:10.1016/j.jbusres.2020.08.019

Díaz-Parralejo, A., Cuerda-Correa, E. M., Macías-García, A., Sánchez-González, J., & Díaz-Díez, M. Á. (2014). The Excellent Mechanical Properties of Cork: A Novel Approach through the Analysis of Contact Stress. *ISRN Materials Science*, *2014*, 1–6. doi:10.1155/2014/898439

Dilmi, S., & Ladjal, M. (2021). A novel approach for water quality classification based on the integration of deep learning and feature extraction techniques. *Chemometrics and Intelligent Laboratory Systems*, *214*, 104329. doi:10.1016/j.chemolab.2021.104329

Dojlido, J., Raniszewski, J., & Woyciechowska, J. (1994). Water quality index–application for rivers in Vistula river basin in Poland. *Water Science and Technology*, *30*(10), 57–64. doi:10.2166/wst.1994.0511

Domakonda, V. K., Farooq, S., Chinthamreddy, S., Puviarasi, R., Sudhakar, M., & Boopathi, S. (2022). Sustainable Developments of Hybrid Floating Solar Power Plants: Photovoltaic System. In Human Agro-Energy Optimization for Business and Industry (pp. 148–167). IGI Global.

Doyle, L., & Weidlich, I. (2021). Sustainable insulation for sustainable DHC. *Energy Reports*, *7*, 150–157. doi:10.1016/j.egyr.2021.08.161

Duan, J., & Xu, C. (2020). Advances in Precision Agriculture Technologies: A Review. *Sustainable Production and Consumption*, *23*, 1–12.

Duan, T., Chapman, S., Guo, Y., & Zheng, B. (2017). Dynamic monitoring of NDVI in wheat agronomy and breeding trials using an unmanned aerial vehicle. *Field Crops Research*, *210*, 71–80. doi:10.1016/j.fcr.2017.05.025

Duchesne, L., Karangelos, E., & Wehenkel, L. (2020). Recent developments in machine learning for energy systems reliability management. *Proceedings of the IEEE*, *108*(9), 1656–1676. doi:10.1109/JPROC.2020.2988715

Durairaj, M., Jayakumar, S., Karpagavalli, V., Maheswari, B. U., Boopathi, S., & ... (2023a). Utilization of Digital Tools in the Indian Higher Education System During Health Crises. In *Multidisciplinary Approaches to Organizational Governance During Health Crises* (pp. 1–21). IGI Global. doi:10.4018/978-1-7998-9213-7.ch001

Ehlers, M. (2008). Geoinformatics and digital earth initiatives: A German perspective. *International Journal of Digital Earth*, *1*(1), 17–30. doi:10.1080/17538940701781975

Ehrlich, P. R., & Harte, J. (2015). To feed the world in 2050 will require a global revolution. *Proceedings of the National Academy of Sciences of the United States of America*, *112*(48), 14743–14744. doi:10.1073/pnas.1519841112 PMID:26627228

Ehsani, M., Falahi, M., & Lotfifard, S. (2012). Vehicle to grid services: Potential and applications. *Energies*, *5*(10), 4076–4090. doi:10.3390/en5104076

Ejarque, J., Badia, R. M., Albertin, L., Aloisio, G., Baglione, E., Becerra, Y., Boschert, S., Berlin, J. R., D'Anca, A., Elia, D., Exertier, F., Fiore, S., Flich, J., Folch, A., Gibbons, S. J., Koldunov, N., Lordan, F., Lorito, S., Løvholt, F., & Volpe, M. (2022). Enabling dynamic and intelligent workflows for HPC, data analytics, and AI convergence. *Future Generation Computer Systems, 134*. doi:10.1016/j.future.2022.04.014

Ekka, V., & Bais, P. (2023). Contribution of Sericulture in the Socio-economic Transformation of the Workers. *Asian Journal of Applied Science and Technology*, *7*(2), 217–224. doi:10.38177/ajast.2023.7223

El Nahry, A. H., & Mohamed, E. S. (2010). Potentiality of land and water resources in African Sahara: A case study of south Egypt. *Environmental Earth Sciences*, *63*(6), 1263–1275. doi:10.1007/s12665-010-0799-5

El-Chakhtoura, J., Prest, E., Saikaly, P., van Loosdrecht, M., Hammes, F., & Vrouwenvelder, H. (2015). *Dynamics of bacterial communities before and after distribution in a full-scale drinking water network. water research, 74, 180-190. GOP (2016).* Government of Punjab. http://epd.punjab.gov.pk/system/files/Punjab%20Environmental%20Quality%20Standards%20for%20%20Drinking%20Water.pdf

Elgenedy, M. A., Massoud, A. M., & Ahmed, S. (2015). Smart grid self-healing: Functions, applications, and developments. In *2015 First Workshop on Smart Grid and Renewable Energy (SGRE)* (pp. 1–6). IEEE. 10.1109/SGRE.2015.7208737

Elias Kaloush, K., Prapoorna Biligiri, K., Abdelaziz Zeiada, W., Carolina Rodezno, M., & Xavier Reed, J. (2010). *Evaluation of Fiber-Reinforced Asphalt Mixtures Using Advanced Material Characterization Tests*. www.astm.org

Elijah, O., Rahman, T. A., Orikumhi, I., Leow, C. Y., & Hindia, M. H. D. N. (2018). An Overview of Internet of Things (IoT) and Data Analytics in Agriculture: Benefits and Challenges. *IEEE Internet of Things Journal*, *5*(5), 3758–3773. doi:10.1109/JIOT.2018.2844296

El-Kassar, A., & Singh, S. K. (2019a). Green innovation and organizational performance: The influence of big data and the moderating role of management commitment and HR practices. *Technological Forecasting and Social Change, 144*, 483–498. doi:10.1016/j.techfore.2017.12.016

Ellis, B. L., Knauth, P., & Djenizian, T. (2014). Three-dimensional self-supported metal oxides for advanced energy storage. In Advanced Materials (Vol. 26, Issue 21, pp. 3368–3397). Wiley-VCH Verlag. doi:10.1002/adma.201306126

Elsisi, M., Tran, M. Q., Mahmoud, K., Lehtonen, M., & Darwish, M. M. (2021). Deep learning-based industry 4.0 and internet of things towards effective energy management for smart buildings. . *Sensors (Basel)*, *21*(4), 1038. doi:10.3390/s21041038 PMID:33546436

Enholm, I. M., Papagiannidis, E., Mikalef, P., & Krogstie, J. (2021). Artificial intelligence and business value: A literature review. *Information Systems Frontiers*.

Fan, F. R., Tang, W., & Wang, Z. L. (2016). Flexible Nanogenerators for Energy Harvesting and Self-Powered Electronics. In Advanced Materials (Vol. 28, Issue 22, pp. 4283–4305). Wiley-VCH Verlag. doi:10.1002/adma.201504299

Fang, B., Yin, X., Tan, Y., Li, C., Gao, Y., Cao, Y., & Li, J. (2016). The contributions of cloud technologies to smart grid. *Renewable & Sustainable Energy Reviews*, *59*, 1326–1331. doi:10.1016/j.rser.2016.01.032

Fan, Z., Yan, Z., & Wen, S. (2023). Deep Learning and Artificial Intelligence in Sustainability: A Review of SDGs, Renewable Energy, and Environmental Health. *Sustainability (Basel)*, *15*(18), 13493. doi:10.3390/su151813493

Farooq, M. S., Riaz, S., Abid, A., Umer, T., & Zikria, Y. B. (2020). Role of IoT Technology in Agriculture: A Systematic Literature Review. *Electronics (Basel)*, *9*(2), 319. doi:10.3390/electronics9020319

Farzaneh, H., Malehmirchegini, L., Bejan, A., Afolabi, T., Mulumba, A., & Daka, P. P. (2021). Artificial intelligence evolution in smart buildings for energy efficiency. *Applied Sciences (Basel, Switzerland)*, *11*(2), 763. doi:10.3390/app11020763

Fatima, S., Alqahtani, H., Naim, A., & Alma'alwi, F. (2022). E-CRM Through Social Media Marketing Activities for Brand Awareness, Brand Image, and Brand Loyalty. In A. Naim & S. Kautish (Eds.), *Building a Brand Image Through Electronic Customer Relationship Management* (pp. 109–138). IGI Global. doi:10.4018/978-1-6684-5386-5.ch006

Feder, G., Just, R. E., & Zilberman, D. (1985). Adoption of agricultural innovations in developing countries: A survey. *Economic Development and Cultural Change*, *33*(2), 255–298. doi:10.1086/451461

Feng, C., Wang, Y., Chen, Q., Ding, Y., Strbac, G., & Kang, C. (2021). Smart grid encounters edge computing: Opportunities and applications. *Advances in Applied Energy*, *1*, 100006. doi:10.1016/j.adapen.2020.100006

Ferguson, R. B., Hergert, G. W., Schepers, J. S., Gotway, C. A., Cahoon, J. E., & Peterson, T. A. (2002). Site-Specific Nitrogen Management of Irrigated Maize. *Soil Science Society of America Journal, 66*(2), 544–553. . doi:10.2136/sssaj2002.5440

Ferrández-Pastor, F. J., García-Chamizo, J. M., Nieto-Hidalgo, M., & Mora-Martínez, J. (2018). Precision Agriculture Design Method Using a Distributed Computing Architecture on Internet of Things Context. *Sensors (Basel)*, *18*(6), 1731. doi:10.3390/s18061731 PMID:29843386

Ferrández-Pastor, F., García-Chamizo, J., Nieto-Hidalgo, M., Mora-Pascual, J., & Mora-Martínez, J. (2016). Developing Ubiquitous Sensor Network Platform Using Internet of Things: Application in Precision Agriculture. *Sensors (Basel)*, *16*(7), 1141. doi:10.3390/s16071141 PMID:27455265

Fic, K., Platek, A., Piwek, J., & Frackowiak, E. (2018). Sustainable materials for electrochemical capacitors. In Materials Today (Vol. 21, Issue 4, pp. 437–454). Elsevier B.V. doi:10.1016/j.mattod.2018.03.005

Filho, W. L., Wall, T., Serafino, A. R. M., Nagy, G. J., Balogun, A.-L., Luetz, J. M., Ng, A. W., Kovaleva, M., Fardous, M. S. A., Alves, F., & Guevara, Z. (2022). Deploying artificial intelligence for climate change adaptation. *Technological Forecasting and Social Change, 180*. doi:10.1016/j.techfore.2022.121662

Fioretto, F., Mak, T. W., & Van Hentenryck, P. (2020). Predicting ac optimal power flows: Combining deep learning and lagrangian dual methods. In *Proceedings of the AAAI Conference on Artificial Intelligence*. Association for the Advancement of Artificial Intelligence. 10.1609/aaai.v34i01.5403

Foster, A. D., & Mark, R. R. (2010). Microeconomics of technology adoption. *Annual Review of Economics*, *2*(1), 395–424. doi:10.1146/annurev.economics.102308.124433 PMID:24386501

Franzo, G., Legnardi, M., Faustini, G., Tucciarone, C. M., & Cecchinato, M. (2023). When Everything Becomes Bigger: Big Data for Big Poultry Production. *Animals (Basel)*, *13*(11), 1804. doi:10.3390/ani13111804 PMID:37889739

Fróna, D., Szenderák, J., & Rákos, M. H. (2019). The Challenge of Feeding the World. *Sustainability (Basel), 11*(20), 5816. doi:10.3390/su11205816

Fuchs, C. (2006). The implications of new information and communication technologies for sustainability. *Environment, Development and Sustainability, 10*(3), 291–309. doi:10.1007/s10668-006-9065-0

Fu, S., Yang, F., & Xiao, Y. (2020). AI inspired intelligent resource management in future wireless network. *IEEE Access : Practical Innovations, Open Solutions, 8*, 22425–22433. doi:10.1109/ACCESS.2020.2968554

Gadzhev, G., Ivanov, V., & Valcheva, R. (2021). HPC Simulations of the Present and Projected Future Climate of the Balkan Region. Springer. doi:10.1007/978-3-030-55347-0_20

Gai, K., Qiu, M., Zhao, H., Tao, L., & Zong, Z. (2016). Dynamic energy-aware cloudlet-based mobile cloud computing model for green computing. *Journal of Network and Computer Applications, 59*, 46–54. doi:10.1016/j.jnca.2015.05.016

Galaz, V., Centeno, M. A., Callahan, P. W., Causevic, A., Patterson, T., Brass, I., Baum, S., Farber, D., Fischer, J., Garcia, D., & ... (2021). Artificial intelligence, systemic risks, and sustainability. *Technology in Society, 67*, 101741.

Gao, B. (2021). The use of machine learning combined with data mining technology in financial risk prevention. *Computational Economics*, 1–21.

Gao, L., & Sun, D. W. (2018). Smart Agriculture: Current Status and Future Perspectives. *Journal of Agricultural Science and Technology, 20*(4), 815–824.

Geneiatakis, D., Kounelis, I., Neisse, R., Nai-Fovino, I., Steri, G., & Baldini, G. (2017). Security and privacy issues for an IoT based smarthome. In *Proceedings of the 2017 40th International Convention on Information and Communication Technology, Electronics and Microelectronics (MIPRO)* (pp. 1292–1297). IEEE.

Ghayoor, F., Fahad, S., Traore, I., Waseem, M., & Goudarzi, A. (2022). A Survey on IoT-Enabled Smart Grids: Emerging, Applications, Challenges, and Outlook. *Energies, 15*(19), 6984. doi:10.3390/en15196984

Gheorghe, G., Tudorache, P., & Roşca, I. M. (2023). The Contribution of Green Marketing in the Development of a Sustainable Destination through Advanced Clustering Methods. *Sustainability (Basel), 15*(18), 13691. doi:10.3390/su151813691

Gnanaprakasam, C., Vankara, J., Sastry, A. S., Prajval, V., Gireesh, N., & Boopathi, S. (2023). Long-Range and Low-Power Automated Soil Irrigation System Using Internet of Things: An Experimental Study. In Contemporary Developments in Agricultural Cyber-Physical Systems (pp. 87–104). IGI Global.

Goel, R. K., Yadav, C. S., Vishnoi, S., & Rastogi, R. (2021). Smart agriculture – Urgent need of the day in developing countries. *Sustainable Computing : Informatics and Systems, 30*, 100512. doi:10.1016/j.suscom.2021.100512

Gonzalez-Dominguez, J., Smith, H., & Li, Z. (2019). Bioelectronics for Plant Monitoring: Opportunities and Challenges. *Sensors (Basel), 19*(21), 4695. PMID:31671733

Goodfellow, I. J., Bengio, Y., & Courville, A. (2016). *Deep learning*. ACM. https://dl.acm.org/citation.cfm?id=3086952

Gotoh, J. Y., Takeda, A., & Yamamoto, R. (2014). Interaction between financial risk measures and machine learning methods. *Computational Management Science, 11*(4), 365–402. doi:10.1007/s10287-013-0175-5

Grbčić, L., Kranjčević, L., & Družeta, S. (2021). Machine learning and simulation-optimization coupling for water distribution network contamination source detection. *Sensors (Basel), 21*(4), 1157. doi:10.3390/s21041157 PMID:33562175

Gulshan, V., Peng, L., Coram, M., Stumpe, M. C., Wu, D., Narayanaswamy, A., Venugopalan, S., Widner, K., Madams, T., Cuadros, J., Ramasamy, K., Raman, R., Nelson, P. C., Mega, J. L., & Webster, D. R. (2016). Development and validation of a deep learning algorithm for detection of diabetic retinopathy in retinal Fundus photographs. *Journal of the American Medical Association, 316*(22), 2402. doi:10.1001/jama.2016.17216 PMID:27898976

Gunasekaran, K., Boopathi, S., & Sureshkumar, M. (2022). Analysis of a Cryogenically Cooled Near-Dry WEDM Process using Different Dielectrics. *Mater-Tehnol.Si. Materials Technology, 56*(2), 179–186.

Guo, Z., Liu, J., Wang, Y., Chen, M., Wang, D., Xu, D., & Cheng, J. (2023). Diffusion models in Bioinformatics: A new wave of deep learning revolution in action. *arXiv (Cornell University)*. https://doi.org//arxiv.2302.10907 doi:10.48550

Guo, B., Yan, S., & Shao, Z. (2009). The redefinition and some discussion of green computing. *Chinese Journal of Computers, 32*(12), 2311–2319. doi:10.3724/sp.j.1016.2009.02311

Gupta, G. (2019). *The Benefits of Bringing Artificial Intelligence to High Performance Computing*. High Performance. https://high-performance-computing.cioreview.com/cxoinsight/the-benefits-of-bringing-artificial-intelligence-to-high-performance-computing-nid-26530-cid-84.html

Gupta, H., Kumar, A., & Wasan, P. (2021). Industry 4.0, cleaner production and circular economy: An integrative framework for evaluating ethical and sustainable business performance of manufacturing organizations. *Journal of Cleaner Production, 295*, 126253. doi:10.1016/j.jclepro.2021.126253

Gupta, J., Pouw, N. R., & Ros-Tonen, M. A. (2015). Towards an elaborated theory of inclusive development. *European Journal of Development Research, 27*(4), 541–559. doi:10.1057/ejdr.2015.30

Gutierrez, A., Boukrami, E., & Lumsden, R. (2015). Technological, organisational and environmental factors influencing managers' decision to adopt cloud computing in the UK. *Journal of Enterprise Information Management, 28*(6), 788–807.

Habib, M., Faris, M., Qaddoura, R., Alomari, M., Alomari, A., & Faris, H. (2021). Toward an automatic quality assessment of voice-based telemedicine consultations: A deep learning approach. *Sensors (Basel), 21*(9), 3279. doi:10.3390/s21093279 PMID:34068602

Hamdaoui, Y., & Maach, A. (2016). *Smart islanding in smart grids. In 2016 IEEE Smart Energy Grid Engineering*. SEGE., doi:10.1109/SEGE.2016.7589521

Haneef, M., Ceseracciu, L., Canale, C., Bayer, I. S., Heredia-Guerrero, J. A., & Athanassiou, A. (2017). Advanced Materials from Fungal Mycelium: Fabrication and Tuning of Physical Properties. *Scientific Reports, 7*(1), 41292. doi:10.1038/srep41292 PMID:28117421

Hanoon, S. K., Abdullah, A. F., Shafri, H. Z., & Wayayok, A. (2022). A Novel Approach Based on Machine Learning and Public Engagement to Predict Water-Scarcity Risk in Urban Areas. *ISPRS International Journal of Geo-Information, 11*(12), 606. doi:10.3390/ijgi11120606

Han, T., Muhammad, K., Hussain, T., Lloret, J., & Baik, S. W. (2020). An efficient deep learning framework for intelligent energy management in IoT networks. *IEEE Internet of Things Journal, 8*(5), 3170–3179. doi:10.1109/JIOT.2020.3013306

Hanumanthakari, S., Gift, M. D. M., Kanimozhi, K. V., Bhavani, M. D., Bamane, K. D., & Boopathi, S. (2023). Biomining Method to Extract Metal Components Using Computer-Printed Circuit Board E-Waste. In P. Srivastava, D. Ramteke, A. K. Bedyal, M. Gupta, & J. K. Sandhu (Eds.), (pp. 123–141). Practice, Progress, and Proficiency in Sustainability. IGI Global. doi:10.4018/978-1-6684-8117-2.ch010

Harikaran, M., Boopathi, S., Gokulakannan, S., & Poonguzhali, M. (2023a). Study on the Source of E-Waste Management and Disposal Methods. In *Sustainable Approaches and Strategies for E-Waste Management and Utilization* (pp. 39–60). IGI Global. doi:10.4018/978-1-6684-7573-7.ch003

Harmon, R. R., & Auseklis, N. (2009). Sustainable IT Services: Assessing the impact of green Computing Practices. IEEE. doi:10.1109/PICMET.2009.5261969

Hassan, M. A., Yang, M., Rasheed, A., Yang, G., Reynolds, M., Xia, X., Xiao, Y., & He, Z. (2019). A rapid monitoring of NDVI across the wheat growth cycle for grain yield prediction using a multi-spectral UAV platform. *Plant Science*, *282*, 95–103. doi:10.1016/j.plantsci.2018.10.022 PMID:31003615

Haydar, S., Arshad, M., & Aziz, J. A. (2016). Evaluation of drinking water quality in urban areas of Pakistan: A case study of Southern Lahore. Pakistan. *Journal of Engineering and Applied Sciences (Asian Research Publishing Network)*.

Hema, N., Krishnamoorthy, N., Chavan, S. M., Kumar, N. M. G., Sabarimuthu, M., & Boopathi, S. (2023). A Study on an Internet of Things (IoT)-Enabled Smart Solar Grid System. In P. Swarnalatha & S. Prabu (Eds.), (pp. 290–308). Advances in Computational Intelligence and Robotics. IGI Global. doi:10.4018/978-1-6684-8098-4.ch017

Hema, N., Krishnamoorthy, N., Chavan, S. M., Kumar, N., Sabarimuthu, M., & Boopathi, S. (2023). A Study on an Internet of Things (IoT)-Enabled Smart Solar Grid System. In *Handbook of Research on Deep Learning Techniques for Cloud-Based Industrial IoT* (pp. 290–308). IGI Global.

Hemanand, D., Mishra, N., Premalatha, G., Mavaluru, D., Vajpayee, A., Kushwaha, S., & Sahile, K. (2022). Applications of intelligent model to analyze the green finance for environmental development in the context of artificial intelligence. *Computational Intelligence and Neuroscience*, *2022*, 2022. doi:10.1155/2022/2977824 PMID:35845917

Henneman, L. R., Liu, C., Mulholland, J. A., & Russell, A. G. (2017). Evaluating the effectiveness of air quality regulations: A review of accountability studies and frameworks. *Journal of the Air & Waste Management Association*, *67*(2), 144–172. doi:10.1080/10962247.2016.1242518 PMID:27715473

Henriksen, A. V., Edwards, T. C. G., Pesonen, L. A., Green, O., & Sørensen, C. A. G. (2019). Internet of Things in arable farming: Implementation, applications, challenges and potential. *Biosystems Engineering*, *191*, 60–84. doi:10.1016/j.biosystemseng.2019.12.013

Hentschel, R., Leyh, C., & Baumhauer, T. (2019). *Critical success factors for the implementation and adoption of cloud services in SMEs.*

He, Q., Wu, D., & Wang, Z. (2020). Advances in Sensors for Agriculture and Environmental Monitoring: A Review. *Sensors (Basel)*, *20*(18), 5157. PMID:32927710

Hermann, E. (2022). Leveraging artificial intelligence in marketing for social good—An ethical perspective. *Journal of Business Ethics*, *179*(1), 43–61. doi:10.1007/s10551-021-04843-y PMID:34054170

Hernández-Ochoa, I. M., Gaiser, T., Kersebaum, K.-C., Webber, H., Seidel, S. J., Grahmann, K., & Ewert, F. (2022). Model-based design of crop diversification through new field arrangements in spatially heterogeneous landscapes. A review. *Agronomy for Sustainable Development*, *42*(4), 74. doi:10.1007/s13593-022-00805-4

Hoda, N., & Naim, A. (Eds.). (2023). *Social Capital in the Age of Online Networking: Genesis, Manifestations, and Implications: Genesis, Manifestations, and Implications*. IGI Global.

Hoepner, A. G., McMillan, D., Vivian, A., & Wese Simen, C. (2021). Significance, relevance and explainability in the machine learning age: An econometrics and financial data science perspective. *European Journal of Finance*, *27*(1-2), 1–7. doi:10.1080/1351847X.2020.1847725

Holzinger, A., Weippl, E., Tjoa, A. M., & Kieseberg, P. (2021). Digital transformation for sustainable development goals — A security, safety and privacy perspective on AI. In A. Holzinger, P. Kieseberg, A. M. Tjoa, & E. Weippl (Eds.), *Machine Learning and Knowledge Extraction* (pp. 1–20). Springer International Publishing. doi:10.1007/978-3-030-84060-0_1

Horton, R. K. (1965). An index number system for rating water quality. *Journal - Water Pollution Control Federation, 37*(3), 300–306.

Hossain, M. A., & Ahmed, K. (2019). Applications of Bioelectronics in Agriculture: A Comprehensive Review. *Frontiers in Plant Science, 10*, 411.

Hu, B., Wang, K., Wu, L., Yu, S. H., Antonietti, M., & Titirici, M. M. (2010). Engineering carbon materials from the hydrothermal carbonization process of biomass. In Advanced Materials, 22(7). doi:10.1002/adma.200902812

Hui, H., Ding, Y., Shi, Q., Li, F., Song, Y., & Yan, J. (2020). 5G network-based Internet of Things for demand response in smart grid: A survey on application potential. *Applied Energy, 257*, 113972. doi:10.1016/j.apenergy.2019.113972

Huntingford, C. (2019). *Machine learning and artificial intelligence to aid climate change research and preparedness.* IOP Publishing. doi:10.1088/1748-9326/ab4e55

Hussain, Z., Babe, M., Saravanan, S., Srimathy, G., Roopa, H., & Boopathi, S. (2023). Optimizing Biomass-to-Biofuel Conversion: IoT and AI Integration for Enhanced Efficiency and Sustainability. In Circular Economy Implementation for Sustainability in the Built Environment (pp. 191–214). IGI Global.

Hussain, Z., Babe, M., Saravanan, S., Srimathy, G., Roopa, H., & Boopathi, S. (2023). Optimizing Biomass-to-Biofuel Conversion: IoT and AI Integration for Enhanced Efficiency and Sustainability. In N. Cobîrzan, R. Muntean, & R.-A. Felseghi (Eds.), (pp. 191–214). Advances in Finance, Accounting, and Economics. IGI Global. doi:10.4018/978-1-6684-8238-4.ch009

Ibrahim, M. S., Dong, W., & Yang, Q. (2020). Machine learning driven smart electric power systems: Current trends and new perspectives. *Applied Energy, 272*, 115237. doi:10.1016/j.apenergy.2020.115237

Ingle, R. B., Swathi, S., Mahendran, G., Senthil, T. S., Muralidharan, N., & Boopathi, S. (2023). Sustainability and Optimization of Green and Lean Manufacturing Processes Using Machine Learning Techniques. In N. Cobîrzan, R. Muntean, & R.-A. Felseghi (Eds.), (pp. 261–285). Advances in Finance, Accounting, and Economics. IGI Global. doi:10.4018/978-1-6684-8238-4.ch012

Ingle, R. B., Swathi, S., Mahendran, G., Senthil, T., Muralidharan, N., & Boopathi, S. (2023). Sustainability and Optimization of Green and Lean Manufacturing Processes Using Machine Learning Techniques. In *Circular Economy Implementation for Sustainability in the Built Environment* (pp. 261–285). IGI Global.

Irimia-Vladu, M. (2014). "Green" electronics: Biodegradable and biocompatible materials and devices for sustainable future. In Chemical Society Reviews, 43(2). Royal Society of Chemistry. doi:10.1039/C3CS60235D

Islam, M. M., & Zhong, R. Y. (2020). Smart Agriculture Technologies for Sustainable Crop Protection: A Comprehensive Review. *Computers and Electronics in Agriculture, 179*, 105827.

Jacobs, H. L., Gabrielson, I. N., Horton, R. K., Lyon, W. A., Hubbard, E. C., & McCallum, G. E. (1965). Water Quality Criteria-Stream vs. Effluent Standards. *Journal - Water Pollution Control Federation*, 292–315.

Jain, S., Parida, A. K., & Sankaranarayanan, S. (2020). Water scarcity prediction for global region using machine learning. *International Journal of Water, 14*(1), 69–88. doi:10.1504/IJW.2020.112746

James, A., Saji, A., Nair, A., & Joseph, D. (2019). CropSense–A Smart Agricultural System using IoT. [Google Scholar]. *J. Electron. Des. Eng., 5*, 1–7.

Janardhana, K., Singh, V., Singh, S. N., Babu, T. R., Bano, S., & Boopathi, S. (2023). Utilization Process for Electronic Waste in Eco-Friendly Concrete: Experimental Study. In Sustainable Approaches and Strategies for E-Waste Management and Utilization (pp. 204–223). IGI Global.

Janardhana, K., Singh, V., Singh, S. N., Babu, T. R., Bano, S., & Boopathi, S. (2023a). Utilization Process for Electronic Waste in Eco-Friendly Concrete: Experimental Study. In Sustainable Approaches and Strategies for E-Waste Management and Utilization (pp. 204–223). IGI Global.

Javaid, M., Haleem, A., Singh, R. P., Suman, R., & Gonzalez, E. S. (2022). Understanding the adoption of Industry 4.0 technologies in improving environmental sustainability. *Sustainable Operations and Computers*, *3*, 203–217.

Jean, N., Burke, M., Xie, M., Davis, W. M., Lobell, D. B., & Ermon, S. (2016). Combining satellite imagery and machine learning to predict poverty. *Science*, *353*(6301), 790–794. doi:10.1126/science.aaf7894 PMID:27540167

Jeevanantham, Y. A., Saravanan, A., Vanitha, V., Boopathi, S., & Kumar, D. P. (2022). Implementation of Internet-of Things (IoT) in Soil Irrigation System. *IEEE Explore*, (pp. 1–5). IEEE.

Jiang, Z., Gao, W., Wang, L., Xiong, X., Zhang, Y., Wen, X., Luo, C., Ye, H., Lu, X., & Zhang, Y. (2018). *HPC AI500: A Benchmark Suite for HPC AI Systems*. Springer. https://link.springer.com/chapter/10.1007/978-3-030-32813-9_2

Jiang, Y., & Li, C. (2019). Applications of Internet of Things in Agriculture: A Systematic Review. *Computers and Electronics in Agriculture*, *157*, 436–447.

Jiao, J. (2020). Title of the Paper. *IOP Conference Series. Earth and Environmental Science*, *510*(2), 022012. doi:10.1088/1755-1315/510/2/022012

Jin, X., Liu, S., Baret, F., Hemerl, M., & Comar, A. (2017). Estimates of plant density of wheat crops at emergence from very low altitude UAV imagery. *Remote Sensing of Environment*, *198*, 105–114. doi:10.1016/j.rse.2017.06.007

Jones, P., & Wynn, M. (2021). The leading digital technology companies and their approach to sustainable development. *Sustainability (Basel)*, *13*(12), 6612. doi:10.3390/su13126612

Joy, J., Jasmin, D. E. A., & John, V. R. (2013). Challenges of Smart Grid. International Journal of Advanced Research in Electrical. *Electronics and Instrumentation Engineering*, *2*, 976–981.

Jung, C. M., Ray, P., & Salkuti, S. R. (2019). Asset management and maintenance: A smart grid perspective. [IJECE]. *Iranian Journal of Electrical and Computer Engineering*, *9*(5), 3391. doi:10.11591/ijece.v9i5.pp3391-3398

Kamal, S., Naim, A., Magd, H., Khan, S. A., & Khan, F. M. (2022). The Relationship Between E-Service Quality, Ease of Use, and E-CRM Performance Referred by Brand Image. In A. Naim & S. Kautish (Eds.), *Building a Brand Image Through Electronic Customer Relationship Management* (pp. 84–108). IGI Global., doi:10.4018/978-1-6684-5386-5.ch005

Kaneko, A., Kennedy, T. W., Mei, L., Sintek, C., Burke, M., Ermon, S., & Lobell, D. B. (2019). *Deep Learning for Crop Yield Prediction in Africa*. In *Proceedings of the International Conference on Machine Learning AI for Social Good Workshop*, LongBeach, CA, USA.

Karanth, S., Benefo, E. O., Patra, D., & Pradhan, A. K. (2023). Importance of artificial intelligence in evaluating climate change and food safety risk. *Journal of Agriculture and Food Research, 11*. doi:10.1016/j.jafr.2022.100485

Karthik, S., Hemalatha, R., Aruna, R., Deivakani, M., Reddy, R. V. K., & Boopathi, S. (2023). Study on Healthcare Security System-Integrated Internet of Things (IoT). In Perspectives and Considerations on the Evolution of Smart Systems (pp. 342–362). IGI Global.

Kaur, K., Dhanoa, I. S., Bhambri, P., & Singh, G. (2020). Energy Saving VM Migration Techniques. *Journal of Critical Reviews*, *7*(9), 2359–2365.

Kavitha, C. R., Varalatchoumy, M., Mithuna, H. R., Bharathi, K., Geethalakshmi, N. M., & Boopathi, S. (2023). Energy Monitoring and Control in the Smart Grid: Integrated Intelligent IoT and ANFIS. In M. Arshad (Ed.), (pp. 290–316). Advances in Bioinformatics and Biomedical Engineering. IGI Global. doi:10.4018/978-1-6684-6577-6.ch014

Kedri, J. (2014). *Simulation and Validation of SPV Micro Grid Comprising 518.2 KWp Distributed Solar Power Plants at Dayalbagh Educational Institute*. [Ph.D. Thesis, Dayalbagh Educational Institute, Dayalbagh, Agra, India].

Kelechi, A. H., Alsharif, M. H., Bameyi, O. J., Ezra, P. J., Joseph, I. K., Atayero, A.-A., Geem, Z. W., & Hong, J. (2020). Artificial Intelligence: An Energy Efficiency Tool for Enhanced High performance computing. *Symmetry*, *12*(6), 1029. doi:10.3390/sym12061029

Kempton, W., & Letendre, S. (1997). Electric vehicles as a new power source for electric utilities. *Transportation Research Part D, Transport and Environment*, *2*(3), 157–175. doi:10.1016/S1361-9209(97)00001-1

Khalid, M. Y., Al Rashid, A., Arif, Z. U., Ahmed, W., Arshad, H., & Zaidi, A. A. (2021). Natural fiber reinforced composites: Sustainable materials for emerging applications. In *Results in Engineering* (Vol. 11). Elsevier B.V. doi:10.1016/j.rineng.2021.100263

Khan, S. A., Magd, H., Al Shamsi, I. R., & Masoom, K. (2022). Social Entrepreneurship Through Innovations in Agriculture. In International Perspectives on Value Creation and Sustainability Through Social Entrepreneurship (pp. 209-222). IGI Global. doi:10.4018/978-1-6684-4666-9.ch010

Khan, A. R., Dubey, M. K., Bisen, P. K., & Saxena, K. K. (2007). Constraints faced by farmers of Narsing Kheda village of Sihore district. [Google Scholar]. *Young*, *8*, 16.

Khanafer, M., & Shirmohammadi, S. (2020). Applied AI in instrumentation and measurement: The deep learning revolution. *IEEE Instrumentation & Measurement Magazine*, *23*(6), 10–17. doi:10.1109/MIM.2020.9200875

Khanna, A., Lamba, B. Y., Jain, S., Bolshev, V., Budnikov, D., Panchenko, V., & Smirnov, A. (2023). Biodiesel Production from Jatropha: A Computational Approach by Means of Artificial Intelligence and Genetic Algorithm. *Sustainability (Basel)*, *15*(12), 9785. doi:10.3390/su15129785

Khan, S. A. R., Ponce, P., Thomas, G., Yu, Z., Al-Ahmadi, M. S., & Tanveer, M. (2021). Digital technologies, circular economy practices and environmental policies in the era of COVID-19. *Sustainability (Basel)*, *13*(22), 12790. doi:10.3390/su132212790

Khan, S., & Yairi, T. (2018). A review on the application of deep learning in system health management. *Mechanical Systems and Signal Processing*, *107*, 241–265. doi:10.1016/j.ymssp.2017.11.024

Khatri, B., Shrimali, H., Khan, S. A., & Naim, A. (2023). Role of HR Analytics in Ensuring Psychological Wellbeing and Job Security: Learnings from COVID-19. In HR Analytics in an Era of Rapid Automation (pp. 36-53). IGI Global.

Khayer, A., Jahan, N., Hossain, M. N., & Hossain, M. Y. (2021). The adoption of cloud computing in small and medium enterprises: A developing country perspective. *VINE Journal of Information and Knowledge Management Systems*, *51*(1), 64–91.

Kimiti, J. M., Odee, D. W., & Vanlauwe, B. (2009). *Area under Grain Legumes Cultivation and Problems Faced by Smallholder Farmers in Legume Production in the Semi-Arid Eastern Kenya*. Academic Press. [Google Scholar]

King, T., Cole, M., Farber, J. M., Eisenbrand, G., Zabaras, D., Fox, E. M., & Hill, J. P. (2017). Food safety for food security: Relationship between global megatrends and developments in food safety. Trends in Food Science &amp. *Trends in Food Science & Technology*, *68*, 160–175. doi:10.1016/j.tifs.2017.08.014

Kodali, R. K., Jain, V., & Karagwal, S. (2016). *IoT based smart greenhouse*. In 2016 IEEE Region 10 Humanitarian Technology Conference (R10-HTC), Agra, India. 10.1109/R10-HTC.2016.7906846

Konečný, P., Teslík, J., & Hamala, M. (2013). Mechanical and physical properties of straw bales. *Advanced Materials Research*, *649*, 250–253. . doi:10.4028/www.scientific.net/AMR.649.250

Konya, A., & Nematzadeh, P. (2024). Recent applications of AI to environmental disciplines: A review. *Science of The Total Environment, 906*. doi:10.1016/j.scitotenv.2023.167705

Koshariya, A. K., Kalaiyarasi, D., Jovith, A. A., Sivakami, T., Hasan, D. S., & Boopathi, S. (2023). AI-Enabled IoT and WSN-Integrated Smart Agriculture System. In R. K. Gupta, A. Jain, J. Wang, S. K. Bharti, & S. Patel (Eds.), (pp. 200–218). Practice, Progress, and Proficiency in Sustainability. IGI Global. doi:10.4018/978-1-6684-8516-3.ch011

Koshariya, A. K., Khatoon, S., Marathe, A. M., Suba, G. M., Baral, D., & Boopathi, S. (2023). Agricultural Waste Management Systems Using Artificial Intelligence Techniques. In S. Kautish, N. K. Chaubey, S. B. Goyal, & P. Whig (Eds.), (pp. 236–258). Advances in Computational Intelligence and Robotics. IGI Global. doi:10.4018/978-1-6684-8171-4.ch009

Kou, G., Chao, X., Peng, Y., Alsaadi, F. E., & Herrera Viedma, E. (2019). *Machine learning methods for systemic risk analysis in financial sectors*.

Kratzke, N. (2018). A brief history of cloud application architectures. *Applied Sciences (Basel, Switzerland)*, *8*(8), 1368.

Kreslins, K., Novik, D., & Vasiljeva, T. (2018). Challenge of cloud computing for SMEs: A case of Baltic countries. *Journal of Innovation Management in Small & Medium Enterprises, 2018*.

Kuang, B., & Mouazen, A. M. (2013). Non-biased prediction of soil organic carbon and total nitrogen with vis-NIR spectroscopy, as affected by soil moisture content and texture. *Biosystems Engineering*, *114*(3), 249–258. doi:10.1016/j.biosystemseng.2013.01.005

Kumalo, N., & Van Der Poll, J. A. (2015). The role of cloud computing in addressing SME challenges in South Africa. *2015 International Conference on Cloud Computing Research and Innovation (Icccri)*, (pp. 139–147). IEEE.

Kumar, M., Kumar, K., Sasikala, P., Sampath, B., Gopi, B., & Sundaram, S. (2023). Sustainable Green Energy Generation From Waste Water: IoT and ML Integration. In Sustainable Science and Intelligent Technologies for Societal Development (pp. 440–463). IGI Global.

Kumar, P. R., Meenakshi, S., Shalini, S., Devi, S. R., & Boopathi, S. (2023). Soil Quality Prediction in Context Learning Approaches Using Deep Learning and Blockchain for Smart Agriculture. In Effective AI, Blockchain, and E-Governance Applications for Knowledge Discovery and Management (pp. 1–26). IGI Global.

Kumar, P. R., Meenakshi, S., Shalini, S., Devi, S. R., & Boopathi, S. (2023). Soil Quality Prediction in Context Learning Approaches Using Deep Learning and Blockchain for Smart Agriculture. In Effective AI, Blockchain, and E-Governance Applications for Knowledge Discovery and Management (pp. 1–26). IGI Global. doi:10.4018/978-1-6684-9151-5.ch001

Kumar, S., & Buyya, R. (2012). Green cloud computing and environmental sustainability. *Harnessing Green IT: Principles and Practices*, 315–339.

Kumara, V., Mohanaprakash, T. A., Fairooz, S., Jamal, K., Babu, T., & B., S. (2023). Experimental Study on a Reliable Smart Hydroponics System: In P. Vasant, R. Rodríguez-Aguilar, I. Litvinchev, & J. A. Marmolejo-Saucedo (Eds.), *Advances in Environmental Engineering and Green Technologies* (pp. 27–45). IGI Global. https://doi.org/ doi:10.4018/978-1-6684-4118-3.ch002

Kumar, B. M., Kumar, K. K., Sasikala, P., Sampath, B., Gopi, B., & Sundaram, S. (2024). Sustainable Green Energy Generation From Waste Water: IoT and ML Integration. In B. K. Mishra (Ed.), (pp. 440–463). Practice, Progress, and Proficiency in Sustainability. IGI Global. doi:10.4018/979-8-3693-1186-8.ch024

Kurp, P. (2008). Green computing. *Communications of the ACM*, *51*(10), 11–13. doi:10.1145/1400181.1400186

Kushwah, J. S., Gupta, M., Shrivastava, S., Saxena, N., Saini, R., & Boopathi, S. (2024). Psychological Impacts, Prevention Strategies, and Intervention Approaches Across Age Groups: Unmasking Cyberbullying. In Change Dynamics in Healthcare, Technological Innovations, and Complex Scenarios (pp. 89–109). IGI Global.

Lahmiri, S., & Bekiros, S. (2019). Can machine learning approaches predict corporate bankruptcy? Evidence from a qualitative experimental design. *Quantitative Finance*, *19*(9), 1569–1577. doi:10.1080/14697688.2019.1588468

Lakshmanan, R., Djama, M., Selvaperumal, S., & Abdulla, R. (2020). Automated smart hydroponics system using internet of things. *Iranian Journal of Electrical and Computer Engineering*, *10*(6), 6389–6398. doi:10.11591/ijece.v10i6.pp6389-6398

Lang, L. (1992). GPS + GIS + remote sensing: An overview. [Google Scholar]. *Earth Observation Magazine*, *1*, 23–26.

Leavy, S., Meaney, G., Wade, K., & Greene, D. (2020). Mitigating gender bias in machine learning data sets. In *Proceedings of the Bias and Social Aspects in Search and Recommendation: First International Workshop*. Springer. 10.1007/978-3-030-52485-2_2

LeCun, Y., Bengio, Y., & Hinton, G. E. (2015). Deep learning. *Nature*, *521*(7553), 436–444. doi:10.1038/nature14539 PMID:26017442

Lee, C. A., Gasster, S. D., Plaza, A., Chang, C.-I., & Huang, B. (2011, September). Recent Developments in High Performance Computing for Remote Sensing: A Review. *IEEE Journal of Selected Topics in Applied Earth Observations and Remote Sensing*, *4*(3), 508–527. doi:10.1109/JSTARS.2011.2162643

Lee, D., & Yoon, S. N. (2021). Application of artificial intelligence-based technologies in the healthcare industry: Opportunities and challenges. *International Journal of Environmental Research and Public Health*, *18*(1), 271. doi:10.3390/ijerph18010271 PMID:33401373

Lee, S., & Lee, D. (2018). Improved prediction of harmful algal blooms in four Major South Korea's Rivers using deep learning models. *International Journal of Environmental Research and Public Health*, *15*(7), 1322. doi:10.3390/ijerph15071322 PMID:29937531

Lehtokunnas, T., Mattila, M., Närvänen, E., & Mesiranta, N. (2022). Towards a circular economy in food consumption: Food waste reduction practices as ethical work. *Journal of Consumer Culture*, *22*(1), 227–245. doi:10.1177/1469540520926252

Lei, Y., Yang, B., Jiang, X., Jia, F., Li, N., & Nandi, A. K. (2020). Applications of machine learning to machine fault diagnosis: A review and roadmap. *Mechanical Systems and Signal Processing*, *138*, 106587. doi:10.1016/j.ymssp.2019.106587

Leo, M., Sharma, S., & Maddulety, K. (2019). Machine learning in banking risk management: A literature review. *Risks*, *7*(1), 29. doi:10.3390/risks7010029

Leong, C., Tan, F., & Ahuja, M. (2020). IS for Good – 10 years to SDG: Where we have been and where we need to go? In *Proceedings of the International Conference on Information Systems*. IEEE.

Leopord, H., Cheruiyot, W. K., & Kimani, S. (2016). A survey and analysis on classification and regression data mining techniques for diseases outbreak prediction in datasets. *International Journal of Engineering Science*, *5*, 1–11.

Li, T., Chen, C., Brozena, A. H., Zhu, J. Y., Xu, L., Driemeier, C., Dai, J., Rojas, O. J., Isogai, A., Wågberg, L., & Hu, L. (2021). Developing fibrillated cellulose as a sustainable technological material. In Nature, 590(7844). doi:10.1038/s41586-020-03167-7

Liakos, K., Busato, P., Moshou, D., Pearson, S., & Bochtis, D. (2018). Machine Learning in Agriculture: A Review. *Sensors (Basel)*, *18*(8), 2674. doi:10.3390/s18082674 PMID:30110960

Li, L., Zhang, Q., & Huang, D. (2018). Development of Agricultural Internet of Things and Its Applications in China. *Computers and Electronics in Agriculture*, *154*, 61–68.

Li, N., Liang, X., Li, X., Wang, C., & Wu, D. D. (2009). Network environment and financial risk using machine learning and sentiment analysis. *Human and Ecological Risk Assessment*, *15*(2), 227–252. doi:10.1080/10807030902761056

Lin, J., Yu, W., Zhang, N., Yang, X., Zhang, H., & Zhao, W. (2017). A Survey on Internet of Things: Architecture, Enabling Technologies, Security and Privacy, and Applications. *IEEE Internet of Things Journal*, *4*(5), 1125–1142. doi:10.1109/JIOT.2017.2683200

Lin, W. Y., Hu, Y. H., & Tsai, C. F. (2011). Machine learning in financial crisis prediction: A survey. *IEEE Transactions on Systems, Man, and Cybernetics. Part C, Applications and Reviews*, *42*(4), 421–436.

Litjens, G., Ciompi, F., Wolterink, J. M., De Vos, B. D., Leiner, T., Teuwen, J., & Išgum, I. (2019). State-of-the-Art deep learning in cardiovascular image analysis. *JACC: Cardiovascular Imaging*, *12*(8), 1549–1565. doi:10.1016/j.jcmg.2019.06.009 PMID:31395244

Liu, C., Li, F., Lai-Peng, M., & Cheng, H. M. (2010). Advanced materials for energy storage. In Advanced Materials, 22(8). doi:10.1002/adma.200903328

Liu, M., & Guo, J. (2021). Integration of Biosensors and Artificial Intelligence for Precision Agriculture. *Frontiers in Plant Science*, *11*, 604524.

Liu, N., Huo, K., McDowell, M. T., Zhao, J., & Cui, Y. (2013). Rice husks as a sustainable source of nanostructured silicon for high performance Li-ion battery anodes. *Scientific Reports*, *3*(1), 1919. doi:10.1038/srep01919 PMID:23715238

Liu, Y., Ni, Z., Karlsson, M., & Gong, S. (2021). Methodology for digital transformation with internet of things and cloud computing: A practical guideline for innovation in small-and medium-sized enterprises. *Sensors (Basel)*, *21*(16), 5355.

Li, Y., & Wang, N. (2019). Precision Agriculture Monitoring System Based on IoT. *IOP Conference Series. Earth and Environmental Science*, *310*, 022015.

Lottes, P., Khanna, R., Pfeifer, J., Siegwart, R., & Stachniss, C. UAV-based crop and weed classification for smart farming. In *Proceedings of the 2017 IEEE International Conference on Robotics and Automation (ICRA)*, (pp. 3024–3031). IEEE. 10.1109/ICRA.2017.7989347

Lundervold, A. S., & Lundervold, A. (2019). An overview of deep learning in medical imaging focusing on MRI. *Zeitschrift für Medizinische Physik*, *29*(2), 102–127. doi:10.1016/j.zemedi.2018.11.002 PMID:30553609

Luo, Y. (2022). A general framework of digitization risks in international business. *Journal of International Business Studies*, *53*(2), 344–361. doi:10.1057/s41267-021-00448-9 PMID:34075261

Lvova, L., & Nadporozhskaya, M. (2017). Chemical sensors for soil analysis: Principles and applications. In A. M. Grumezescu (Ed.), *Series Nanotechnology in the Agri-Food Industry; New Pesticides and Soil Sensors* (Vol. 10, pp. 637–678). Elsevier. doi:10.1016/B978-0-12-804299-1.00018-7

Maghsudi, S., Lan, A., Xu, J., & van Der Schaar, M. (2021). Personalized education in the artificial intelligence era: What to expect next. *IEEE Signal Processing Magazine*, *38*(3), 37–50. doi:10.1109/MSP.2021.3055032

Maguluri, L. P., Ananth, J., Hariram, S., Geetha, C., Bhaskar, A., & Boopathi, S. (2023). Smart Vehicle-Emissions Monitoring System Using Internet of Things (IoT). In Handbook of Research on Safe Disposal Methods of Municipal Solid Wastes for a Sustainable Environment (pp. 191–211). IGI Global.

Maguluri, L. P., Arularasan, A. N., & Boopathi, S. (2023). Assessing Security Concerns for AI-Based Drones in Smart Cities. In R. Kumar, A. B. Abdul Hamid, & N. I. Binti Ya'akub (Eds.), (pp. 27–47). Advances in Computational Intelligence and Robotics. IGI Global. doi:10.4018/978-1-6684-9151-5.ch002

Maheswari, B. U., Imambi, S. S., Hasan, D., Meenakshi, S., Pratheep, V. G., & Boopathi, S. (2023). Internet of Things and Machine Learning-Integrated Smart Robotics. In M. K. Habib (Ed.), (pp. 240–258). Advances in Computational Intelligence and Robotics. IGI Global. doi:10.4018/978-1-6684-7791-5.ch010

Maier, A., Syben, C., Lasser, T., & Rieß, C. (2019). A gentle introduction to deep learning in medical image processing. *Zeitschrift für Medizinische Physik*, *29*(2), 86–101. doi:10.1016/j.zemedi.2018.12.003 PMID:30686613

Maiurova, A., Kurniawan, T. A., Kustikova, M., Bykovskaia, E., Othman, M. H. D., Singh, D., & Goh, H. H. (2022). Promoting digital transformation in waste collection service and waste recycling in Moscow (Russia): Applying a circular economy paradigm to mitigate climate change impacts on the environment. *Journal of Cleaner Production*, *354*, 131604. doi:10.1016/j.jclepro.2022.131604

Makhija, P., & Chacko, E. (2021). Efficiency and advancement of artificial intelligence in service sector with special reference to banking industry. Maleh, Y. (eds) Computational Intelligence Techniques for Green Smart Cities. Green Energy and Technology. Springer, Cham. doi:10.1007/978-981-16-3250-1_2

Malathi, J., Kusha, K., Isaac, S., Ramesh, A., Rajendiran, M., & Boopathi, S. (2024). IoT-Enabled Remote Patient Monitoring for Chronic Disease Management and Cost Savings: Transforming Healthcare. In Advances in Explainable AI Applications for Smart Cities (pp. 371–388). IGI Global.

Maleki, M. R., Van Holm, L., Ramon, H., Merckx, R., De Baerdemaeker, J., & Mouazen, A. M. (2006). Phosphorus Sensing for Fresh Soils using Visible and Near Infrared Spectroscopy. *Biosystems Engineering*, *95*(3), 425–436. doi:10.1016/j.biosystemseng.2006.07.015

Malik, P. K., Naim, A., & Singh, R. (Eds.). (2022). *Printed Antennas: Design and Challenges*. CRC Press. doi:10.1201/9781003347057

Mallarino, A., & Wittry, D. (1997). Use of DGPs, Yield Monitors, Soil Testing, and Variable Rate Technology to Improve Phosphorus and Potassium Management. *Proceedings of the Integrated Crop Management Conference*. IEEE. 10.31274/icm-180809-589

Manubens-Gil, D., Vegas-Regidor, J., Prodhomme, C., Mula-Valls, O., & Doblas-Reyes, F. J. (2016). Seamless management of ensemble climate prediction experiments on HPC platforms. *2016 International Conference on High Performance Computing & Simulation (HPCS)*, Innsbruck, Austria. 10.1109/HPCSim.2016.7568429

Mao, Y., You, C., Zhang, J., Huang, K., & Letaief, K. B. (2017). A survey on Mobile Edge Computing: The Communication Perspective. *IEEE Communications Surveys and Tutorials*, *19*(4), 2322–2358. doi:10.1109/COMST.2017.2745201

Marques, M. F. V., Lunz, J. N., Aguiar, V. O., Grafova, I., Kemell, M., Visentin, F., Sartori, A., & Grafov, A. (2015). Thermal and Mechanical Properties of Sustainable Composites Reinforced with Natural Fibers. *Journal of Polymers and the Environment*, *23*(2), 251–260. doi:10.1007/s10924-014-0687-2

Martos, V., Ahmad, A., Cartujo, P., & Ordoñez, J. (2021). Ensuring agricultural sustainability through remote sensing in the era of agriculture 5.0. *Applied Sciences (Basel, Switzerland)*, *11*(13), 5911. doi:10.3390/app11135911

Mashrur, A., Luo, W., Zaidi, N. A., & Robles-Kelly, A. (2020). Machine learning for financial risk management: A survey. *IEEE Access : Practical Innovations, Open Solutions*, *8*, 203203–203223. doi:10.1109/ACCESS.2020.3036322

Mathew, A., Amudha, P., & Sivakumari, S. (2020). Deep Learning Techniques: An Overview. In Advances in intelligent systems and computing (pp. 599–608). Springer. doi:10.1007/978-981-15-3383-9_54

Medaglia, R., Rukanova, B., & Zhang, Z. (2024). Digital government and the circular economy transition: An analytical framework and a research agenda. *Government Information Quarterly*, *41*(1), 101904. doi:10.1016/j.giq.2023.101904

Mekala, M. S., & Viswanathan, P. (2017). A Survey: Smart agriculture IoT with cloud computing. *2017 International Conference on Microelectronic Devices, Circuits and Systems (ICMDCS)*. IEEE. 10.1109/ICMDCS.2017.8211551

Meliani, M., el Barkany, A., el Abbassi, I., Darcherif, A. M., & Mahmoudi, M. (2020). Control system in the smart grid: State of the art and opportunities. In *2020 IEEE 13th International Colloquium of Logistics and Supply Chain Management (LOGISTIQUA)* (pp. 1–6). IEEE. 10.1109/LOGISTIQUA49782.2020.9353878

Melita, L., Gumrah, F., & Amareanu, M. (2014). *Issue 2, 2014* (Vol. 5). "Porous polymer membranes used for wastewater treatment", Membrane Water Treatment. Techno-Press., doi:10.12989/mwt.2014.5.2.147

Mhlanga, D. (2021). Artificial intelligence in the industry 4.0, and its impact on poverty, innovation, infrastructure development, and the sustainable development goals: Lessons from emerging economies? *Sustainability (Basel)*, *13*(11), 5788. doi:10.3390/su13115788

Michalakes, J. (2020). HPC for Weather Forecasting. In A. Grama & A. Sameh (Eds.), *Parallel Algorithms in Computational Science and Engineering*. Modeling and Simulation in Science, Engineering and Technology. Birkhäuser. doi:10.1007/978-3-030-43736-7_10

Mies, A., & Gold, S. (2021). Mapping the social dimension of the circular economy. *Journal of Cleaner Production*, *321*, 128960. doi:10.1016/j.jclepro.2021.128960

Milano, M., O'Sullivan, B., & Gavanelli, M. (2014). Sustainable policy making: A strategic challenge for artificial intelligence. *AI Magazine*, *35*(3), 22–35. doi:10.1609/aimag.v35i3.2534

Milojicic, D., Faraboschi, P., Dube, N., & Roweth, D. (2021). Future of HPC: Diversifying Heterogeneity. 2021 Design, Automation & Test in Europe Conference & Exhibition. IEEE. doi:10.23919/DATE51398.2021.9474063

Minevich, M. (2022). How To Fight Climate Change Using A. I. *Forbes*. https://www.forbes.com/sites/markminevich/2022/07/08/how-to-fight-climate-change-using-ai/?sh=5682c32a838e

Mishra, H., & Mishra, D. (2023). Artificial Intelligence and Machine Learning in Agriculture: Transforming Farming Systems. *Res. Trends Agric. Sci*, *1*, 1–16.

Mittal, A., & Singh, A. (2007). Microcontroller based pest management system. *Second International Conference on Systems (ICONS'07)*. IEEE. 10.1109/ICONS.2007.35

Modisane, P., & Jokonya, O. (2021). Evaluating the benefits of cloud computing in small, medium and micro-sized enterprises (SMMEs). *Procedia Computer Science*, *181*, 784–792.

Mohammed, A. (2023). *Fraud Prevention, Confidentiality, and Data Security for Implementation of Big Data in Modern Business: Fraud Detection and Prevention Techniques*. Academic Press.

Mohammed, A. B. (2023). *AS Syed - Global Applications of the Internet of Things in Digital*. Applications and Impact of Internet of Things in Digital Marketing.

Mohanty, A., Venkateswaran, N., Ranjit, P., Tripathi, M. A., & Boopathi, S. (2023). Innovative Strategy for Profitable Automobile Industries: Working Capital Management. In Handbook of Research on Designing Sustainable Supply Chains to Achieve a Circular Economy (pp. 412–428). IGI Global.

Mohanty, A., Jothi, B., Jeyasudha, J., Ranjit, P., Isaac, J. S., & Boopathi, S. (2023). Additive Manufacturing Using Robotic Programming. In *AI-Enabled Social Robotics in Human Care Services* (pp. 259–282). IGI Global. doi:10.4018/978-1-6684-8171-4.ch010

Mondal, S., Singh, S., & Gupta, H. (2023). Green entrepreneurship and digitalization enabling the circular economy through sustainable waste management-An exploratory study of emerging economy. *Journal of Cleaner Production*, *422*, 138433. doi:10.1016/j.jclepro.2023.138433

Moreno Escobar, J. J., Morales Matamoros, O., Tejeida Padilla, R., Lina Reyes, I., & Quintana Espinosa, H. (2021). A Comprehensive Review on Smart Grids: Challenges and Opportunities. *Sensors (Basel)*, *21*(21), 6978. doi:10.3390/s21216978 PMID:34770285

Mujinga, M. (2020). Cloud Computing Inhibitors Among Small and Medium Enterprises. *2020 3rd International Conference on Intelligent Sustainable Systems (ICISS)*, 1385–1391.

Muniasamy, A., Tabassam, S., Hussain, M. A., Sultana, H., Muniasamy, V., & Bhatnagar, R. Deep Learning for Predictive Analytics in Healthcare. In: A H, A A, T G, R B, M FT, eds. The International Conference on Advanced Machine Learning Technologies and Applications (AMLTA2019). Cairo, Egypt: Springer International Publishing; 2019: 32–42.

Muthumanickam, D., Kannan, P., Kumaraperumal, R., Natarajan, S., Sivasamy, R., & Poongodi, C. (2011). Drought assessment and monitoring through remote sensing and GIS in western tracts of Tamil Nadu, India. *International Journal of Remote Sensing*, *32*(18), 5157–5176. doi:10.1080/01431161.2010.494642

Mwasilu, F., Justo, J. J., Kim, E.-K., Do, T. D., & Jung, J.-W. (2014). Electric vehicles and smart grid interaction: A review on vehicle to grid and renewable energy sources integration. *Renewable & Sustainable Energy Reviews*, *34*, 501–516. doi:10.1016/j.rser.2014.03.031

Myilsamy, S., & Sampath, B. (2017). Grey Relational Optimization of Powder Mixed Near-Dry Wire Cut Electrical Discharge Machining of Inconel 718 Alloy. *Asian Journal of Research in Social Sciences and Humanities*, *7*(3), 18–25.

Myilsamy, S., & Sampath, B. (2021). Experimental comparison of near-dry and cryogenically cooled near-dry machining in wire-cut electrical discharge machining processes. *IOP:Surface Topography: Metrology and Properties*, *9*(3), 035015.

Nahorniak, H., & Hannouf, K. (2023). Features of the challenges and opportunities of formation of intellectual capital of domestic enterprises in the engineering sector ukraine in the conditions of digitalization. *Probl. Transp. Telecommun.*, *24*, 1–15.

Naim, A. (2022). Benefits of Artificial Intelligence (AI) in Financial Practices. *International Journal of Studies in Business Management, Economics and Strategies, 1*, 1–12. https://scholarsdigest.org/index.php/bmes/article/view/7

Naim, A., & Khan, S. A. (2023). Impact and Assessment of Electronic Commerce on Consumer Buying Behaviour. In Global Applications of the Internet of Things in Digital Marketing (pp. 264-289). IGI Global. doi:10.4018/978-1-6684-8166-0.ch014

Naim, A., Alqahtani, H., Fatima, S., & Khan, M. F. (2023). Applications of Artificial Intelligence of Things in Green Supply Chain Management: Challenges and Future Directions. In Blockchain, IoT, and AI Technologies for Supply Chain Management (pp. 99-126). CRC Press.

Naim, A., Muniasamy, A., & Alqahtani, H. (2022). Application of Internet of Things in Marketing Management. *Ambient Intelligence and Internet of Things: Convergent Technologies*, 273-299.

Naim, A., Syed, A., Bilfaqih, S. M., & Muniasamy, A. (2023). Impact of Individual and Organizational Social Capital of Marketing in a Global Network. In Social Capital in the Age of Online Networking: Genesis, Manifestations, and Implications (pp. 1-17). IGI Global. doi:10.4018/978-1-6684-8953-6.ch001

Naim, A. (2021). Applications of MIS in building Electronic Relationship with customers: A case-based study. Periodica Journal of Modern Philosophy. *Social Sciences and Humanities*, *1*, 1–8.

Naim, A. (2021). Green Information Technologies in Business Operations. Periodica Journal of Modern Philosophy. *Social Sciences and Humanities*, *1*, 36–49.

Naim, A. (2021). New Trends in Business Process Management: Applications of Green Information Technologies. *British Journal of Environmental Studies*, *1*(1), 12–23.

Naim, A. (2023). Application of HPC techniques to identify the business financial risks. *Academy of Marketing Studies Journal*, *27*(S5), 1–11.

Naim, A. (2023). Application of machine learning techniques to identify the business financial risks. *Academy of Marketing Studies Journal*, *27*(S5), 1–11.

Naim, A., Alahmari, F., & Rahim, A. (2021). Role of Artificial Intelligence in Market Development and Vehicular Communication. Smart Antennas. *Recent Trends in Design and Applications*, *2*, 28–39. doi:10.2174/9781681088594121020006

Naim, A., & Alqahtani, H. (2023). Role of IoT in Industrial Transformation and Marketing Management. In *Global Applications of the Internet of Things in Digital Marketing* (pp. 67–93). IGI Global. doi:10.4018/978-1-6684-8166-0.ch005

Naim, A., Alqahtani, H., Muniasamy, A., Bilfaqih, S. M., Mahveen, R., & Mahjabeen, R. (2023). Applications of Information Systems and Data Security in Marketing Management. In A. Naim, P. Malik, & F. Zaidi (Eds.), *Fraud Prevention, Confidentiality, and Data Security for Modern Businesses* (pp. 57–83). IGI Global. doi:10.4018/978-1-6684-6581-3.ch003

Naim, A., & Alqahtani, K. (2021). Role of Information Systems in Customer Relationship Management. *Pulse*, *2*(2).

Naim, A., Alshawaf, S. M., Kumar Malik, P., & Singh, R. (2023). Effective E-Learning Practices by HPC and Artificial Intelligence. *2023 International Conference on Artificial Intelligence and Smart Communication (AISC)*, Greater Noida, India. 10.1109/AISC56616.2023.10085391

Naim, A., & Kautish, S. K. (Eds.). (2022). *Building a Brand Image Through Electronic Customer Relationship Management*. IGI Global., doi:10.4018/978-1-6684-5386-5

Naim, A., Malik, P. K., & Zaidi, F. A. (Eds.). (2023). *Fraud Prevention, Confidentiality, and Data Security for Modern Businesses*. IGI Global. doi:10.4018/978-1-6684-6581-3

Naim, A., Muniasamy, A., Clementking, A., & Rajkumar, R. (2022). Relevance of Green Manufacturing and IoT in Industrial Transformation and Marketing Management. In M. Lahby, A. Al-Fuqaha, & Y. Maleh (Eds.), *Computational Intelligence Techniques for Green Smart Cities. Green Energy and Technology*. Springer. doi:10.1007/978-3-030-96429-0_19

Nanda, A. K., Sharma, A., Augustine, P. J., Cyril, B. R., Kiran, V., & Sampath, B. (2024). Securing Cloud Infrastructure in IaaS and PaaS Environments. In Improving Security, Privacy, and Trust in Cloud Computing (pp. 1–33). IGI Global.

Nanda, A. K., Sharma, A., Augustine, P. J., Cyril, B. R., Kiran, V., & Sampath, B. (2024). Securing Cloud Infrastructure in IaaS and PaaS Environments. In Improving Security, Privacy, and Trust in Cloud Computing (pp. 1–33). IGI Global. doi:10.4018/979-8-3693-1431-9.ch001

Nanda, A. K., Sharma, A., Augustine, P. J., Cyril, B. R., Kiran, V., & Sampath, B. (2024a). Securing Cloud Infrastructure in IaaS and PaaS Environments. In Improving Security, Privacy, and Trust in Cloud Computing (pp. 1–33). IGI Global.

Nanda, A. K., Sharma, A., Augustine, P. J., Cyril, B. R., Kiran, V., & Sampath, B. (2024b). Securing Cloud Infrastructure in IaaS and PaaS Environments. In Improving Security, Privacy, and Trust in Cloud Computing (pp. 1–33). IGI Global.

Nascimento, D. L. M., Alencastro, V., Quelhas, O. L. G., Caiado, R. G. G., Garza-Reyes, J. A., Rocha-Lona, L., & Tortorella, G. (2019). Exploring Industry 4.0 technologies to enable circular economy practices in a manufacturing context: A business model proposal. *Journal of Manufacturing Technology Management*, *30*(3), 607–627. doi:10.1108/JMTM-03-2018-0071

Naval, N., & Yusta, J. M. (2021). Virtual power plant models and electricity markets-A review. *Renewable & Sustainable Energy Reviews*, *149*, 111393.

Naveeenkumar, N., Rallapalli, S., Sasikala, K., Priya, P. V., Husain, J., & Boopathi, S. (2024). Enhancing Consumer Behavior and Experience Through AI-Driven Insights Optimization. In *AI Impacts in Digital Consumer Behavior* (pp. 1–35). IGI Global. doi:10.4018/979-8-3693-1918-5.ch001

Navulur, S., & Sastry, S. C. S., A., & N. Giri Prasad, M. (. (2017). Agricultural Management through Wireless Sensors and Internet of Things. [IJECE]. *Iranian Journal of Electrical and Computer Engineering*, *7*(6), 3492. doi:10.11591/ijece.v7i6.pp3492-3499

Nghiêm-Phú, B., & Suter, J. R. (2022). The Application of Anthropomorphism in Marketing: Implications for Green Economy in a digital world. In *Handbook of Research on Building Greener Economics and Adopting Digital Tools in the Era of Climate Change* (pp. 13–42). IGI Global. doi:10.4018/978-1-6684-4610-2.ch002

Nishanth, J. R., Deshmukh, M. A., Kushwah, R., Kushwaha, K. K., Balaji, S., & Sampath, B. (2023). Particle Swarm Optimization of Hybrid Renewable Energy Systems. In B. K. Mishra (Ed.), (pp. 291–308). Advances in Civil and Industrial Engineering. IGI Global., doi:10.4018/979-8-3693-0044-2.ch016

Nishanth, J., Deshmukh, M. A., Kushwah, R., Kushwaha, K. K., Balaji, S., & Sampath, B. (2023). Particle Swarm Optimization of Hybrid Renewable Energy Systems. In *Intelligent Engineering Applications and Applied Sciences for Sustainability* (pp. 291–308). IGI Global.

Nishant, R., Kennedy, M., & Corbett, J. (2020). Artificial intelligence for sustainability: Challenges, opportunities, and a research agenda. *International Journal of Information Management*, *53*, 102104.

Noh, S. K. (2021). Recycled clothing classification system using intelligent IoT and deep learning with AlexNet. . *Computational Intelligence and Neuroscience*, *2021*, 5544784. doi:10.1155/2021/5544784 PMID:33854541

Nozari, H., Szmelter-Jarosz, A., & Ghahremani-Nahr, J. (2021). The Ideas of Sustainable and Green Marketing Based on the Internet of Everything—The Case of the Dairy Industry. *Future Internet*, *13*(10), 266. doi:10.3390/fi13100266

O'Grady, M. J., & O'Hare, G. M. P. (2017). Modelling the smart farm. *Information Processing in Agriculture*, *4*(3), 179–187. doi:10.1016/j.inpa.2017.05.001

Obaisi, A. I., Adegbeye, M. J., Elghandour, M. M. M. Y., Barbabosa-Pliego, A., & Salem, A. Z. M. (2022). Natural Resource Management and Sustainable Agriculture. Handbook of Climate Change Mitigation and Adaptation, (pp. 2577–2613). Springer. doi:10.1007/978-3-030-72579-2_133

Ojo, A. O., Raman, M., & Downe, A. G. (2019). Toward green computing practices: A Malaysian study of green belief and attitude among Information Technology professionals. *Journal of Cleaner Production*, *224*, 246–255. doi:10.1016/j.jclepro.2019.03.237

Ojo, O. I., & Ilunga, F. (2018). *Geospatial Analysis for Irrigated Land Assessment, Modeling and Mapping*. Multi-Purposeful Application of Geospatial Data. doi:10.5772/intechopen.73314

Oltean, V. A., Renault, S., Valvo, M., & Brandell, D. (2016). Sustainable materials for sustainable energy storage: Organic Na electrodes. In Materials, 9(3). MDPI AG. doi:10.3390/ma9030142

Omitaomu, O. A., & Niu, H. (2021). Artificial Intelligence Techniques in Smart Grid: A Survey. *Smart Cities*, *4*(2), 548–568. doi:10.3390/smartcities4020029

Ott, W. (1978). *Water quality indices: a survey of indices used in the United States* (Vol. 1). Environmental Protection Agency, Office of Research and Development, Office of Monitoring and Technical Support.

Otuka, R., Preston, D., & Pimenidis, E. (2014). The use and challenges of cloud computing services in SMEs in Nigeria. *Proceedings of the European Conference on Information Management*, 325. IEEE.

Oualmakran, Y., Melendez, J., & Herraiz, S. (2012). *Self-healing for smart grids: Problem formulation and considerations. In 2012 3rd IEEE PES Innovative Smart Grid Technologies Europe (ISGT Europe)*. IEEE., doi:10.1109/ISGTEurope.2012.6465856

Ouyang, H., Wei, X., & Wu, Q. (2019). Agricultural commodity futures prices prediction via long- and short-term time series network. *Journal of Applied Econometrics*, *22*, 468–483.

Ozdemir, E., Ozdemir, S., Erhan, K., & Aktas, A. (2016). Energy storage technologies opportunities and challenges in smart grids. In *2016 International Smart Grid Workshop and Certificate Program (ISGWCP)* (pp. 1–6). IEEE. 10.1109/ISGWCP.2016.7548263

Pachiappan, K., Anitha, K., Pitchai, R., Sangeetha, S., Satyanarayana, T. V. V., & Boopathi, S. (2023). Intelligent Machines, IoT, and AI in Revolutionizing Agriculture for Water Processing. In B. B. Gupta & F. Colace (Eds.), (pp. 374–399). Advances in Computational Intelligence and Robotics. IGI Global., doi:10.4018/978-1-6684-9999-3.ch015

Pachiappan, K., Anitha, K., Pitchai, R., Sangeetha, S., Satyanarayana, T., & Boopathi, S. (2024a). Intelligent Machines, IoT, and AI in Revolutionizing Agriculture for Water Processing. In *Handbook of Research on AI and ML for Intelligent Machines and Systems* (pp. 374–399). IGI Global.

Palaniappan, M., Tirlangi, S., Mohamed, M. J. S., Moorthy, R. S., Valeti, S. V., & Boopathi, S. (2023). Fused Deposition Modelling of Polylactic Acid (PLA)-Based Polymer Composites: A Case Study. In Development, Properties, and Industrial Applications of 3D Printed Polymer Composites (pp. 66–85). IGI Global.

Pallathadka, H., Mustafa, M., Sanchez, D. T., Sajja, G. S., Gour, S., & Naved, M. (2023). Impact of machine learning on management, healthcare and agriculture. *Materials Today: Proceedings*, *80*, 2803–2806.

Palmer, T. (2014). Climate forecasting: Build high-resolution global climate models. *Nature*, *515*(7527), 338–339. doi:10.1038/515338a PMID:25409812

Patel, V., & Lee, W. M. (2020). Artificial Intelligence in Agriculture: A Comprehensive Review. *Computers and Electronics in Agriculture*, *176*, 105554.

Patil, K. A., & Kale, N. R. (2016). A model for smart agriculture using IoT. *2016 International Conference on Global Trends in Signal Processing, Information Computing and Communication (ICGTSPICC)*. IEEE. 10.1109/ICGTSPICC.2016.7955360

Paul, A., & Thilagham, K. KG, J.-, Reddy, P. R., Sathyamurthy, R., & Boopathi, S. (2024). Multi-criteria Optimization on Friction Stir Welding of Aluminum Composite (AA5052-H32/B4C) using Titanium Nitride Coated Tool. Engineering Research Express.

Paul, K., Sorrentino, M., Lucini, L., Rouphael, Y., Cardarelli, M., Bonini, P., Reynaud, H., Canaguier, R., Trtílek, M., Panzarová, K., & Colla, G. (2019). Understanding the Biostimulant Action of Vegetal-Derived Protein Hydrolysates by High-Throughput Plant Phenotyping and Metabolomics: A Case Study on Tomato. *Frontiers in Plant Science*, *10*, 47. doi:10.3389/fpls.2019.00047 PMID:30800134

Paul, S. G., Saha, A., Arefin, M. S., Bhuiyan, T., Biswas, A. A., Reza, A. W., Alotaibi, N. M., Alyami, S. A., & Moni, M. A. (2023). A comprehensive review of green computing: Past, present, and future research. *IEEE Access : Practical Innovations, Open Solutions*, *11*, 87445–87494. doi:10.1109/ACCESS.2023.3304332

Pedrycz, W. (2022). Towards green machine learning: Challenges, opportunities, and developments. *Journal of Smart Environments and Green Computing*, *2*(4), 163–174. doi:10.20517/jsegc.2022.16

Persis, D. J., Venkatesh, V., Sreedharan, V. R., Shi, Y., & Sankaranarayanan, B. (2021). Modelling and analysing the impact of Circular Economy; Internet of Things and ethical business practices in the VUCA world: Evidence from the food processing industry. *Journal of Cleaner Production*, *301*, 126871. doi:10.1016/j.jclepro.2021.126871

Pierdicca, R., Paolanti, M., Felicetti, A., Piccinini, F., & Zingaretti, P. (2020). Automatic faults detection of photovoltaic farms: SolAIr, a deep learning-based system for thermal images. *Energies*, *13*(24), 6496. doi:10.3390/en13246496

Pimentel, D., & Burgess, M. (2013). Soil erosion threatens food production. *Agriculture*, *3*(3), 443–463. doi:10.3390/agriculture3030443

Potamitis, I., Rigakis, I., Tatlas, N. A., & Potirakis, S. (2019). In-Vivo Vibroacoustic Surveillance of Trees in the Context of the IoT. *Sensors (Basel)*, *19*(6), 1366. doi:10.3390/s19061366 PMID:30893798

Prabhuswamy, M., Tripathi, R., Vijayakumar, M., Thulasimani, T., Sundharesalingam, P., & Sampath, B. (2024). A Study on the Complex Nature of Higher Education Leadership: An Innovative Approach. In *Challenges of Globalization and Inclusivity in Academic Research* (pp. 202–223). IGI Global. doi:10.4018/979-8-3693-1371-8.ch013

Prakash, B., Jayashri, S., & Karthik, T. (2019). A hybrid genetic artificial neural network (G-ANN) algorithm for optimization of energy component in a wireless mesh network toward green computing. *Soft Computing*, *23*(8), 2789–2798. doi:10.1007/s00500-019-03789-8

Prasad, R., Rohokale, V., Prasad, R., & Rohokale, V. (2020). Artificial intelligence and machine learning in cyber security. *Cyber security: the lifeline of information and communication technology*, 231-247.

Priya, E. D., Sharmila, A., Rajarajeshwari, K. C., Anand, K. G., & Naim, A. (2023). Wearable Proximity Coupled Antenna for IoT Applications. In *Internet of Things Enabled Antennas for Biomedical Devices and Systems: Impact, Challenges and Applications* (pp. 249–258). Springer Nature Singapore. doi:10.1007/978-981-99-0212-5_20

Provis, J. L. (2018). Alkali-activated materials. In *Cement and Concrete Research* (Vol. 114, pp. 40–48). Elsevier Ltd., doi:10.1016/j.cemconres.2017.02.009

Puranik, T. A., Shaik, N., Vankudoth, R., Kolhe, M. R., Yadav, N., & Boopathi, S. (2024). Study on Harmonizing Human-Robot (Drone) Collaboration: Navigating Seamless Interactions in Collaborative Environments. In Cybersecurity Issues and Challenges in the Drone Industry (pp. 1–26). IGI Global.

Qureshi, S., & Zia, A. (2019). Big Data in Agriculture: A Survey. *Computers and Electronics in Agriculture*, *156*, 133–146.

Quy, V. K., Hau, N. V., Anh, D. V., Quy, N. M., Ban, N. T., Lanza, S., Randazzo, G., & Muzirafuti, A. (2022). IoT-Enabled Smart Agriculture: Architecture, Applications, and Challenges. *Applied Sciences (Basel, Switzerland), 12*(7), 3396. doi:10.3390/app12073396

Ragazou, K., Garefalakis, A., Zafeiriou, E., & Passas, I. (2022). Agriculture 5.0: A New Strategic Management Mode for a Cut Cost and an Energy Efficient Agriculture Sector. [s Note: MDPI stays neutral with regard to jurisdictional claims in published....]. *Energies, 2022*(15), 3113.

Rahamathunnisa, U., Sudhakar, K., Padhi, S., Bhattacharya, S., Shashibhushan, G., & Boopathi, S. (2024). Sustainable Energy Generation From Waste Water: IoT Integrated Technologies. In Adoption and Use of Technology Tools and Services by Economically Disadvantaged Communities: Implications for Growth and Sustainability (pp. 225–256). IGI Global.

Rahamathunnisa, U., Sudhakar, K., Padhi, S., Bhattacharya, S., Shashibhushan, G., & Boopathi, S. (2024a). Sustainable Energy Generation From Waste Water: IoT Integrated Technologies. In Adoption and Use of Technology Tools and Services by Economically Disadvantaged Communities: Implications for Growth and Sustainability (pp. 225–256). IGI Global.

Rahamathunnisa, U., Subhashini, P., Aancy, H. M., Meenakshi, S., Boopathi, S., & ... (2023). Solutions for Software Requirement Risks Using Artificial Intelligence Techniques. In *Handbook of Research on Data Science and Cybersecurity Innovations in Industry 4.0 Technologies* (pp. 45–64). IGI Global.

Rahamathunnisa, U., Sudhakar, K., Murugan, T. K., Thivaharan, S., Rajkumar, M., & Boopathi, S. (2023). Cloud Computing Principles for Optimizing Robot Task Offloading Processes. In *AI-Enabled Social Robotics in Human Care Services* (pp. 188–211). IGI Global.

Rahmanian, N., Ali, S. H. B., Homayoonfard, M., Ali, N. J., Rehan, M., Sadef, Y., & Nizami, A. S. (2015). Analysis of physiochemical parameters to evaluate the drinking water quality in the State of Perak, Malaysia. *Journal of Chemistry, 2015*, 2015. doi:10.1155/2015/716125

Raj, R. (2020). *High-Performance Computing Education: Current Challenges and Future Directions_ITiCSE-WGR '20: Proceedings of the Working Group Reports on Innovation and Technology in Computer Science Education*. ACM. doi:10.1145/3437800.3439203

Ramudu, K., Mohan, V. M., Jyothirmai, D., Prasad, D., Agrawal, R., & Boopathi, S. (2023). Machine Learning and Artificial Intelligence in Disease Prediction: Applications, Challenges, Limitations, Case Studies, and Future Directions. In Contemporary Applications of Data Fusion for Advanced Healthcare Informatics (pp. 297–318). IGI Global.

Ramudu, K., Mohan, V. M., Jyothirmai, D., Prasad, D. V. S. S. S. V., Agrawal, R., & Boopathi, S. (2023). Machine Learning and Artificial Intelligence in Disease Prediction: Applications, Challenges, Limitations, Case Studies, and Future Directions. In G. S. Karthick & S. Karupusamy (Eds.), (pp. 297–318). Advances in Healthcare Information Systems and Administration. IGI Global. doi:10.4018/978-1-6684-8913-0.ch013

Rangel-Martinez, D., Nigam, K., & Ricardez-Sandoval, L. A. (2021). Machine learning on sustainable energy: A review and outlook on renewable energy systems, catalysis, smart grid and energy storage. *Chemical Engineering Research & Design, 174*, 414–441.

Rathore, B. (2019). From Trendy to Green: Exploring AI's Role in Sustainable Fashion Marketing. *International Journal of New Media Studies: International Peer Reviewed Scholarly Indexed Journal, 6*(2), 12–22. doi:10.58972/eiprmj.v6i2y19.120

Raut, R. D., Gardas, B. B., Narkhede, B. E., & Narwane, V. S. (2019). To investigate the determinants of cloud computing adoption in the manufacturing micro, small and medium enterprises: A DEMATEL-based approach. *Benchmarking, 26*(3), 990–1019.

Ravisankar, A., Sampath, B., & Asif, M. M. (2023). Economic Studies on Automobile Management: Working Capital and Investment Analysis. In Multidisciplinary Approaches to Organizational Governance During Health Crises (pp. 169–198). IGI Global.

Ravisankar, A., Shanthi, A., Lavanya, S., Ramaratnam, M., Krishnamoorthy, V., & Boopathi, S. (2024). Harnessing 6G for Consumer-Centric Business Strategies Across Electronic Industries. In AI Impacts in Digital Consumer Behavior (pp. 241–270). IGI Global.

Raza, K., Patle, V., & Arya, S. (2012). A review on green computing for eco-friendly and sustainable it. *Journal of Computational Intelligence and Electronic Systems*, *1*(1), 3–16. doi:10.1166/jcies.2012.1023

Rebecca, B., Kumar, K. P. M., Padmini, S., Srivastava, B. K., Halder, S., & Boopathi, S. (2023). Convergence of Data Science-AI-Green Chemistry-Affordable Medicine: Transforming Drug Discovery. In B. B. Gupta & F. Colace (Eds.), (pp. 348–373). Advances in Computational Intelligence and Robotics. IGI Global. doi:10.4018/978-1-6684-9999-3.ch014

Rebecca, B., Kumar, K. P. M., Padmini, S., Srivastava, B. K., Halder, S., & Boopathi, S. (2024a). Convergence of Data Science-AI-Green Chemistry-Affordable Medicine: Transforming Drug Discovery. In *Handbook of Research on AI and ML for Intelligent Machines and Systems* (pp. 348–373). IGI Global.

Reddy, M. A., Gaurav, A., Ushasukhanya, S., Rao, V. C. S., Bhattacharya, S., & Boopathi, S. (2023). Bio-Medical Wastes Handling Strategies During the COVID-19 Pandemic. In C. S. V. Negrão, I. G. P. Maia, & J. A. F. Brito (Eds.), (pp. 90–111). Advances in Logistics, Operations, and Management Science. IGI Global. doi:10.4018/978-1-7998-9213-7.ch006

Reddy, M. A., Reddy, B. M., Mukund, C., Venneti, K., Preethi, D., & Boopathi, S. (2023). Social Health Protection During the COVID-Pandemic Using IoT. In *The COVID-19 Pandemic and the Digitalization of Diplomacy* (pp. 204–235). IGI Global.

Reddy, T. K., & Dutta, M. (2018). Impact of Agricultural Inputs on Agricultural GDP in Indian Economy. *Theoretical Economics Letters*, *08*(10), 1840–1853. doi:10.4236/tel.2018.810121

Rembold, F., Atzberger, C., Savin, I., & Rojas, O. (2013). Using Low Resolution Satellite Imagery for Yield Prediction and Yield Anomaly Detection. *Remote Sensing (Basel)*, *5*(4), 1704–1733. doi:10.3390/rs5041704

Revathi, S., Babu, M., Rajkumar, N., Meti, V. K. V., Kandavalli, S. R., & Boopathi, S. (2024). Unleashing the Future Potential of 4D Printing: Exploring Applications in Wearable Technology, Robotics, Energy, Transportation, and Fashion. In Human-Centered Approaches in Industry 5.0: Human-Machine Interaction, Virtual Reality Training, and Customer Sentiment Analysis (pp. 131–153). IGI Global.

Revathi, S., Babu, M., Rajkumar, N., Meti, V. K. V., Kandavalli, S. R., & Boopathi, S. (2024a). Unleashing the Future Potential of 4D Printing: Exploring Applications in Wearable Technology, Robotics, Energy, Transportation, and Fashion. In Human-Centered Approaches in Industry 5.0: Human-Machine Interaction, Virtual Reality Training, and Customer Sentiment Analysis (pp. 131–153). IGI Global.

Revathi, S., Babu, M., Rajkumar, N., Meti, V. K. V., Kandavalli, S. R., & Boopathi, S. (2024b). Unleashing the Future Potential of 4D Printing: Exploring Applications in Wearable Technology, Robotics, Energy, Transportation, and Fashion. In Human-Centered Approaches in Industry 5.0: Human-Machine Interaction, Virtual Reality Training, and Customer Sentiment Analysis (pp. 131–153). IGI Global.

Roberts, H., Zhang, J., Bariach, B., Cowls, J., Gilburt, B., Juneja, P., Tsamados, A., Ziosi, M., Taddeo, M., & Floridi, L. (2022). Artificial intelligence in support of the circular economy: Ethical considerations and a path forward. *AI & Society*, 1–14.

Romero, H. F. M., Hernández-Callejo, L., Rebollo, M. Á. G., Cardeñoso-Payo, V., Gómez, V. A., Bello, H. J., Moyo, R. T., & Aragonés, J. I. M. (2023). Synthetic Dataset of Electroluminescence Images of Photovoltaic Cells by Deep Convolutional Generative Adversarial Networks. *Sustainability (Basel)*, *15*(9), 7175. doi:10.3390/su15097175

Roth, L., Aasen, H., Walter, A., & Liebisch, F. (2018). Extracting leaf area index using viewing geometry effects new perspective on high-resolution unmanned aerial system photography. *ISPRS Journal of Photogrammetry and Remote Sensing*, *141*, 161–175. doi:10.1016/j.isprsjprs.2018.04.012

Rouphael, Y., Spíchal, L., Panzarová, K., Casa, R., & Colla, G. (2018). High-Throughput Plant Phenotyping for Developing Novel Biostimulants: From Lab to Field or From Field to Lab? *Frontiers in Plant Science*, *9*, 1197. doi:10.3389/fpls.2018.01197 PMID:30154818

Ruohomaa, H., & Ivanova, N. (2019). From solid waste management towards the circular economy and digital driven symbiosis. *IOP Conference Series. Earth and Environmental Science*, *337*(1), 012032. doi:10.1088/1755-1315/337/1/012032

Rusch, M., Schöggl, J.-P., & Baumgartner, R. J. (2023). Application of digital technologies for sustainable product management in a circular economy: A review. *Business Strategy and the Environment*, *32*(3), 1159–1174. doi:10.1002/bse.3099

Saeed, M. A., Ahmed, Z., Yang, J., & Zhang, W. (2020). An optimal approach of wind power assessment using Chebyshev metric for determining the Weibull distribution parameters. *Sustainable Energy Technologies and Assessments*, *37*, 100612.

Saeed, S., Shezan, S. K. A., Arbab, M. I., & Rana, S. (2017). Battery monitoring system for the smart grid applications. In *2017 Australasian Universities Power Engineering Conference (AUPEC)* (pp. 1–5). IEEE. 10.1109/AUPEC.2017.8282490

Saha, B. (2014). Green computing. [IJCTT]. *International Journal of Computer Trends and Technology*, *14*(2), 46–50. doi:10.14445/22312803/IJCTT-V14P112

Saha, B. (2018). Green computing: Current research trends. *International Journal on Computer Science and Engineering*, *6*(3), 467–469.

Saini, H., & Tarkar, P. (2022). Artificial intelligence in human resource practices with challenges and future directions. In *Handbook of Research on Innovative Management Using AI in Industry 5.0* (pp. 222–230). IGI Global. doi:10.4018/978-1-7998-8497-2.ch015

Sambo, P., Nicoletto, C., Giro, A., Pii, Y., Valentinuzzi, F., Mimmo, T., Lugli, P., Orzes, G., Mazzetto, F., Astolfi, S., Terzano, R., & Cesco, S. (2019). Hydroponic Solutions for Soilless Production Systems: Issues and Opportunities in a Smart Agriculture Perspective. *Frontiers in Plant Science*, *10*, 923. doi:10.3389/fpls.2019.00923 PMID:31396245

Samikannu, R., Koshariya, A. K., Poornima, E., Ramesh, S., Kumar, A., & Boopathi, S. (2022). Sustainable Development in Modern Aquaponics Cultivation Systems Using IoT Technologies. In *Human Agro-Energy Optimization for Business and Industry* (pp. 105–127). IGI Global.

Sampath, B. (2021). *Sustainable Eco-Friendly Wire-Cut Electrical Discharge Machining: Gas Emission Analysis.*

Sampath, B., Sureshkumar, T., Yuvaraj, M., & Velmurugan, D. (2021). Experimental Investigations on Eco-Friendly Helium-Mist Near-Dry Wire-Cut EDM of M2-HSS Material. *Materials Research Proceedings*, *19*, 175–180.

Sangeetha, M., Kannan, S. R., Boopathi, S., Ramya, J., Ishrat, M., & Sabarinathan, G. (2023). Prediction of Fruit Texture Features Using Deep Learning Techniques. *2023 4th International Conference on Smart Electronics and Communication (ICOSEC)*, 762–768.

Sankar, K. M., Booba, B., & Boopathi, S. (2023). Smart Agriculture Irrigation Monitoring System Using Internet of Things. In G. S. Karthick (Ed.), (pp. 105–121). Advances in Environmental Engineering and Green Technologies. IGI Global. doi:10.4018/978-1-6684-7879-0.ch006

Sanni-Anibire, M. O., Mohamad Zin, R., & Olatunji, S. O. (2021). Developing a preliminary cost estimation model for tall buildings based on machine learning. *International Journal of Management Science and Engineering Management*, *16*(2), 134–142. doi:10.1080/17509653.2021.1905568

Saravanan, A., Venkatasubramanian, R., Khare, R., Surakasi, R., Boopathi, S., Ray, S., & Sudhakar, M. (2022). POLICY TRENDS OF RENEWABLE ENERGY AND NON. *Renewable Energy*.

Sarkar, S., & Misra, S. (2016). Theoretical modelling of fog computing: A green computing paradigm to support IoT applications. *IET Networks*, *5*(2), 23–29. doi:10.1049/iet-net.2015.0034

Sarker, I. H. (2022). AI-Based Modeling: Techniques, Applications and Research Issues Towards Automation, Intelligent and Smart Systems. *SN Computer Science*, *3*(2), 158. doi:10.1007/s42979-022-01043-x PMID:35194580

Satav, S. D., Lamani, D., Harsha, K., Kumar, N., Manikandan, S., & Sampath, B. (2023). Energy and Battery Management in the Era of Cloud Computing: Sustainable Wireless Systems and Networks. In Sustainable Science and Intelligent Technologies for Societal Development (pp. 141–166). IGI Global.

Satav, S. D., Lamani, D., Harsha, K., Kumar, N., Manikandan, S., & Sampath, B. (2023a). Energy and Battery Management in the Era of Cloud Computing: Sustainable Wireless Systems and Networks. In Sustainable Science and Intelligent Technologies for Societal Development (pp. 141–166). IGI Global.

Satav, S. D., Lamani, D., Harsha, K., Kumar, N., Manikandan, S., & Sampath, B. (2023b). Energy and Battery Management in the Era of Cloud Computing: Sustainable Wireless Systems and Networks. In Sustainable Science and Intelligent Technologies for Societal Development (pp. 141–166). IGI Global.

Satav, S. D., Hasan, D. S., Pitchai, R., Mohanaprakash, T., Sultanuddin, S., & Boopathi, S. (2023). Next generation of internet of things (ngiot) in healthcare systems. In *Sustainable Science and Intelligent Technologies for Societal Development* (pp. 307–330). IGI Global.

Sathish, T., Sunagar, P., Singh, V., Boopathi, S., Al-Enizi, A. M., Pandit, B., Gupta, M., & Sehgal, S. S. (2023). Characteristics estimation of natural fibre reinforced plastic composites using deep multi-layer perceptron (MLP) technique. *Chemosphere*, *337*, 139346. PMID:37379988

Sauvé, S., Bernard, S., & Sloan, P. (2016). Environmental sciences, sustainable development and circular economy: Alternative concepts for trans-disciplinary research. *Environmental Development*, *17*, 48–56. doi:10.1016/j.envdev.2015.09.002

Savazzi, S., Nicoli, M., & Rampa, V. (2020). Federated learning with cooperating devices: A consensus approach for massive IoT networks. *IEEE Internet of Things Journal*, *7*(5), 4641–4654. doi:10.1109/JIOT.2020.2964162

Sayed, K., Gabbar, H. A., & Krishnamoorthy, S. (2017). SCADA and smart energy grid control automation. In *Smart Energy Grid Engineering* (pp. 481–514). Academic Press. doi:10.1016/B978-0-12-805343-0.00018-8

Schieweck, A., & Bock, M. C. (2015). Emissions from low-VOC and zero-VOC paints - Valuable alternatives to conventional formulations also for use in sensitive environments? *Building and Environment*, *85*, 243–252. doi:10.1016/j.buildenv.2014.12.001

Schniederjans, D. G., & Hales, D. N. (2016). Cloud computing and its impact on economic and environmental performance: A transaction cost economics perspective. *Decision Support Systems*, *86*, 73–82.

See, S., & Adie, J. (2021). Challenges and opportunities for a hybrid modelling approach to earth system science. *CCF Transactions on High Performance Computing*, *3*(3), 320–329. doi:10.1007/s42514-021-00071-y

Sejnowski, T. J. (2018). The deep learning revolution. The MIT Press eBooks. doi:10.7551/mitpress/11474.001.0001

Selvakumar, S., Adithe, S., Isaac, J. S., Pradhan, R., Venkatesh, V., & Sampath, B. (2023a). A Study of the Printed Circuit Board (PCB) E-Waste Recycling Process. In Sustainable Approaches and Strategies for E-Waste Management and Utilization (pp. 159–184). IGI Global.

Selvakumar, S., Adithe, S., Isaac, J. S., Pradhan, R., Venkatesh, V., & Sampath, B. (2023b). A Study of the Printed Circuit Board (PCB) E-Waste Recycling Process. In Sustainable Approaches and Strategies for E-Waste Management and Utilization (pp. 159–184). IGI Global.

Selvakumar, S., Shankar, R., Ranjit, P. S., Bhattacharya, S., Gopala Gupta, A. S. A. L. G., & Boopathi, S. (2023). E-Waste Recovery and Utilization Processes for Mobile Phone Waste. In P. Srivastava, D. Ramteke, A. K. Bedyal, M. Gupta, & J. K. Sandhu (Eds.), (pp. 222–240). Practice, Progress, and Proficiency in Sustainability. IGI Global. doi:10.4018/978-1-6684-8117-2.ch016

Semiromi, F. B., Hassani, A. H., Torabian, A., Karbassi, A. R., & Lotfi, F. H. (2011). Water quality index development using fuzzy logic: A case study of the Karoon River of Iran. *African Journal of Biotechnology*, *10*(50), 10125–10133. doi:10.5897/AJB11.1608

Sengeni, D., Padmapriya, G., Imambi, S. S., Suganthi, D., Suri, A., & Boopathi, S. (2023). Biomedical waste handling method using artificial intelligence techniques. In *Handbook of Research on Safe Disposal Methods of Municipal Solid Wastes for a Sustainable Environment* (pp. 306–323). IGI Global. doi:10.4018/978-1-6684-8117-2.ch022

Senthil, T., Puviyarasan, M., Babu, S. R., Surakasi, R., Sampath, B., & Associates. (2023). Industrial Robot-Integrated Fused Deposition Modelling for the 3D Printing Process. In Development, Properties, and Industrial Applications of 3D Printed Polymer Composites (pp. 188–210). IGI Global.

Senthil, T. S., Ohmsakthi Vel, R., Puviyarasan, M., Babu, S. R., Surakasi, R., & Sampath, B. (2023). Industrial Robot-Integrated Fused Deposition Modelling for the 3D Printing Process. In R. Keshavamurthy, V. Tambrallimath, & J. P. Davim (Eds.), (pp. 188–210). Advances in Chemical and Materials Engineering. IGI Global. doi:10.4018/978-1-6684-6009-2.ch011

Serradilla, O., Zugasti, E., Rodriguez, J., & Zurutuza, U. (2022). Deep learning models for predictive maintenance: A survey, comparison, challenges and prospects. *Applied Intelligence*, *52*(10), 10934–10964. doi:10.1007/s10489-021-03004-y

Shafieian, H., & Le, D. (2019). The Application of IoT in Agriculture: A Systematic Literature Review. *Computers and Electronics in Agriculture*, *157*, 436–447.

Shahid, M. K., Pyo, M., & Choi, Y. G. (2017). Carbonate scale reduction in reverse osmosis membrane by CO2 in wastewater reclamation. *Membrane Water Treatment*, *8*(2), 125–136. doi:10.12989/mwt.2017.8.2.125

Shaikh, T. A., Rasool, T., & Lone, F. R. (2022). Towards leveraging the role of machine learning and artificial intelligence in precision agriculture and smart farming. *Computers and Electronics in Agriculture*, *198*, 107119.

Shamshiri, R. R., Kalantari, F., Ting, K. C., Thorp, K. R., Hameed, I. A., Weltzien, C., Ahmad, D., & Shad, Z. (2018). Advances in greenhouse automation and controlled environment agriculture: A transition to plant factories and urban agriculture. *International Journal of Agricultural and Biological Engineering*, *11*, 1–22. doi:10.25165/j.ijabe.20181101.3210

Sharma, D. M., Ramana, K. V., Jothilakshmi, R., Verma, R., Maheswari, B. U., & Boopathi, S. (2024). Integrating Generative AI Into K-12 Curriculums and Pedagogies in India: Opportunities and Challenges. *Facilitating Global Collaboration and Knowledge Sharing in Higher Education With Generative AI*, (pp. 133–161). IEEE.

Sharma, D. M., Ramana, K. V., Jothilakshmi, R., Verma, R., Maheswari, B. U., & Boopathi, S. (2024a). Integrating Generative AI Into K-12 Curriculums and Pedagogies in India: Opportunities and Challenges. *Facilitating Global Collaboration and Knowledge Sharing in Higher Education With Generative AI*, 133–161.

Sharma, D. M., Ramana, K. V., Jothilakshmi, R., Verma, R., Maheswari, B. U., & Boopathi, S. (2024b). Integrating Generative AI Into K-12 Curriculums and Pedagogies in India: Opportunities and Challenges. *Facilitating Global Collaboration and Knowledge Sharing in Higher Education With Generative AI*, 133–161.

Sharma, A., Jain, A., Gupta, P., & Chowdary, V. (2020). Machine learning applications for precision agriculture: A comprehensive review. *IEEE Access : Practical Innovations, Open Solutions*, *9*, 4843–4873.

Sharma, G. D., Yadav, A., & Chopra, R. (2020). Artificial intelligence and effective governance: A review, critique and research agenda. *Sustainable Futures : An Applied Journal of Technology, Environment and Society*, *2*, 100004.

Sharma, M., Sharma, M., Sharma, N., & Boopathi, S. (2023). Building Sustainable Smart Cities Through Cloud and Intelligent Parking System. In B. B. Gupta & F. Colace (Eds.), (pp. 195–222). Advances in Computational Intelligence and Robotics. IGI Global. doi:10.4018/978-1-6684-9999-3.ch009

Sharma, M., Sharma, M., Sharma, N., & Boopathi, S. (2024). Building Sustainable Smart Cities Through Cloud and Intelligent Parking System. In *Handbook of Research on AI and ML for Intelligent Machines and Systems* (pp. 195–222). IGI Global.

Sharma, P., & Khamparia, A. (2020). Internet of Things (IoT) and Artificial Intelligence (AI) in Agriculture: A Comprehensive Review. *Journal of King Saud University. Computer and Information Sciences.*

Sharma, P., Said, Z., Kumar, A., Nizetic, S., Pandey, A., Hoang, A. T., Huang, Z., Afzal, A., Li, C., Le, A. T., Nguyen, X. P., & Tran, V. D. (2022). Recent advances in machine learning research for nanofluid-based heat transfer in renewable energy system. *Energy & Fuels*, *36*(13), 6626–6658. doi:10.1021/acs.energyfuels.2c01006

Sharma, R., Bhambri, P., & Sohal, A. K. (2020). Mobile Adhoc Networks. *Journal of Composition Theory*, *13*(2), 982–985.

Shen, D., Wu, G., & Suk, H. (2017). Deep learning in medical image analysis. *Annual Review of Biomedical Engineering*, *19*(1), 221–248. doi:10.1146/annurev-bioeng-071516-044442 PMID:28301734

Shetty, J. P., & Panda, R. (2021). An overview of cloud computing in SMEs. *Journal of Global Entrepreneurship Research*, *11*(1), 175–188.

Shivanna, K. R. (2022). Climate change and its impact on biodiversity and human welfare. *Proceedings of the Indian National Science Academy. Part A, Physical Sciences*, *88*(2), 160–171. doi:10.1007/s43538-022-00073-6

Shi, X., An, X., Zhao, Q., Liu, H., Xia, L., Sun, X., & Guo, Y. (2019). State-of-the-Art Internet of Things in Protected Agriculture. *Sensors (Basel)*, *19*(8), 1833. doi:10.3390/s19081833 PMID:30999637

Shorten, C., & Khoshgoftaar, T. M. (2019). A survey on Image Data Augmentation for Deep Learning. *Journal of Big Data*, *6*(1), 60. doi:10.1186/s40537-019-0197-0

Siano, P. (2014). Demand response and smart grids—A survey. *Renewable & Sustainable Energy Reviews*, *30*, 461–478. doi:10.1016/j.rser.2013.10.022

Si, D. (2022). A Framework to analyze the Impacts of AI with the Sustainable Development Goals. *Highlights Sci. Eng. Technol.*, *17*, 313–323. doi:10.54097/hset.v17i.2621

Siddiquee, S. M. S., Howard, B., O'Sullivan, D. T. J., & Bruton, K. (2020). Demand Response in Smart Grid – A Systematic Mapping Study. In 2020 2nd International Conference on Smart Power & Internet Energy Systems (SPIES) (pp. 1-6). IEEE. 10.1109/SPIES48661.2020.9243107

Sii, H. I., Sherrard, J. H., & Wilson, T. E. (1993). A water quality index based on fuzzy set theory. In *ENVIRONMENTAL ENGINEERING-CONFERENCE* (pp. 1727–1727). American Society Of Civil Engineers.

Silva, N. H. C. S., Vilela, C., Marrucho, I. M., Freire, C. S. R., Pascoal Neto, C., & Silvestre, A. J. D. (2014). Protein-based materials: From sources to innovative sustainable materials for biomedical applications. *Journal of Materials Chemistry. B*, *2*(24), 3715–3740. doi:10.1039/c4tb00168k PMID:32261720

Singh, M. (2022). *Quantum Artificial Intelligence for the Science of Climate Change*. Springer. . doi:10.1201/9781003250357-13

Singh, M., Bhambri, P., Dhanoa, I. S., Jain, A., & Kaur, K. (2021). Data Mining Model for Predicting Diabetes. *Annals of the Romanian Society for Cell Biology*, *25*(4), 6702–6712.

Singh, R., & Singh, R. (2023). Applications of sentiment analysis and machine learning techniques in disease outbreak prediction–A review. *Materials Today: Proceedings*, *81*, 1006–1011. doi:10.1016/j.matpr.2021.04.356

Singh, R., Singh, R., Gehlot, A., Akram, S. V., Priyadarshi, N., & Twala, B. (2022). Horticulture 4.0: Adoption of Industry 4.0 Technologies in Horticulture for Meeting Sustainable Farming. *Applied Sciences (Basel, Switzerland)*, *12*(24), 12557. doi:10.3390/app122412557

Sisinni, E., Saifullah, A., Han, S., Jennehag, U., & Gidlund, M. (2018). Industrial Internet of Things: Challenges, Opportunities, and Directions. *IEEE Transactions on Industrial Informatics*, *14*(11), 4724–4734. doi:10.1109/TII.2018.2852491

Smith, J. A., & Brown, R. W. (2018). Advances in Biosensors for Agricultural Applications. *Biosensors & Bioelectronics*, *102*, 292–305.

Solanki, A., Agrawal, H., & Khare, K. (2015). Predictive analysis of water quality parameters using deep learning. *Int. J. Comput. Appl.*, *125*.

Soltani, N., Bahrami, A., Pech-Canul, M. I., & González, L. A. (2015). Review on the physicochemical treatments of rice husk for production of advanced materials. In *Chemical Engineering Journal* (Vol. 264, pp. 899–935). Elsevier. doi:10.1016/j.cej.2014.11.056

Song, X. P., Hu, Z. H., Du, J. G., & Sheng, Z. H. (2014). Application of machine learning methods to risk assessment of financial statement fraud: Evidence from China. *Journal of Forecasting*, *33*(8), 611–626. doi:10.1002/for.2294

Sonia, R., Gupta, N., Manikandan, K., Hemalatha, R., Kumar, M. J., & Boopathi, S. (2024). Strengthening Security, Privacy, and Trust in Artificial Intelligence Drones for Smart Cities. In *Analyzing and Mitigating Security Risks in Cloud Computing* (pp. 214–242). IGI Global. doi:10.4018/979-8-3693-3249-8.ch011

Soomro, T. R., & Sarwar, M. (2012). Green Computing: From current to future trends. Zenodo *(CERN European Organization for Nuclear Research)*. doi:10.5281/zenodo.1062818

Soori, M. (2023). Artificial intelligence, machine learning, and deep learning in advanced robotics, a review. *Cognitive Robotics, 3*. doi:10.1016/j.cogr.2023.04.001

Spalding, R. A. (2016). Fault Location, Isolation and service restoration (FLISR) functionalities tests in a Smart Grids laboratory for evaluation of the quality of service. In *2016 17th International Conference on Harmonics and Quality of Power (ICHQP)* (pp. 879–884). IEEE. 10.1109/ICHQP.2016.7783370

Spanaki, K., Sivarajah, U., Fakhimi, M., Despoudi, S., & Irani, Z. (2022). Disruptive technologies in agricultural operations: A systematic review of AI-driven AgriTech research. *Annals of Operations Research*, *308*(1–2), 491–524.

Srinivas, B., Maguluri, L. P., Naidu, K. V., Reddy, L. C. S., Deivakani, M., & Boopathi, S. (2023). Architecture and Framework for Interfacing Cloud-Enabled Robots. In *Handbook of Research on Data Science and Cybersecurity Innovations in Industry 4.0 Technologies* (pp. 542–560). IGI Global.

Srisruthi, S., Swarna, N., Ros, G. M. S., & Elizabeth, E. (2016). Sustainable agriculture using eco-friendly and energy efficient sensor technology. *2016 IEEE International Conference on Recent Trends in Electronics, Information & Communication Technology (RTEICT)*. IEEE.

Stroppiana, D., Migliazzi, M., Chiarabini, V., Crema, A., Musanti, M., Franchino, C., & Villa, P. Rice yield estimation using multispectral data from UAV: A preliminary experiment in northern Italy. In *Proceedings of the 2015 IEEE International Geoscience and Remote Sensing Symposium (IGARSS)*, Milan, Italy. 10.1109/IGARSS.2015.7326869

Strubell, E., Ganesh, A., & McCallum, A. (2020). Energy and policy considerations for modern deep learning research. *Proceedings of the AAAI Conference on Artificial Intelligence*. 10.1609/aaai.v34i09.7123

Suardana, N. P. G., Piao, Y., & Lim, J. K. (2011). *Mechanical Properties Of Hemp Fibers And HEMP/PP Composites: Effects Of Chemical Surface Treatment*.

Subha, S., Inbamalar, T., Komala, C., Suresh, L. R., Boopathi, S., & Alaskar, K. (2023). A Remote Health Care Monitoring system using internet of medical things (IoMT). *IEEE Explore*, (pp. 1–6). IEEE.

Su, J., Liu, C., Coombes, M., Hu, X., Wang, C., Xu, Z., Li, Q., Guo, L., & Chen, W. H. (2018). Wheat yellow rust monitoring by learning from multispectral UAV aerial imagery. *Computers and Electronics in Agriculture*, *155*, 157–166. doi:10.1016/j.compag.2018.10.017

Sukeshini, S. P., Ved, M., Chintalapti, J., & Pal, S.N. (2021). Big Data Analytics and Machine Learning Technologies for HPC Applications. In Evolving Technologies for Computing, Communication and Smart World. Lecture Notes in Electrical Engineering. Springer. doi:10.1007/978-981-15-7804-5_31

Sultan, V., Aryal, A., Chang, H., & Kral, J. (2022). Integration of EVs into the smart grid: A systematic literature review. *Energy Informatics*, *5*(1), 65. doi:10.1186/s42162-022-00251-2

Sun, B., Zhang, L., & Chen, Z. (2019). Smart Farming: Agriculture's Future. *Precision Agriculture*, *20*(6), 1114–1124.

Sundaramoorthy, K., Singh, A., Sumathy, G., Maheshwari, A., Arunarani, A., & Boopathi, S. (2024). A Study on AI and Blockchain-Powered Smart Parking Models for Urban Mobility. In *Handbook of Research on AI and ML for Intelligent Machines and Systems* (pp. 223–250). IGI Global.

Suryani, W., & Luthfiyyah, A. (2023). Artificial Intelligence (AI) Opportunities and Challenges of Green Marketing Tourism. 5G. *Artificial Intelligence, and Next Generation Internet of Things: Digital Innovation for Green and Sustainable Economies*, 178-188.

Syamala, M. C. R., K., Pramila, P. V., Dash, S., Meenakshi, S., & Boopathi, S. (2023). Machine Learning-Integrated IoT-Based Smart Home Energy Management System: In P. Swarnalatha & S. Prabu (Eds.), Advances in Computational Intelligence and Robotics (pp. 219–235). IGI Global. doi:10.4018/978-1-6684-8098-4.ch013

Syamala, M., Komala, C., Pramila, P., Dash, S., Meenakshi, S., & Boopathi, S. (2023a). Machine Learning-Integrated IoT-Based Smart Home Energy Management System. In *Handbook of Research on Deep Learning Techniques for Cloud-Based Industrial IoT* (pp. 219–235). IGI Global.

Tabassam, S., & Al-Qahtane, E. 21. S. Tabassam and E. Al-Qahtane, "Comparative Analysis on Requirement Engineering Modelling Techniques Case Study of Personal Al-Haj E-Guide," 2019 2nd International Conference on Computer Applications & Information Security (ICCAIS), 2019, pp. 1-8, 10.1109/CAIS.2019.8769549

Tabassam, S., Al-Saeed, W., Almughram, O., & Alghamdi, K. (2019). Scalable Data Analysis and Query Processing. *International Journal on Engineering Applications*, *7*(3), 81–87. doi:10.15866/irea.v7i3.17012

Tabassam, S., Hassan, O., Al-Qahtnae, E., & Al-Ahmary, N. (2019). Goal Question Metrics and Its Application to Process Management and Improvement. *International Journal on Engineering Applications*, *7*(2), 52–58. doi:10.15866/irea.v7i2.17013

Tabassam, S., Shah, H., Alghamdi, K., & Badshah, A. (2019). Social Networks and Digital Security. *2019 International Conference on Electrical, Communication, and Computer Engineering (ICECCE)*, Swat, Pakistan. 10.1109/ICECCE47252.2019.8940808

Tamburri, D. A., Miglierina, M., & Di Nitto, E. (2020). Cloud applications monitoring: An industrial study. *Information and Software Technology*, *127*, 106376.

Tang, H., Huang, H., Liu, J., Zhu, J., Gou, F., & Wu, J. (2022). AI-Assisted Diagnosis and Decision-Making Method in Developing Countries for Osteosarcoma. *Health Care*, *10*, 2313. PMID:36421636

Tao, Y., & Zhang, Y. (2018). Research on the Application of IoT Technology in Agriculture. *IOP Conference Series. Earth and Environmental Science*, *170*, 012007.

Terence, S., & Purushothaman, G. (2020). Systematic review of Internet of Things in smart farming. *Transactions on Emerging Telecommunications Technologies*, *31*(6), e3958. doi:10.1002/ett.3958

Theopoulos, A., Boursianis, A., Koukounaras, A., & Samaras, T. (2018). *Prototype wireless sensor network for real-time measurements in hydroponics cultivation.* In *Proceedings of the 7th International Conference on Modern Circuits and Systems Technologies (MOCAST)*, Thessaloniki, Greece. 10.1109/MOCAST.2018.8376576

Titirici, M. M., White, R. J., Brun, N., Budarin, V. L., Su, D. S., Del Monte, F., Clark, J. H., & MacLachlan, M. J. (2015a). Sustainable carbon materials. In Chemical Society Reviews, 44(1). Royal Society of Chemistry. doi:10.1039/C4CS00232F

Tiwari, A. K., De Maio, M., Singh, P. K., & Mahato, M. K. (2015). Evaluation of surface water quality by using GIS and a heavy metal pollution index (HPI) model in a coal mining area, India. *Bulletin of Environmental Contamination and Toxicology*, *95*(3), 304–310. doi:10.1007/s00128-015-1558-9 PMID:25962697

Tripathy, K. P., & Mishra, A. K. (2023). Deep Learning in Hydrology and Water Resources Disciplines: Concepts, Methods, Applications, and Research Directions. *Journal of Hydrology*. doi:10.1016/j.jhydrol.2023.130458

Tripathy, P. K., Tripathy, A. K., Agarwal, A., & Mohanty, S. P. (2021). MyGreen: An IoT-Enabled Smart Greenhouse for Sustainable Agriculture. *IEEE Consumer Electronics Magazine*, *10*(4), 57–62. doi:10.1109/MCE.2021.3055930

Tripodi, P., Massa, D., Venezia, A., & Cardi, T. (2018). Sensing Technologies for Precision Phenotyping in Vegetable Crops: Current Status and Future Challenges. *Agronomy (Basel)*, *8*(4), 57. doi:10.3390/agronomy8040057

Truong, H.-L., & Dustdar, S. (2015). Principles for engineering IoT cloud systems. *IEEE Cloud Computing*, *2*(2), 68–76.

Tsai, S. B., Xue, Y., Zhang, J., Chen, Q., Liu, Y., Zhou, J., & Dong, W. (2017). Models for forecasting growth trends in renewable energy. *Renewable & Sustainable Energy Reviews*, *77*, 1169–1178. doi:10.1016/j.rser.2016.06.001

Tucker, C. J., Elgin, J. H., & McMurtrey, J. E. (1980). Relationship of crop radiance to alfalfa agronomic values. *International Journal of Remote Sensing*, *1*(1), 69–75. doi:10.1080/01431168008948211

Tzachor, A., Devare, M., King, B., & Avin, S., & Ó hÉigeartaigh, S. (2022). Responsible artificial intelligence in agriculture requires systemic understanding of risks and externalities. *Nature Machine Intelligence*, *4*(2), 104–109.

Tzounisa, A., Katsoulasa, N., Bartzanasb, T., & Kittas, C. (2017). Internet of Things in agriculture, recent advances and future challenges. *Biosystems Engineering*, *164*, 31–48. doi:10.1016/j.biosystemseng.2017.09.007

Ugandar, R., Rahamathunnisa, U., Sajithra, S., Christiana, M. B. V., Palai, B. K., & Boopathi, S. (2023). Hospital Waste Management Using Internet of Things and Deep Learning: Enhanced Efficiency and Sustainability. In Applications of Synthetic Biology in Health, Energy, and Environment (pp. 317–343). IGI Global.

Ugandar, R. E., Rahamathunnisa, U., Sajithra, S., Christiana, M. B. V., Palai, B. K., & Boopathi, S. (2023). Hospital Waste Management Using Internet of Things and Deep Learning: Enhanced Efficiency and Sustainability. In M. Arshad (Ed.), (pp. 317–343). Advances in Bioinformatics and Biomedical Engineering. IGI Global. doi:10.4018/978-1-6684-6577-6.ch015

Usman, N., Arief, P., Gilang, L., Erfan, R., & Hendra, P. (2018). Implementation IoT in System Monitoring Hydroponic Plant Water Circulation and Control. [Google Scholar]. *Int. J. Eng. Technol.*, *7*, 122–126.

Vajjhala, N. R., & Ramollari, E. (2021). *Big data using cloud computing-opportunities for small and medium-sized enterprises.*

Vandelook, S., Elsacker, E., Van Wylick, A., De Laet, L., & Peeters, E. (2021). Current state and future prospects of pure mycelium materials. In Fungal Biology and Biotechnology, 8(1). BioMed Central Ltd. doi:10.1186/s40694-021-00128-1

Vanitha, S. K. R., & Boopathi, S. (2023). Artificial Intelligence Techniques in Water Purification and Utilization. In P. Vasant, R. Rodríguez-Aguilar, I. Litvinchev, & J. A. Marmolejo-Saucedo (Eds.), (pp. 202–218). Advances in Environmental Engineering and Green Technologies. IGI Global. doi:10.4018/978-1-6684-4118-3.ch010

Vasiljeva, T., Shaikhulina, S., & Kreslins, K. (2017). Cloud computing: Business perspectives, benefits and challenges for small and medium enterprises (case of Latvia). *Procedia Engineering*, *178*, 443–451.

Vasirani, M., Kota, R., Cavalcante, R. L. G., Ossowski, S., & Jennings, N. R. (2013). An agent-based approach to virtual power plants of wind power generators and electric vehicles. *IEEE Transactions on Smart Grid*, *4*(3), 1314–1322. doi:10.1109/TSG.2013.2259270

Velte, P. (2022). Meta-analyses on corporate social responsibility (CSR): A literature review. *Management Review Quarterly*, *72*(3), 627–675. doi:10.1007/s11301-021-00211-2

Venkateswaran, N., Vidhya, K., Ayyannan, M., Chavan, S. M., Sekar, K., & Boopathi, S. (2023a). A Study on Smart Energy Management Framework Using Cloud Computing. In 5G, Artificial Intelligence, and Next Generation Internet of Things: Digital Innovation for Green and Sustainable Economies (pp. 189–212). IGI Global.

Venkateswaran, N., Vidhya, K., Ayyannan, M., Chavan, S. M., Sekar, K., & Boopathi, S. (2023b). A Study on Smart Energy Management Framework Using Cloud Computing. In 5G, Artificial Intelligence, and Next Generation Internet of Things: Digital Innovation for Green and Sustainable Economies (pp. 189–212). IGI Global.

Venkateswaran, N., Kumar, S. S., Diwakar, G., Gnanasangeetha, D., & Boopathi, S. (2023). Synthetic Biology for Waste Water to Energy Conversion: IoT and AI Approaches. *Applications of Synthetic Biology in Health. Energy & Environment*, 360–384.

Venkateswaran, N., Vidhya, K., Ayyannan, M., Chavan, S. M., Sekar, K., & Boopathi, S. (2023). A Study on Smart Energy Management Framework Using Cloud Computing. In P. Ordóñez De Pablos & X. Zhang (Eds.), (pp. 189–212). Practice, Progress, and Proficiency in Sustainability. IGI Global. doi:10.4018/978-1-6684-8634-4.ch009

Venkateswaran, N., Vidhya, R., Naik, D. A., Raj, T. M., Munjal, N., & Boopathi, S. (2023). Study on Sentence and Question Formation Using Deep Learning Techniques. In *Digital Natives as a Disruptive Force in Asian Businesses and Societies* (pp. 252–273). IGI Global. doi:10.4018/978-1-6684-6782-4.ch015

Vennila, T., Karuna, M., Srivastava, B. K., Venugopal, J., Surakasi, R., & Sampath, B. (2022). New Strategies in Treatment and Enzymatic Processes: Ethanol Production From Sugarcane Bagasse. In Human Agro-Energy Optimization for Business and Industry (pp. 219–240). IGI Global.

Verma, R., Christiana, M. B. V., Maheswari, M., Srinivasan, V., Patro, P., Dari, S. S., & Boopathi, S. (2024). Intelligent Physarum Solver for Profit Maximization in Oligopolistic Supply Chain Networks. In *AI and Machine Learning Impacts in Intelligent Supply Chain* (pp. 156–179). IGI Global.

Verma, R., & Gupta, B. D. (2015). Detection of heavy metal ions in contaminated water by surface plasmon resonance based optical fibre sensor using conducting polymer and chitosan. *Food Chemistry*, *166*, 568–575. doi:10.1016/j.foodchem.2014.06.045 PMID:25053095

Vidhyalakshmi, R., & Kumar, V. (2016). Determinants of cloud computing adoption by SMEs. *International Journal of Business Information Systems*, *22*(3), 375–395.

Vignesh, S., Arulshri, K., SyedSajith, S., Kathiresan, S., Boopathi, S., & Dinesh Babu, P. (2018). Design and development of ornithopter and experimental analysis of flapping rate under various operating conditions. *Materials Today: Proceedings*, *5*(11), 25185–25194. doi:10.1016/j.matpr.2018.10.320

Vijayakumar, G. N. S., Domakonda, V. K., Farooq, S., Kumar, B. S., Pradeep, N., & Boopathi, S. (2024). Sustainable Developments in Nano-Fluid Synthesis for Various Industrial Applications. In Adoption and Use of Technology Tools and Services by Economically Disadvantaged Communities: Implications for Growth and Sustainability (pp. 48–81). IGI Global.

Vinuesa, R., Azizpour, H., Leite, I., Balaam, M., Dignum, V., Domisch, S., Felländer, A., Langhans, S. D., Tegmark, M., & Fuso Nerini, F. (2020). The role of artificial intelligence in achieving the Sustainable Development Goals. . *Nature Communications*, *11*(1), 233. doi:10.1038/s41467-019-14108-y PMID:31932590

Vinuesa, R., & Sirmacek, B. (2021). Interpretable deep-learning models to help achieve the Sustainable Development Goals. *Nature Machine Intelligence*, *3*(11), 926. doi:10.1038/s42256-021-00414-y

Vital, J.-A., Gaurut, M., Lardy, R., Viovy, N., Soussana, J.-F., Bellocchi, G., & Martin, R. (2013). High-Performance computing for climate change impact studies with the Pasture Simulation model. *Computers and Electronics in Agriculture*, *98*, 131–135. doi:10.1016/j.compag.2013.08.004

VoPham, T. (2018). Emerging trends in geospatial artificial intelligence (geoAI): Potential applications for environmental epidemiology. *Environmental Health*, *17*, 1–6.

Wagle, D. V., Zhao, H., & Baker, G. A. (2014). Deep eutectic solvents: Sustainable media for nanoscale and functional materials. *Accounts of Chemical Research*, *47*(8), 2299–2308. doi:10.1021/ar5000488 PMID:24892971

Wang, D. & Yuan, F. (2020). *High-Performance Computing for Earth System Modeling*. Springer. . doi:10.1007/978-3-030-47998-5_10

Wang, D., Wu, Z., & Zhu, B. (2022). Controlling Shareholder Characteristics and Corporate Debt Default Risk: Evidence Based on Machine Learning. *Emerging Markets Finance & Trade*, *58*(12), 3324–3339. doi:10.1080/1540496X.2022.2037416

Wang, D., & Zhang, Y. (2018). Precision Agriculture—An Opportunity for IoT Technologies in Agriculture: A Review. *Sustainable Production and Consumption*, *13*, 37–47.

Wang, H., Fu, T., Du, Y., Gao, W., Huang, K., Liu, Z., Chandak, P., Liu, S., Van Katwyk, P., Deac, A., Anandkumar, A., Bergen, K., Gomes, C. P., Ho, S., Kohli, P., Lasenby, J., Leskovec, J., Liu, T.-Y., Manrai, A, & Zitnik, M. (2023). Scientific discovery in the age of artificial intelligence. *Nature*, *620*(7972), 47–60. doi:10.1038/s41586-023-06221-2 PMID:37532811

Wang, J., Zhang, C., & Zhang, L. (2021). A Review on the Use of Artificial Intelligence in Agriculture. *Precision Agriculture*, *22*(4), 701–722.

Wang, Z., Li, P., & Zhang, H. (2017). Internet of Things and Artificial Intelligence in Agriculture: A Review. *Computers and Electronics in Agriculture*, *142*, 283–297.

Widyastuti, D., & Irwansyah, I. (2018a). Benefits and challenges of cloud computing technology adoption in small and medium enterprises (SMEs). *Bandung Creative Movement (BCM)*, *4*(1).

Widyastuti, D., & Irwansyah, I. (2018b). Benefits and challenges of cloud computing technology adoption in small and medium enterprises (SMEs). *Bandung Creative Movement (BCM)*, *4*(1).

Willard, J., Jia, X., Xu, S., Steinbach, M., & Kumar, V. (2022). Integrating scientific knowledge with machine learning for engineering and environmental systems. *ACM Computing Surveys*, *55*(4), 1–37. doi:10.1145/3514228

Wolfert, S., Ge, L., Verdouw, C., & Bogaardt, M.-J. (2017). Big Data in Smart Farming – A review. *Agricultural Systems*, *153*, 69–80. doi:10.1016/j.agsy.2017.01.023

Wong, L. W., Tan, G. W. H., Ooi, K. B., Lin, B., & Dwivedi, Y. K. (2022). Artificial intelligence-driven risk management for enhancing supply chain agility: A deep-learning-based dual-stage PLS-SEM-ANN analysis. *International Journal of Production Research*, 1–21. doi:10.1080/00207543.2022.2063089

Wu, C.-J., Raghavendra, R., Gupta, U., Acun, B., Ardalani, N., Maeng, K., Chang, G., Aga, F., Huang, J., Bai, C., & ... (2022). Sustainable ai: Environmental implications, challenges and opportunities. *Proceedings of Machine Learning and Systems*, *4*, 795–813.

Wu, L., & Gao, S. (2019). A Survey of IoT Applications in Agricultural Field. *Computers and Electronics in Agriculture*, *157*, 218–229.

Wurtz, F., & Delinchant, B. (2017). "Smart buildings" integrated in "smart grids": A key challenge for the energy transition by using physical models and optimization with a "human-in-the-loop" approach. *Comptes Rendus Physique*, *18*(7–8), 428–444. doi:10.1016/j.crhy.2017.09.007

Wu, Z. Y., & Rahman, A. (2017). Optimized deep learning framework for water distribution data-driven modeling. *Procedia Engineering*, *186*, 261–268. doi:10.1016/j.proeng.2017.03.240

Xuan, F. (2021). Regression analysis of supply chain financial risk based on machine learning and fuzzy decision model. *Journal of Intelligent & Fuzzy Systems*, *40*(4), 6925–6935. doi:10.3233/JIFS-189523

Xu, X., Fu, S., Cai, Q., Tian, W., Liu, W., Dou, W., Sun, X., & Liu, A. X. (2018). Dynamic resource allocation for load balancing in fog environment. *Wireless Communications and Mobile Computing*, *2018*, 1–15. doi:10.1155/2018/6421607

Yang, W., Li, X., Li, Y., Zhu, R., & Pang, H. (2019). Applications of Metal–Organic-Framework-Derived Carbon Materials. In Advanced Materials, 31(6). Wiley-VCH Verlag. doi:10.1002/adma.201804740

Yang, H., Ma, M., Thompson, J. R., & Flower, R. J. (2018). Waste management, informal recycling, environmental pollution and public health. *Journal of Epidemiology and Community Health*, *72*(3), 237–243. doi:10.1136/jech-2016-208597 PMID:29222091

Yang, H., & Zhu, K. (2022). Research on AI-assisted grading of math questions based on deep learning. In *International Conference on Internet of Things and Machine Learning (IoTML 2021)*. SPIE. 10.1117/12.2628649

Yang, T., Zhao, L., Li, W., & Zomaya, A. Y. (2020). Reinforcement learning in sustainable energy and electric systems: A survey. *Annual Reviews in Control*, *49*, 145–163. doi:10.1016/j.arcontrol.2020.03.001

Yang, W., Feng, H., Zhang, X., Zhang, J., Doonan, J. H., Batchelor, W. D., Xiong, L., & Yan, J. (2020). Crop Phenomics and High-Throughput Phenotyping: Past Decades, Current Challenges, and Future Perspectives. *Molecular Plant*, *13*(2), 187–214. doi:10.1016/j.molp.2020.01.008 PMID:31981735

Yang, Z., Hu, J., Ai, X., Wu, J., & Yang, G. (2020). Transactive energy supported economic operation for multi-energy complementary microgrids. *IEEE Transactions on Smart Grid*, *12*(1), 4–17.

Yeğin, T., & Ikram, M. (2022). Performance evaluation of green furniture brands in the marketing 4.0 period: An integrated MCDM approach. *Sustainability (Basel)*, *14*(17), 10644. doi:10.3390/su141710644

Yi, G., & Loia, V. (2019). High-performance computing systems and applications for AI. *The Journal of Supercomputing*, *75*(8), 4248–4251. doi:10.1007/s11227-019-02937-z

Yoldaş, Y., Önen, A., Muyeen, S. M., Vasilakos, A. V., & Alan, İ. (2017). Enhancing smart grid with microgrids: Challenges and opportunities. *Renewable & Sustainable Energy Reviews*, *72*, 205–214. doi:10.1016/j.rser.2017.01.064

Yong, W., Shuaishuai, L., Li, L., Minzan, L., Ming, L., Arvanitis, K. G., Georgieva, C., & Sigrimis, N. (2018). Smart Sensors from Ground to Cloud and Web Intelligence. *IFAC-PapersOnLine*, *51*(17), 31–38. doi:10.1016/j.ifacol.2018.08.057

Yu, L., Qin, S., Zhang, M., Shen, C., Jiang, T., & Guan, X. (2021). A review of deep reinforcement learning for smart building energy management. *IEEE Internet of Things Journal*, *8*(15), 12046–12063. doi:10.1109/JIOT.2021.3078462

Yupapin, P., Trabelsi, Y., Nattappan, A., & Boopathi, S. (2023). Performance improvement of wire-cut electrical discharge machining process using cryogenically treated super-conductive state of Monel-K500 alloy. *Iranian Journal of Science and Technology. Transaction of Mechanical Engineering*, *47*(1), 267–283. doi:10.1007/s40997-022-00513-0

Yu, X., & Xue, Y. (2016). Smart Grids: A Cyber–Physical Systems Perspective. *Proceedings of the IEEE*, *104*(5), 1058–1070. doi:10.1109/JPROC.2015.2503119

Zekrifa, D. M. S., Kulkarni, M., Bhagyalakshmi, A., Devireddy, N., Gupta, S., & Boopathi, S. (2023). Integrating Machine Learning and AI for Improved Hydrological Modeling and Water Resource Management. In V. Shikuku (Ed.), (pp. 46–70). Advances in Environmental Engineering and Green Technologies. IGI Global., doi:10.4018/978-1-6684-6791-6.ch003

Zeng, W., Shu, L., Li, Q., Chen, S., Wang, F., & Tao, X. M. (2014). Fiber-based wearable electronics: A review of materials, fabrication, devices, and applications. In Advanced Materials, 26(31). Wiley-VCH Verlag. doi:10.1002/adma.201400633

Zhang, J., Zhu, Y., Zhang, X., Ye, M., & Yang, J. (2018). Developing a Long Short-Term Memory (LSTM) based model for predicting water table depth in agricultural areas. *Journal of Hydrology (Amsterdam)*, *561*, 918–929. doi:10.1016/j.jhydrol.2018.04.065

Zhang, L., & Sun, B. (2019). An Overview of Smart Agriculture in China. *Agricultural Systems*, *172*, 1–8.

Zhang, L., Wen, J., Li, Y., Chen, J., Ye, Y., Fu, Y., & Livingood, W. (2021). A review of machine learning in building load prediction. *Applied Energy*, *285*, 116452. doi:10.1016/j.apenergy.2021.116452

Zhao, C., Liu, B., Piao, S., Wang, X., Lobell, D. B., Huang, Y., Huang, M., Yao, Y., Bassu, S., Ciais, P., Durand, J.-L., Elliott, J., Ewert, F., Janssens, I. A., Li, T., Lin, E., Liu, Q., Martre, P., Müller, C., & Asseng, S. (2017). Temperature increase reduces global yields of major crops in four independent estimates. *Proceedings of the National Academy of Sciences of the United States of America, 114*(35), 9326–9331. doi:10.1073/pnas.1701762114 PMID:28811375

Zhen, W., & Yan, L. (2018). Development of a Wireless Biosensor System for Agricultural Applications. *IEEE Transactions on Instrumentation and Measurement, 67*(10), 2402–2411.

ZhouJ.ReynoldsD.WebsdaleD.Le CornuT.Gonzalez NavarroO.ListerC.OrfordS.LaycockS.FinlaysonG.StittT. (2017). Cropquant: An automated and scalable field phenotyping platform for crop monitoring and trait measurements to facilitate breeding and digital agriculture. bioRxiv. doi:10.1101/161547

Zhou, Y., Zheng, S., Liu, Z., Wen, T., Ding, Z., Yan, J., & Zhang, G. (2020). Passive and active phase change materials integrated building energy systems with advanced machine-learning based climate-adaptive designs, intelligent operations, uncertainty-based analysis and optimisations: A state-of-the-art review. *Renewable & Sustainable Energy Reviews, 130*, 109889. doi:10.1016/j.rser.2020.109889

Zhu, Y., Romain, C., & Williams, C. K. (2016). Sustainable polymers from renewable resources. In Nature, 540(7633). doi:10.1038/nature21001

About the Contributors

Arshi Naim is an Associate Professor and Quality Programme head in Business Management, KSA She has completed her PHD in business management Singhania University Rajasthan, MBA in marketing management from HNB University and economics honours from AMU, Aligarh, India. Also, she has accomplished some certificates like QM peer reviewer and E-Learning expert from quality matters, US, ICDL from Australia and certificate on digital skills and digital marketing from UK. She is expert in curriculum and programme development, quality and accreditation process, E-learning and Quality Matters Peer Reviewer. She has research publications in Web of Sceince, IEEE, ACM, and Scopus Indexed Journals. She has also published books with Taylor and Francis, Nova Science Publications, IGI, and Springer. She is also closely associated with international and national universities for supervising PHD students.

***

Pankaj Bhambri is affiliated with the Department of Information Technology at Guru Nanak Dev Engineering College in Ludhiana. Additionally, he fulfills the role of the Convener for his Departmental Board of Studies. He possesses nearly two decades of teaching experience. He is an active member of IE India, ISTE New Delhi, IIIE Navi Mumbai, IETE New Delhi and CSI Mumbai. He has contributed to the various research activities while publishing articles in the renowned SCIE and Scopus journals and conference proceedings. He has also published several international patents. Dr. Bhambri has garnered extensive experience in the realm of academic publishing, having served as an editor/author for a multitude of books in collaboration with esteemed publishing houses such as CRC Press, Elsevier, Scrivener, and Bentham Science. Dr. Bhambri has been honored with several prestigious accolades, including the ISTE Best Teacher Award in 2023 and 2022, the I2OR National Award in 2020, the Green ThinkerZ Top 100 International Distinguished Educators award in 2020, the I2OR Outstanding Educator Award in 2019, the SAA Distinguished Alumni Award in 2012, the CIPS Rashtriya Rattan Award in 2008, the LCHC Best Teacher Award in 2007, and numerous other commendations from various government and non-profit organizations. He has provided guidance and oversight for numerous research projects and dissertations at the postgraduate and Ph.D. levels. He successfully organized a diverse range of educational programmes, securing financial backing from esteemed institutions such as the AICTE, the TEQIP, among others. Dr. Bhambri's areas of interest encompass machine learning, bioinformatics, wireless sensor networks, and network security.

Vijay Bhutani is currently completing his Ph.D. in Mechanical Engineering at Amity University, UP, India. Concurrently, he serves as an Assistant Professor in the Department of Mechanical Engineering at Amity University, Greater Noida, U.P., India. His research contributions include numerous publications in esteemed journals indexed in SCI/SCOPUS and other reputable indexing platforms. His expertise lies in thermal engineering, renewable and sustainable energy, green energy, interdisciplinary technology integration, and the design of solar energy systems.

Sampath Boopathi is an accomplished individual with a strong academic background and extensive research experience. He completed his undergraduate studies in Mechanical Engineering and pursued his postgraduate studies in the field of Computer-Aided Design. Dr. Boopathi obtained his Ph.D. from Anna University, focusing his research on Manufacturing and optimization. Throughout his career, Dr. Boopathi has made significant contributions to the field of engineering. He has authored and published over 190 research articles in internationally peer-reviewed journals, highlighting his expertise and dedication to advancing knowledge in his area of specialization. His research output demonstrates his commitment to conducting rigorous and impactful research. In addition to his research publications, Dr. Boopathi has also been granted one patent and has three published patents to his name. This indicates his innovative thinking and ability to develop practical solutions to real-world engineering challenges. With 17 years of academic and research experience, Dr. Boopathi has enriched the engineering community through his teaching and mentorship roles. He has served in various engineering colleges in Tamilnadu, India, where he has imparted knowledge, guided students, and contributed to the overall academic development of the institutions. Dr. Sampath Boopathi's diverse background, ranging from mechanical engineering to computer-aided design, along with his specialization in manufacturing and optimization, positions him as a valuable asset in the field of engineering. His research contributions, patents, and extensive teaching experience exemplify his expertise and dedication to advancing engineering knowledge and fostering innovation.

Vishal Goar is currently working as Dean of Research at Bikaner Technical University, Bikaner. He has wide experience of 17 years as an academician, researcher and in Industry. He has published 46 research papers in international journals (SCI/Scopus/WoS) and presented 45 papers in national/international conferences. He has been editor/authored 13 books including reference, text and edited books of Springer, Singapore, ACM USA and Scholars-Press, Germany. He has published two patents and 3 patents grants in his credit. He supervised 6 Ph.d research scholars in area of cloud computing, software development, Neural Networks and Deep Learning. He also served as Organizing Chair, Convener, Publicity Chair of various Springer/IEEE/ACM internationally renowned conferences.

Santosh K C is an accomplished scholar and educator with over 14 years of experience in the field of Computer Science and Engineering (CSE). Holding the position of Associate Professor within the Department of Computer Science and Engineering. Dr. Santosh K C has made significant contributions to both academia and industry through research, teaching, and mentorship. He completed his doctoral studies in Computer Science from Visveshvaraya Technological University, Belagavi, specializing in Image Processing. As a researcher, He has a diverse portfolio of publications in top-tier conferences and journals. In addition to his research contributions, Dr. Santosh K C is deeply committed to teaching and mentorship. He has developed and taught numerous undergraduate and graduate courses in Computer Science.

M. Sureshkumar completed his undergraduate in Mechanical Engineering and postgraduate in the field of Engineering Design. He completed his Ph.D. from Anna University, Chennai, Tamil Nādu, India.

S.K. Sansaniwal is an expert in multidisciplinary research related to sustainable cooling, energy efficiency, built environment, renewable energy, and policy development. He has over 11 years of experience in research and academia and has worked on various national and international collaborative research projects. Dr. Sansaniwal has published peer-reviewed research and serves as a reviewer and member of the editorial board for several publishers.

Preeti Singh Bahadur completed her Ph.D Degree in Physics from Barkatullah University, Bhopal, M.P., India and is currently working as a Associate Professor in the Department of Applied Sciences, Amity University, Greater Noida, U.P., India. She has published many research papers and chapters in various reputed journals, indexed in SCI/SCOPUS and other indexing. She has presented numerous research papers in various international and national conferences.

Jitendra Singh Kushwah is an Associate Professor Department of Information Technology, Institute of Technology and Management

Nagendra Singh Yadav is a Software Testing professional with experience managing remote teams and leading project teams. His background in software testing, technical writing, Research, Author, blogger, and Project management informs his mindful but competitive approach. Nagendra is fueled by his passion for understanding the project requirements and asking from the stakeholders. He considers himself a 'forever student,' eager to build on his academic foundations as a researcher, technical writer, and blogger, and stays in tune with the latest software testing & project management strategies through continued work experience. He is the Global Assistant General Secretary, International Council of Computer Science (ICCS) at Eudoxia Research Centre and holds degrees in M.C.A, B.C.A & a diploma in Blogging and Content Marketing. He has worked with Infosys Limited for 3 years in the field of software validation solutions as a Test Engineer (2016- 2019).

Index

A

Adoption 13, 15, 23, 25, 40-42, 47-48, 91, 100, 112, 115, 143, 153, 156, 158, 160, 173, 191, 193, 215-216, 224, 228-229, 233, 252, 254, 258, 264, 274, 280, 292-299, 305-306, 310-315, 322, 324, 326-327, 329, 331, 333
Agriculture 1-3, 7-14, 19, 22-45, 48, 66, 96, 101-102, 111, 139-148, 151, 153-165, 186-187, 189, 192, 247, 276-277, 328, 330
AI-driven solutions 2-3, 9-11, 16, 19, 21
Applications 4-5, 12, 14-19, 21-28, 33-34, 42, 44-45, 48-49, 57-58, 60, 62, 64, 66, 71-72, 75-77, 86-87, 90, 94-96, 101, 106-108, 112-115, 124, 135-137, 145-147, 149-150, 152, 156, 158-159, 161-163, 169, 186-199, 201-202, 210, 212-213, 218, 224, 231-234, 236, 238-244, 246, 248, 250-254, 257, 261-263, 269, 271, 275-278, 282, 286-289, 294, 296-298, 301-302, 305-318, 322-323, 326, 328, 330-333, 351
Artificial Intelligence 1-2, 4, 22-24, 26-29, 33-38, 41, 43-45, 56, 75, 94-96, 99-101, 104, 108-113, 136-139, 143-144, 154, 178, 184, 186, 188-189, 191-195, 211-213, 228, 230-231, 233-234, 266-272, 274-277, 299, 302, 305, 327-328, 330, 333
AWQ 334

B

Best Practices 18, 41, 69, 166, 172, 175, 182, 216, 226, 229, 305-306, 315, 322-324, 327
Blockchain 23-24, 114-116, 134-135, 137, 152, 167, 177-178, 187, 189-190, 195, 209, 212, 228, 231, 298, 328
Building Energy Management 74, 76-77, 82, 85-86, 94, 100, 107-109, 113
Business Financial Risks 137, 194, 196, 200, 212

C

Carbon Emissions 46, 48-49, 61, 66, 68-69, 134, 167, 171, 214-222, 226, 229, 246, 254-255, 260, 305, 321-322, 326-327
Challenges 2-4, 6, 12, 14, 18, 20, 22, 24, 27-28, 30-34, 36-37, 41-43, 46-48, 53, 59-60, 62, 64-67, 70, 72, 75-77, 94, 96, 99, 102, 105, 108-109, 111-112, 117, 136-138, 150-156, 159-161, 163-165, 167, 169-171, 173-175, 183-185, 191-192, 194, 197, 212-213, 216-218, 222-223, 226-227, 229, 233, 236-237, 243, 252-253, 255-260, 264-269, 272-274, 276-277, 279-280, 282, 289-303, 305-307, 313-315, 326, 329-333
circular economy 3, 12, 19, 21, 23-24, 95, 166-167, 172-173, 175-177, 179-180, 183, 185, 188-191, 214-216, 224-229, 231-232, 237, 262
Cloud computing 24-25, 42, 47, 60-61, 71, 94, 97, 139, 143, 162, 186-187, 193, 214-220, 230-234, 282, 297, 299-300, 305-311, 313, 315-316, 318, 320-322, 325-333
consumer behavior change 166

D

Data Analytics 2-3, 10, 23, 33, 48, 79, 95, 144, 156, 160, 188, 226, 231, 270, 275, 277, 279, 292, 295, 297-300, 303, 318, 326
Deep Learning 15, 23-24, 27, 46, 49-50, 52, 56-57, 59, 63-67, 70-72, 97, 99, 101-102, 104, 106, 108-113, 161, 186, 189, 193-195, 197, 200, 234, 270, 272-273, 275, 277, 285-287, 328-329, 332, 350-351
Deep Reinforcement Learning 55, 62-63, 99-100, 104, 107, 113, 286
Development 2, 15-16, 20, 23-24, 29, 31, 36, 39, 41-42, 44-45, 47-48, 59, 69-71, 74, 83, 87, 91-93, 95, 101-103, 107-113, 120-126, 128, 130, 136-137, 139,

141, 143, 145-146, 149, 151, 154, 157, 160-161, 170, 173, 175, 177, 183, 185-190, 192-193, 212, 216, 227, 230-232, 234, 236-237, 241, 252, 255-256, 258-260, 262, 264, 267, 269-270, 273-274, 280, 282, 289, 293, 295-297, 299, 307, 309-310, 315, 317, 321-322, 324, 326, 328-329, 331, 351
Digital Practices 166-168, 172-176, 180, 182, 184-185, 214-216, 218, 224-228, 230
digital waste management 166-168, 172, 175, 182-183, 185, 231

E

Eco-Friendly Technologies 70, 177, 214-216
Economics 96, 138-139, 155, 158, 160, 163, 199, 211, 331
Electronic Waste 46-47, 59, 63, 68, 167-171, 175, 177-179, 184-185, 189, 214-218, 222, 225, 232, 254, 307
Email Waste Management 214-216, 220-221, 224
energy distribution 1-7, 22, 61, 75-76, 81-83, 92
Energy Efficiency 4, 6, 12, 47-48, 58, 66-70, 76, 82, 85-86, 106, 108, 110, 174, 182-183, 216-220, 226, 228, 251-252, 276, 290-291, 293, 296, 298, 300, 318-320, 322
Energy Systems 3, 6, 25, 74-76, 82-83, 95-96, 98, 100, 108, 157, 232, 278-279, 298, 302, 330
Engineering 1, 23-28, 43, 46, 51, 57, 59-60, 72, 74-79, 84, 86, 88, 93-98, 101, 111-113, 151-152, 161-166, 187-188, 193, 213-214, 231-233, 235-237, 259, 261-262, 276-277, 301-302, 305-307, 315-318, 326, 328-330, 332, 350-351
Environmental Concepts 305-306
environmental conservation 1-2, 14-18, 20, 22, 29, 40, 65, 92
Environmental impact 2-3, 7, 9-10, 14-15, 19-20, 32, 35-37, 39-40, 46-48, 58, 60-61, 63-68, 70, 75, 90-92, 94, 108, 156, 168-172, 175, 177, 179-180, 184, 215-227, 229, 237, 239, 241, 244, 246, 248, 250-253, 282, 290, 297, 305-306, 318, 320-322, 326-327
Environmental sustainability 2-3, 23, 28, 31, 38, 42, 48, 62, 70, 114-115, 128, 134-135, 158, 167-168, 172, 175-177, 225-228, 253, 266, 289, 306, 322, 330
Essential-Inorganic 334, 336
ethical considerations 14, 22, 39-42, 94, 166-168, 172, 185, 191

F

Financial Consultancies 194
Financial Institutions 194, 197, 201
Financial Sustainability 194, 211
Future Trends 22, 72, 94, 184-185, 227-229, 231, 305, 326-327

G

Green Computing 43, 46-49, 57-59, 61-63, 65, 67, 70-72, 214-220, 224, 229, 231, 233

H

harnessed 12, 49, 266, 282, 285-286
High-performance computing 266, 268-269, 272, 274-278

I

Industrial Revolution 236-237
Information Technology 46-47, 59-61, 68, 72, 166, 201, 230, 327, 329
Innovation 2-3, 6, 8, 12, 22, 33-34, 47, 56, 69-71, 76, 94, 106, 111, 126, 134-135, 138, 167, 173-175, 177, 183, 185, 187, 193, 215, 226, 229, 231, 234, 236, 244, 251, 254-256, 259-260, 265, 276, 305-309, 315, 317-318, 324, 326, 328-330, 333
Internet of Things (IoT) 28, 38, 44, 107, 139, 144-145, 160, 178, 184, 189, 228, 232, 278
IoT 10, 22-26, 28, 35, 38-45, 56, 60, 72, 79, 94-96, 100, 102, 107, 110, 112, 137-139, 144-146, 149-150, 152, 155-157, 159-163, 165, 167, 178, 184, 186-187, 189, 191, 193, 199, 212-213, 228, 230-234, 278, 280, 282-283, 299, 301-302, 326, 328-332

L

Leveraging 4, 7, 9, 20, 26, 49, 53, 64, 77, 83, 87, 108, 136, 194, 215, 219, 222, 224, 226, 266-269, 273, 278, 287, 290, 293, 297-298, 300, 311

M

Machine Learning 1-10, 22-28, 33-34, 36-38, 41-42, 44, 46, 48-61, 65-68, 70, 72, 74-83, 85-86, 88, 90-91, 93-102, 107-113, 149, 156, 159, 161-162, 186, 189, 191, 194-195, 198, 200, 211-213, 226, 231, 233, 270, 272, 276-277, 299, 318, 328, 333, 351
Manufacturing 23, 47-48, 61, 64, 69, 80, 89-90, 95-96, 99, 138, 169-171, 175, 177-180, 183, 185-190, 222, 224-225, 227, 231, 236, 243, 251-253, 255, 316, 326-328, 331, 335

Marketing Framework 114, 116, 127-135
Marketing Models 114, 116, 135
Materials Science 74-77, 86, 88, 94, 183, 185, 260-261
MFIS 334-336
mitigation 18, 22, 90, 163, 194, 266, 268, 271, 273-274, 285, 298, 300
ML 1-4, 7-9, 11-20, 22, 24-27, 49, 61, 74-94, 96, 100-101, 107, 156, 191, 193-207, 209-211, 232-234, 267, 285-286, 293, 298, 300, 329-332, 334

O

Organic 9, 129, 134, 148, 159-160, 162, 238, 244, 248, 262, 334-336, 343, 347

P

PCSIR 334
Pioneering Sustainability 236
policy interventions 166, 173, 176
Precision Agriculture 2-3, 12, 22, 26, 28-30, 32, 35, 37-38, 42-45, 144, 148, 151, 158-160, 186
precision farming 1-2, 7-8, 19, 30-32, 34-35, 39, 66, 152, 159
Predictive Maintenance 5-6, 10, 12, 59-60, 63-65, 67, 74-80, 82, 88, 93, 112, 178, 297-299, 303

R

Reference model 114
Remote Sensing 20, 28, 38, 65, 85, 90, 111, 143, 146, 148, 152, 159-165, 270, 275-276
renewable energy 1-8, 11-14, 25-26, 47-48, 58, 60-63, 65-66, 70, 74-76, 79, 82-86, 91, 96-97, 99-100, 104-108, 113, 134, 157, 175-176, 183, 192, 215, 217, 220, 226, 228, 232, 259, 275, 278-281, 291-299, 301-303, 306-307, 319-322, 326-327, 330-331
Renewable Energy Forecasting 74, 76, 83-85
renewable energy integration 4-5, 58, 60-62, 91, 278-279, 292-293, 320-321, 326

S

Small and Medium Enterprises (SMEs) 305, 307, 333
smart grid management 7, 62, 297
Smart Grids 2-4, 22, 62, 74-76, 82-83, 157-158, 278-283, 285-289, 291-304
Supply Chain Optimization 21, 37, 61, 74, 76, 88, 90
Sustainability 2-4, 9-12, 14, 20, 23-25, 27-33, 35, 38, 42, 46, 48, 59-60, 62, 65-66, 68-71, 74-77, 80, 82, 90-97, 99-101, 103-104, 108, 111-112, 114-115, 128, 134-136, 138, 140, 142, 146, 158-161, 167-169, 172, 175-177, 180-185, 188-189, 191, 193-194, 211, 215, 217-218, 220, 224-233, 237-239, 244, 248-250, 252-254, 256-260, 264, 266, 275, 278, 280, 289, 296, 299, 306-307, 318, 320-322, 326-327, 329-331, 333
Sustainable 1-4, 6-10, 14-16, 18-19, 21-27, 29-30, 32-34, 36, 38-40, 42-46, 48, 63-64, 66, 68-71, 74-76, 80, 82-83, 85-86, 90-94, 96-104, 108-113, 134-136, 138-140, 146, 154-155, 157-158, 161, 163-168, 170, 172-177, 179-193, 214-216, 218-220, 222, 225-234, 236-248, 250-265, 291-294, 296-299, 301-303, 305-306, 321-322, 326-333
sustainable agriculture 1, 7, 9, 19, 22, 38-39, 48, 139-140, 155, 157, 163-165, 247
Sustainable Development Goals 101, 103, 108, 111
Sustainable materials 177, 225, 237-238, 244, 250-252, 255-262, 264-265

T

technological innovations 166, 184-185, 189
Technology 1, 6, 10, 16, 23, 25-26, 28, 30-31, 35-37, 39-41, 43-44, 46-48, 59-61, 64-70, 72, 74-75, 79-80, 85, 94, 99, 103, 114-115, 133-135, 138, 141-142, 144-145, 147-148, 151-153, 155, 157, 159-162, 164, 166-168, 170, 173, 177-178, 185-191, 193-194, 201, 211-214, 216, 226-228, 230, 233, 235, 243, 249, 253-254, 259, 265, 276, 279-280, 282-283, 286, 289-292, 294-295, 297, 299-301, 305-306, 310, 314, 319, 322, 327-329, 331-333, 350
Toxic-Inorganic 334, 336

W

WHO 125, 133, 152, 272, 309, 334-336, 349
Wireless sensors 142, 147, 155, 163

www.ingramcontent.com/pod-product-compliance
Lightning Source LLC
LaVergne TN
LVHW080629011026
849160LV00014B/1161